AF662622

Mathematics *for* JEE

Main and Advanced

Trigonometry

Rejaul Makshud
M.Sc. (Calcutta University)

UNICORN BOOKS

Publisher
UNICORN BOOKS

F-2/16, Ansari Road, Daryaganj, New Delhi-110002
E-mail: info@unicornbooks.in • Website: www.unicornbooks.in
☎ 011-23275434, 23262683

Branch : Mumbai
23-25, Zaoba Wadi Thakurdwar, Mumbai-401002
☎ 022-22010941, 022-22053387
E-mail: rapidex@bom5.vsnl.net.in

Showroom :
➤ **PM Publications, New Delhi**
- 10-B, Netaji Subhash Marg, Daryaganj, New Delhi-110002
- 6686, Khari Baoli, Delhi-110006

© Copyright: Author
ISBN 978-81-7806-342-3

Edition: 2014

No part of this publication may be reproduced, stored in a retrieval system or transmitted, in any form or by any means, electronic, mechanical, photocopying, recording or otherwise, without the prior permission in writing from the Publisher or author Anybody doing so shall face legal action and will be liable for damages.

Printed at : *Param Offsetters, Okhla, New Delhi-110020*

Preface

It gives me great pleasure and satisfaction to express myself before the students through a comprehensive text book on Trigonometry after having taught for many years at a national level institute.

Trigonometry is a vital topic under mathematics and constitutes nearly 15 to 20% of all questions asked in the IIT-JEE and other engineering entrance exams. After having taught for so many years I realized that it was necessary to design a book in a systematic way, using which a reader could build their base, improve their level and also enjoy the subject.

This book has been designed after considering the new pattern of questions in JEE Main and Advanced exams. It contains six chapters and each of them contains theory, followed by a large number of worked out problems and exercise based problems, namely

- Level – I : Questions based on Fundamentals
- Level – II : Objective type questions
- Level – III : Subjective type questions for JEE Advanced

Further included are **Linked Comprehension type, Match Matrix type, Assertion and Reason type questions** along with previous years questions from Roorkee and IIT-JEE.

Friends, no problem in mathematics is difficult. Once you understand the concept, every problem becomes easy. So do not jump directly to the exercise without going through the theory, the worked-out problems and the objective type questions. The exercise problems are so arranged that they gradually require more thinking.

I hope this book will help the students in enjoying the subject and improve their confidence when they try to solve any type of questions.

My special thanks go to Mr Rakesh Singh (IISc, Bangalore), Mr Ravi Bhushan (IIT, Delhi), Mr. M.P. Singh (IISc Bangalore) Mr Manoj Kumar (IIT, Delhi), Mr Vivek Bhatt (BIET) who have helped, inspired and motivated me to accomplish this task and finally my sincere thanks to the publisher Dr. Ashok Gupta of Unicorn Books for publishing this book.

I owe a special debt of gratitude to my early teachers Mr Swapan Halder, Mr. Jadunandan Mishra, Mr Mahadev Roy and Mr Dilip Bhattacharya for showing me, what clear and engaging mathematical teaching is all about.

I have tried my best to keep errors out of this book. I shall be grateful to the readers who point out any inadvertent errors and make constructive suggestions.

Rejaul Makshud

Contents

Pages

Chapters

1. Trigonometric Ratios and Identities .... 1 – 78
2. Graphs of Trigonometric Functions .... 79 – 102
3. Trigonometric Equation .... 103 – 148
4. Trigonometric Inequation .... 149 – 174
5. Inverse Trigonometric Functions .... 175 – 254
6. Properties of Triangles .... 255 – 320

CHAPTER 1 TRIGONOMETRIC RATIOS AND IDENTITIES

1.1 INTRODUCTION

Trigonometry (from Greek trigonon "triangle" + metron "measure") is a branch of mathematics that stud ies triangles and the relationships between the lengths of their sides and the angles between those sides.

Trigonometry defines the trigonometric functions, which describe those relationships and have applicability to cyclical phenomena, such as waves. The field evolved during the third century BC as a branch of geometry used extensively for astronomical studies. It is also the foundation of the practical art of surveying.

Trigonometry basics are often taught in school either as a separate course or as part of a precalculus course. The trigonometric functions are pervasive in parts of pure mathematics and applied mathematics such as Fourier analysis and the wave equation, which are, in turn, essential to many branches of science and technology.

1.2 HISTORY OF TRIGONOMETRY :

Sumerian astronomers introduced angle measure, using a division of circles into 360 degrees. They and their successors the Babylonians studied the ratios of the sides of similar triangles and discovered some properties of these ratios, but did not turn that into a systematic method for finding sides and angles of triangles. The ancient Nubians used a similar method. The ancient Greeks transformed trigonometry into an ordered science.

Classical Greek mathematicians (such as Euclid and Archimedes) studied the properties of chords and inscribed angles in circles, and proved theorems that are equivalent to modern trigonometric formulae, although they presented them geometrically, rather than algebraically. Claudius Ptolemy expanded upon Hipparchus' Chords in a Circle in his Almagest. The modern sine function was first defined in the Surya Siddhanta, and its properties were further documented by the 5th century Indian mathematician and as tronomer Aryabhata. These Greek and Indian works were translated and expanded by medieval Islamic mathematicians.

By the 10th century, Islamic mathematicians were using all six trigonometric functions, had tabulated their values, and were applying them to problems in spherical geometry.[citation needed] At about the same time, Chinese mathematicians developed trigonometry independently, although it was not a major field of study for them. Knowledge of trigonometric functions and methods reached Europe via Latin translations of the works of Persian and Arabic astronomers such as Al Battani and Nasir al-Din al-Tusi. One of the earliest works on trigonometry by a European mathematician is De Triangulis by the 15th century German mathematician Regiomontanus. Trigonometry was still so little known in 16th century Europe that Nicolaus Copernicus devoted two chapters of De revolutionibus orbium coelestium to explaining its basic concepts. Driven by the demands of navigation and the growing need for accurate maps of large areas, trigonometry grew into a major branch of mathematics. Bartholomaeus Pitiscus was the first to use the word, publishing his Trigonometria in 1595. Gemma Frisius described for the first time the method of triangulation still used today in surveying. It was Leonhard Euler who fully incorporated complex numbers into trigonometry. The works of James Gregory in the 17th century and Colin Maclaurin in the 18th century were influential in the development of trigonometric series. Also in the 18th century, Brook Taylor defined the general Taylor series..

1.3 APPLICATION OF TRIGONOMETRY:

There are an enormous number of uses of trigonometry and trigonometric functions. For instance, the technique of triangulation is used in astronomy to measure the distance to nearby stars, in geography to measure distances between landmarks, and in satellite navigation systems. The sine and cosine functions are fundamental to the theory of periodic functions such as those that describe sound and light waves.

Fields that use trigonometry or trigonometric functions include astronomy (especially for locating apparent positions of celestial objects, in which spherical trigonometry is essential) and hence navigation (on the oceans, in aircraft, and in space), music theory, acoustics, optics, analysis of financial markets, electronics, probability theory, statistics, biology, medical imaging (CAT scans and ultrasound), pharmacy, chemistry, number theory (and hence cryptology), seismology, meteorology, oceanography, many physical sciences, land surveying and geodesy, architecture, phonetics, economics, electrical engineering, mechanical engineering, civil engineering, computer graphics, cartography, crystallography and game development.

1.4. TRIGONOMETRICAL FUNCTIONS:

In mathematics, the trigonometric functions (also called the **circular functions**) are functions of an angle. They are used to relate the angles of a triangle to the lengths of the sides of a triangle. Trigonometric unctions are important in the study of triangles and modeling periodic phenomena, among many other applications. The most familiar trigonometric functions are the sine, cosine, and tangent. In the context of the standard unit circle with radius 1, where a triangle is formed by a ray originating at the origin and making some angle with the x-axis, the sine of the angle gives the length of the y-component (rise) of the triangle, the cosine gives the length of the x-component (run), and the tangent function gives the slope (y-component divided by the x-component). More precise definitions are detailed below. Trigonometric functions are commonly defined as ratios of two sides of a right triangle containing the angle, and can equivalently be defined as the lengths of various line segments from a unit circle. More modern definitions express them as infinite series or as solutions of certain differential equations, allowing their extension to arbitrary positive and negative values and even to complex numbers.

Trigonometric functions have a wide range of uses including computing unknown lengths and angles in triangles (often right triangles). In this use, trigonometric functions are used, for instance, in navigation, engineering, and physics. A common use in elementary physics is resolving a vector into Cartesian coordinates. The sine and cosine functions are also commonly used to model periodic function phenomena such as sound and light waves, the position and velocity of harmonic oscillators, sunlight intensity and day length, and average temperature variations through the year.

In modern usage, there are six basic trigonometric functions, tabulated here with equations that relate them to one another. Especially with the last four, these relations are often taken as the definitions of those functions, but one can define them equally well geometrically, or by other means, and then derive these relations.

1.5 MEASUREMENT OF ANGLES :

1. **Angle** : The measurement of an angle is the amount of rotation from the initial side to the terminal side.

2. **Sense of an Angle** : The sense of an angle is +ve or –ve according as the initial side rotates in anti-clock-wise or clockwise direction to get the terminal side.

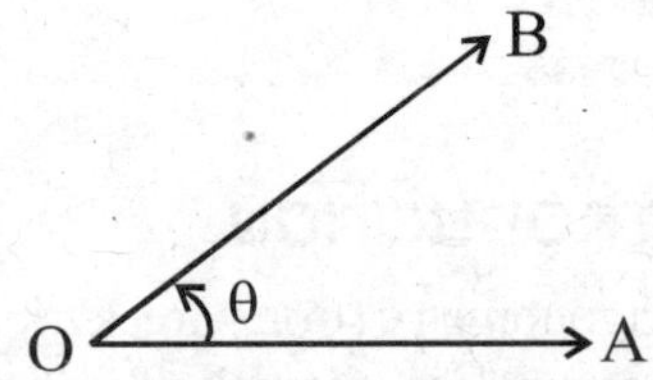

positive angle

-ve angle

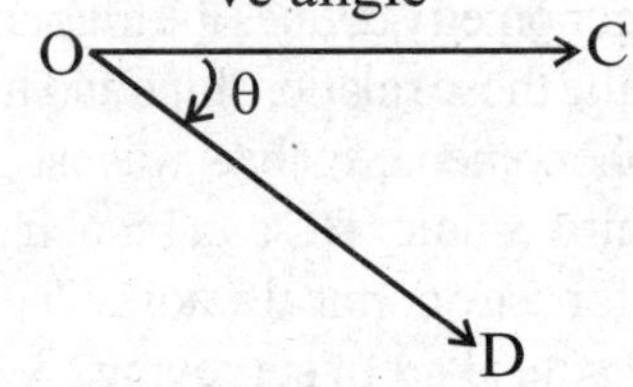

3. **System of measuring angles** :
There are three systems of measuring angles such as
i) Sexagesimal system
ii) Centisimal system
iii) Circular system

4. In sexagesimal system, we have

$$1 \text{ right angle} = 90^\circ$$

$$1^\circ = 60'$$

$$1' = 60''$$

5. In centasimal system, we have

$$1 \text{ right angle} = 100^g$$

$$1^g = 100'$$

$$1' = 100''$$

6. In circular system, the unit of measurement is radian

Radian : One radian is the measure of an angle subtended at the centre of a circle by an arc of length equal to the radius of the circle.

Here, $\angle AOB = 1 \text{ radian} = 1^c$.

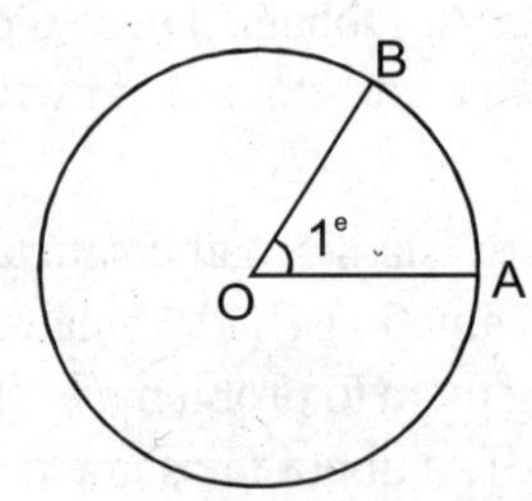

Notes :

1. When an angle is expressed in radi ans, the word radian is omitted.

2. Since $180^\circ = \pi$ radian $= \left(\frac{22}{7\times180}\right)$ radian $= 0.01746$ radian

3. $1 \text{ radian} = \frac{180^\circ}{\pi} = \left(\frac{180}{22}\times 7\right)$

$= 57^\circ\ 16'\ 22'$

4. The angle between two consecutive digits is $30^\circ \left(\frac{\pi}{6} \text{ radians}\right)$

5. The hour hand rotates through an angle of 30° in one hour (i.e. $\left(\frac{1}{2}\right)$ in one minute).

6. The minute hand rotates through an angle of 6° in one minute.

7. The relation amongst three systems of measurement of an angle is

$$\frac{D}{90^\circ} = \frac{G}{100} = \frac{2R}{\pi}$$

8. The number of radians in an angle subtended by an arc of a circle at the centre is $\frac{\text{Arc}}{\text{Radius}}$

i.e. $\theta = \frac{s}{r}$

Exercise - 1

1. Find the length of an arc of a circle of radius 5 cm. subtending a central angle of measuring 15°.
2. In a circle of diameter 40 cm. the length of a chord is 20 cm. Find the length of minor arc corresponding to the chord.
3. If the arcs of same length in the circles subtended angles of 60° and 75° at their centres. Find the ratio of their radii.
4. A horse is tied to a post by a rope. If the horse moves along a circular path always keeping the rope tight and discribes 88 meters when it has traced out 72° at the centre, find the length of the rope.
5. The Moon's distance from the Earth is 36000 kms. and its diameter subtends an angle of 31° at the eye of the observer. Find the diameter of the Moon.
6. The difference between the acute angles of a right angled triangle is $\frac{2\pi}{3}$ radians. Express the angles in degrees.
7. The angles of a quadrilateral are in A.P. and the greatest angle is 120°. Find the angles in radians.
8. The angles of a triangle are in A.P. such that the greatest is 5-times the least. Find the angles in radians.
9. A wheel makes 180 revolutions per minute through how many radians does it turn in 1 second?
10. Find the distance from the eye at which a coin of 2 cm. diameter should be held so as to conceal the full Moon whose angular diameter is $31'$.
11. The interrior angles of a triangle are in A.P. The smallest angle is 120° and the common difference is 5°. Find the number of sides of the polygon.
12. A wheel makes 30 revolutions per minute. Find the circular measure of the angle described by a spoke in 1/2 second.
13. A man running along a circular track at the rate of 10 miles per hour travels in 36 seconds, an arc which subtends 56° at the centre. Find the diameter of the circle.
14. At what distance does a man $5\frac{1}{2}$ ft in height, subtend an angle of $15''$?
15. Find the angle between thehour hand and minute - hand in circular measure at 4 O' clock.

Answers

1. $\frac{5}{12}\pi$ cm.
2. $\frac{20\pi}{3}$ cm.
3. 5 : 4
4. 70 meters
5. 3247.62 kms.
6. 81°, 9°
7. $\frac{\pi}{3}, \frac{4\pi}{9}, \frac{5\pi}{9}, \frac{2\pi}{3}$.
8. $\frac{\pi}{9}, \frac{\pi}{3}, \frac{5\pi}{9}$

9. 6π 10. 2.217 m.

11. No. of sides = 9. 12. $\pi/2$

13. 1080 ft 14. 14.32 miles

15. $\frac{4\pi}{3}$.

1.6 TRIGONOMETRICAL RATIOS

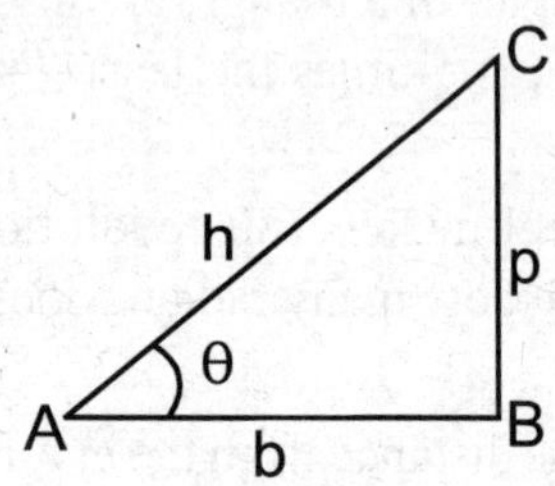

1.6.1 Definitions of Trigonometric Ratios:

1. $\text{Sin}\,\theta = \frac{p}{h}$ 2 $\text{cosec}\,\theta = \frac{h}{p}$

3. $\cos\theta = \frac{b}{h}$ 4. $\sec\theta = \frac{h}{b}$

5. $\tan\theta = \frac{p}{b}$ 6. $\cot\theta = \frac{b}{p}$

1.6.2 Signs of Trigonometrical Ratios :

The signs of the trigonometrical ratios in different quadrants are remembered by the following chart.

sin, cosec are +ve	All are +ve
tan, cot are +ve	cos, sec are +ve

It is also known as all, sin, tan, cos formula.

1.6.3 Relation between the Trigonometrical Ratios of an Angle.

Step-I . (i) $\sin\theta \cdot \text{cosec}\,\theta = 1$

(ii). $\cos\theta \cdot \sec\theta = 1$

(iii). $\tan\theta \cdot \cot\theta = 1$

Step - II (i) $\tan\theta = \frac{\sin\theta}{\cos\theta}$

(ii) $\cot\theta = \frac{\cos\theta}{\sin\theta}$

Step - III (i) $\sin^2\theta + \cos^2\theta = 1$

(ii). $\sec^2\theta = 1 + \tan^2\theta$

(iii) $\text{cosec}^2\theta = 1 + \cot^2\theta$

1.7 Limits of the values of Trigonometrical Functions.

1. $-1 \le \sin\theta \le 1$.
2. $-1 \le \cos\theta \le 1$
3. $\text{cosec}\,\theta \ge 1$ and $\text{cosec}\,\theta \le -1$
4. $\text{Sec}\,\theta \ge 1$ and $\sec\theta \le -1$
5. $-\infty < \tan\theta < \infty$
6. $-\infty < \cot\theta < \infty$

1.8 Some solved examples :

Ex-1. If $\sec\theta + \tan\theta = 3$, where θ lies in the first quadrant, then find the value of $\cos\theta$

Soln. Given $\sec\theta + \tan\theta = 3$ ....(i)

$\Rightarrow (\sec\theta - \tan\theta) = \frac{1}{(\sec\theta + \tan\theta)} = \frac{1}{3}$ ....(ii)

Adding (i) and (ii), we get, $2\sec\theta = 3 + \frac{1}{3} = \frac{10}{3}$

$\Rightarrow \sec\theta = \frac{5}{3}$

$\Rightarrow \cos\theta = \frac{3}{5}$.

Ex-2. If $\text{cosec}\,\theta - \cot\theta = \frac{1}{5}$, then find the value of $\sin\theta$

Soln. Given $\text{cosec}\,\theta - \cot\theta = \frac{1}{5}$ ....(i)

$\Rightarrow \text{cosec}\,\theta - \cot\theta = \frac{1}{\text{cosec}\,\theta + \cot\theta} = 5$ ....(ii)

Adding (i) and (ii), we get, $2\,\text{cosec}\,\theta = 5 + \frac{1}{5} = \frac{26}{5}$

$\Rightarrow \text{cosec}\theta = \frac{13}{5} \Rightarrow \sin\theta = \frac{5}{13}$.

Ex-3. If $a = c\cos\theta + d\sin\theta$ and $b = c\sin\theta - d\cos\theta$ such that $a^m + b^n = c^p + d^q$, where m, n, p, q $\in$ N then find the value of $m+n+p+q+42$.

Soln. Given $a = c\cos\theta + d\sin\theta$ ....(i) and

$b = c\sin\theta - d\cos\theta$ ....(ii)

Squaring and adding (i) and (ii), we get,

$$a^2 + b^2 = (c\cos\theta + d\sin\theta)^2 + (c\sin\theta - d\cos\theta)^2$$

$$\Rightarrow \quad a^2 + b^2 = (c^2\cos^2\theta + d^2\sin^2\theta) + (c^2\sin^2\theta + d^2\cos^2\theta)$$

$$\Rightarrow \quad a^2 + b^2 = c^2 + d^2$$

$$\Rightarrow \quad a^2 + b^2 = c^2 + d^2$$

$$\Rightarrow \quad m = 2,\ n = 2,\ p = 2,\ q = 2$$

Hence, the value of $m+n+p+q+42 = 50$

Ex-4. If $3\sin\theta + 4\cos\theta = 5$, then find the value of $3\cos\theta - 4\sin\theta$.

Soln. Let $x = 3\cos\theta - 4\sin\theta$ ....(i) and

$5 = 3\sin\theta - 4\cos\theta$ ....(ii)

Squaring and adding (i) and (ii), we get

$$x^2 + 5^2 = (3\cos\theta + 4\sin\theta)^2 + (3\sin\theta - 4\cos\theta)^2$$

$$\Rightarrow \quad x^2 + 5^2 = (9\cos^2\theta + 16\sin^2\theta + 24\sin\theta\cos\theta) + +(9\sin^2\theta + 16\cos^2\theta - 24\sin\theta\cos\theta)$$

$$\Rightarrow \quad x^2 + 5^2 = (9\cos^2\theta + 16\sin^2\theta) + (9\sin^2\theta + 16\cos^2\theta)$$

$$\Rightarrow \quad x^2 + 5^2 = 9(\cos^2\theta + \sin^2\theta) + 16(\cos^2\theta + \sin^2\theta)$$

$$\Rightarrow \quad x^2 + 25 = 25$$

$$\Rightarrow \quad x^2 = 0$$

$$\Rightarrow \quad x = 0$$

$$\Rightarrow \quad 3\cos\theta - 4\sin\theta = 0$$

Ex-5. If $x = r\cos\theta\sin\varphi$, $y = r\cos\theta\cos\varphi$ and $z = r\sin\theta$ such that $x^m + y^n + z^p = r^2$, where m,n,p $\in$ N, m,n,p $\in$ N, then find the value of $(m+n+p-4)^{m+n+p+4}$.

Soln. We have, $x^2 + y^2 + z^2$

$$= (r\cos\theta\cos\varphi)^2 + (r\cos\theta\sin\varphi)^2 + +(r\sin\theta)^2$$

$$\Rightarrow \quad x^2 + y^2 + z^2 = (r^2\cos^2\theta\cos^2\varphi) + (r^2\cos^2\theta\sin^2\varphi) + +(r^2\sin^2\theta)$$

$$\Rightarrow \quad x^2 + y^2 + z^2 = r^2\cos^2\theta(\cos^2\varphi + \sin^2\varphi) + (r^2\sin^2\theta)$$

$$= r^2\cos^2\theta + r^2\sin^2\theta$$

$$= r^2(\cos^2\theta + \sin^2\theta) = r^2$$

$$\Rightarrow \quad x^2 + y^2 + z^2 = r^2$$

$$\Rightarrow \quad m = 2,\ n = 2,\ p = 2$$

Thus, the value of $(m+n+p-4)^{(m+n+p+4)}$

$$= 2^{10} = 1024$$

Ex-6. If $x = \dfrac{2\sin\alpha}{1+\cos\alpha+3\sin\alpha}$, then find the value of $\dfrac{\sin\alpha - 3\cos\alpha + 3}{2 - 2\cos\alpha}$

Soln. Given $x = \dfrac{2\sin\alpha}{1+\cos\alpha+3\sin\alpha}$

We have, $\dfrac{\sin\alpha - 3\cos\alpha + 3}{2 - 2\cos\alpha}$

$$= \frac{\sin\alpha + 3(1-\cos\alpha)}{2(1-\cos\alpha)}$$

$$= \frac{\sin\alpha}{2(1-\cos\alpha)} + \frac{3}{2}$$

$$= \frac{\sin\alpha(1+\cos\alpha)}{2(1-\cos^2\alpha)} + \frac{3}{2}$$

$$= \frac{\sin\alpha(1+\cos\alpha)}{2\sin^2\alpha} + \frac{3}{2}$$

$$= \frac{(1+\cos\alpha)}{2\sin\alpha} + \frac{3}{2}$$

$$= \frac{(1+\cos\alpha+3\sin\alpha)}{2\sin\alpha}$$

$$= \frac{1}{x}$$

Ex-7. If $P = \sec^6\theta - \tan^6\theta - 3\sec^2\theta\tan^2\theta$, $Q = \text{cosec}^6\theta - \cot^6\theta - 3\,\text{cosec}^2\theta\cot^2\theta$ and $R = \sin^6\theta + \cos^6\theta + 3\sin^2\theta\cos^2\theta$, then find the value of $(P+Q+R)^{(P+Q+R)}$

Soln. We have, $P = \sec^6\theta - \tan^6\theta - 3\sec^2\theta\tan^2\theta$

$$= (\sec^2\theta - \tan^2\theta)^3 = 1$$

$Q = \text{cosec}^6\theta - \cot^6\theta - 3\,\text{cosec}^2\theta\cot^2\theta$

$$= (\text{cosec}^2\theta - \cot^2\theta)^3 = 1$$

and $R = \sin^6\theta + \cos^6\theta + 3\sin^2\theta\cos^2\theta$

$$= (\sin^2\theta + \cos^2\theta)^3 = 1$$

Hence, the value of $(P+Q+R)^{(P+Q+R)}$

$$= 3^3 = 27$$

Ex-8. If $3\sin x + 4\cos x = 5$, for all x in $\left(0, \frac{\pi}{2}\right)$, then find the value of $2\sin x + \cos x + 4\tan x$

Soln. We have $3\sin x + 4\cos x = 5$

Let $y = 3\cos x - 4\sin x$

Now, $y^2 + 5^2 = (3\cos x - 4\sin x)^2 + (3\sin x + 4\cos x)^2$

$\Rightarrow$ $y^2 + 25 = 9\cos^2 x + 16\sin^2 x - 24\sin x\cos x + 9\sin^2 x + 16\cos^2 x + 24\sin x\cos x$

$\Rightarrow$ $y^2 + 25 = 25(\cos^2 x + \sin^2 x) = 25$

$\Rightarrow$ $y^2 = 0$

$\Rightarrow$ $y = 0$

$\Rightarrow$ $3\cos x - 4\sin x = 0$

$\Rightarrow$ $3\cos x = 4\sin x$

$\Rightarrow$ $\tan x = 3/4$

Hence, the value of $2\sin x + \cos x + 4\tan x$

$$= 2\left(\frac{3}{5}\right) + \left(\frac{4}{5}\right) + 4\left(\frac{3}{4}\right) = 2 + 3 = 5.$$

Ex-9. If $\sin A + \sin B + \sin C + 3 = 0$, then find the value of $\cos A + \cos B + \cos C + 10$

Soln. Given $\sin A + \sin B + \sin C = -3$

$\Rightarrow$ $\sin A = -1, \sin B = -1, \sin C = -1$

$\Rightarrow$ $A = -\frac{\pi}{2}, B = -\frac{\pi}{2}, C = -\frac{\pi}{2}$

Hence, the value of $\cos A + \cos B + \cos C + 10$

$= 0 + 0 + 0 + 10 = 10.$

Ex-10. If $(1+\sin\theta)(1+\cos\theta) = \frac{5}{4}$, then find the value of $(1-\sin\theta)(1-\cos\theta)$

Soln. We have $(1+\sin\theta)(1+\cos\theta) = \frac{5}{4}$

$\Rightarrow$ $1 + \sin\theta + \cos\theta + \sin\theta\cos\theta = \frac{5}{4}$

$\Rightarrow$ $1 + t + \left(\frac{t^2-1}{2}\right) = \frac{5}{4}$ $(\sin\theta + \cos\theta = t, \text{ say})$

$\Rightarrow$ $t + \left(\frac{t^2-1}{2}\right) = \frac{1}{4}$

$\Rightarrow$ $t^2 + 2t - 1 = \frac{1}{2}$

$\Rightarrow$ $2t^2 + 4t - 3 = 0$

$\Rightarrow$ $t = \frac{-4 \pm \sqrt{16+24}}{4} = \frac{-4 \pm 2\sqrt{10}}{4} = -1 \pm \frac{1}{2}\sqrt{10}$

$\Rightarrow \quad t=-1+\frac{1}{2}\sqrt{10}$

$\Rightarrow \quad \sin\theta+\cos\theta=-1+\frac{1}{2}\sqrt{10}$

Now, $(1-\sin\theta)(1-\cos\theta)$

$= 1-\sin\theta-\cos\theta+\sin\theta\cos\theta$

$= 1-(\sin\theta+\cos\theta)+\sin\theta\cos\theta$

$= 1-\left(-1+\frac{\sqrt{10}}{2}\right)+\frac{1}{2}\left(\frac{10}{4}-\sqrt{10}\right)$

$= \left(2+\frac{5}{4}\right)-\sqrt{10}$

$= \left(\frac{13}{4}-\sqrt{10}\right)$

Ex-11. Find the minimum value of the expression $f(x)=\frac{9x^2\sin^2x+4}{x\sin x}$, for all x in $(0, \pi)$

Soln. Given $f(x)=\frac{9x^2\sin^2x+4}{x\sin x}$

$= 9x\sin x+\frac{4}{x\sin x}$

Applying, A.M $\geq$ G.M, we get,

$$\left(\frac{9x\sin x+\frac{4}{x\sin x}}{2}\right)\geq\sqrt{9x\sin x\times\frac{4}{x\sin x}}$$

$$\Rightarrow \left(\frac{9x\sin x+\frac{4}{x\sin x}}{2}\right)\geq 6$$

$$\Rightarrow \left(9x\sin x+\frac{4}{x\sin x}\right)\geq 12$$

Hence, the minimum value of f(x) is 12.

Ex-12. If $\cos\theta + \sin\theta = \sqrt{2}\cos\theta$, than prove that $\cos\theta - \sin\theta = \sqrt{2}\sin\theta$.

Soln. We have, $\cos\theta+\sin\theta=\sqrt{2}\cos\theta$

$\Rightarrow \quad \sin\theta=\left(\sqrt{2}-1\right)\cos\theta$

$\Rightarrow \quad \cos\theta=\frac{\sin\theta}{\left(\sqrt{2}-1\right)}$

$\Rightarrow \quad \cos\theta=\left(\sqrt{2}+1\right)\sin\theta$

$\Rightarrow \quad \cos\theta-\sin\theta=\sqrt{2}\sin\theta$

Ex-13. If $\tan^2\theta = 1-e^2$, then prove that

$$\sec\theta+\tan^3\theta.\text{cosec}\,\theta=\left(2-e^2\right)^{3/2}$$

Soln. We have, $\tan^2\theta=1-e^2$

$\Rightarrow \quad 1+\tan^2\theta=1+1-e^2=2-e^2$

$\Rightarrow \quad \sec^2\theta=2-e^2$

$\Rightarrow \quad \sec\theta=\sqrt{2-e^2}$

Now, $\sec\theta+\tan^3\theta.\text{cosec}\,\theta$

$= \sec\theta+\frac{\sin^3\theta}{\cos^3\theta}.\frac{1}{\sin\theta}$

$= \sec\theta+\frac{\sin^2\theta}{\cos^3\theta}$

$= \sec\theta+\frac{\sin^2\theta}{\cos^2\theta}.\frac{1}{\cos\theta}$

$= \sec\theta+\tan^2\theta.\sec\theta$

$= \sec\theta\left(1+\tan^2\theta\right)$

$= \sec^3\theta$

$= \left(2-e^2\right)^{3/2}$

Ex-14. If $\sin\theta+\sin^2\theta+\sin^3\theta=1$, then prove that, $\cos^6\theta-4\cos^4\theta+8\cos^2\theta=4$

Soln. Given $\sin\theta+\sin^2\theta+\sin^3\theta=1$

$\Rightarrow \quad \left(\sin\theta+\sin^3\theta\right)=1-\sin^2\theta=\cos^2\theta$

$\Rightarrow \quad \left(\sin\theta+\sin^3\theta\right)^2=\left(\cos^2\theta\right)^2$

$\Rightarrow \quad \left(\sin\theta + \sin^3\theta\right)^2 = \left(\cos^2\theta\right)^2$

$\Rightarrow \quad \left(1 - \cos^2\theta\right)\left(2 - \cos^2\theta\right)^2 = \cos^4\theta$

$\Rightarrow \quad \left(1 - \cos^2\theta\right)\left(4 - 4\cos^2\theta + \cos^4\theta\right) = \cos^4\theta$

$\Rightarrow \quad 4 - 4\cos^2\theta + \cos^4\theta - 4\cos^2\theta + 4\cos^4\theta$

$- \cos^6\theta = \cos^4\theta$

$\Rightarrow \quad \cos^6\theta - 4\cos^4\theta + 8\cos^2\theta = 4$

Ex-15. If $x = \dfrac{2\sin\theta}{1+\cos\theta+\sin\theta}$, then prove that $\dfrac{1-\cos\theta+\sin\theta}{1+\sin\theta} = x$.

Soln. Given $x = \dfrac{2\sin\theta}{1+\cos\theta+\sin\theta}$

$$= \frac{2\sin\theta}{(1+\sin\theta)+\cos\theta}$$

$$= \frac{2\sin\theta\left((1+\sin\theta)-\cos\theta\right)}{\left((1+\sin\theta)+\cos\theta\right)\left((1+\sin\theta)-\cos\theta\right)}$$

$$= \frac{2\sin\theta\left((1+\sin\theta)-\cos\theta\right)}{\left((1+\sin\theta)^2-\cos^2\theta\right)}$$

$$= \frac{2\sin\theta\left((1+\sin\theta)-\cos\theta\right)}{\left(1+\sin^2\theta+2\sin\theta-\cos^2\theta\right)}$$

$$= \frac{2\sin\theta\left((1+\sin\theta)-\cos\theta\right)}{\left(\sin^2\theta+2\sin\theta+\left(1-\cos^2\theta\right)\right)}$$

$$= \frac{2\sin\theta\left((1+\sin\theta)-\cos\theta\right)}{\left(2\sin\theta+2\sin^2\theta\right)}$$

$$= \frac{2\sin\theta\left((1+\sin\theta)-\cos\theta\right)}{2\sin\theta(1+\sin\theta)}$$

$$= \frac{\left((1+\sin\theta)-\cos\theta\right)}{(1+\sin\theta)}$$

$$= \frac{(1-\cos\theta+\sin\theta)}{(1+\sin\theta)}$$

Ex-16. If $\dfrac{\sin^4\alpha}{a} + \dfrac{\cos^4\alpha}{b} = \dfrac{1}{a+b}$, then prove that

$$\frac{\sin^8\alpha}{a^3} + \frac{\cos^8\alpha}{b^3} = \frac{1}{(a+b)^3}$$

Soln. We have, $\dfrac{\sin^4\alpha}{a} + \dfrac{\cos^4\alpha}{b} = \dfrac{1}{a+b}$

$\Rightarrow \quad \left(\dfrac{a+b}{a}\right)\sin^4\alpha + \left(\dfrac{a+b}{b}\right)\cos^4\alpha = 1$

$\Rightarrow \quad \left(1+\dfrac{b}{a}\right)\sin^4\alpha + \left(1+\dfrac{a}{b}\right)\cos^4\alpha = 1$

$\Rightarrow \quad \left(\dfrac{b}{a}\sin^4\alpha + \dfrac{a}{b}\cos^4\alpha\right) + \left(\sin^4\alpha + \cos^4\alpha\right) = 1$

$\Rightarrow \quad \left(\dfrac{b}{a}\sin^4\alpha + \dfrac{a}{b}\cos^4\alpha\right) + \left(1 - 2\sin^2\alpha.\cos^2\alpha\right) = 1$

$\Rightarrow \quad \left(\dfrac{b}{a}\sin^4\alpha + \dfrac{a}{b}\cos^4\alpha - 2\sin^2\alpha.\cos^2\alpha\right) = 0$

$\Rightarrow \quad \left(\sqrt{\dfrac{b}{a}}\sin^2\alpha\right)^2 + \left(\sqrt{\dfrac{a}{b}}\cos^2\alpha\right)^2$

$- 2\sqrt{\dfrac{b}{a}}\sin^2\alpha.\sqrt{\dfrac{a}{b}}\cos^2\alpha = 0$

$\Rightarrow \quad \left(\sqrt{\dfrac{b}{a}}\sin^2\alpha - \sqrt{\dfrac{a}{b}}\cos^2\alpha\right)^2 = 0$

$\Rightarrow \quad \left(\sqrt{\dfrac{b}{a}}\sin^2\alpha - \sqrt{\dfrac{a}{b}}\cos^2\alpha\right) = 0$

$\Rightarrow \quad \sqrt{\dfrac{b}{a}}\sin^2\alpha = \sqrt{\dfrac{a}{b}}\cos^2\alpha$

$\Rightarrow \quad \dfrac{\sin^2\alpha}{a} = \dfrac{\cos^2\alpha}{b} = \dfrac{1}{a+b}$

$\Rightarrow \quad \sin^2\alpha = \dfrac{a}{a+b}, \ \cos^2\alpha = \dfrac{b}{a+b}$

Now, $\dfrac{\sin^8\alpha}{a^3}+\dfrac{\cos^8\alpha}{b^3}$

$= \dfrac{(\sin^2\alpha)^4}{a^3}+\dfrac{(\cos^2\alpha)^4}{b^3}$

$= \dfrac{\left(\dfrac{a}{a+b}\right)^4}{a^3}+\dfrac{\left(\dfrac{b}{a+b}\right)^4}{b^3}$

$= \dfrac{a^4}{a^3(a+b)^4}+\dfrac{b^4}{b^3(a+b)^4}$

$= \dfrac{a}{(a+b)^4}+\dfrac{b}{(a+b)^4}$

$= \dfrac{a+b}{(a+b)^4}$

$= \dfrac{1}{(a+b)^3}$

Ex-17. Prove that,

$3(\sin x-\cos x)^4+6(\sin x+\cos x)^2$

$+4(\sin^6 x+\cos^6 x)=13$

Soln. We have, $3(\sin x - \cos x)^4 + 6(\sin x + \cos x)^2 + 4(\sin^6 x + \cos^6 x)$

$= 3(\sin^4 x - 4\sin^3 x\cos x + 6\sin^2 x\cos^2 x - 4\sin x\cos^3 x + \cos^4 x)$

$+ 6(\sin^2 x + \cos^2 x + 2\sin x\cos x) + 4\{(\sin^2 x)^3 + (\cos^2 x)^3\}$

$= 3(\sin^4 x + \cos^4 x - 4\sin x\cos x(\sin^2 x + \cos^2 x) + 6\sin^2 x\cos^2 x)$

$+ 6(1 + 2\sin x\cos x) + 4(\sin^2 x + \cos^2 x)^2 - 12\sin^2 x\cos^2 x.$

$= 3 - 6\sin^2 x\cos^2 x - 12\sin x\cos x + 18\sin^2 x\cos^2 x + 6 + 12\sin x\cos x$

$+ 4 - 12\sin^2 x\cos^2 x$

$= 3+6+4$

$= 13.$

Ex-18. If $\sin x + \sin^2 x = 1$, then find the value of $\cos^8 x + 2\cos^6 x + \cos^4 x$

Soln. We have, $\sin x + \sin^2 x = 1$

$\Rightarrow \sin x = 1 - \sin^2 x = \cos^2 x$

Now, $\cos^8 x + 2\cos^6 x + \cos^4 x$

$= (\cos^4 x)^2 + 2.\cos^4 x.\cos^2 x + (\cos^2 x)^2$

$= (\cos^4 x + \cos^2 x)^2$

$= (\sin^2 x + \sin x)^2$

$= (1)^2 = 1.$

Ex-19. If $0 \le \theta \le 180^\circ$ and $81^{\sin^2\theta}+81^{\cos^2\theta}=30$, then find the value of θ

Soln. We have, $81^{\sin^2\theta}+81^{\cos^2\theta}=30$

$\Rightarrow 81^{\sin^2\theta}+81^{1-\sin^2\theta}=30$

$\Rightarrow 81^{\sin^2\theta}+\dfrac{81}{81^{\sin^2\theta}}=30$

$\Rightarrow a+\dfrac{81}{a}=30,\quad a=81^{\sin^2\theta}$

$\Rightarrow a^2-30a+81=0$

$\Rightarrow (a-27)(a-3)=0$

$\Rightarrow a=3, 27$

When $a=3 \Rightarrow 81^{\sin^2\theta}=3$

$\Rightarrow 3^{4\sin^2\theta}=3$

$\Rightarrow 4\sin^2\theta=1$

$\Rightarrow \sin^2\theta=\left(\dfrac{1}{2}\right)^2$

$\Rightarrow \sin^2\theta=\sin^2\left(\dfrac{\pi}{6}\right)$

$\Rightarrow \theta=\left(n\pi\pm\dfrac{\pi}{6}\right)$

$\Rightarrow \theta=\dfrac{\pi}{6},\dfrac{5\pi}{6}$

When $a=27 \Rightarrow 81^{\sin^2\theta}=27$

$\Rightarrow 3^{4\sin^2\theta}=3^3$

$\Rightarrow 4\sin^2\theta=3$

$$\Rightarrow \quad \sin^2\theta = \left(\frac{\sqrt{3}}{2}\right)^2$$

$$\Rightarrow \quad \sin^2\theta = \sin^2\left(\frac{\pi}{3}\right)$$

$$\Rightarrow \quad \theta = \left(n\pi \pm \frac{\pi}{3}\right)$$

$$\Rightarrow \quad \theta = \frac{\pi}{3}, \frac{2\pi}{3}$$

Hence, the values of θ are $\frac{\pi}{6}, \frac{\pi}{3}, \frac{2\pi}{3}, \frac{5\pi}{6}$.

Exercise - 2

1. Prove that $\frac{\cot\theta - \tan\theta}{1 - 2\sin^2\theta} = \sec\theta \operatorname{cosec}\theta$

2. Prove that $\tan\theta + \sqrt{\frac{1-\sin\theta}{1+\sin\theta}} = \sec\theta$.

3. Prove that $\frac{\tan\alpha + \sec\alpha - 1}{\tan\alpha - \sec\alpha + 1} = \frac{1+\sin\alpha}{\cos\alpha}$

4. Prove that $\sin^8 A - \cos^8 A = (\sin^2 A - \cos^2 A) \times (1 - 2\sin^2 A \cos^2 A)$

5. If $U_n = \sin^n\theta + \cos^n\theta$, prove that $2U_6 - 3U_4 + 1 = 0$

6. Prove that $\frac{1}{\operatorname{cosec}\theta - \cot\theta} - \frac{1}{\sin\theta}$ $= \frac{1}{\sin\theta} - \frac{1}{\operatorname{cosec}\theta + \cot\theta}$

7. Prove that the equation $\frac{(a+b)^2}{4ab} = \sin^2\theta$ is possible only when a = b

8. Prove that $\sin^2\theta + \operatorname{cosec}^2\theta \geq 2$

9. Prove that $\sec^2\theta + \operatorname{cosec}^2\theta \geq 4$

10. If $\sec\theta + \tan\theta = 3$, then find the value of $\cos\theta$.

11. If $\operatorname{cosec}\theta - \cot\theta = 5$, then find the value of $\sin\theta$.

12. If $x = a\cos\theta + b\sin\theta$ and $y = a\sin\theta - b\cos\theta$, then prove that $x^2 + y^2 = a^2 + b^2$

13. If $3\sin\theta + 5\cos\theta = 5$, then prove that $5\sin\theta - 3\cos\theta = \pm 3$.

14. If $x = r\cos\theta\cos\phi, y = r\cos\theta\sin\phi, z = r\sin\theta$, then prove that $x^2 + y^2 + z^2 = r^2$.

15. If $\cos\theta + \sin\theta = \sqrt{2}\cos\theta$, then prove that $\cos\theta - \sin\theta = \sqrt{2}\sin\theta$.

16. If $\tan^2 A = 1 + 2\tan^2 B$, then prove that $\cos^2 B = 2\cos^2 A$.

17.. If $\tan^2\theta = 1 - e^2$, then prove that $\sec\theta + \tan^3\theta.\operatorname{cosec}\theta = (2 - e^2)^{3/2}$

18. If $\sin^2\theta + \sin\theta = 1$, then prove that $\cos^4\theta + \cos^2\theta = 1$

19. If $\sin^2\theta + \sin\theta = 1$, then find the value of $\cos^{12}\theta + 3\cos^{10}\theta + 3\cos^8\theta + \cos^6\theta$.

20. If $\cos^4\theta + \cos^2\theta = 1$, then prove that $\tan^4\theta + \tan^2\theta = 1$

21. If $\sin^4\theta + \sin^2\theta = 1$, then prove that $\cos^8\theta + 2\cos^6\theta + \cos^4\theta = 1$

22. If $\sin\theta + \sin^2\theta + \sin^3\theta = 1$, then prove that, $\cos^6\theta - 4\cos^4\theta + 8\cos^2\theta = 1$

23. If $\sin\theta_1 + \sin\theta_2 + \sin\theta_3 = 1$, then prove that, $\cos\theta_1 + \cos\theta_2 + \cos\theta_3 = 1$

24. If $x = \frac{2\sin\theta}{1+\cos\theta+\sin\theta}$, then prove that, $\frac{1-\cos\theta+\sin\theta}{1+\sin\theta} = x$.

25. Prove

$$\text{tal} - \frac{\sin^2\theta}{1+\cos\theta} + \frac{1+\cos\theta}{\sin\theta} - \frac{\sin\theta}{1-\cos\theta} = \cos\theta.$$

26. Prove that,

$$3(\sin x - \cos x)^4 + 6(\sin x + \cos x)^2 + 4(\sin^6 x + \cos^6 x) = 13$$

27. If $a^3 = \cos ec\theta - \sin\theta$ and $b^3 = \sec\theta - \cos\theta$, then prove that $a^2b^2(a^2 + b^2) = 1.$

28. If $x\sin^3\alpha + y\cos^3\alpha = \sin\alpha\cos\alpha$ and $x\sin\alpha = y\cos\alpha$ then prove that $x^2 + y^2 = 1$.

29.. If $\tan\theta + \sin\theta = m$, $\tan\theta - \sin\theta = n$, then prove that $m^2 - n^2 = 4\sqrt{mn}$.

30. If $\dfrac{\cos^4 x}{\cos^2 y} + \dfrac{\sin^4 x}{\sin^2 y} = 1$, then prove that $\dfrac{\cos^4 y}{\cos^2 x} + \dfrac{\sin^4 y}{\sin^2 x} = 1$

31. If $\dfrac{\sin^4\alpha}{a} + \dfrac{\cos^4\alpha}{b} = \dfrac{1}{a+b}$, then prove that $\dfrac{\sin^8\alpha}{a^3} + \dfrac{\cos^8\alpha}{b^3} = \dfrac{1}{(a+b)^3}$

32.. If $\dfrac{\sin^4\theta}{2} + \dfrac{\cos^4\theta}{3} = \dfrac{1}{5}$, then prove that $\dfrac{\sin^8\theta}{8} + \dfrac{\cos^8\theta}{27} = \dfrac{1}{125}$.

Answers

10. $\frac{3}{5}$ 11. $\frac{5}{13}$

1.9 MEASUREMENT OF THE ANGLES OF DIFFERENT T-RATIOS:

1.9.1 Recognization of the quadrants :

We have introdcuced six t- ratios. Signs of these t-ratios depend upon the quadrant in which the terminal side of the angle lies. We take always the length of OP vector is denoted by 'r', which is always positive.

Thus, $\sin\theta = \dfrac{y}{r}$ has the sign of y, $\cos\theta = \dfrac{x}{r}$ has the sign of x and $\tan\theta = \dfrac{y}{x}$ depends on the signs of both x and y.

Similarly, the signs of other trigonometric functions can be obtained by the signs of x and y.

(A) In first quadrant, we have $x > 0, y > 0$

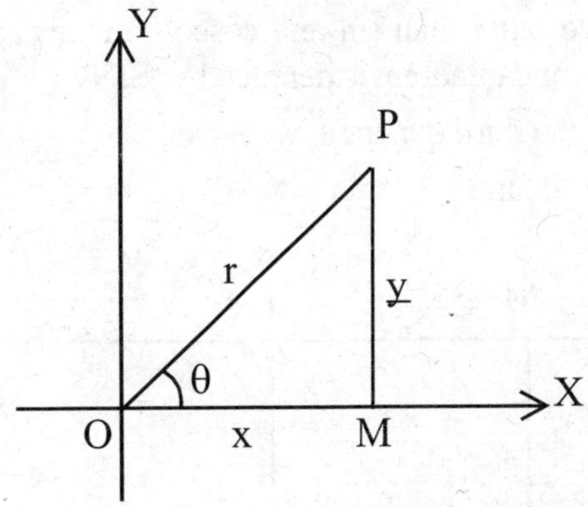

(i) $\sin\theta = \dfrac{y}{r} > 0$, $\text{cosec}\ \theta = \dfrac{r}{y} > 0$

(ii) $\cos\theta = \dfrac{y}{r} > 0$, $\sec\theta = \dfrac{r}{y} > 0$

(iii) $\tan\theta = \dfrac{y}{x} > 0$, $\cot\theta = \dfrac{y}{x} > 0$

Thus, in the first quadrant all trigonometric ratios are positive. Due to this fact, and for this reason, the first quadrant is represented by 'ALL'

(B) In the second quadrant, we have $x < 0$ and $y > 0$

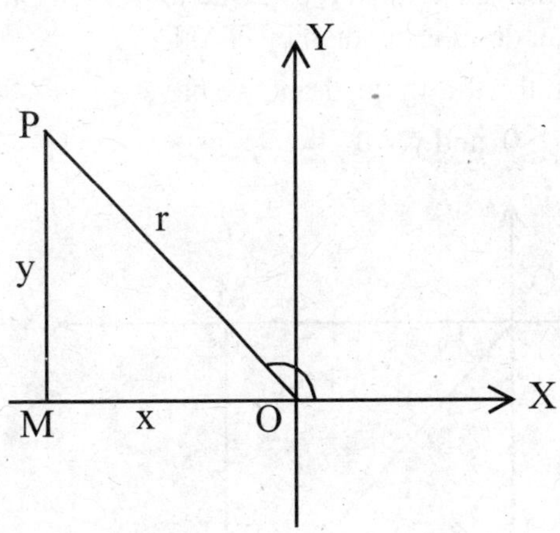

(i) $\sin\theta = \dfrac{y}{r} > 0$, $\text{cosec}\theta = \dfrac{y}{r} > 0$

(ii) $\cos\theta = \dfrac{x}{r} < 0$, $\sec\theta = \dfrac{r}{x} < 0$

(iii) $\tan\theta = \frac{y}{x} < 0$, $\cot\theta = \frac{y}{x} < 0$

Thus, in the second quadrant all t-raios are negative other than sin and cosec. For this reason, second quadrant is denoted by 'SIN'

(C) In the third quadrant, we have

$x < 0$ and $y < 0$

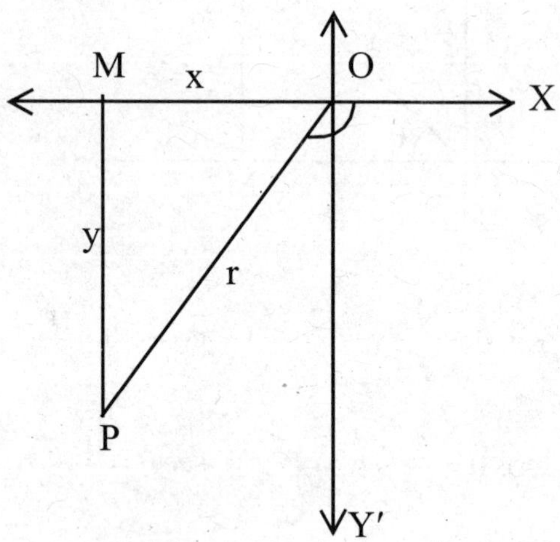

(i) $\sin\theta = \frac{y}{r} < 0\ 0$, $\operatorname{cosec}\theta = \frac{r}{y} < 0$

(ii) $\cos\theta = \frac{x}{r} < 0$, $\sec\theta = \frac{r}{x} < 0$

(iii) $\tan\theta = \frac{y}{x} > 0$, $\cot\theta = \frac{y}{x} > 0$

Thus, in the third quadrant all t-raios are negative other than tan and cot. Due to this reason, third quadrant is denoted by 'TAN'

(D) In the fourth quadrant, we have

$x > 0$ and $y < 0$

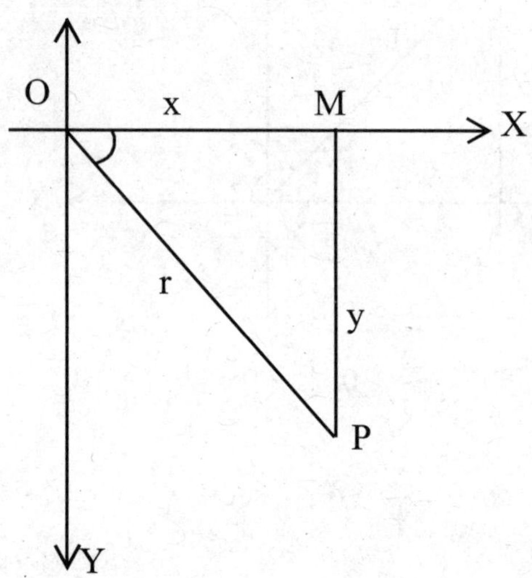

(i) $\sin\theta = \frac{y}{r} < 0$, $\operatorname{cosec}\theta = \frac{r}{y} < 0$

(ii) $\cos\theta = \frac{x}{r} > 0$, $\sec\theta = \frac{r}{x} > 0$

(iii) $\tan\theta = \frac{y}{x} < 0$, $\cot\theta = \frac{y}{x} < 0$

Thus, in the fourth quadrant all t-raios are negative other than cos and sec. For this reason, fourth quadrant is denoted by 'COS'.

(E)

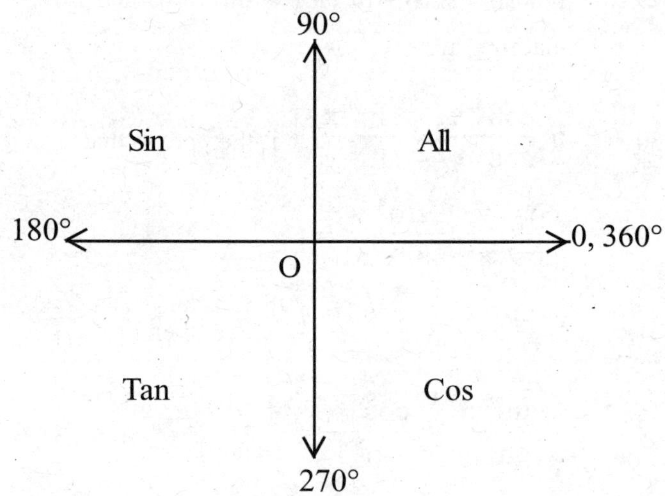

1.9.2 T-ratios of the angle $(-\theta)$, in terms of θ, for all values of θ.

1. (i) $\sin(-\theta) = -\sin\theta$

(ii). $\cos(-\theta) = \cos\theta$

(iii) $\tan(-\theta) = -\tan\theta$

(iv). $\operatorname{cosec}(-\theta) = -\operatorname{cosec}\theta$

(v) $\sec(-\theta) = \sec\theta$

(vi) $\cot(-\theta) = -\cot\theta$

1.9.3 T-ratios of the different angle in terms of θ, for all values of θ.

2 (i). $\sin(90-\theta)$

$= \sin\left(90^\circ \times 1 - \theta\right) = \cos\theta$

(ii) $\sin(90+\theta)$

$= \sin\left(90^\circ \times 1 + \theta\right) = \cos\theta$

(iii) $\sin(180-\theta)$

$= \sin(90^\circ \times 2 - \theta) = \sin\theta$

(iv) $\sin(180 + \theta)$

$= \sin(90^\circ \times 2 + \theta) = -\sin\theta$

(v) $\sin(270 - \theta)$

$= \sin(90^\circ \times 3 - \theta) = -\cos\theta$

(vi) $\sin(270 + \theta)$

$= \sin(90^\circ \times 3 + \theta) = -\cos\theta$

(vii) $\sin(360 - \theta)$

$= \sin(90^\circ \times 4 - \theta) = -\sin\theta$

(viii) $\sin(360 + \theta)$

$= \sin(90^\circ \times 4 + \theta) = \sin\theta$

3 (i). $\cos(90 - \theta)$

$= \cos(90^\circ \times 1 - \theta) = \sin\theta$

(ii) $\cos(90 + \theta)$

$= \cos(90^\circ \times 1 + \theta) = -\sin\theta$

(iii) $\cos(180 - \theta)$

$= \cos(90^\circ \times 2 - \theta) = -\cos\theta$

(iv) $\cos(180 + \theta)$

$= \cos(90^\circ \times 2 + \theta) = -\cos\theta$

(v) $\cos(270 - \theta)$

$= \cos(90^\circ \times 3 - \theta) = -\sin\theta$

(vi) $\cos(270 + \theta)$

$= \cos(90^\circ \times 3 + \theta) = -\sin\theta$

(vii) $\cos(360 - \theta)$

$= \cos(90^\circ \times 4 - \theta) = \cos\theta$

(viii) $\cos(360 + \theta)$

$= \cos(90^\circ \times 4 + \theta) = \cos\theta$.

4. (i). $\tan(90 - \theta)$

$= \tan(90^\circ \times 1 - \theta) = \cot\theta$

(ii) $\tan(90 + \theta)$

$= \tan(90^\circ \times 1 + \theta) = -\cot\theta$

(iii) $\tan(180 - \theta)$

$= \tan(90^\circ \times 2 - \theta) = -\tan\theta$

(iv) $\tan(180 + \theta)$

$= \tan(90^\circ \times 2 + \theta) = \tan\theta$

(v) $\tan(270 - \theta)$

$= \tan(90^\circ \times 3 - \theta) = \cot\theta$

(vi) $\tan(270 + \theta)$

$= \tan(90^\circ \times 3 + \theta) = -\cot\theta$

(vii) $\tan(360 - \theta)$

$= \tan(90^\circ \times 4 - \theta) = -\tan\theta$

(viii) $\tan(360 + \theta)$

$= \tan(90^\circ \times 4 + \theta) = \tan\theta$.

Note : All the above results can be remembered by the following simple rule.

1. If θ be measured with an even multiple of 90° by + or – sign, then the T-ratios remains unaltered (i.e. sine remains sine and cosine remains cosine etc.) and treating θ as an acute angle, the quadrant in which the associated angle lies, is determined and then the sign of the T-ratio is determined by the All – Sin – Tan – Cos formula.

2. If θ be associated with an odd multiple of 90 by +ve or –ve sign, then the T-ratios is altered in form (i.e. sine becomes cosine and cosine becomes sine, tangent becomes cotangent and, conversely, etc.) and the sign of the ratio is determined as in the previous paragraph.

3. If the multiple of 90 is more than 4, then divide it by 4 and find out remainder. If remainder is 0, then the degree lies on right of x-axis, if remainder is 1, then the degree lies on the +ve y-axis, if remainder is 2, then the degree lies on -ve of x-axis and if the remainder is 3, then the degree lies on the -ve of y-axis respectively.

For examples :

i) $\sin(570^\circ)$

$= \sin(90 \times 6 + 30^\circ)$

$= -\sin 30^\circ = -\frac{1}{2}$

ii) $\tan(1950^\circ)$

$= \tan(90 \times 22 - 30^\circ)$

$= -\tan(30°) = -\frac{1}{\sqrt{3}}$.

iii) $\cos(2310°)$

$= \cos(90\times 25 + 60°)$

$= -\sin(60°) = -\frac{\sqrt{3}}{2}$.

1.10 SOME SOLVED EXAMPLES:

Ex-1. Find the value each of

(i) $\sin 120^{o}$

(ii) $\sin 150^{o}$

(iii) $\sin 210^{o}$

(iv) $\sin 225^{o}$

(v) $\sin 300^{o}$

(vi) $\sin 330^{o}$

(vii) $\sin 405^{o}$

(viii) $\sin 650^{o}$

(ix) $\sin 1500^{o}$

(x) $\sin 2013^{o}$

Soln. (i) $\sin(120^{o}) = \sin(90\times 1 + 30^{o})$

$= \cos(30^{o}) = \frac{\sqrt{3}}{2}$

(ii) $\sin(150^{o}) = \sin(90\times 2 - 30^{o})$

$= \sin(30^{o}) = \frac{1}{2}$

(iii) $\sin(210^{o}) = \sin(90\times 2 + 30^{o})$

$= -\sin(30^{o}) = -\frac{1}{2}$

(iv) $\sin(225^{o}) = \sin(90\times 2 + 45^{o})$

$= -\sin(45^{o}) = -\frac{1}{\sqrt{2}}$

(v) $\sin(300^{o}) = \sin(90\times 3 + 30^{o})$

$= -\cos(30^{o}) = -\frac{\sqrt{3}}{2}$

(vi) $\sin(330^{o}) = \sin(90\times 3 + 60^{o})$

$= -\cos(60^{o}) = -\frac{1}{2}$

(vii) $\sin(405^{o}) = \sin(90\times 4 + 45^{o})$

$= \sin(45^{o}) = \frac{1}{\sqrt{2}}$

(viii) $\sin(660^{o}) = \sin(90\times 7 + 30^{o})$

$= \sin(30^{o}) = \frac{1}{2}$

(ix) $\sin(1500^{o}) = \sin(90\times 16 + 60^{o})$

$= \sin(60^{o}) = \frac{\sqrt{3}}{2}$

(x) $\sin(2013^{o}) = \sin(90\times 22 + 33^{o})$

$= -\sin(33^{o})$

Ex-2. Find the value of

$\cos(1^{o})\cos(2^{o})\cos(3^{o})\ldots\ldots\cos(189^{o})$

Soln. We have ,

$\cos(1^{o})\cos(2^{o})\cos(3^{o})\ldots\ldots\cos(189^{o})$

$= \cos(1^{o})\cos(2^{o})\cos(3^{o})\ldots\ldots\cos(89^{o})$
$\cos(90^{o})\cos(91^{o})\ldots\ldots\cos(189^{o})$

$= \cos(1^{o})\cos(2^{o})\cos(3^{o})\ldots\cos(89^{o})$
$\times 0 \times \cos(91^{o})\ldots\ldots\cos(189^{o})$

$= 0$

Ex-3. Find the value of

$\tan(1^{o})\tan(2^{o})\tan(3^{o})\ldots\ldots\tan(89^{o})$

Soln. We have,

$\tan(1^{o})\tan(2^{o})\tan(3^{o})\ldots\ldots\tan(89^{o})$

$= \tan(1^{o})\tan(2^{o})\tan(3^{o})\ldots\ldots\tan(44^{o})$
$\tan(45^{o})\tan(46^{o})\ldots\ldots\tan(87^{o})\tan(88^{o})\tan(89^{o})$

$= \{\tan(1^{o})\times\tan(89^{o})\}.\{\tan(2^{o})\times\tan(88^{o})\}$
$\ldots\ldots\{\tan(44^{o})\times\tan(46^{o})\}.\tan(45^{o})$

$= 1.$

Ex-4. Find the value of

$\tan 35^o.\tan 40^o.\tan 45^o.\tan 50^o.\tan 55^o$

Soln. We have,

$\tan 35^o.\tan 40^o.\tan 45^o.\tan 50^o.\tan 55^o$

$= \{\tan 35^o \times \tan 55^o\}\{\tan 40^o \times \tan 50^o\}$

$.\times \tan 45^o$

$= \{\tan 35^o \times \cot 35^o\}.\{\tan 40^o \times \cot 40^o\}$

$\times \tan 45^o$

$= 1$

Ex-5. Find the value of

$\sin(10^o)+\sin(20^o)+\sin(30^o)$

$+\sin(40^o)+..........+\sin(360^o)$

Soln. We have, $\sin(10^o)+\sin(20^o)+\sin(30^o)$

$+\sin(40^o)+..........+\sin(360^o)$

$= \sin(10^o)+\sin(20^o)+\sin(30^o)$

$+\sin(40^o)+.........+\sin(150^o)$

$+\sin(340^o)+\sin(350^o)+\sin(360^o)$

$= \sin(10^o)+\sin(20^o)+\sin(30^o)$

$+\sin(40^o)+.........+\sin(80^o)$

$+\sin(90^o)+\sin(100^o)$

$+\sin(360^o-40^o)+\sin(360^o-30^o)$

$+\sin(360^o-20^o)+\sin(360^o-10^o)$

$+\sin(360^o)$

$= \sin(10^o)+\sin(20^o)+\sin(30^o)$

$+\sin(40^o)+.........+\sin(80^o)$

$+\sin(90^o)+\sin(100^o)$

$-\sin(40^o)-\sin(30^o)$

$-\sin(20^o)-\sin(10^o)+\sin(180^o)$

$= 0$

Ex-6. Find the value of

$\cos(10^o)+\cos(20^o)+\cos(30^o)$

$+\cos(40^o)+.........+\cos(360^o)$

Soln. We have, $\cos(10^o)+\cos(20^o)+\cos(30^o)$

$+\cos(40^o)+.........+\cos(360^o)$

$= \cos 20^o + \cos 30^o + \cos 40^o +.........+$
$\cos 140^o + \cos 150^o +$
$\cos 160^o + \cos 170^o$
$+ \cos 180^o + (\cos 190^o +$
$\cos 200^o + \cos 210^o + \cos 220^o$
$+..............+ \cos 360^o)$

$= \cos 10^o + \cos 20^o + \cos 30^o + \cos 40^o$
$+......- \cos 40^o - \cos 50^o - \cos 60 -$
$\cos 70^o + \cos 180^o + (\cos 190^o +$
$\cos 200^o + \cos 210^o + \cos 220^o$
$+...........+ \cos 360^o)$

$= \cos 180^o + \cos 360^o$

$= -1 + 1$

$= 0$

Ex-7. Find the value of

$\sin^2 5^o + \sin^2 10^o + \sin^2 15^o + ... + \sin^2 90^o$

Soln. We have,

$\sin^2 5^o + \sin^2 10^o + \sin^2 15^o + ... + \sin^2 90^o$

$= \sin^2 5^o + \sin^2 10^o + \sin^2 15^o +$

$...... + \sin^2 40 + \sin^2 45$

$+ \sin^2 50 + \sin^2 80 + \sin^2 85 + \sin^2 90^o$

$= (\sin^2 5^o + \sin^2 85^o)$

$+(\sin^2 10^o + \sin^2 80^o)$

$+(\sin^2 15^o + \sin^2 75^o)$

$+ ....+(\sin^2 40^o + \sin^2 50^o)$

$+ (\sin^2 45^o + \sin^2 90^o)$

$= (\sin^2 5^o + \cos^2 5^o)$

$+(\sin^2 10^o + \cos^2 10^o)$

$+(\sin^2 15^o + \cos^2 15^o)+ ....+$

$(\sin^2 40^o + \cos^2 40^o)$

$+(\sin^2 45^o + \sin^2 90^o)$

$= (1+1+\ldots\ldots\ldots\text{8 times}) + \left(\frac{1}{2}+1\right)$

$= \left(8+1+\frac{1}{2}\right)$

$= 9\frac{1}{2}$

Ex-8. Find the value of

$$\sin^2\left(\frac{\pi}{18}\right)+\sin^2\left(\frac{\pi}{9}\right)+\sin^2\left(\frac{4\pi}{9}\right)+\sin^2\left(\frac{7\pi}{18}\right)$$

Soln. We have, $\sin^2\left(\frac{\pi}{18}\right)+\sin^2\left(\frac{\pi}{9}\right)$

$+\sin^2\left(\frac{4\pi}{9}\right)+\sin^2\left(\frac{7\pi}{18}\right)$

$= \sin^2\left(\frac{\pi}{18}\right)+\sin^2\left(\frac{2\pi}{18}\right)$

$+\sin^2\left(\frac{8\pi}{18}\right)+\sin^2\left(\frac{7\pi}{18}\right)$

$= \sin^2\left(\frac{\pi}{18}\right)+\sin^2\left(\frac{\pi}{2}-\frac{7\pi}{18}\right)$

$+\sin^2\left(\frac{\pi}{2}-\frac{\pi}{18}\right)+\sin^2\left(\frac{7\pi}{18}\right)$

$= \sin^2\left(\frac{\pi}{18}\right)+\cos^2\left(\frac{7\pi}{18}\right)$

$+\cos^2\left(\frac{\pi}{18}\right)+\sin^2\left(\frac{7\pi}{18}\right)$

$= \left\{\sin^2\left(\frac{\pi}{18}\right)+\cos^2\left(\frac{\pi}{18}\right)\right\}$

$+\left\{\cos^2\left(\frac{7\pi}{18}\right)+\sin^2\left(\frac{7\pi}{18}\right)\right\}$

$= 1+1 = 2$

Exercise -3

1. Find the values of

i)	sin (135°)	ii)	cos (150°)
iii)	tan (120°)	iv)	sin (225°)
v)	sec (240°)	vi)	tan (300°)
vii)	sin (330°)	viii)	tan (315°)
ix)	cos (315°)	x)	sin (405°)

2. Find the values of

i)	sin (675°)	ii)	cos (1230°)
iii)	tan 1020°	iv)	cosec (1305°)
v)	sec (-1035°)	vi)	tan (–1755°)
vii)	sin (1410°)	viii)	cos (1450°)
ix)	tan (2010°)	x)	sin (1950°)

3. Express in terms of ratios of the smallest +ve angles.

i) sin 240°

ii) cos 780°

iii) sin (–1358°)

iv) cosec (–1150°)

v) tan (–1750°)

4. If $\theta = \frac{23\pi}{6}$, then find the value of $\sec\theta - \tan\theta$.

5. Simplify : $\frac{\sin(270+A)\cos(90-A)}{\sin(180-A)\cos(180-A)}$

6. Simplify : tan 25°. tan 35°. tan 45° tan 55°. tan 65°

7. Prove that $\sin^2\frac{\pi}{8}+\sin^2\frac{3\pi}{8}+\sin^2\frac{5\pi}{8}+\sin^2\frac{7\pi}{8} = 2$

8. Prove that $\sin^2\frac{\pi}{4}+\sin^2\frac{3\pi}{4}+\sin^2\frac{5\pi}{4}+\sin^2\frac{7\pi}{4} = 2.$

9. Prove that

i) $\sin^2 5° + \sin^2 10° + \sin^2 15° + \ldots\ldots\ldots + \sin^2 90° = 9\frac{1}{2}$.

ii) $\sin^2 6° + \sin^2 12° + \sin^2 18° + \ldots\ldots\ldots + \sin^2 90° = 8$

iii) $\sin^2 10° + \sin^2 20° + \sin^2 30° + \ldots\ldots\ldots + \sin^2 90° = 5$

10. Prove that

i) $\tan 1°. \tan 2°. \tan 3°..... \tan 89° = 1.$

ii) $\cos 1°. \cos 2°. \cos 3°......\cos 189°. = 0$

11. Solve for θ; $2\sin^2\theta + 3\cos\theta = 0$, where $0 < \theta < 360°$.

12. Solve for θ; $\cos\theta + \sqrt{3}\sin\theta = 2$, where $0 < \theta < 360°$.

13. If $4n\alpha = \pi$, then prove that $\tan\alpha \tan 2\alpha \tan 3\alpha .... \tan(2n-1)\alpha = 1.$

14. If A, B, C, D be the angles of a cyclic quadrilateral ABCD, then prove that

i) $\cos A + \cos B + \cos C + \cos D = 0$

ii) $\tan A + \tan B + \tan C + \tan D = 0.$

iii) $\sin 2A + \sin 2B + \sin 2C + \sin 2D = 0$

15. Prove that $\cos 18° + \cos 162° + \cos 234° + \cos 306° = 0.$

Answers

4. $\sqrt{3}$ **5.** 1 **6.** 1 **7.** $\theta = 120°, 240°$ **8.** $\theta = 60°$

Graph of Trigonometric functions :

1. Graph of f(x) = sin x.

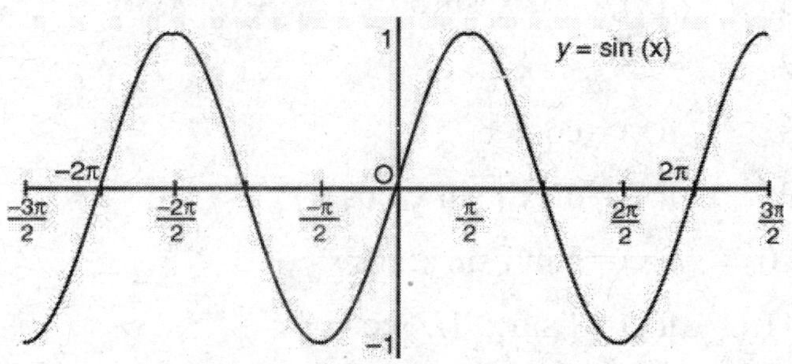

Characteristics of sine function:

1 It is an odd function, since $\sin(-x) = -\sin x$

2. It is a periodic function with period 2π

3. $\sin x = 1 \Rightarrow x = (4n+1)\frac{\pi}{2}, \ n \in I$

4. $\sin x = 0 \Rightarrow x = n\pi, \ n \in I$

5. $\sin x = -1 \Rightarrow x = (4n-1)\frac{\pi}{2}, \ n \in I$

Graph of f(x) = cos x :

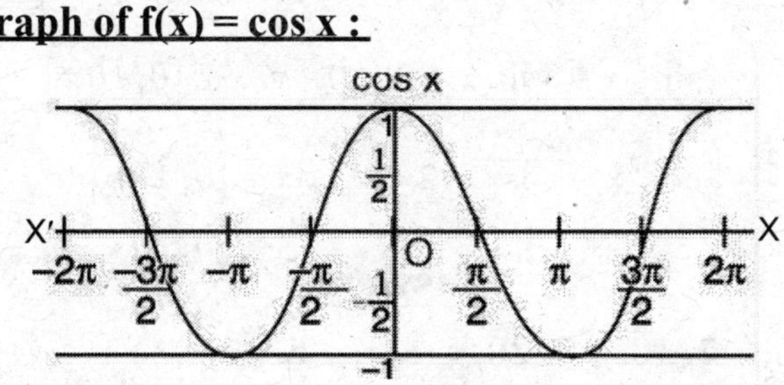

Characteristics of cosine function :

1. It is an even function, since $\cos(-x) = \cos x$

2. It is a periodic function with period 2π.

3. $\cos x = 1 \Rightarrow x = 2n\pi, \ n \in I$

4. $\cos x = 0 \Rightarrow x = (2n+1)\frac{\pi}{2}, \ n \in I$

5. $\cos x = -1 \Rightarrow x = (2n+1)\pi, \ n \in I$.

3. Graph of f(x) = tan x.

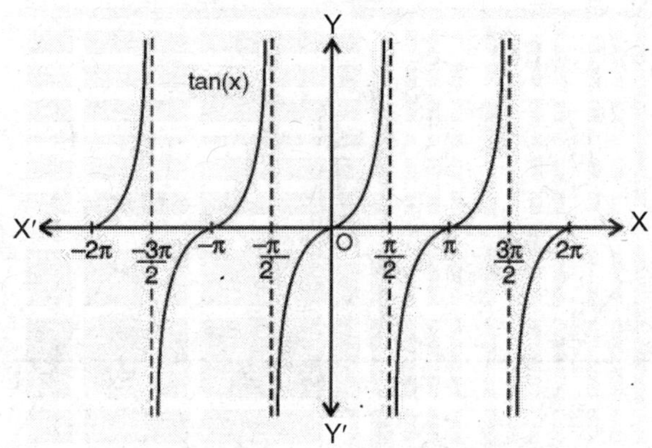

Characteristics of sine function:

1. It is an odd function, since $\tan(-x) = -\tan x$

2. It is a periodic function with period π

3 $\tan x = 1 \Rightarrow x = (4n+1)\frac{\pi}{4}, \ n \in I$

4. $\tan x = 0 \Rightarrow x = n\pi, \ n \in I$

5. $\tan x = -1 \Rightarrow x = (4n-1)\frac{\pi}{4}, \ n \in I$

4. Graph of f(x) = cot x

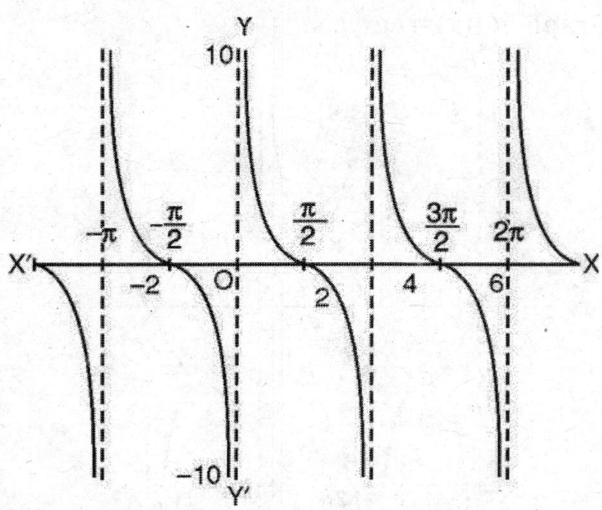

Characteristics of cotangent function :

1. It is an odd function, since cot (-x) = - cotx
2. It is a periodic function with period π
3. $\cot x = 1 \Rightarrow x = (4n+1)\frac{\pi}{4}$, $n \in I$
4. $\cot x = 0 \Rightarrow x = (2n+1)\frac{\pi}{2}$, $n \in I$
5. $\cot x = -1 \Rightarrow x = (4n-1)\frac{\pi}{4}$, $n \in I$

Graph of f(x) = sec x

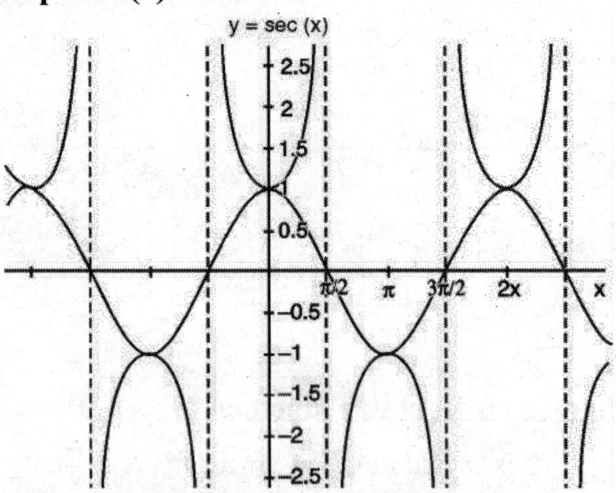

Characteristics of secant function

1. It is an even function, sec(-x) = sec x
2. It is a periodic function with period 2π
3. sec x can never be zero.
4. $\sec x = 1 \Rightarrow x = 2n\pi$, $n \in I$
5. $\sec x = -1 \Rightarrow x = (2n+1)\pi$, $n \in I$

Graph of f(x) = cosec x

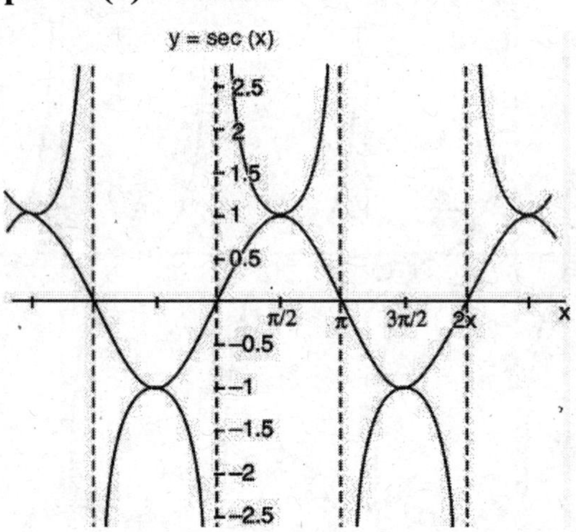

Characteristics of co-secant function :

1 It is an odd function, since cosec(-x) = - cosec x
2. It is a periodic function with period 2π
3. $\operatorname{cosec} x = 1 \Rightarrow x = (4n+1)\frac{\pi}{2}$, $n \in I$
4. cosecx can never be zero.
5. $\sec x = -1 \Rightarrow x = (4n-1)\frac{\pi}{2}$, $n \in I$

Exercise - 4

Q.A. Draw the graphs of

1. f(x) = sin x + 1
2. f(x) = sin x - 1
3. f(x) = - sin x
4. f(x) = 1 - sin x
5. f(x) -1- sinx.
6. f(x) = sin2 x, sin 3 x
7. . . . . ^{2}x
8. $f(x) = \cos^2 x$
9. f(x) = max { sin x, cos x }
10. f(x) = min { sin x, cos x }
11. f(x) = { sin x, 1/2, cos x }
12. f(x) = max { tan x, cot x }
13. f(x) = min { tan x, cot x }

Q. B. Find the number of solutions of

1. $\sin x = \frac{1}{2}$, $\forall\ x \in [0, 6\pi]$
2. $\cos x = \frac{\sqrt{3}}{2}$, $\forall\ x \in [0, 10\pi]$
3. $4\sin^2 x - 1 = 0$, $\forall\ x \in [0, 10\pi]$
4. $\sin^2 x - 3\sin x + 2 = 0$, $\forall\ x \in [0, 10\pi]$
5. $\cos^2 x - \cos x - 2 = 0$, $\forall x \in [0, 10\pi]$

Answers

1. 6, **2.** 10, **3.** 20, **4.** 5, **5.** 10

T-RATIOS OF COMPOUND ANGLES

1.11 DEFINITION :

The algebraic sum or difference of two or more angles is called a compound angle such as $A+B,\ A-B, A+B+C, A+B-C$ etc.

1.11.1 The Addition Formula :

1. $\sin(A+B) = \sin A\cos B + \cos A\sin B$

2. $\cos(A+B) = \cos A\cos B - \sin A\sin B$

3. $\tan(A+B) = \dfrac{\tan A + \tan B}{1-\tan A\tan B}$

Proof: Let $\angle XOP = A$ and $\angle POQ = B$

Then $\angle QOX = A+B$.

Draw PM, QS perpendiculars on OX and PR is parallel to OX.

Clearly, $\angle RPO = A$ and $\angle PQR = A$

In ΔQOS

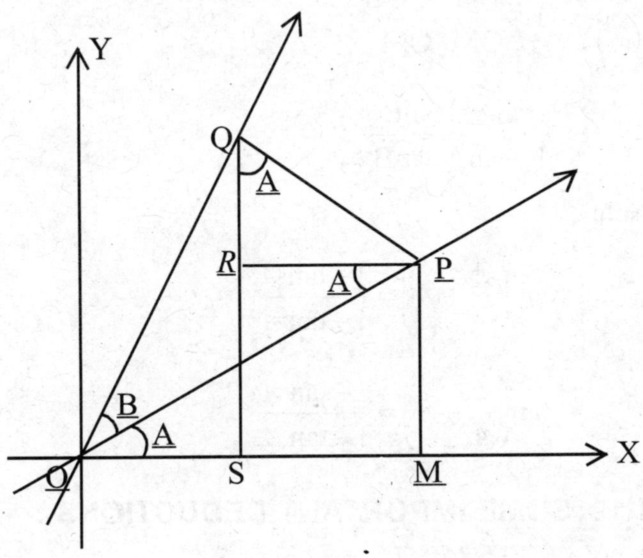

1. $\sin(A+B) = \dfrac{QS}{OQ} = \dfrac{RS+RQ}{OQ}$

$= \dfrac{RS}{OQ} + \dfrac{RQ}{OQ}$

$= \dfrac{PM}{OQ} + \dfrac{RQ}{OQ}$

$= \dfrac{PM}{OP}\cdot\dfrac{OP}{OQ} + \dfrac{QR}{PQ}\cdot\dfrac{PQ}{OQ}$

$=$ sin A cos B + cos A. sin B.

2. $\cos(A+B) = \dfrac{OS}{OQ} = \dfrac{OM-MS}{OQ}$

$= \dfrac{OM}{OQ} - \dfrac{MS}{OQ}$

$= \dfrac{OM}{OQ} - \dfrac{MS}{OQ}$

$= \dfrac{OM}{OP}\cdot\dfrac{OP}{OQ} - \dfrac{PR}{PQ}\cdot\dfrac{PQ}{OQ}$

$=$ cos A.cos B - sin A. sin B.

3. $\tan(A+B) = \dfrac{QS}{OS} = \dfrac{QR+RS}{OS}$

$= \dfrac{QR+RS}{OM-SM}$

$= \dfrac{\dfrac{QR}{OM}+\dfrac{RS}{OM}}{1-\dfrac{SM}{OM}}$

$= \dfrac{\dfrac{QR}{OM}+\dfrac{RS}{OM}}{1-\dfrac{PR}{PQ}\cdot\dfrac{PQ}{OM}}$

$= \dfrac{\dfrac{PM}{OM}+\dfrac{PQ}{OP}}{1-\dfrac{PM}{OQ}\cdot\dfrac{PQ}{OP}}$

$= \dfrac{\tan A + \tan B}{1-\tan A.\tan B}$

1.11.2 Subtraction formula :

1. sin (A – B) = sin A cos B – cos A sin B
2. cos (A – B) = cos A cos B + sin A sin B
3. $\tan(A-B)=\dfrac{\tan A-\tan B}{1+\tan A\tan B}$

Proof: Let $\angle POX = A$ and $\angle QOX = B$

Clearly, $\angle POQ = A - B$

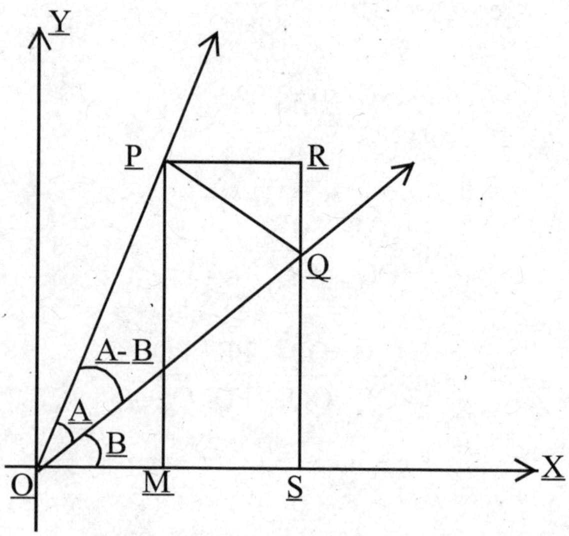

Draw PM, RS perpendiculars on OX and PR is parallel to OX.

From geometry, we can say that, $\angle PQR = A$

In ΔQOS,

1. $\sin(A-B)$

$=\dfrac{QS}{OQ}=\dfrac{RS-RQ}{OQ}=\dfrac{PM-RQ}{OQ}$

$=\dfrac{PM}{OQ}-\dfrac{RQ}{OQ}$

$=\dfrac{PM}{OP}\cdot\dfrac{OP}{OQ}-\dfrac{RQ}{PQ}\cdot\dfrac{PQ}{OQ}$

= sin A. cos B - cos A. sin B

2. $\cos(A-B)$

$=\dfrac{OQ}{OQ}=\dfrac{OM+MS}{OQ}=\dfrac{OM+PR}{OQ}$

$=\dfrac{OM}{OQ}+\dfrac{PR}{OQ}$

$=\dfrac{OM}{OP}\cdot\dfrac{OP}{OQ}+\dfrac{PR}{PQ}\cdot\dfrac{PQ}{OQ}$

= cos A. cos B + sin A. sin B.

3. $\tan(A-B)$

$=\dfrac{QS}{OS}=\dfrac{RS-QR}{OM+MS}=\dfrac{RS-QR}{OM+PR}$

$=\dfrac{\dfrac{PL}{OM}-\dfrac{QR}{OM}}{1+\dfrac{PR}{OM}}$

$=\dfrac{\dfrac{PL}{OM}-\dfrac{QR}{OM}}{1+\dfrac{PR}{PM}\cdot\dfrac{PM}{OM}}$

$=\dfrac{\dfrac{PM}{OM}-\dfrac{PQ}{OM}}{1+\dfrac{PM}{OM}\cdot\dfrac{PQ}{OM}}$

$=\dfrac{\tan A-\tan B}{1+\tan A.\tan B}$.

Note :

1. $\tan\left(\dfrac{\pi}{4}+A\right)=\dfrac{1+\tan A}{1-\tan A}$

2. $\tan\left(\dfrac{\pi}{4}-A\right)=\dfrac{1-\tan A}{1+\tan A}$

1.12 SOME IMPORTANT DEDUCTIONS :

Deduction 1. sin (A + B) sin (A – B)
= sin² A – sin² B = cos² B – cos² A

Proof: We have $\sin(A+B)\sin(A-B)$

$=\{\sin A\cos B+\cos A\sin B\}$

$\times\{\sin A\cos B-\cos A\sin B\}$

$=\{\sin^2 A\cos^2 B-\cos^2 A\sin^2 B\}$

$= \{\sin^2A(1-\sin^2B)-(1-\sin^2A)\sin^2B\}$

$= \{\sin^2A - \sin^2A\sin^2B - \sin^2B - \sin^2A\sin^2B\}$

$= \sin^2A - \sin^2B$

$= (1-\cos^2A)-(1-\cos^2B)$

$= \cos^2B - \cos^2A$

Deduction -2. cos (A + B) cos (A – B)
$= \cos^2A - \sin^2B = \cos^2B - \sin^2A$

Proof: We have, $\cos(A+B)\cos(A-B)$

$= \{\cos A\cos B + \sin A\sin B\}$
$\times\{\cos A\cos B - \sin A\sin B\}$

$= \{\cos^2A\cos^2B - \sin^2A\sin^2B\}$

$= \{\cos^2A(1-\sin^2B)-(1-\cos^2A)\sin^2B\}$

$= \{\cos^2A - \cos^2A\sin^2B - \sin^2B + \cos^2A\sin^2B\}$

$= \cos^2A - \sin^2B$

Deduction-3. $\cot(A+B) = \dfrac{\cot A\cot B - 1}{\cot B + \cot A}$

Proof: We have, $\cot(A+B)$

$$= \frac{\cos(A+B)}{\sin(A+B)}$$

$$= \frac{\cos A\cos B - \sin A\sin B}{\sin A\cos B + \cos A\sin B}$$

$$= \frac{\dfrac{\cos A\cos B}{\sin A\sin B} - \dfrac{\sin A\sin B}{\sin A\sin B}}{\dfrac{\sin A\cos B}{\sin A\sin B} + \dfrac{\cos A\sin B}{\sin A\sin B}}$$

$$= \frac{\cot A\cot B - 1}{\cot B + \cot A}$$

Deduction -4. $\cot(A-B) = \dfrac{\cot A\cot B + 1}{\cot B - \cot A}$

Proof: We have, $\cot(A-B)$

$$= \frac{\cos(A-B)}{\sin(A-B)}$$

$$= \frac{\cos A\cos B + \sin A\sin B}{\sin A\cos B - \cos A\sin B}$$

$$= \frac{\dfrac{\cos A\cos B}{\sin A\sin B} + \dfrac{\sin A\sin B}{\sin A\sin B}}{\dfrac{\sin A\cos B}{\sin A\sin B} - \dfrac{\cos A\sin B}{\sin A\sin B}}$$

$$= \frac{\cot A\cot B + 1}{\cot B - \cot A}$$

Deduction -5. sin(A + B + C)
= cos A cos B cos C (tan A + tan B + tan C
– tan A tan B tan C)

Proof: We have $\sin(A+B+C)$

$= \sin(A+B)\cos C + \cos(A+B)\sin C$

$= \{\sin A.\cos B + \cos A.\sin B\}\cos C$
$+ \{\cos A\cos B - \sin A\sin B\}\sin C$

$= \sin A.\cos B.\cos C + \sin B.\cos A.\cos C$
$+ \sin C.\cos A\cos B - \sin A.\sin B.\sin C$

$= \cos A.\cos B.\cos C$
$\{\tan A + \tan B + \tan C - \tan A.\tan B.\tan C\}$

Deduction-6. cos(A + B + C)
= cos A cos B cos C (1–tan A tan B –
tan B tan C – tan C tan A)

Proof: We have, $\cos(A+B+C)$

$= \cos(A+B)\cos C - \sin(A+B)\sin C$

$= \{\cos A\cos B - \sin A\sin B\}\cos C$
$- \{\sin A\cos B + \cos A\sin B\}\sin C$

$= \cos A\cos B\cos C - \sin A\sin B\cos C$
$\{-\sin A\sin C\cos B - \cos A\sin B\sin C\}$

$= \cos A\cos B\cos C(1 - \tan A - \tan B - \tan C)$

Deduction -7. tan (A + B + C)

$$= \frac{\tan A + \tan B + \tan C - \tan A\tan B\tan C}{1 - \tan A\tan B - \tan B\tan C - \tan C\tan A}$$

Proof: We have, $\tan(A+B+C)$

$$= \frac{\sin(A+B+C)}{\cos(A+B+C)}$$

$$\frac{\cos A \cos B \cos C\{\tan A + \tan B + \tan C - \tan A \tan B \tan C\}}{\cos A \cos B \cos C\{1 - \tan A \tan B - \tan B \tan C - \tan C \tan A\}}$$

$$= \left(\frac{\tan A + \tan B + \tan C - \tan A \tan B \tan C}{1 - \tan A \tan B - \tan B \tan C - \tan C \tan A}\right)$$

1.13 SOME SOLVED EXAMPLES :

Ex-1. Find the values of

(i) $\sin(15^\circ)$,

(ii) $\cos(15^\circ)$,

(iii) $\tan(15^\circ)$

Soln. We have, (i)

$$\sin(15^\circ) = \sin(45^\circ - 30^\circ)$$

$$= \sin(45^\circ)\cos(30^\circ) - \cos(45^\circ)\sin(30^\circ)$$

$$= \frac{1}{2}.\frac{\sqrt{3}}{2} - \frac{1}{2}.\frac{1}{\sqrt{2}}$$

$$= \frac{\sqrt{3}-1}{2\sqrt{2}}$$

(ii) $$\cos(15^\circ) = \cos(45^\circ - 30^\circ)$$

$$= \cos(45^\circ)\cos(30^\circ) + \sin(45^\circ)\sin(30^\circ)$$

$$= \frac{1}{\sqrt{2}}.\frac{\sqrt{3}}{2} + \frac{1}{\sqrt{2}}.\frac{1}{2}$$

$$= \frac{\sqrt{3}+1}{2\sqrt{2}}$$

(iii) $$\tan(15^\circ) = \tan(45^\circ - 30^\circ)$$

$$= \frac{\tan 45^\circ - \tan 30^\circ}{1 + \tan 45^\circ . \tan 30^\circ}$$

$$= \frac{1 - \frac{1}{\sqrt{3}}}{1 + \frac{1}{\sqrt{3}}}$$

$$= \frac{\sqrt{3}-1}{\sqrt{3}+1}$$

$$= \frac{\left(\sqrt{3}-1\right)^2}{3-1}$$

$$= \frac{3+1-2\sqrt{3}}{2}$$

$$= \frac{4-2\sqrt{3}}{2}$$

$$= 2-\sqrt{3}$$

Note :

(i) $$\cot(15^\circ) = \frac{1}{\tan(15^\circ)} = \frac{1}{2-\sqrt{3}} = 2+\sqrt{3}$$

(ii) $$\tan(105^\circ) = -\cot(15^\circ) = -\left(2+\sqrt{3}\right)$$

(iii) $$\cot(105^\circ) = -\tan(15^\circ) = -\left(2-\sqrt{3}\right)$$

$$= \sqrt{3}-2$$

Ex-2. Find the value of $\tan(75^\circ) + \cot(75^\circ)$

Soln. We have, $\tan(75^\circ) + \cot(75^\circ)$

$$= \cot(15^\circ) + \tan(15^\circ)$$

$$= \left(2+\sqrt{3}\right) + \left(2-\sqrt{3}\right)$$

$$= 4$$

Ex-3. Prove that

$$\cos(9^\circ) + \sin(9^\circ) = \sqrt{2}\sin(54^\circ)$$

Soln. We have, $\cos(9^\circ) + \sin(9^\circ)$

$$= \sqrt{2}\left(\frac{1}{\sqrt{2}}\cos(9^\circ) + \frac{1}{\sqrt{2}}\sin(9^\circ)\right)$$

$$= \sqrt{2}\left(\sin(45^\circ)\cos(9^\circ) + \cos(45^\circ)\sin(9^\circ)\right)$$

$$= \sqrt{2}\left(\sin(45^\circ + 9^\circ)\right)$$

$$= \sqrt{2}\sin(54^\circ)$$

Ex-4. Prove that

$$\tan(70^\circ) = 2\tan(50^\circ) + \tan(20^\circ)$$

Soln. We have, $\tan(70^\circ) = \tan(50^\circ + 20^\circ)$

$\Rightarrow \quad \tan(70^\circ) = \dfrac{\tan 50^\circ + \tan 20^\circ}{1 - \tan 50^\circ \tan 20^\circ}$

$\Rightarrow \quad \tan(70^\circ) - \tan(70^\circ)\tan(50^\circ)\tan(20^\circ)$

$= \tan 50^\circ + \tan 20^\circ$

$\Rightarrow \quad \tan(70^\circ) - \tan(70^\circ)\tan(50^\circ)\cot(70^\circ)$

$= \tan 50^\circ + \tan 20^\circ$

$\Rightarrow \quad \tan(70^\circ) - \tan(50^\circ) = \tan 50^\circ + \tan 20^\circ$

$\Rightarrow \quad \tan(70^\circ) = 2\tan 50^\circ + \tan 20^\circ$

Ex-5. If $A + B = 45^\circ$, then find the value of $(1 + \tan A)(1 + \tan B)$

Soln. We have, $A + B = 45^\circ$

$\Rightarrow \quad \tan(A + B) = \tan(45^\circ)$

$\Rightarrow \quad \tan(A + B) = 1$

$\Rightarrow \quad \dfrac{\tan A + \tan B}{1 - \tan A . \tan B} = 1$

$\Rightarrow \quad \tan A + \tan B = 1 - \tan A . \tan B$

$\Rightarrow \quad \tan A + \tan B + \tan A . \tan B = 1$

$\Rightarrow \quad 1 + \tan A + \tan B + \tan A . \tan B = 1 + 1 = 2$

$\Rightarrow \quad (1 + \tan A) + \tan B(1 + \tan A) = 2$

$\Rightarrow \quad (1 + \tan A)(1 + \tan B) = 2$

Ex-6. Find the value of

$(1 + \tan 20^\circ)(1 + \tan 24^\circ)$

$(1 + \tan 25^\circ)(1 + \tan 21^\circ)$

Soln. We have,

$(1 + \tan 20^\circ)(1 + \tan 24^\circ)(1 + \tan 25^\circ)$

$(1 + \tan 21^\circ)$

$= \{(1 + \tan 20^\circ)(1 + \tan 25^\circ)\}$

$\times \{(1 + \tan 24^\circ)(1 + \tan 21^\circ)\}$

$= 2 \times 2$

$= 4$

Ex-7. Find the value of

$(1 + \tan 225^\circ)(1 + \tan 230^\circ)$

$(1 + \tan 240^\circ)(1 - \tan 200^\circ)$

$(1 - \tan 205^\circ)(1 - \tan 215^\circ)$

Soln. We have, $(1 + \tan 225^\circ)(1 + \tan 230^\circ)$

$(1 + \tan 240^\circ)(1 - \tan 200^\circ)$

$(1 - \tan 205^\circ)(1 - \tan 215^\circ)$

$= \{(1 + \tan 225^\circ)(1 + \tan(-200^\circ))\}$

$\{(1 + \tan 230^\circ)(1 + \tan(-205^\circ))\}$

$\times \{(1 + \tan 240^\circ)(1 + \tan(-215^\circ))\}$

$= 2 \times 2 \times 2$

$= 8$

Ex-8. If $\tan\alpha + \tan\beta = a$, $\cot\alpha + \cot\beta = b$ and $\tan(\alpha + \beta) = c$ then find a relation in a, b and c

Soln. .We have, $\tan(\alpha + \beta) = c$

$\Rightarrow \quad \dfrac{\tan\alpha + \tan\beta}{1 - \tan\alpha \tan\beta} = c$

$\Rightarrow \quad \dfrac{a}{1 - \tan\alpha \tan\beta} = c$

$\Rightarrow \quad \tan\alpha \tan\beta = \dfrac{a}{c} - 1 = \dfrac{a - c}{c}$ ....(i)

Now, $\cot\alpha + \cot\beta = b$

$\Rightarrow \quad \dfrac{1}{\tan\alpha} + \dfrac{1}{\tan\beta} = b$

$\Rightarrow \quad \dfrac{\tan\alpha + \tan\beta}{\tan\alpha . \tan\beta} = b$

$\Rightarrow \quad \tan\alpha . \tan\beta = \dfrac{a}{b}$ ....(ii)

From (i) and (ii), we get, $\dfrac{a - c}{c} = \dfrac{a}{b}$

$\Rightarrow \quad ac + bc = ab$

Which is the required relation.

Ex-9. If $\tan\beta = \dfrac{n\sin\alpha\cos\alpha}{1-n\sin^2\alpha}$, then prove that

$\tan(\alpha-\beta) = (1-n)\tan\alpha$

Soln. We have, $\tan(\alpha-\beta) = \dfrac{\tan\alpha - \tan\beta}{1+\tan\alpha\tan\beta}$

$$= \frac{\dfrac{\sin\alpha}{\cos\alpha} - \left(\dfrac{n\sin\alpha\cos\alpha}{1-n\sin^2\alpha}\right)}{1+\dfrac{\sin\alpha}{\cos\alpha}\times\left(\dfrac{n\sin\alpha\cos\alpha}{1-n\sin^2\alpha}\right)}$$

$$= \frac{\sin\alpha(1-n\sin^2\alpha) - n\sin\alpha\cos^2\alpha}{\cos\alpha(1-n\sin^2\alpha)+n\sin^2\alpha.\cos\alpha}$$

$$= \frac{\sin\alpha - n\sin\alpha(\sin^2\alpha+\cos^2\alpha)}{\cos\alpha - n\sin^2\alpha.\cos\alpha + n\sin^2\alpha.\cos\alpha}$$

$$= \frac{\sin\alpha - n\sin\alpha}{\cos\alpha}$$

$$= \frac{(1-n)\sin\alpha}{\cos\alpha}$$

$$= (1-n)\tan\alpha$$

Ex-10. If $\cos(\alpha-\beta)+\cos(\beta-\gamma)+\cos(\gamma-\alpha) = -\dfrac{3}{2}$, then prove that, $\cos\alpha+\cos\beta+\cos\gamma = 0$ and $\sin\alpha+\sin\beta+\sin\gamma = 0$

Soln. We have,

$$\cos(\alpha-\beta)+\cos(\beta-\gamma)+\cos(\gamma-\alpha) = -\frac{3}{2}$$

$$\Rightarrow 2(\cos(\alpha-\beta)+\cos(\beta-\gamma)+\cos(\gamma-\alpha))+3 = 0$$

$$\Rightarrow 2(\cos\alpha\cos\beta+\cos\beta\cos\gamma+\cos\gamma\cos\alpha) +2(\sin\alpha\sin\beta+\sin\beta\sin\gamma+\sin\gamma\sin\alpha)+3 = 0$$

$$\Rightarrow (\cos^2\alpha+\sin^2\alpha)+(\cos^2\beta+\sin^2\beta) +(\cos^2\gamma+\sin^2\gamma)+ 2(\cos\alpha\cos\beta+\cos\beta\cos\gamma+\cos\gamma\cos\alpha) +2(\sin\alpha\sin\beta+\sin\beta\sin\gamma+\sin\gamma\sin\alpha) = 0$$

$$\Rightarrow \cos^2\alpha+\cos^2\beta+\cos^2\gamma +2\cos\alpha\cos\beta+2\cos\beta\cos\gamma + 2\cos\gamma\cos\alpha + \sin^2\alpha+\sin^2\beta+\sin^2\gamma +2\sin\alpha\sin\beta+2\sin\beta\sin\gamma +2\sin\gamma\sin\alpha = 0$$

$$\Rightarrow (\cos\alpha+\cos\beta+\cos\gamma)^2 + (\sin\alpha+\sin\beta+\sin\gamma)^2 = 0$$

$$\Rightarrow (\cos\alpha+\cos\beta+\cos\gamma) = 0$$

and $(\sin\alpha+\sin\beta+\sin\gamma) = 0$

Ex-11. If $\sin\alpha\sin\beta - \cos\alpha\cos\beta + 1 = 0$, then prove that $1+\cot\alpha\tan\beta = 0$

Soln. We have, $\sin\alpha\sin\beta - \cos\alpha\cos\beta + 1 = 0$

$$\Rightarrow \cos\alpha\cos\beta - \sin\alpha\sin\beta = 1$$

$$\Rightarrow \cos(\alpha+\beta) = 1$$

Therefore,

$$\sin(\alpha+\beta) = \sqrt{1-\cos^2(\alpha+\beta)} = \sqrt{1-1} = 0$$

Now, $1+\cot\alpha.\tan\beta$

$$= 1+\frac{\cos\alpha.\sin\beta}{\sin\alpha.\cos\beta}$$

$$= \frac{\sin\alpha.\cos\beta+\cos\alpha.\sin\beta}{\sin\alpha.\cos\beta}$$

$$= \frac{\sin(\alpha+\beta)}{\sin\alpha.\cos\beta}$$

$$= 0.$$

Ex-12: If $\alpha+\beta = 90^{\circ}$ and $\beta+\gamma = \alpha$, then prove that $\tan\alpha = \tan\beta + 2\tan\gamma$

Soln. Given $\beta+\gamma = \alpha$

$$\Rightarrow \tan(\beta+\gamma) = \tan\alpha$$

$$\Rightarrow \tan(\beta+\gamma) = \tan(90^{\circ}-\alpha)$$

$$\Rightarrow \tan(\beta+\gamma) = \cot\alpha$$

$$\Rightarrow \frac{\tan\beta+\tan\gamma}{1-\tan\beta\tan\gamma} = \cot\alpha$$

$\Rightarrow \quad \dfrac{\tan\beta + \tan\gamma}{1 - \tan\beta\tan\gamma} = \cot\beta$

$\Rightarrow \quad \tan\beta + \tan\gamma = \cot\beta - \tan\beta.\cot\beta.\tan\gamma$

$\Rightarrow \quad \tan\beta + \tan\gamma = \cot\beta - \tan\gamma$

$\Rightarrow \quad \tan\beta + \tan\gamma = \cot(90^o - \alpha) - \tan\gamma$

$\Rightarrow \quad \tan\beta + \tan\gamma = \tan\alpha - \tan\gamma$

$\Rightarrow \quad \tan\alpha = \tan\beta + 2\tan\gamma$

Ex-13. If $x + y + z = 0$, then prove that

$$\cot(x+y-z).\cot(y+z-x) + \cot(y+z-x).\cot(z+x-y) + \cot(z+x-y).\cot(x+y-z) = 1$$

Soln. Let $A = x + y - z$, $B = y + z - x$, $C = x + y - z$

Then, $A + B + C$

$= (x+y-z) + (y+z-x) + (x+y-z)$

$= (x+y+z) = 0$

$\Rightarrow \quad A + B = -C$

$\Rightarrow \quad \cot(A+B) = \cot(-C)$

$\Rightarrow \quad \dfrac{\cot A\cot B - 1}{\cot A + \cot B} = \cot(-C)$

$\Rightarrow \quad \dfrac{\cot A\cot B - 1}{\cot A + \cot B} = -\cot C$

$\Rightarrow \quad \cot A\cot B - 1 = -\cot A\cot C - \cot B\cot C$

$\Rightarrow \quad \cot A\cot B + \cot A\cot C + \cot B\cot C = 1$

$\Rightarrow \quad \cot(x+y-z)\cot(y+z-x) + \cot(y+z-x)\cot(z+x-y) + \cot(z+x-y)\cot(x+y-z) = 1$

Ex-14.: If $2\tan\alpha = 3\tan\beta$, then show that,

$$\tan(\alpha-\beta) = \frac{\sin 2\beta}{5 - \cos 2\beta}$$

Soln. We have,

$$\tan(\alpha-\beta) = \frac{\tan\alpha - \tan\beta}{1 + \tan\alpha.\tan\beta}$$

$$= \frac{\frac{3}{2}\tan\beta - \tan\beta}{1 + \frac{3}{2}\tan\beta\tan\beta}$$

$$= \frac{\tan\beta}{2 + 3\tan^2\beta}$$

$$= \frac{\frac{\sin\beta}{\cos\beta}}{2 + 3\frac{\sin^2\beta}{\cos^2\beta}}$$

$$= \frac{\sin\beta\cos\beta}{2\cos^2\beta + 3\sin^2\beta}$$

$$= \frac{\sin\beta\cos\beta}{2 + \sin^2\beta}$$

$$= \frac{2\sin\beta\cos\beta}{4 + 2\sin^2\beta}$$

$$= \frac{\sin 2\beta}{4 + 1 - \cos 2\beta}$$

$$= \frac{\sin 2\beta}{5 - \cos 2\beta}$$

Ex-15. If $\dfrac{x}{\cos\theta} = \dfrac{y}{\cos\left(\theta + \frac{2\pi}{3}\right)} = \dfrac{z}{\cos\left(\theta - \frac{2\pi}{3}\right)}$, then find the value of $x + y + z$.

Soln. Given

$$\frac{x}{\cos\theta} = \frac{y}{\cos\left(\theta + \frac{2\pi}{3}\right)} = \frac{z}{\cos\left(\theta - \frac{2\pi}{3}\right)} = m \text{ (say)}$$

Now, $x + y + z$

$$= m\times\left(\cos\theta + \cos\left(\theta + \frac{2\pi}{3}\right) + \cos\left(\theta - \frac{2\pi}{3}\right)\right)$$

$$= m\times\left(\cos\theta + 2\cos\theta.\cos\left(\frac{2\pi}{3}\right)\right)$$

$$= m\times\left(\cos\theta + 2\cos\theta.\left(-\frac{1}{2}\right)\right)$$

$= m\times(\cos\theta-\cos\theta)$

$= 0$

Hence, the value of x + y + z is zero.

Exercise - 5

1. Find the values of sin15°, cos15°, sin 75°, cos 75°.
2. Find the values of tan 15°, cot 15°, tan 75°, cot 75°.
3. Prove that
   (i) $\tan 15° + \cot 15° = 4$
   (ii) $\tan 75° + \cot 75° = 4$
   (iii) $\cot 15° - \tan 15° = 2\sqrt{3}$
   (iv) $\tan 75° - \cot 75° = 2\sqrt{3}$.
4. Prove that $\cos 18° - \sin 18° = \sqrt{2}\sin 27°$.
5. Prove that $\sin(n+1)x\,\sin(n+2)x + \cos(n+1)x\,\cos(n+2)x = \cos x$
6. Prove that $1+\tan x.\tan\left(\frac{x}{2}\right)=\sec x$.
7. Prove that $\cot x-\cot 2x=\operatorname{cosec} 2x$.
8. Prove that $\frac{\tan 69^o+\tan 66^o}{1-\tan 69^o\tan 66^o} = -1$
9. Prove that tan 70° = 2 tan 50° + tan 20°.
10. If $A+B=\frac{\pi}{4}$, then prove that (1 + tanA) (1 + tanB)= 2
11. Prove that $\frac{\cos 20°-\sin 20°}{\cos 20°+\sin 20°}=\tan 25°$
12. Prove that $\frac{\cos 7°+\sin 7°}{\cos 7°-\sin 7°}=\tan 52°$.
13. Prove that tan20° + tan 25° + tan 20° tan 25° = 1.
14. Prove that tan 13 A – tan 9 A – tan 4A = tan 4A. tan9A. tan13A
15. Prove that tan 9A - tan 7A - tan 2A = tan 2A. tan 7A. tan 9A
16. Prove that cot x cot 2 x - cot 2 x cot 3 x - cot 3x cot 2 x = 1
17. If $\tan\alpha=\frac{m}{m+1}$, $\tan\beta=\frac{1}{2m+1}$, then prove that $\alpha+\beta=\frac{\pi}{4}$.
18. Find the value of
    (i) $\sin^2 75^o-\sin^2 15^o$
    (ii) $\cos^2 75^o-\sin^2 15^o$.
19. Prove that $\cos^2\left(\frac{\pi}{4}+\theta\right)-\sin^2\left(\frac{\pi}{4}-\theta\right)=0$
20. Prove that $\sin^2 B=\sin^2 A+\sin^2(A-B) - 2\sin A\cos A\sin(A-B)$.
21. Prove that $\cos 2x\cos 2y+\cos^2(x+y) - \cos^2(x-y)=\cos(2x+2y)$
22. If $\sin\theta=\frac{x-y}{x+y}$, then prove that $\tan\left(\frac{\pi}{4}+\frac{\theta}{2}\right)=\pm\sqrt{\frac{x}{y}}$.
23. If $\tan\beta=\frac{n\sin\alpha\cos\alpha}{1-n\sin^2\alpha}$, prove that $\tan(\alpha-\beta)=(1-n)\tan\alpha$.
24. If $\tan\alpha=\frac{Q\sin\beta}{P+Q\cos\beta}$, prove that $\tan(\beta-\alpha)=\frac{P\sin\alpha}{Q+P\cos\beta}$
25. If $\cos(\alpha-\beta)+\cos(\beta-\gamma)+\cos(\gamma-\alpha)=-\frac{3}{2}$ prove that $\cos\alpha+\cos\beta+\cos\gamma=0$ and $\sin\alpha+\sin\beta+\sin\gamma=0$
26. If $\tan(\alpha+\theta)=n\tan(\alpha-\theta)$, show that $\frac{\sin 2\theta}{\sin 2\alpha}=\frac{n-1}{n+1}$
27. If $\sin\alpha+\sin\beta=a$ and $\cos\alpha+\cos\beta=b$, then show that

(i) $\cos(\alpha+\beta)=\dfrac{b^2-a^2}{b^2+a^2}$

(ii) $\sin(\alpha+\beta)=\dfrac{2ab}{b^2+a^2}$

28. If α and β are the roots of $a\cos\theta+b\sin\theta=c$, then prove that

(i) $\cos(\alpha+\beta)=\dfrac{a^2-b^2}{a^2+b^2}$

(ii) $\cos(\alpha-\beta)=\dfrac{2c-(a^2+b^2)}{a^2+b^2}$

29. If α and β are the roots of $a\tan\theta+b\sec\theta=c$, then show that

$\tan(\alpha+\beta)=\dfrac{2ac}{a^2-c^2}$

30. If $\tan(\pi\cos\theta)=\cot(\pi\sin\theta)$, prove that

$\cos\left(\theta-\dfrac{\pi}{4}\right)=\pm\dfrac{1}{2\sqrt{2}}$.

31. If $\tan\theta=\dfrac{x\sin\varphi}{1-x\cos\varphi}$ and

$\tan\varphi=\dfrac{y\sin\theta}{1-y\cos\theta}$,

then prove that, $x\sin\varphi=y\sin\theta$

TRANSFORMATION FORMULAE

➤ **Transformation of products into sums or differences :**

1. $2\sin A\cos B=\sin(A+B)+\sin(A-B)$
2. $2\cos A\sin B=\sin(A+B)-\sin(A-B)$
3. $2\cos A\cos B=\cos(A+B)+\cos(A-B)$
4. $2\sin A\sin B=\cos(A-B)-\cos(A+B)$

Proof: As we know that

$\sin(A+B)=\sin A.\cos B+\cos A.\sin B$ ....(i)

$\sin(A-B)=\sin A.\cos B-\cos A.\sin B$ ......(ii)

Adding (i) and (ii), we get

1. $2\sin A.\cos B=\sin(A+B)+\sin(A-B)$

Subtracting (i) and (ii), we get

2. $2\cos A.\sin B=\sin(A+B)-\sin(A-B)$

Also, we have,

$\cos(A+B)=\cos A.\cos B-\sin A.\sin B$......(iii)

$\cos(A-B)=\cos A.\cos B+\sin A.\sin B$..........(iv)

Adding, (iii) and (iv), we get,

3. $2\cos A.\cos B=\cos(A+B)+\cos(A-B)$

Subtracting (iv) from (iii), we get,

4. $2\sin A.\sin B=\cos(A-B)-\cos(A+B)$.

➤ **Transformations of sums or differences into Products :**

1. $\sin C+\sin D=2\sin\dfrac{C+D}{2}\cos\dfrac{C-D}{2}$
2. $\sin C-\sin B=2\cos\dfrac{C+D}{2}\sin\dfrac{C-D}{2}$
3. $\cos C+\cos D=2\cos\dfrac{C+D}{2}\cos\dfrac{C-D}{2}$
4. $\cos C-\tan D=-2\sin\dfrac{C+D}{2}\sin\dfrac{C-D}{2}$.

Proof: As we know that,

$\sin(A+B)+\sin(A-B)=2\sin A\cos B$ ....(i)

$\sin(A+B)-\sin(A-B)=2\cos A\sin B$ ....(ii)

$\cos(A+B)+\cos(A-B)=2\cos A\cos B$ ....(iii)

$\cos(A+B)-\cos(A-B)=-2\sin A\sin B$ ....(iv)

Put $A+B=C$ & $A-B=D$

$\Rightarrow\quad A=\dfrac{C+D}{2}$ & $B=\dfrac{C-D}{2}$

From (i), we get

1. $\sin C+\sin D=2\sin\left(\dfrac{C+D}{2}\right)\cos\left(\dfrac{C-D}{2}\right)$

From (ii), we get

2. $\sin C-\sin D=2\cos\left(\dfrac{C+D}{2}\right)\sin\left(\dfrac{C-D}{2}\right)$

From (iii), we get

3. $\cos C+\cos D=2\cos\left(\frac{C+D}{2}\right)\cos\left(\frac{C-D}{2}\right)$

From (iv), we get,

4. $\cos C-\cos D=-2\sin\left(\frac{C+D}{2}\right)\sin\left(\frac{C-D}{2}\right)$

Some solved examples :

Ex-1. Prove that $\sin(47^o)+\cos(77^o)=\cos(17^o)$

Soln. We have,

$\sin(47^o)+\cos(77^o)=\sin(47^o)+\sin(13^o)$

$= 2\sin\left(\frac{47^o+13^o}{2}\right)\cos\left(\frac{47^o-13^o}{2}\right)$

$= 2\sin(30^o)\cos(17^o)$

$= 2\times\frac{1}{2}\times\cos(17^o)$

$= \cos(17^o)$

Ex-2. Prove that

$\cos(80^o)+\cos(40^o)-\cos(20^o)=0$

Soln. We have, $\cos(80^o)+\cos(40^o)-\cos(20^o)$

$= 2\cos\left(\frac{80^o+40^o}{2}\right)\cos\left(\frac{80^o-40^o}{2}\right)-\cos(20^o)$

$= 2\cos(60^o)\cos(20^o)-\cos(20^o)$

$= 2\times\frac{1}{2}\times\cos(20^o)-\cos(20^o)$

$= \cos(20^o)-\cos(20^o)$

$= 0$

Ex-3. Prove that

$\sin(10^o)+\sin(20^o)+\sin(40^o)+\sin(50^o)-\sin(70^o)-\sin(80^o)=0$

Soln. We have, $\sin(10^o)+\sin(20^o)+\sin(40^o)+\sin(50^o)-\sin(70^o)-\sin(80^o)$

$= \{\sin(50^o)+\sin(10^o)\}+\{\sin(40^o)+\sin(20^o)\}-\sin(70^o)-\sin(80^o)$

$= 2\sin\left(\frac{50^o+10^o}{2}\right)\cos\left(\frac{50^o-10^o}{2}\right)+2\sin\left(\frac{40^o+20^o}{2}\right)\cos\left(\frac{40^o-20^o}{2}\right)-\sin(70^o)-\sin(80^o)$

$= 2\sin(30^o)\cos(20^o)+2\sin(30^o)\cos(10^o)-\sin(70^o)-\sin(80^o)$

$= \cos(20^o)+\cos(10^o)-\sin(70^o)-\sin(80^o)$

$= \cos(20^o)+\cos(10^o)-\cos(20^o)-\cos(10^o)$

$= 0$

Ex-4. Prove $\frac{\sin A+\sin 2A+\sin 4A+\sin 5A}{\cos A+\cos 2A+\cos 4A+\cos 5A}=\tan 3A$

Soln. We have,

$\frac{\sin A+\sin 2A+\sin 4A+\sin 5A}{\cos A+\cos 2A+\cos 4A+\cos 5A}$

$= \frac{(\sin 5A+\sin A)+(\sin 4A+\sin 2A)}{(\cos 5A+\cos A)+(\cos 4A+\cos 2A)}$

$= \frac{2\sin 3A\cos 2A+2\sin 3A\cos A}{2\cos 3A\cos 2A+2\cos 3A\cos A}$

$= \frac{\sin 3A(\cos 2A+\cos A)}{\cos 3A(\cos 2A+\cos A)}$

$= \frac{\sin 3A}{\cos 3A}$

$= \tan 3A.$

Hence, the result.

Ex-5. Prove that

$\cos\alpha+\cos\beta+\cos\gamma+\cos(\alpha+\beta+\gamma)$

$$= 4\cos\left(\frac{\alpha+\beta}{2}\right)\cos\left(\frac{\beta+\gamma}{2}\right)\cos\left(\frac{\gamma+\alpha}{2}\right)$$

Soln. We have

$$\cos\alpha+\cos\beta+\cos\gamma+\cos(\alpha+\beta+\gamma)$$

$$= (\cos\alpha+\cos\beta)+(\cos(\alpha+\beta+\gamma)+\cos\gamma)$$

$$= 2\cos\left(\frac{\alpha+\beta}{2}\right)\cos\left(\frac{\alpha-\beta}{2}\right) + 2\cos\left(\frac{\alpha+\beta+\gamma+\gamma}{2}\right)\cos\left(\frac{\alpha+\beta+\gamma-\gamma}{2}\right)$$

$$= 2\cos\left(\frac{\alpha+\beta}{2}\right)\cos\left(\frac{\alpha-\beta}{2}\right) + 2\cos\left(\frac{\alpha+\beta+2\gamma}{2}\right)\cos\left(\frac{\alpha+\beta}{2}\right)$$

$$= 2\cos\left(\frac{\alpha+\beta}{2}\right)\times \left\{\cos\left(\frac{\alpha-\beta}{2}\right)+2\cos\left(\frac{\alpha+\beta+2\gamma}{2}\right)\right\}$$

$$= 2\cos\left(\frac{\alpha+\beta}{2}\right)\times \cos\left(\frac{\frac{\alpha-\beta}{2}+\frac{\alpha+\beta+2\gamma}{2}}{2}\right)$$

$$\cos\left(\frac{\frac{\alpha-\beta}{2}-\frac{\alpha+\beta+2\gamma}{2}}{2}\right)$$

$$= 2\cos\left(\frac{\alpha+\beta}{2}\right)\times\left\{\cos\left(\frac{\alpha+\gamma}{2}\right)\cos\left(\frac{\beta+\gamma}{2}\right)\right\}$$

Ex-6. If $\sin A-\sin B=\frac{1}{2}$ and $\cos A-\cos B=\frac{1}{3}$, then find $\tan\left(\frac{A+B}{2}\right)$

Soln. Given $\sin A-\sin B=\frac{1}{2}$ ....(i)

and $\cos A-\cos B=\frac{1}{3}$ ....(ii)

Dividing (i) by (ii), we get,

$$\frac{\sin A-\sin B}{\cos A-\cos B}=\frac{1/2}{1/3}=\frac{3}{2}$$

$$\Rightarrow \frac{2\cos\left(\frac{A+B}{2}\right)\sin\left(\frac{A-B}{2}\right)}{-2\sin\left(\frac{A+B}{2}\right)\sin\left(\frac{A-B}{2}\right)}=\frac{3}{2}$$

$$\Rightarrow \frac{\cos\left(\frac{A+B}{2}\right)}{\sin\left(\frac{A+B}{2}\right)}=-\frac{3}{2}$$

$$\Rightarrow \cot\left(\frac{A+B}{2}\right)=-\frac{3}{2}$$

$$\Rightarrow \tan\left(\frac{A+B}{2}\right)=-\frac{2}{3}.$$

Exercise - 6

1. Express as a sum or difference :

i) $2\sin3\alpha\ \cos2\beta$

ii) $\frac{1}{4}\cos10\alpha\ \cos20\beta$

iii) $2\sin5\theta\cos\theta$

iv) $\sin75^\circ\cos15^\circ$

v) $\cos75^\circ\cos15^\circ$

2. Express as a product :

i) $\cos\phi-\cos5\phi$

ii) $\cos45^\circ+\sin75^\circ$

iii) $\cos6\theta-\cos8\theta$

iv) $\cos4\theta+\cos8\theta$

v) $\sin6\theta-\sin2\theta$

3. Prove that :

i) $\frac{\sin5A-\sin3A}{\cos5A+\cos3A}=\tan A$

ii) $\frac{\sin A+\sin3A}{\cos A+\cos3A}=\tan2A$

iii) $\dfrac{\sin A+\sin B}{\cos A+\cos B}=\tan\left(\dfrac{A+B}{2}\right)$

4. Prove that :

i) $\sin 38°+\sin 22°=\sin 82°$

ii) $\sin 105°+\cos 105°=\cos 45°$

iii) $\cos 55°+\cos 65°+\cos 175°=0$.

iv) $\cos 20°+\cos 100°+\cos 120°=0$

v) $\sin 50°-\sin 70+\sin 10°=0$

5. Prove that : $(\cos\alpha+\cos\beta)^2$

$$+(\sin\alpha+\sin\beta)^2=4\cos^2\left(\frac{\alpha-\beta}{2}\right)$$

6. Prove that : $(\cos\alpha-\cos\beta)^2$

$$+(\sin\alpha-\sin\beta)^2=4\sin^2\left(\frac{\alpha-\beta}{2}\right)$$

7. Prove that

i) $\cos 20 \text{ co } 40 \cos 80=\dfrac{1}{8}$

ii) $\sin 20.\ \sin 40 \sin 80=\dfrac{\sqrt{3}}{8}$

iii) $\sin 10 \sin 50.\ \sin 60.\ \sin 70=\dfrac{\sqrt{3}}{16}$

iv) $\cos 10.\ \cos 30.\ \cos 50.\ \cos 70°=\dfrac{3}{16}$

8. Prove that

(i) $(\sin\alpha-\sin\beta)^2+(\cos\alpha-\cos\beta)^2$

$=4\sin^2\left(\dfrac{\alpha-\beta}{2}\right)$

(ii) If $\cos A+\cos B=\dfrac{1}{2}$, $\sin A+\sin B=\dfrac{1}{4}$, then

prove that $\tan\left(\dfrac{A+B}{2}\right)=\dfrac{1}{2}$.

9. Prove that :

i) $\dfrac{\sin A+\sin 3A+\sin 5A}{\cos A+\cos 3A+\cos 5A}=\tan 3A$

ii) $\dfrac{\cos 4x+\cos 3x+\cos 12x}{\sin 4x+\sin 3x+\sin 2x}=\cot 3x$

iii) $\dfrac{\sin A+\sin 3A+\sin 5A+\sin 7A}{\cos A+\cos 3A+\cos 5A+\cos 7A}$

$=\tan 4A$

10. If $\operatorname{cosec} A+\sec A=\operatorname{cosec} B+\sec B$, then prove that $\tan A\tan B=\cot\left(\dfrac{A+B}{2}\right)$

11. If $\sin 2A=\lambda\sin 2B$, then prove that

$$\frac{\tan(A+B)}{\tan(A-B)}=\frac{\lambda+1}{\lambda-1}$$

12. Find the value of $\sqrt{3}\cot 20°-4\cos 20°$.

13. If $\sin A+\sin B=a$ and $\cos A+\cos B=b$, then find $\cos(A+B)$.

14. If $2\cos A=x+\dfrac{1}{x}$, $2\cos B=y+\dfrac{1}{y}$, then find $\cos(A-B)$.

15. Prove that $\sin 47°+\sin 61°-\sin 11°-\sin 25°$. $=\cos 7$.

16. If $\tan\alpha=\dfrac{m}{m+1}$, and $\tan\beta=\dfrac{1}{2m+1}$, then find $\tan(\alpha+\beta)$.

17. Find the number of integral values of k for which $7\cos x+5\sin x=2k+1$ has a solution.

Answers

12. –1, **13.** $\left(\dfrac{b^v-\lambda^v}{b^v+a^v}\right)$, **14.** $\dfrac{1}{2}\left(xy+\dfrac{1}{\lambda y}\right)$

16. 1, **17.** 7

1.14 MULTIPLE ANGLES

1.14.1 Definition :

An angle, of the form nA, $n\in Z$, is called a multiple angle of A. 2A, 3A, 4A etc are multiple angles of A.

1.14.2 Trigonometrical ratios of 2A in terms of t-ratio of A :

1. $\sin 2A=2\sin A\cos A$
2. $\cos 2A=\cos^2 A-\sin^2 A=2\cos^2 A-1=1-2\sin^2 A$.
3. $\tan 2A=\dfrac{2\tan A}{1-\tan^2 A}$

Proof: 1. We know that,

$$\sin(A+B)=\sin A\cos B+\cos A\sin B$$

Put B = A, we get,

sin 2 A = sin A. cos A. + sin A. cos A

= 2 sin A. cos A.

2. As we know that,

$$\cos(A+B) = \cos A.\cos B - \sin A.\sin B$$

Put B = A, we get,

cos 2 A = cosA. cos A - sin A. sinA

= $\cos^2 A - \sin^2 A$.

= $\cos^2 A - (1 - \cos^2 A)$

= $2\cos^2 A - 1$

= $2(1 - \sin^2 A) - 1$

= $1 - 2\sin^2 A$.

3. As we know that,

$$\tan(A+B) = \frac{\tan A + \tan B}{1 - \tan A.\tan B}$$

Put B = A, we get,

$$\tan 2A = \frac{\tan A + \tan A}{1 - \tan A.\tan A} = \frac{2\tan A}{1 - \tan^2 A}$$

1.14.3 T-ratios of angle 2A

4. (i) $\sin 2A = \dfrac{2\tan A}{1+\tan^2 A}$,

(ii) $\cos 2A = \dfrac{1-\tan^2 A}{1+\tan^2 A}$

5. (i) $1 - \cos 2A = 2\sin^2 A$,

(ii) $1 + \cos 2A = 2\cos^2 A$

6. (i) $\tan A = \dfrac{\sin 2A}{1+\cos 2A}$,

(ii) $\tan A = \dfrac{1-\cos 2A}{\sin 2A}$

Proof: 4.(i) As we know that, sin 2A

= 2 sin A cos A

$$= \frac{2\sin A\cos A}{1}$$

$$= \frac{2\sin A\cos A}{\cos^2 A + \sin^2 A}$$

$$= \frac{\dfrac{2\sin A\cos A}{\cos^2 A}}{1 + \dfrac{\sin^2 A}{\cos^2 A}}$$

$$= \frac{2\tan A}{1+\tan^2 A}.$$

(ii) Also, we have, cos 2 A

= $\cos^2 A - \sin^2 A$

$$= \frac{\cos^2 A - \sin^2 A}{1}$$

$$= \frac{\cos^2 A - \sin^2 A}{\cos^2 A + \sin^2 A}$$

$$= \frac{1 - \dfrac{\sin^2 A}{\cos^2 A}}{1 + \dfrac{\sin^2 A}{\cos^2 A}}$$

$$= \frac{1-\tan^2 A}{1+\tan^2 A}.$$

5 .(i) we have, 1 - cos 2 A

= $1 - (1 - 2\sin^2 A)$

= $2\sin^2 A$

(ii) Also, we have, 1 + cos 2 A

= $1 + (2\cos^2 A - 1)$

= $2\cos^2 A$.

6. (i) tan A

$$= \frac{\sin A}{\cos A} = \frac{2\sin A\cos A}{2\cos^2 A} = \frac{\sin 2A}{1+\cos 2A}$$

(ii) tan A

$$= \frac{\sin A}{\cos A} = \frac{2\sin^2 A}{2\sin A\cos A} = \frac{1-\cos 2A}{\sin 2A}$$

1.14.4 Trigonometrical ratios of 3A in terms of t-ratio of A.

7. $\sin 3A = 3\sin A - 4\sin^3 A$

8. $\cos 3A = 4\cos^3 A - 3\cos A$

9. $\tan 3A = \dfrac{3\tan A - \tan^3 A}{1 - 3\tan^2 A}$

Proof: 7. We have, sin 3A

= sin (2A + A)

= sin 2 A cos A + cos 2A. sin A

= 2 sin A.cos A.cos A

+ $(1 - 2\sin^2 A)\sin A$

$= 2\sin A.\cos^2 A + (1 - 2\sin^2 A).\sin A$

$= 2\sin A(1 - \sin^2 A) + (1 - 2\sin^2 A).\sin A$

$= 2\sin A - 2\sin^3 A + \sin A - 2\sin^3 A$

$= 3\sin A - 4\sin^3 A.$

8. We have, $\cos 3A$

$= \cos(2A + A)$

$= \cos 2A.\cos A - \sin 2A.\sin A$

$= (2\cos^2 A - 1)\cos A - 2\sin^2 A.\cos A$

$= (2\cos^2 A - 1)\cos A - 2\cos A(1 - \cos^2 A)$

$= 2\cos^3 A - \cos A - 2\cos A + 2\cos^3 A.$

$= 4\cos^3 A - 3\cos A.$

9. we have, $\tan 3A$

$= \tan(2A + A)$

$$= \frac{\tan 2A + \tan A}{1 - \tan 2A.\tan A}$$

$$= \frac{\dfrac{2\tan A}{1 - \tan^2 A} + \tan A}{1 - \dfrac{2\tan A}{1 - \tan^2 A}.\tan A}$$

$$= \frac{2\tan A + \tan A - \tan^3 A}{1 - \tan^2 A - 2\tan^2 A}$$

$$= \frac{3\tan A - \tan^3 A}{1 - 3\tan^2 A}$$

1.15 SOME IMPORTANT DEDUCTIONS

Deduction-1 $\sin^2 A = \frac{1}{2}(1 - \cos 2A)$

Proof: We have,

$$\sin^2 A = \frac{1}{2}(2\sin^2 A) = \frac{1}{2}(1 - \cos 2A)$$

Deduction -2. $\cos^2 A = \frac{1}{2}(1 + \cos 2A)$

Proof: We have,

$$\cos^2 A = \frac{1}{2}(2\cos^2 A) = \frac{1}{2}(1 + \cos 2A)$$

Deduction -3. $\sin^3 A = \frac{1}{4}(3\sin A - \sin 3A)$

Proof: We have, $\sin 3A = 3\sin A - 4\sin^3 A$

$\Rightarrow \quad 4\sin^3 A = 3\sin A - \sin 3A$

$\Rightarrow \quad \sin^3 A = \frac{1}{4}(3\sin A - \sin 3A)$

Deduction-4. $\cos^3 A = \frac{1}{4}(\cos 3A + 3\cos A)$

Proof: We have, $\cos 3A = 4\cos^3 A - 3\cos A$

$\Rightarrow \quad 4\cos^3 A = \cos 3A + 3\cos A$

$\Rightarrow \quad \cos^3 A = \frac{1}{4}(\cos 3A + 3\cos A)$

Deduction-5. $\sin A \sin(60 - A).\sin(60 + A)$

$= \frac{1}{4}\sin 3A$

Proof: We have, $\sin A.\sin(60^\circ - A)\sin(60^\circ + A)$

$= \sin A.(\sin^2 60^\circ - \sin^2 A)$

$= \sin A.\left(\frac{3}{4} - \sin^2 A\right)$

$= \frac{\sin A.}{4}(3 - 4\sin^2 A)$

$= \frac{1}{4}(3\sin A - 4\sin^3 A)$

$= \frac{1}{4} \times \sin 3A$

Deduction-6. $\cos A.\cos(60 - A).\cos(60 + A) = \frac{1}{4}\cos 3A$

Proof: We have,

$\cos A.\cos(60^\circ - A)\cos(60^\circ + A)$

$= \cos A.(\cos^2 60^\circ - \sin^2 A)$

$= \cos A.\left(\frac{1}{4} - 1 + \cos^2 A\right)$

$= \cos A.\left(-\frac{3}{4} + \cos^2 A\right)$

$= \frac{\cos A}{4}.\left(-3+4\cos^2 A\right)$

$= \frac{1}{4}.\left(-3\cos A+4\cos^3 A\right)$

$= \frac{1}{4}.\left(4\cos^3 A - 3\cos A\right)$

$= \frac{1}{4}\times \cos 3A$

Deduction-7. tan A. tan (60 – A). tan (60 + A) = tan 3A

Proof: We have, $\tan A.\tan\left(60^{\circ}-A\right)\tan\left(60^{\circ}+A\right)$

$= \frac{\sin A.\sin\left(60^{\circ}-A\right)\sin\left(60^{\circ}+A\right)}{\cos A.\cos\left(60^{\circ}-A\right)\cos\left(60^{\circ}+A\right)}$

$= \frac{\frac{1}{4}\sin 3A}{\frac{1}{4}\cos 3A}$

$= \frac{\sin 3A}{\cos 3A}$

$=$ tan 3A .

Deduction -8. sin 4A = 4 sin cos A – 8 cos A $\sin^3$A

Proof: We have, sin 4A

$=$ 2 sin 2A. cos 2A

$=$ 2 (2 sin A cos A) (1 - 2 $\sin^2$ A)

$=$ 4 sin A.cos A (1 - 2 $\sin^2$ A)

$=$ 4 sin A. cos A - 8 $\sin^3$ A. cos A

Deduction-9. cos 4 A = 1 – 8 $\sin^2$ A + 8 $\sin^4$ A

Proof: We have, cos 4A

$=$ cos 2 (2 A)

$=$ 1- 2 $\sin^2$ (2A)

$=$ 1 - 2 (2 sin A. cos A) ^{2}

$=$ 1 - 8 $\sin^2$ A. $\cos^2$ A

$=$ 1 - 8 $\sin^2$ A (1 - $\sin^2$ A)

$=$ 1 - 8 $\sin^2$ A + 8 $\sin^4$ A

Deduction-10. $\tan 4A = \frac{4\tan A - 4\tan^3 A}{1-6\tan^2 A+\tan^4 A}$

Proof: We have, tan 4 A

$= \tan 2.(2A)$

$= \frac{2\tan 2A}{1+\tan^2 2A}$

$= \frac{\frac{4\tan A}{1-\tan^2 A}}{1-\left(\frac{2\tan A}{1-\tan^2 A}\right)^2}$

$= \frac{4\tan A\left(1-\tan^2 A\right)}{\left(1-\tan^2 A\right)^2-4\tan^2 A}$

$= \frac{4\tan A-4\tan^3 A}{1-6\tan^2 A+\tan^4 A}$

Deduction -11: sin 5 A = 16 $\sin^5$ A – 20 $\sin^3$ A + 5 sin A

Proof: We have sin 5A

$= \sin\left(3A+2A\right)$

$=$ sin 3A cos 2A + cos 3A. sin 2A

$= \left(3\sin A-4\sin^3 A\right)\left(1-2\sin^2 A\right)$
$+2\left(4\cos^3 A-3\cos A\right)\sin A\cos A$

$= \left(3\sin A-4\sin^3 A\right)\left(1-2\sin^2 A\right)$
$+2\left(4\cos^2 A-3\right)\sin A\cos^2 A$

$= \left(3\sin A-4\sin^3 A\right)\left(1-2\sin^2 A\right)$
$+2\left(1-4\sin^2 A\right)\left(\sin A-\sin^3 A\right)$

$= \left(3\sin A-4\sin^3 A-6\sin^3 A+8\sin^5 A\right)$
$+2\left(\sin A-4\sin^3 A-\sin^3 A+4\sin^5 A\right)$

$=$ 5 sin A - 20 $\sin^3$ A + 16 $\sin^5$A

$=$ 16 $\sin^5$ A - 20 $\sin^3$ A + 5 sin A.

Deduction-12: cos 5 A = 16 $\cos^5$ A – 20 $\cos^3$ A + 5 cos A

Proof: We have cos 5 A

$=$ cos (3A + 2 A)

$=$ cos 3A cos 2A - sin 3A sin 2 A

$=$ (4 $\cos^3$ A - 3 cos A) (2 $\cos^2$ A - 1)
- (3 sinA - 4 $\sin^3$A) (2 sin A cos A)

$=$ 8 $\cos^5$ A - 6 $\cos^3$ A - 4 $\cos^3$ A + 3 cos A

$- (3 - 4\sin^2 A)\, 2\cos A\, (1 - \cos^2 A)$

$= 8\cos^5 A - 10\cos^3 A + 3\cos A - (4\cos^2 A - 1)(2\cos A - 2\cos^3 A)$

$= 8\cos^5 A - 10\cos^3 A + 3\cos A - 8\cos^3 A + 2\cos A + 8\cos^5 A - 2\cos^3 A$

$= 16\cos^5 A - 20\cos^3 A + 5\cos A.$

Deduction -13: $\sin 6A = (6\sin A - 32\sin^3 A + 32\sin^5 A)\cos A.$

Proof: We have $\sin 6A$

$= \sin 2(3A)$

$= 2\sin 3A.\cos 3A$

$= 2(3\sin A - 4\,\text{Sin}^3 A)(4\cos^3 A - 3\cos A)$

$= 2(3\sin A - 4\sin^3 A)(1 - 4\sin^2 A)\cos A$

$= 2(3\sin A - 4\sin^3 A - 12\sin^3 A + 16\sin 5A)\cos A$

$= 2(3\sin A - 16\sin^3 A + 16\sin^5 A)\cos A$

$= (6\sin A - 32\sin^3 A + 32\sin^5 A)\cos A$

Deduction-14. $\cos 6A = 32\cos^6 A - 48\cos^4 A + 18\cos^2 A - 1$

Proof: We have $\cos 6A$

$= \cos 2(3A)$

$= 2\cos^2(3A) - 1$

$= 2(4\cos^3 A - 3\cos A)^2 - 1$

$= 2(16\cos^6 A - 24\cos^4 A + 9\cos^2 A) - 1$

$= 32\cos^6 A - 48\cos^4 A + 18\cos^2 A - 1$

1.15 SOME SOLVED EXAMPLES

Ex-1. Prove that $\dfrac{1 - \cos 2\theta}{\sin 2\theta} = \tan\theta$

Soln. We have,

$$\frac{1 - \cos 2\theta}{\sin 2\theta} = \frac{2\sin^2\theta}{2\sin\theta\cos\theta} = \frac{\sin\theta}{\cos\theta} = \tan\theta$$

Ex-2. Prove that $\dfrac{1 + \cos 2\theta}{\sin 2\theta} = \cot\theta$

Soln. We have,

$$\frac{1 + \cos 2\theta}{\sin 2\theta} = \frac{2\cos^2\theta}{2\sin\theta\cos\theta} = \frac{\cos\theta}{\sin\theta} = \cot\theta$$

Ex-3. Prove that $\cot\theta - \tan\theta = 2\cot(2\theta)$

Soln. We have,

$$\cot\theta - \tan\theta = \frac{\cos\theta}{\sin\theta} - \frac{\sin\theta}{\cos\theta}$$

$$= \frac{\cos^2\theta - \sin^2\theta}{\sin\theta\cos\theta}$$

$$= \frac{2\cos 2\theta}{2\sin\theta\cos\theta}$$

$$= \frac{2\cos 2\theta}{\sin 2\theta}$$

$$= 2\cot 2\theta$$

Ex-4. If $\tan\theta = \dfrac{b}{a}$, prove that

$a\cos(2\theta) + b\sin(2\theta) = a$

Soln. We have, $a\cos(2\theta) + b\sin(2\theta)$

$$= a\left(\frac{1 - \tan^2\theta}{1 + \tan^2\theta}\right) + b\left(\frac{2\tan\theta}{1 + \tan^2\theta}\right)$$

$$= a\left(\frac{1 - \frac{b^2}{a^2}}{1 + \frac{b^2}{a^2}}\right) + b\left(\frac{2\frac{b}{a}}{1 + \frac{b^2}{a^2}}\right)$$

$$= a\left(\frac{a^2 - b^2}{a^2 + b^2}\right) + b\left(\frac{2ab}{a^2 + b^2}\right)$$

$$= \frac{a(a^2 - b^2 + 2b^2)}{a^2 + b^2}$$

$$= \frac{a(a^2 + b^2)}{a^2 + b^2}$$

$= a.$

Ex-5. Prove that $\sqrt{3}\,\text{cosec}(20^\circ) - \sec(20^\circ) = 4$

Soln. We have, $\sqrt{3}\,\text{cosec}(20^\circ) - \sec(20^\circ)$

$$= \frac{\sqrt{3}}{\sin(20^\circ)} - \frac{1}{\cos(20^\circ)}$$

$$= \frac{\sqrt{3}\cos(20^\circ) - \sin(20^\circ)}{\sin(20^\circ)\cos(20^\circ)}$$

$$= \frac{4\left(\frac{\sqrt{3}}{2}\cos(20^\circ) - \frac{1}{2}\sin(20^\circ)\right)}{2\sin(20^\circ)\cos(20^\circ)}$$

$$= \frac{4(\sin(60^\circ)\cos(20^\circ) - \cos(60^\circ)\sin(20^\circ))}{2\sin(20^\circ)\cos(20^\circ)}$$

$$= \frac{4(\sin(60^\circ - 20^\circ))}{\sin(40^\circ)}$$

$$= 4$$

Ex-6. Prove that $\tan(9^\circ) - \tan(27^\circ) - \tan(63^\circ) + \tan(81^\circ) = 4$

Soln. We have,

$$\tan(9^\circ) - \tan(27^\circ) - \tan(63^\circ) + \tan(81^\circ)$$

$$= \{\tan(9^\circ) + \tan(81^\circ)\} - \{\tan(27^\circ) + \tan(63^\circ)\}$$

$$= \{\tan(9^\circ) + \cot(9^\circ)\} - \{\tan(27^\circ) + \cot(27^\circ)\}$$

$$= \left\{\frac{\sin(9^\circ)}{\cos(9^\circ)} + \frac{\cos(9^\circ)}{\sin(9^\circ)}\right\} - \left\{\frac{\sin(27^\circ)}{\cos(27^\circ)} + \frac{\cos(27^\circ)}{\sin(27^\circ)}\right\}$$

$$= \left\{\frac{\sin^2(9^\circ) + \cos^2(9^\circ)}{\sin(9^\circ)\cos(9^\circ)}\right\} - \left\{\frac{\sin^2(27^\circ) + \cos^2(27^\circ)}{\sin(27^\circ)\cos(27^\circ)}\right\}$$

$$= \left\{\frac{2}{2\sin(9^\circ)\cos(9^\circ)}\right\} - \left\{\frac{2}{2\sin(27^\circ)\cos(27^\circ)}\right\}$$

$$= \left\{\frac{2}{\sin(18^\circ)}\right\} - \left\{\frac{2}{\sin(54^\circ)}\right\}$$

$$= \left\{\frac{2}{\left(\frac{\sqrt{5}-1}{4}\right)}\right\} - \left\{\frac{2}{\frac{\sqrt{5}+1}{4}}\right\}$$

$$= \left\{\frac{8}{\sqrt{5}-1}\right\} - \left\{\frac{8}{\sqrt{5}+1}\right\}$$

$$= \left\{\frac{8(\sqrt{5}+1-\sqrt{5}+1)}{5-1}\right\}$$

$$= 4$$

Ex-7. Prove that $\frac{\sec 8A - 1}{\sec 4A - 1} = \frac{\tan 8A}{\tan 2A}$

Soln. We have, $\frac{\sec 8A - 1}{\sec 4A - 1}$

$$= \frac{\frac{1}{\cos 8A} - 1}{\frac{1}{\cos 4A} - 1}$$

$$= \frac{1 - \cos 8A}{1 - \cos 4A} \times \frac{\cos 4A}{\cos 8A}$$

$$= \frac{2\sin^2 4A}{2\sin^2 2A} \times \frac{\cos 4A}{\cos 8A}$$

$$= \frac{2\sin 4A\cos 4A}{\cos 8A} \times \frac{\sin 4A}{2\sin^2 2A}$$

$$= \frac{\sin 8A}{\cos 8A} \times \frac{2\sin 2A\cos 2A}{2\sin^2 2A}$$

$$= \frac{\sin 8A}{\cos 8A} \times \frac{\cos 2A}{\sin 2A}$$

$$= \tan 8A \times \cot 2A$$

$$= \frac{\tan 8A}{\tan 2A}.$$

Ex-8. Prove that $\tan\theta + 2\tan(2\theta) + 4\tan(4\theta) + 8\cot(8\theta) = \cot\theta$

Soln. We have, $\tan\theta + 2\tan(2\theta) + 4\tan(4\theta) + 8\cot(8\theta)$

$$= \cot\theta - (\cot\theta - \tan\theta) + 2\tan 2\theta + 4\tan 4\theta + 8\cot 8\theta$$

$$= \cot\theta - 2\cot 2\theta + 2\tan 2\theta + 4\tan 4\theta + 8\cot 8\theta$$

$= \cot\theta - 2(\cot 2\theta - \tan 2\theta)$

$+ 4\tan 4\theta + 8\cot 8\theta$

$= \cot\theta - 4\cot 4\theta + 4\tan 4\theta + 8\cot 8\theta$

$= \cot\theta - 4(\cot 4\theta - \tan 4\theta) + 8\cot 8\theta$

$= \cot\theta - 8\cot 8\theta + 8\cot 8\theta$

$= \cot\theta$

Ex-9. Prove that $\cos^2(\theta) + \cos^2\left(\frac{2\pi}{3} - \theta\right) + \cos^2\left(\frac{2\pi}{3} + \theta\right) = \frac{3}{2}$

Soln. We have,

$\cos^2(\theta) + \cos^2\left(\frac{2\pi}{3} - \theta\right) + \cos^2\left(\frac{2\pi}{3} + \theta\right)$

$= \frac{1}{2}\left(2\cos^2(\theta) + 2\cos^2\left(\frac{2\pi}{3} - \theta\right) + 2\cos^2\left(\frac{2\pi}{3} + \theta\right)\right)$

$= \frac{1}{2}(1 + \cos(2\theta)) + \frac{1}{2}\left(1 + \cos\left(\frac{4\pi}{3} - 2\theta\right)\right)$

$+ \frac{1}{2}\left(1 + \cos\left(\frac{4\pi}{3} + 2\theta\right)\right)$

$= \frac{1}{2}\left(3 + \left(\cos 2\theta + \cos\left(\frac{4\pi}{3} - 2\theta\right) + \cos\left(\frac{4\pi}{3} + 2\theta\right)\right)\right)$

$= \frac{1}{2}\left(3 + \left(\cos 2\theta + 2\cos\left(\frac{4\pi}{3}\right)\cos(2\theta)\right)\right)$

$= \frac{1}{2}\left(3 + \left(\cos 2\theta + 2\left(-\frac{1}{2}\right)\cos(2\theta)\right)\right)$

$= \frac{3}{2}$

Ex-10 Prove that

$\sin^2\theta + \sin^2(120^\circ - \theta) + \sin^2(240^\circ + \theta)$

$= (1 + \sin^2\theta)$

Soln. We have,

$\sin^2\theta + \sin^2(120^\circ - \theta) + \sin^2(240^\circ + \theta)$

$= \frac{1}{2}\left(2\sin^2\theta + 2\sin^2\left(\frac{2\pi}{3} - \theta\right)\right)$

$+ \frac{1}{2}\left(2\sin^2\left(\frac{4\pi}{3} + \theta\right)\right)$

$= \frac{1}{2}\left((1 - \cos 2\theta) + \left(1 - \cos\left(\frac{4\pi}{3} - 2\theta\right)\right)\right)$

$+ \frac{1}{2}\left(\left(1 - \cos\left(\frac{8\pi}{3} + 2\theta\right)\right)\right)$

$= \frac{3}{2} + \frac{1}{2}\cos 2\theta + \frac{1}{2}\cos\left(2\pi - \left(\frac{2\pi}{3} + 2\theta\right)\right)$

$+ \left(\frac{1}{2}\left(\cos\left(2\pi + \left(\frac{2\pi}{3} + 2\theta\right)\right)\right)\right)$

$= \frac{1}{2}\left(3 - \left(\cos 2\theta + \cos\left(\frac{4\pi}{3} - 2\theta\right) + \cos\left(\frac{8\pi}{3} + 2\theta\right)\right)\right)$

$= \frac{1}{2}\left(3 - \left(\cos 2\theta - \cos\left(\frac{2\pi}{3} + 2\theta\right) + \cos\left(\frac{2\pi}{3} + 2\theta\right)\right)\right)$

$= \frac{1}{2}(3 - \cos 2\theta)$

$= \frac{1}{2}(2 + (1 - \cos 2\theta))$

$= \frac{1}{2}(2 + 2\sin^2\theta)$

$= (1 + \sin^2\theta)$

Ex-11. Prove that

$4\sin(\theta)\sin(60^\circ + \theta)\sin(60^\circ - \theta) = \sin 3\theta$

Soln. We have, $4\sin(\theta)\sin\left(\frac{\pi}{3} - \theta\right)\sin\left(\frac{\pi}{3} + \theta\right)$

$= 4\sin(\theta) \times \left(\sin\left(\frac{\pi}{3} - \theta\right)\sin\left(\frac{\pi}{3} + \theta\right)\right)$

$$= 4\sin(\theta)\times\left(\sin^2\left(\frac{\pi}{3}\right)-\sin^2\theta\right)$$

$$= 4\sin(\theta)\times\left(\frac{3}{4}-\sin^2\theta\right)$$

$$= 4\times\left(\frac{3}{4}\sin\theta-\sin^3\theta\right)$$

$$= \left(3\sin\theta-4\sin^3\theta\right)$$

$$= \sin(3\theta)$$

Ex-12. Prove that

$$\sin(20^\circ)\sin(40^\circ)\sin(80^\circ)=\frac{\sqrt{3}}{8}$$

Soln. We have, $\sin(20^\circ)\sin(40^\circ)\sin(80^\circ)$

$$= \frac{1}{4}\left(\sin\left(3.20^\circ\right)\right)$$

$$= \frac{1}{4}\left(\sin\left(60^\circ\right)\right)$$

$$= \frac{\sqrt{3}}{8}$$

Ex-13. Prove that

$$4\cos(\theta).\cos(60^\circ-\theta).\cos(60^\circ+\theta)$$

$$=\cos(3\theta)$$

Soln. We have,

$$4\cos(\theta).\cos(60^\circ-\theta).\cos(60^\circ+\theta)$$

$$= 4\cos(\theta).\left(\cos(60^\circ-\theta).\cos(60^\circ+\theta)\right)$$

$$= 4\cos(\theta).\left(\cos^2(60^\circ)-\sin^2\theta\right)$$

$$= 4\cos(\theta).\left(\frac{1}{4}-1+\cos^2\theta\right)$$

$$= 4\cos(\theta).\left(-\frac{3}{4}+\cos^2\theta\right)$$

$$= \cos(\theta).\left(-3+4\cos^2\theta\right)$$

$$= \left(4\cos^3\theta-3\cos(\theta)\right)$$

$$= \cos(3\theta)$$

Ex-14. Prove that

$$\cos(10^\circ).\cos(20^\circ).\cos(40^\circ)=\frac{1}{8}$$

Soln. We have, $\cos(10^\circ).\cos(20^\circ).\cos(40^\circ)$

$$= \cos(10^\circ).\cos(30^\circ-10^\circ).\cos(30^\circ+10^\circ)$$

$$= \frac{1}{4}\left(4\cos(10^\circ).\cos(30^\circ-10^\circ).\cos(30^\circ+10^\circ)\right)$$

$$= \frac{1}{4}\left(\cos\left(3.10^\circ\right)\right)$$

$$= \frac{\sqrt{3}}{8}$$

Ex-15. Prove that

$$\tan(\theta)+\tan(60^\circ+\theta)-\tan(60^\circ-\theta)$$

$$=3\tan(3\theta)$$

Soln. We have,

$$\tan(\theta)-\tan(60^\circ-\theta)+\tan(60^\circ+\theta)$$

$$= \tan(\theta)-\frac{\tan(60^\circ)-\tan(\theta)}{1+\tan(60^\circ).\tan(\theta)}$$

$$+\frac{\tan(60^\circ)+\tan(\theta)}{1-\tan(60^\circ).\tan(\theta)}$$

$$= \tan(\theta)-\frac{\sqrt{3}-\tan(\theta)}{1+\sqrt{3}.\tan(\theta)}+\frac{\sqrt{3}+\tan(\theta)}{1-\sqrt{3}.\tan(\theta)}$$

$$= \tan(\theta)+$$

$$\frac{-\sqrt{3}+3\tan(\theta)+\tan(\theta)-\sqrt{3}\tan^2(\theta)+\sqrt{3}+3\tan(\theta)+\tan(\theta)+\sqrt{3}\tan^2(\theta)}{\left(1-3\tan^2\theta\right)}$$

$$= \tan(\theta)+\frac{8\tan(\theta)}{1-3\tan^2(\theta)}$$

$$= \frac{\tan(\theta)-3\tan^3(\theta)+8\tan(\theta)}{1-3\tan^2(\theta)}$$

$$= 3\times\left(\frac{3\tan(\theta)-\tan^3(\theta)}{1-3\tan^2(\theta)}\right)$$

$= \quad 3\times\tan(3\theta)$

Ex-16 Find the value of $\sqrt{3}\cot(20^{o})-4\cos(20^{o})$

Soln. We have, $\sqrt{3}\cot(20^{o})-4\cos(20^{o})$

$$= \frac{\sqrt{3}\cos(20^{o})-4\sin(20^{o})\cos(20^{o})}{\sin(20^{o})}$$

$$= \frac{2\left(\frac{\sqrt{3}}{2}\cos(20^{o})-2\sin(20^{o})\cos(20^{o})\right)}{\sin(20^{o})}$$

$$= \frac{2(\sin(60^{o})\cos(20^{o})-\sin(40^{o}))}{\sin(20^{o})}$$

$$= \frac{(2\sin(60^{o})\cos(20^{o})-2\sin(40^{o}))}{\sin(20^{o})}$$

$$= \frac{(\sin(80^{o})+\sin(40^{o})-2\sin(40^{o}))}{\sin(20^{o})}$$

$$= \frac{(\sin(80^{o})-\sin(40^{o}))}{\sin(20^{o})}$$

$$= \frac{2\cos(60^{o})\sin(20^{o})}{\sin(20^{o})}$$

$= \quad 1$

Ex-17. Prove that

$\sin(2^{o})+\sin(4^{o})+\sin(6^{o})+$

$\sin(8^{o})+\ldots\ldots\ldots\ldots+\sin(180^{o})=\cot(1^{o})$

Soln. We have, $\sin(2^{o})+\sin(4^{o})+\sin(6^{o})$

$+\sin(8^{o})+\ldots\ldots\ldots\ldots+\sin(180^{o})$

$= \quad \sin(2^{o})+\sin(2^{o}+2^{o})+\sin(2^{o}+2.2^{o})$

$+\sin(2^{o}+3.2^{o})$

$+\ldots.+\sin(2^{o}+(90-1)2^{o})$

$$= \frac{\sin\left(\frac{90^{o}.2^{o}}{2}\right)}{\sin\left(\frac{2^{o}}{2}\right)}\times\sin\left(2^{o}+(90^{o}-1)\frac{2^{o}}{2}\right)$$

$$= \frac{1}{\sin(1^{o})}\times\sin(91^{o})$$

$$= \frac{\cos(1^{o})}{\sin(1^{o})}$$

$= \quad \cot(1^{o})$

Ex-18. Find the value of

$$\sin\left(\frac{\pi}{2013}\right)+\sin\left(\frac{3\pi}{2013}\right)+\sin\left(\frac{5\pi}{2013}\right)$$

$$+\sin\left(\frac{7\pi}{2013}\right)+\ldots\text{upto }(2013)\text{trems}$$

Soln. We have, $\sin\left(\frac{\pi}{2013}\right)+\sin\left(\frac{\pi}{2013}+\frac{2\pi}{2013}\right)$

$+\sin\left(\frac{\pi}{2013}+\frac{2.2\pi}{2013}\right)$

$+\ldots\ldots+\sin\left(\frac{\pi}{2013}+\frac{1006.2\pi}{2013}\right)$

$$= \frac{\sin\left(2013.\left(\frac{\frac{2\pi}{2013}}{2}\right)\right)}{\sin\left(\frac{\frac{2\pi}{2013}}{2}\right)}\times\sin\left(\frac{\pi}{2013}+(2012)\left(\frac{\frac{2\pi}{2013}}{2}\right)\right)$$

$= \quad 0$

Ex-19. If $\tan y=\left(\frac{1+\sqrt{1+y}}{1+\sqrt{1-y}}\right)$, then prove that

$\sin(4y)=y$

Soln. Put $y=\sin(4\theta)$

Then $\tan(y)=\dfrac{1+\sqrt{1+\sin(4\theta)}}{1+\sqrt{1-\sin(4\theta)}}$

$\Rightarrow \quad \tan(y) = \dfrac{1+\sqrt{(\cos(2\theta)+\sin(2\theta))^2}}{1+\sqrt{(\cos(2\theta)-\sin(2\theta))^2}}$

$\Rightarrow \quad \tan(y) = \dfrac{1+\cos(2\theta)+\sin(2\theta)}{1+(\cos(2\theta)-\sin(2\theta))}$

$\Rightarrow \quad \tan(y) = \dfrac{2\cos^2(\theta)+\sin(2\theta)}{2\cos^2(\theta)-\sin(2\theta)}$

$\Rightarrow \quad \tan(y) = \dfrac{2\cos^2(\theta)+2\sin(\theta)\cos(\theta)}{2\cos^2(\theta)-2\sin(\theta)\cos(\theta)}$

$\Rightarrow \quad \tan(y) = \dfrac{\cos(\theta)+\sin(\theta)}{\cos(\theta)-\sin(\theta)}$

$\Rightarrow \quad \tan(y) = \tan\left(\dfrac{\pi}{4}-\theta\right)$

$\Rightarrow \quad y = \dfrac{\pi}{4}-\theta$

$\Rightarrow \quad 4y = \pi - 4\theta$

$\Rightarrow \quad \sin(4y) = \sin(\pi-4\theta) = \sin(4\theta) = y$

Hence, the result.

Ex-20. If $M = \sqrt{a^2\cos^2\theta + b^2\sin^2\theta}$ $+\sqrt{a^2\sin^2\theta + b^2\cos^2\theta}$ such that max (M^2) = m_1 and min (M^2) = m_2, then find the value of m_1 - m_2.

Soln. Given $M = \sqrt{a^2\cos^2\theta + b^2\sin^2\theta}$ $+\sqrt{a^2\sin^2\theta + b^2\cos^2\theta}$

$\Rightarrow \quad M^2 = a^2\cos^2\theta + b^2\sin^2\theta + a^2\sin^2\theta + b^2\cos^2\theta + 2\sqrt{(a^2\cos^2\theta + b^2\sin^2\theta)(a^2\sin^2\theta + b^2\cos^2\theta)}$

$\Rightarrow \quad M^2 = a^2 + b^2 + 2\sqrt{(a^2\cos^2\theta + b^2\sin^2\theta)}\sqrt{(a^2\sin^2\theta + b^2\cos^2\theta)}$

$\Rightarrow \quad M^2 = a^2 + b^2 + 2\sqrt{(a^4+b^4)\sin^2\theta\cos^2\theta + a^2b^2(\cos^4\theta + \sin^4\theta)}$

$\Rightarrow \quad M^2 = a^2 + b^2 + 2\sqrt{(a^4+b^4)\sin^2\theta\cos^2\theta + a^2b^2(1-2\sin^2\theta\cos^2\theta)}$

$\Rightarrow \quad M^2 = a^2 + b^2 + 2\sqrt{(a^4+b^4-2a^2b^2)\sin^2\theta\cos^2\theta + a^2b^2}$

$\Rightarrow \quad M^2 = a^2 + b^2 + \sqrt{\left(4(a^4+b^4-2a^2b^2)\sin^2\theta\cos^2\theta + 4a^2b^2\right)}$

$\Rightarrow \quad M^2 = a^2 + b^2 + \sqrt{\left((a^2-b^2)^2 \ldots (\sin(2\theta))^2 + 4a^2b^2\right)}$

Thus,

$\max(M^2) = a^2 + b^2 + (a^2 + b^2)$

$= 2(a^2 + b^2)$

and $\min(M^2) = a^2 + b^2 + 2ab = (a+b)^2$

Hence, the value of $m_1 - m_2 = \max(M^2) - \min(M^2) = 2(a^2+b^2) - (a+b)^2$.

Exercise - 7

1. **Prove that :**

i) $\dfrac{\sin 2\theta}{1+\cos 2\theta} = \tan\theta$

ii) $\dfrac{\cos 2\theta}{1+\sin 2\theta} = \tan\left(\dfrac{\pi}{4}-\theta\right)$

iii) $\dfrac{1+\sin 2\theta + \cos 2\theta}{1+\sin 2\theta - \cos 2\theta} = \cot\theta$

iv) $\cot\theta - \tan\theta = 2\cot 2\theta$

v) $\sqrt{2+\sqrt{2+2\cos 4\theta}} = 2\cos\theta$

2. Prove that :

cosec 10° – $\sqrt{3}$ sec 10° = 4.

3. If $\tan\theta = \dfrac{b}{a}$, then prove that

$a\cos 2\theta + b\sin 2\theta = a$

4. Prove that :

i) $\cos^2\dfrac{\pi}{8} + \cos^2\dfrac{3\pi}{8} + \cos^2\dfrac{5\pi}{8} + \cos^2\dfrac{7\pi}{8} = 2.$

ii) $\sin^2\frac{\pi}{8}+\sin^2\frac{3\pi}{8}+\sin^2\frac{5\pi}{8}+\sin^2\frac{7\pi}{8}=2.$

5. Prove that : $\frac{\sec 8A-1}{\sec 4A-1}=\frac{\tan 8A}{\tan 2A}$

6. If cosec 2A + cosec 2 B + cosec 2 C = 0, then prove that tan A + tan B + tan C + cot A + cot B + cot C = 0

7. If tan 25° = a, then prove that : $\frac{\tan 155°-\tan 115°}{1+\tan 155°.\tan 115°}=\frac{1-a^2}{2a}$

8. If $\tan^2\theta=2\tan^2\phi+1$, then prove that $\cos 2\theta+\sin^2\phi=0$

9. If $2\tan\alpha=3\tan\beta$, then prove that :

$$\tan(\alpha-\beta)=\frac{\sin 2\beta}{5-\cos 2\beta}$$

10. If $a+\frac{1}{a}=2\cos\theta$, then prove that :

$$a^4+\frac{1}{a^4}=2\cos 4\theta.$$

11. Prove that $\tan\alpha+2\tan 2\alpha+4\tan 4\alpha+8\cot 8\alpha.=\cot\alpha$.

12. Prove that $3(\sin x-\cos x)^4+6(\sin x+\cos x)^2+4(\sin^6 x+\cos^6 x)=13$.

13. If α and β are the roots of $3\cos x+4\sin x=5$, then find $\cot(\alpha+\beta)$.

14. If α and β are the solutions of the equation $a\tan\theta+b\sec\theta=c$, then find $\tan(\alpha+\beta)$.

15. If $\tan\beta=3\tan\alpha$, then prove that

$$\tan(\alpha+\beta)=\frac{2\sin 2\beta}{1+2\cos 2\beta}$$

16. Find the value of cot 91. cot 92. cot 93 cot 94..............cot 178. cot 179.

17. Find the value of

$$\left(1+\cos\left(\frac{\pi}{8}\right)\right)\left(1+\cos\left(\frac{3\pi}{8}\right)\right)\left(1+\cos\left(\frac{5\pi}{8}\right)\right)\left(1+\cos\left(\frac{7\pi}{8}\right)\right)$$

18. If $\alpha+\beta=90$, then find the maximum value of $\cos\alpha\cos\beta$.

19. Prove that $\cos^2\theta+\cos^2\left(\frac{2\pi}{3}-\theta\right)+\cos^2\left(\frac{2\pi}{3}+\theta\right)=\frac{3}{2}$

20. Prove that $\cos^2\theta+\cos^2\left(\frac{\pi}{3}-\theta\right)+\cos^2\left(\frac{\pi}{3}+\theta\right)=\frac{3}{2}$

21. Prove that $\sqrt{3}\,\mathrm{cosec}\,2o^o-\sec 2o^o=4$

22. Prove that

$$\sin\theta\cos^3\theta-\sin^3\theta\cos\theta=\frac{1}{4}\sin 3\theta$$

23. Prove that

$$\tan\theta+\tan\left(\frac{\pi}{3}+\theta\right)+\tan\left(\frac{\pi}{3}-\theta\right)=3\tan(3\theta)$$

24. Prove that

$$\cot\theta+\cot\left(\frac{\pi}{3}+\theta\right)+\cot\left(\frac{2\pi}{3}+\theta\right)=3\cot 3\theta$$

25. Prove that $\tan 4\theta=\frac{4\tan\theta-4\tan^3\theta}{1-6\tan^2\theta+\tan^4\theta}$.

26. Prove that

$$\sin^3\alpha+\sin^3\left(\frac{2\pi}{3}+\alpha\right)+\left(\frac{4\pi}{3}+\alpha\right)=-\frac{3}{4}\sin 3\alpha$$

27. Prove that $2^n\cos\theta os\theta.co.\cos 2^2\theta.\cos 2^3\theta.$

$$......\cos 2^{n-1}\theta=\frac{\sin(2^n\theta)}{\sin\theta}$$

28. If $\theta=\frac{\pi}{2^n+1}$, then prove that

$2^n \cos\theta os\theta.co.\cos 2^2\theta$

$\cos 2^3\theta ....\cos 2^{n-1}\theta = 1$

29. If $\cos 6\theta = A\cos^6\theta + B\cos^4\theta$

$+ C\cos^2\theta + D$,

then find the vale of $A + B + C + D + 2$.

30. If $\sin 5\theta = A\sin\theta + B\sin^3\theta + C\sin^5\theta$, then find the value of $A + B + C + 10$.

31. Prove that

$$\frac{\sin x}{\cos 3x} + \frac{\sin 3x}{\cos 9x} + \frac{\sin 9x}{\cos 27x}$$

$$= \frac{1}{2}(\tan 27x - \tan x)$$

32. Prove that

$$\cos\left(\frac{2\pi}{7}\right) + \cos\left(\frac{4\pi}{7}\right) + \cos\left(\frac{6\pi}{7}\right) = -\frac{1}{2}.$$

33. Prove that

$$\sin\left(\frac{2\pi}{7}\right) + \sin\left(\frac{4\pi}{7}\right) + \sin\left(\frac{8\pi}{7}\right) = \frac{\sqrt{7}}{2}.$$

34. Prove that

$$\frac{\tan^{2n}(\theta)}{\tan(\theta)} = (1 + \sec 2\theta)(1 + \sec 2^2\theta)$$

$$(1 + \sec 2^3\theta).....(1 + \sec 2^n\theta)$$

Answers

13. $\frac{5}{12}$, **14.** $\frac{2c}{1-c^v}$, **16.** 0, **17.** $\frac{1}{8}$, **18.** $\frac{1}{2}$

29. 3, **30.** 11.

1.16 THE MAXIMUM AND MINIMUM VALUES OF F(X) = A COS X + B SIN X + C

We have $f(x) = a\cos x + b\sin x + c$

Let $a = r\sin\theta$ & $b = r\cos\theta$

Then $r = \sqrt{a^2 + b^2}$ and $\tan(\theta) = \frac{a}{b}$

Now, $f(x) = a\cos x + b\sin x + c$

$= r(\sin\theta\cos x + \cos\theta\sin x)$

$= r\sin(\theta + x)$

As we know that, $-1 \le \sin(\theta + x) \le 1$

$\Rightarrow \quad -r + c \le r\sin(\theta + x) + c \le r + c$

$\Rightarrow \quad -r + c \le f(x) \le r + c$

$\Rightarrow \quad -\sqrt{a^2 + b^2} + c \le f(x) \le \sqrt{a^2 + b^2} + c$

Thus, the maximum value of f(x) is

$\sqrt{a^2 + b^2} + c$

and the minimum values of f(x) is

$-\sqrt{a^2 + b^2} + c$.

Ex-1. Find the max and min values of $f(x) = 3\sin x + 4\cos x + 10$

Soln. Here, $a = 3, b = 4$ and $c = 10$

Thus, the minimum values of f(x) =

$-\sqrt{a^2 + b^2} + c = -5 + 10 = 5$

and the maximum values of f(x) =

$\sqrt{a^2 + b^2} + c = 5 + 10 = 15$.

Ex-2. Find the range of $f(x) = \sin x + \cos x + 3$

Soln. $R_f = [\min f(x), \ \max f(x)]$

$= [-\sqrt{2} + 3, \ \sqrt{2} + 3]$

Ex-3. Let $A = \sin^4\theta + \cos^4\theta$, then find A

Soln. We have, $A = \sin^4\theta + \cos^4\theta$

$= (\sin^2\theta)^2 + (\cos^2\theta)^2$

$= 1 - 2\sin^2\theta.\cos^2\theta$

$= 1 - \frac{(\sin 2\theta)^2}{2}$

$= 1 + \frac{\{-\sin^2(2\theta)\}}{2}$

As we know that $-1 \le \{-\sin^2(2\theta)\} \le 0$

$\Rightarrow \quad -\frac{1}{2} \le \frac{\{-\sin^2(2\theta)\}}{2} \le 0$

$\Rightarrow \quad -\frac{1}{2} + 1 \le \frac{\{-\sin^2(2\theta)\}}{2} \le 0 + 1$

$\Rightarrow \quad \frac{1}{2} \le A \le 1$

Ex-4. Find the max and min values of

$f(\theta)=\sin^6\theta+\cos^6\theta$

Soln. We have $f(\theta)=\sin^6\theta+\cos^6\theta$

$= (\sin^2\theta)^3+(\cos^2\theta)^3$

$= (\sin^2\theta+\cos^2\theta)^2-3\sin^2\theta\cos^2\theta(\sin^2\theta+\cos^2\theta)$

$= 1-3\sin^2\theta\cos^2\theta$

$= 1-\frac{3}{4}(4\sin^2\theta\cos^2\theta)$

$= 1-\frac{3}{4}(\sin^2 2\theta)$

$= 1+\frac{3}{4}(-\sin^2 2\theta)$

As we know that $-1\le(-\sin^2 2\theta)\le 0$

$\Rightarrow -\frac{3}{4}\le\frac{3(-\sin^2 2\theta)}{4}\le 0$

$\Rightarrow 1-\frac{3}{4}\le 1+\frac{3(-\sin^2 2\theta)}{4}\le 1$

$\Rightarrow \frac{1}{4}\le f(\theta)\le 1$

Hence, the maximum value = 1 and

the minimum value $=\frac{1}{4}$

Ex-5. If $A=\cos^2\theta+\sin^4\theta$ and

$B=\cos^4\theta+\sin^2\theta$ such that

m_1 = Max of A and m_2 = Min of B

then find the value of $m_1^2+m_2^2+m_1m_2$

Soln. Now, A

$= \cos^2\theta+\sin^4\theta$

$= \frac{1}{2}(2\cos^2\theta)+\frac{1}{4}(2\sin^2\theta)^2$

$= \frac{1}{2}(1+\cos(2\theta))+\frac{1}{4}(1-\cos(2\theta))^2$

$= \frac{1}{2}(1+\cos(2\theta))+\frac{1}{4}(1-2\cos(2\theta)+\cos^2(2\theta))$

$= \frac{1}{2}+\frac{1}{2}\cos(2\theta)+\frac{1}{4}-\frac{1}{2}\cos(2\theta)+\frac{1}{4}(\cos^2(2\theta))$

$= \frac{3}{4}+\frac{1}{4}(\cos^2(2\theta))$

Max value of A $= m_1=\frac{3}{4}+\frac{1}{4}.1=1$

Also, B

$= \sin^2\theta+\cos^4\theta$

$= \frac{1}{2}(2\sin^2\theta)+\frac{1}{4}(4\cos^4\theta)$

$= \frac{1}{2}(2\sin^2\theta)+\frac{1}{4}(2\cos^2\theta)^2$

$= \frac{1}{2}(1-\cos(2\theta))+\frac{1}{4}(1+\cos(2\theta))^2$

$= \frac{1}{2}(1-\cos(2\theta))+\frac{1}{4}(1+2\cos(2\theta)+\cos^2(2\theta))$

$= \frac{1}{2}+\frac{1}{4}+\frac{1}{4}\cos^2(2\theta)$

$= \frac{3}{4}+\frac{1}{4}\cos^2(2\theta)$

Thus, the minimum value of B

$= m_2=\frac{3}{4}+\frac{1}{4}.0=\frac{3}{4}$

Now, the value of

$m_1^2+m_2^2+m_1m_2$

$= 1+\frac{9}{16}+\frac{3}{4}$

$= \frac{37}{16}$

Ex-6. Find the max and min values of

$f(\theta)=\sin^2(\sin\theta)+\cos^2(\cos\theta)$

Soln. We have, $\sin^2(\sin\theta)+\cos^2(\cos\theta)$

$= \sin^2(\cos\theta)+\cos^2(\cos\theta)$

$+\sin^2(\sin\theta)-\sin^2(\cos\theta)$

$= (\sin^2(\cos\theta)+\cos^2(\cos\theta))$

$+ +\sin^2(\sin\theta)-\sin^2(\cos\theta)$

$= 1+(\sin^2(\sin\theta)-\sin^2(\cos\theta))$

Max value of $f(\theta)$

$= 1+\left(\sin^2\left(\sin\left(\frac{\pi}{2}\right)\right)-\sin^2\left(\cos\left(\frac{\pi}{2}\right)\right)\right)$

$= \quad 1+\sin^2(1)$

Min value of $f(\theta)$

$= \quad 1+\left(\sin^2(\sin(0))-\sin^2(\cos(0))\right)$

$= \quad 1-\sin^2(1)$

Ex-7. Find the minimum value of

$$f(\theta)=(3\sin(\theta)-4\cos(\theta)-10)(3\sin(\theta)+4\cos(\theta)-10)$$

Soln. We have $f(\theta)=(3\sin(\theta)-4\cos(\theta)-10)(3\sin(\theta)+4\cos(\theta)-10)$

$= \left(9\sin^2(\theta)-16\cos^2(\theta)\right)-10(3\sin\theta+4\cos\theta)-10(3\sin\theta-4\cos\theta)$

$= \left(9\sin^2(\theta)-16\cos^2(\theta)\right)-10(3\sin\theta+4\cos\theta+3\sin\theta-4\cos\theta)$

$= \left(9\sin^2(\theta)-16\cos^2(\theta)\right)-60\sin(\theta)$

$= 25\sin^2\theta-60\sin(\theta)-16$

$= (5\sin\theta-6)^2-36-16$

$= (5\sin\theta-6)^2-52$

Hence, the minimum value of $f(\theta)=121-52=69$

Ex-8. Find the range of $A=\sin^{2010}\theta+\cos^{2014}\theta$

Soln. Now, $0<\sin^{2010}\theta\le\sin^2\theta$ ....(i) and

$0<\cos^{2014}\theta\le\cos^2\theta$ ....(ii)

Adding (i) and (ii), we get,

$0<\sin^{2012}\theta+\cos^{2014}\theta\le\sin^2\theta+\cos^2\theta$

$\Rightarrow \quad 0<A\le 1$

Thus, the range of A = $(0, 1]$

Ex-9. Find the minimum value of

$$f(x)=\frac{x^2\sin^2x+4}{x\sin x}, \text{ where } x\in\left(0,\frac{\pi}{2}\right)$$

Soln. We have $f(x)=\dfrac{x^2\sin^2x+4}{x\sin x}$

$= x\sin x+\dfrac{4}{x\sin x}\ge 4$

Hence, the minimum values of $f(x)$ is 4

Exercise - 8.

1. Find the Maximum and Minimum values of
   (i) f(x) = 3 sin x + 4 cos x + 5
   (ii) f(x) = 3 sin (100) x + 4 cos (100) x + 10
   (iii) f(x) = 3 sin x + 4
   (iv) f(x) = 2 cos x + 5
   (v) f(x) = sin x + cos x
   (vi) f(x) = sin x - cos x
   (vii) f(x) = sin (sin x)
   (viii) f(x) = cos (cos x)
   (ix) f(x) = sin (sin x) + cos (sinx)
   (x) f(x) = cos (sin x) + sin (cos x).
2. Find the max and min value of 3 sin2x + 4cos 2x + 3
3. Prove that
   $$-4\le 5\cos\theta+3\cos\left(\theta+\frac{\pi}{3}\right)+3\le 10.$$
4. Find the max. and min. values of $\cos^2\theta-6\sin\theta\cos\theta+3\sin^2\theta+2$
5. Find the greatest and the least values of $2\sin^2\theta+3\cos^2\theta$.
6. Find the least value of $\text{cosec}^2 x+25\sec^2 x$.
7. Find the ratio of the greatest value of $2-\cos x+\sin^2 x$ to its least value.
8. If $y=4\sin^2\theta-\cos 2\theta$, then y lies in the interval............
9. If $x\sin^3\theta+y\cos^3\theta=\sin\theta\cos\theta$ and $x\sin\theta=y\cos\theta$, then prove that $x^2+y^2=1$.
10. Find the maximum and minimum values of $f(x)=\sin^4x+\cos^4x$.
11. Find the maximum and minimum values of $f(x)=\sin^6x+\cos^6x$.
12. Find the max and min values of
    $$f(x)=(\sin x+\cos x+\text{cosec}2x)^3$$
13. Find the max and min values of
    $$f(x)=\log_x y+\log_y x \quad x, y \ \&\ +\text{ve}$$
14. Find the max and min values of
    $$f(x)=\frac{5}{\sin^2\theta-6\sin\theta\cos\theta+3\cos^2\theta}$$
15. Find the maximum and the minimum values of
    $$f(x)=\sin^2x+\cos^4x$$

16. Find the maximum and the minimum values of $f(x) = \cos^2 x + \sin^4 x$

Answers

1. (*i*) Max V = 10, Min V = 0
   (*ii*) Max V = 15, Min V = 5
   (*iii*) Max V = 7, Min V = 1
   (*iv*) Max V = 7, Min V = 3
   (*v*) Max V = $\sqrt{2}$, Min V = $-\sqrt{2}$
   (*vi*) Max V = $\sqrt{2}$, Min V = $-\sqrt{2}$
   (*vii*) Max V = sin 1, Min V = – sin 1
   (*viii*) Max V = cos 1, Min V = 0
   (*ix*) Max V = $\sqrt{2}$, Min V = $-\sqrt{2}$
   (*x*) Max V = $\sqrt{2} - \sin 1$, Min V = $-\sqrt{2} + \sin 1$
2. Max V = 8, Min V = –2
4. Max V = $4 + \sqrt{10}$, Min V = $4 - \sqrt{10}$
5. Max V = 3, Min V = 2
6. L.V. = 10
7. 13 : 12
8. [–1, 5]
10. Max V = 1, Min V = $\frac{1}{2}$
11. Max V = 1, Min V = $\frac{1}{4}$
12. Min V = $\frac{27}{2}$, Max V = Not defined.
13. Min V = 2, Max V = Not defined
14. Max V = $\frac{5}{2+\sqrt{10}}$, Min V = $\frac{5}{2-\sqrt{10}}$
15. Max V = 1, Min V = $\frac{3}{4}$
16. Max V = 1, Min V = $\frac{3}{4}$

1.17 SUB–MULTIPLE ANGLES

1.17.1 Definition :

An angle, of the form $\frac{A}{n}$, $n \in Z\ (\neq 0)$, is called a sub-multiple angle of A. Thus $\frac{A}{2}, \frac{A}{3}, \frac{A}{4}, \frac{A}{5}$ etc. are sub-multiple angles of A.

1.17.2 T-ratios of angle $\frac{A}{2}$ and $\frac{A}{3}$:

1. $\sin A = 2 \sin \frac{A}{2} \cos \frac{A}{2} = \dfrac{2\tan\frac{A}{2}}{1+\tan^2\frac{A}{2}}$

2. $\cos A = \cos^2 \frac{A}{2} - \sin^2 \frac{A}{2}$

   $= 2\cos^2 \frac{A}{2} - 1$

   $= 1 - 2\sin^2 \frac{A}{2} = \dfrac{1-\tan^2\frac{A}{2}}{1+\tan^2\frac{A}{2}}$

3. $\tan A = \dfrac{2\tan\frac{A}{2}}{1-\tan^2\frac{A}{2}}$.

4. $\sin A = 3\sin\left(\frac{A}{3}\right) - 4\sin^3\left(\frac{A}{3}\right)$

5. $\cos A = 4\cos^3\left(\frac{A}{3}\right) - 3\cos\left(\frac{A}{3}\right)$

.6. $\tan A = \dfrac{3\tan\left(\frac{A}{3}\right) - \tan^3\left(\frac{A}{3}\right)}{1-3\tan^2\left(\frac{A}{3}\right)}$

1.17.3 Values of sin 18°, cos 18° and tan 18°:

1. $\sin 18° = \dfrac{\sqrt{5}-1}{4}$

Proof : Let A = 18°

$\Rightarrow\ 5A = 90^{\circ}$

$\Rightarrow\ 2A = 90^{\circ} - 3A$

$\Rightarrow\ \sin 2A = \sin(90^{\circ} - 3A) = \cos 3A$

$\Rightarrow\ 2\sin A \cos A = 4\cos^3 A - 3\cos A$

$\Rightarrow\ 2\sin A = 4\cos^2 A - 3$

$\Rightarrow\ 2\sin A = 4 - 4\sin^2 A - 3 = 1 - 4\sin^2 A$

$\Rightarrow\ 4\sin^2 A + 2\sin A - 1 = 0$

$\Rightarrow\ \sin A = \dfrac{-2 \pm \sqrt{20}}{8} = \dfrac{-2 \pm 2\sqrt{5}}{8}$

$\Rightarrow \quad \sin A = \dfrac{-1 \pm \sqrt{5}}{4}$

$\Rightarrow \quad \sin A = \dfrac{\sqrt{5}-1}{4}, \dfrac{-\sqrt{5}-1}{4}$

$\Rightarrow \quad \sin(18^\circ) = \dfrac{\sqrt{5}-1}{4}$, since 18° lies on the first quadrant.

2, $\quad \cos 18^\circ = \dfrac{1}{4}\sqrt{10+2\sqrt{5}}$

Proof: We have, $\cos(18^\circ) = \sqrt{1-\sin^2(18^\circ)}$

$$= \sqrt{1-\left(\frac{\sqrt{5}-1}{4}\right)^2}$$

$$= \sqrt{1-\left(\frac{5+1-2\sqrt{5}}{16}\right)}$$

$$= \sqrt{\left(\frac{16-5-1+2\sqrt{5}}{16}\right)}$$

$$= \frac{1}{4}\sqrt{10+2\sqrt{5}}$$

3. $\quad \tan 18^\circ = \dfrac{\sqrt{5}-1}{\sqrt{10+2\sqrt{5}}}$

Proof: We have, $\tan(18^\circ) = \dfrac{\sin(18^\circ)}{\cos(18^\circ)}$

$$= \frac{\left(\frac{\sqrt{5}-1}{4}\right)}{\frac{\sqrt{10+2\sqrt{5}}}{4}}$$

$$= \left(\frac{\sqrt{5}-1}{\sqrt{10+2\sqrt{5}}}\right)$$

Notes :

i) $\quad \sin 72^\circ = \cos 18^\circ = \dfrac{1}{4}\sqrt{10+2\sqrt{5}}$

ii) $\quad \cos 72^\circ = \sin 18^\circ = \dfrac{\sqrt{5}-1}{4}$

➢ **Values of sin 36°, cos 36° and tan 36° :**

1. $\quad \cos 36^\circ = \dfrac{\sqrt{5}+1}{4}$

Proof: We have, $\cos(36^\circ)$

$$= \cos 2(18^\circ)$$

$$= 1-2\sin^2(18^\circ)$$

$$= 1-2\left(\frac{\sqrt{5}-1}{4}\right)^2$$

$$= 1-2\left(\frac{5+1-2\sqrt{5}}{16}\right)$$

$$= \left(\frac{8-5-1+2\sqrt{5}}{8}\right)$$

$$= \left(\frac{2+2\sqrt{5}}{8}\right)$$

$$= \left(\frac{\sqrt{5}+1}{4}\right)$$

2. $\quad \sin 36^\circ = \dfrac{1}{4}\sqrt{10-2\sqrt{5}}$

Proof: We have, $\sin(36^\circ) = \sqrt{1-\cos^2(36^\circ)}$

$$= \sqrt{1-\left(\frac{\sqrt{5}+1}{4}\right)^2}$$

$$= \sqrt{1-\left(\frac{5+1+2\sqrt{5}}{16}\right)}$$

$$= \sqrt{\left(\frac{16-5-1-2\sqrt{5}}{16}\right)}$$

$$= \sqrt{\left(\frac{10-2\sqrt{5}}{16}\right)}$$

$$= \frac{1}{4}\sqrt{10-2\sqrt{5}}$$

3. $\quad \tan 36^\circ = \dfrac{1}{4} \times (\sqrt{5}-1) \times \sqrt{10-2\sqrt{5}}$.

Proof: We have $\tan(36^\circ) = \dfrac{\sin(36^\circ)}{\cos(36^\circ)}$

$$= \frac{\frac{\sqrt{10-2\sqrt{5}}}{4}}{\frac{\sqrt{5}+1}{4}}$$

$$= \frac{\sqrt{10-2\sqrt{5}}}{\left(\sqrt{5}+1\right)}$$

$$= \frac{\left(\sqrt{5}-1\right)\times\left(\sqrt{10-2\sqrt{5}}\right)}{4}$$

Notes :

i) $\sin 54° = \cos 36° = \frac{\sqrt{5}+1}{4}$

ii) $\cos 54° = \sin 36° = \frac{1}{4}\sqrt{10-2\sqrt{5}}$.

1.17.4 Some Important Deductions :

Deduction -1: $\tan 7\frac{1}{2}^{\circ} = \sqrt{6} - \sqrt{4} - \sqrt{3} + \sqrt{2}$

Proof: As we know that, $\tan\theta = \frac{1-\cos(2\theta)}{\sin(2\theta)}$

Put $\theta = 7\frac{1}{2}^{\circ}$, then

$$\tan\left(7\frac{1}{2}^{\circ}\right) = \frac{1-\cos(15^{\circ})}{\sin(15^{\circ})}$$

$$= \frac{1-\frac{\sqrt{3}+1}{2\sqrt{2}}}{\frac{\sqrt{3}-1}{2\sqrt{2}}}$$

$$= \frac{2\sqrt{2}-\sqrt{3}-1}{\sqrt{3}-1}$$

$$= \frac{\left(2\sqrt{2}-\sqrt{3}-1\right)\left(\sqrt{3}+1\right)}{\left(\sqrt{3}-1\right)\left(\sqrt{3}+1\right)}$$

$$= \frac{\left(2\sqrt{2}-\sqrt{3}-1\right)\left(\sqrt{3}+1\right)}{2}$$

$$= \frac{2\sqrt{6}-3-\sqrt{3}+2\sqrt{2}-\sqrt{3}-1}{2}$$

$$= \frac{2\left(\sqrt{6}-\sqrt{4}-\sqrt{3}+\sqrt{2}\right)}{2}$$

$$= \left(\sqrt{6}-\sqrt{4}-\sqrt{3}+\sqrt{2}\right)$$

Deduction-2:. $\cot 7\frac{1}{2}^{\circ} = \sqrt{6} + \sqrt{4} + \sqrt{3} + \sqrt{2}$

Proof: As we know that, $\cot(\theta) = \frac{1+\cos(2\theta)}{\sin(2\theta)}$

Put $\theta = 7\frac{1}{2}^{\circ}$,

$$\cot\left(7\frac{1}{2}^{\circ}\right) = \frac{1+\cos(15^{\circ})}{\sin(15^{\circ})}$$

$$= \frac{1+\frac{\sqrt{3}+1}{2\sqrt{2}}}{\frac{\sqrt{3}-1}{2\sqrt{2}}}$$

$$= \frac{2\sqrt{2}+\sqrt{3}+1}{\sqrt{3}-1}$$

$$= \frac{\left(2\sqrt{2}+\sqrt{3}+1\right)\left(\sqrt{3}+1\right)}{\left(\sqrt{3}-1\right)\left(\sqrt{3}+1\right)}$$

$$= \frac{\left(2\sqrt{6}+3+\sqrt{3}+2\sqrt{2}+\sqrt{3}+1\right)}{2}$$

$$= \frac{2\left(\sqrt{6}+\sqrt{4}+\sqrt{3}+\sqrt{2}\right)}{2}$$

$$= \left(\sqrt{6}+\sqrt{4}+\sqrt{3}+\sqrt{2}\right)$$

Deduction -3:. $\sin 22\frac{1}{2}^{\circ} = \frac{1}{2}\sqrt{2-\sqrt{2}}$

Proof: As we know that, $2\sin^2(\theta) = 1-\cos 2\theta$

Put, $\theta = 22\frac{1}{2}^{\circ}$,

$$2\sin^2\left(22\frac{1}{2}^\circ\right) = 1-\cos(45^\circ)$$

$$= 1-\left(\frac{1}{\sqrt{2}}\right)$$

$$= \frac{\sqrt{2}-1}{\sqrt{2}}$$

$$\Rightarrow \sin\left(22\frac{1}{2}^\circ\right) = \pm\sqrt{\frac{\sqrt{2}-1}{2\sqrt{2}}}$$

$$\Rightarrow \sin\left(22\frac{1}{2}^\circ\right) = \sqrt{\frac{\sqrt{2}-1}{2\sqrt{2}}}$$ since, $22\frac{1}{2}^\circ$ lies in the first quadrant.

$$\Rightarrow \sin\left(22\frac{1}{2}^\circ\right) = \frac{1}{2}\sqrt{2-\sqrt{2}}.$$

Deduction-4:. $\cos 22\frac{1}{2}^\circ = \frac{1}{2}\sqrt{2+\sqrt{2}}$

Proof: As we know that $2\cos^2(\theta) = 1-\cos 2\theta$

Put, $\theta = 22\frac{1}{2}^\circ$,

$$2\cos^2\left(22\frac{1}{2}^\circ\right) = 1+\cos(45^\circ)$$

$$= 1+\left(\frac{1}{\sqrt{2}}\right)$$

$$= \frac{\sqrt{2}+1}{\sqrt{2}}$$

$$\Rightarrow \cos\left(22\frac{1}{2}^\circ\right) = \pm\sqrt{\frac{\sqrt{2}-1}{2\sqrt{2}}}$$

$$\Rightarrow \cos\left(22\frac{1}{2}^\circ\right) = \sqrt{\frac{\sqrt{2}+1}{2\sqrt{2}}}$$ since, $22\frac{1}{2}^\circ$ lies in the first quadrant.

$$\Rightarrow \cos\left(22\frac{1}{2}^\circ\right) = \frac{1}{2}\sqrt{2+\sqrt{2}}.$$

Deduction- 5: $\tan 22\frac{1}{2}^\circ = \sqrt{2}-1.$

Proof: As we know that, $\tan\theta = \frac{1-\cos(2\theta)}{\sin(2\theta)}$

Put $\theta = 22\frac{1}{2}^\circ$,

$$\tan\left(22\frac{1}{2}^\circ\right) = \frac{1-\cos(45^\circ)}{\sin(45^\circ)}$$

$$= \frac{1-\frac{1}{\sqrt{2}}}{\frac{1}{\sqrt{2}}} = \sqrt{2}-1$$

Deduction-6:. $\cot 22\frac{1}{2}^\circ = \sqrt{2}+1.$

Proof: As we know that $\cot(\theta) = \frac{1+\cos(2\theta)}{\sin(2\theta)}$

Put $\theta = 22\frac{1}{2}^\circ$,

$$\cot\left(22\frac{1}{2}^\circ\right) = \frac{1+\cos(45^\circ)}{\sin(45^\circ)}$$

Deduction-7:. $\sin 112\frac{1}{2}^\circ = \frac{1}{2}\sqrt{2+\sqrt{2}}$

Proof: We have, $\sin\left(112\frac{1}{2}^\circ\right)$

$$= \sin\left(90^\circ\times 1+22\frac{1}{2}^\circ\right)$$

$$= \cos\left(22\frac{1}{2}^\circ\right)$$

$$= \frac{1}{2}\sqrt{2+\sqrt{2}}$$

Deduction -8:: $\cos 112\frac{1}{2}^\circ = \frac{1}{2}\sqrt{2-\sqrt{2}}$

Proof: We have, $\cos\left(112\frac{1}{2}^{\circ}\right)$

$= \cos\left(90^{\circ}\times 1+22\frac{1}{2}^{\circ}\right)$

$= -\sin\left(22\frac{1}{2}^{\circ}\right)$

$= -\frac{1}{2}\sqrt{2+\sqrt{2}}$

Deduction -9: $\tan 112\frac{1}{2}^{\circ} = -(\sqrt{2}+1)$

Proof: We have, $\tan\left(112\frac{1}{2}^{\circ}\right)$

$= \tan\left(90^{\circ}\times 1+22\frac{1}{2}^{\circ}\right)$

$= -\cot\left(22\frac{1}{2}^{\circ}\right)$

$= -\left(\sqrt{2}+1\right)$

1.18 SOME SOLVED EXAMPLES :

Ex-1. Prove that $\sin^2(24^{\circ})-\sin^2(6^{\circ})=\frac{(\sqrt{5}-1)}{8}$

Soln. We have, $\sin^2(24^{\circ})-\sin^2(6^{\circ})$

$= \sin(24^{\circ}+6^{\circ})\times\sin(24^{\circ}-6^{\circ})$

$= \sin(30^{\circ})\times\sin(18^{\circ})$

$= \frac{1}{2}\times\sin(18^{\circ})$

$= \frac{1}{2}\times\frac{\sqrt{5}-1}{4}$

$= \frac{(\sqrt{5}-1)}{8}$

Ex-2. Prove that $\sin^2(48^{\circ})-\cos^2(12^{\circ})=\frac{(\sqrt{5}+1)}{8}$

Soln. We have, $\sin^2(48^{\circ})-\cos^2(12^{\circ})$

$= \cos(48^{\circ}+12^{\circ})\times\cos(48^{\circ}-12^{\circ})$

$= \cos(60^{\circ})\times\cos(36^{\circ})$

$= \frac{1}{2}\times\frac{\sqrt{5}+1}{4}$

$= \frac{\sqrt{5}+1}{8}$

Ex-3. Prove that $\sin(12^{\circ})\sin(48^{\circ})\sin(54^{\circ})=\frac{1}{8}$

Soln. We have, $\sin(12^{\circ})\sin(48^{\circ})\sin(54^{\circ})$

$= \frac{1}{\sin(72^{\circ})}(\sin(12^{\circ})\sin(48^{\circ})\sin(72^{\circ}))(\sin(54^{\circ}))$

$= \frac{1}{4\sin(72^{\circ})}$
$(4\sin(60^{\circ}-12^{\circ})\sin(12^{\circ})\sin(60^{\circ}+12^{\circ}))$
$(\cos(36^{\circ}))$

$= \frac{1}{4\sin(72^{\circ})}(.\sin(36^{\circ})\cos(36^{\circ}))$

$= \frac{1}{8\sin(72^{\circ})}(2\sin(36^{\circ})\cos(36^{\circ}))$

$= \frac{1}{8\sin(72^{\circ})}(\sin(72^{\circ}))$

$= \frac{1}{8}$

Ex-4. Prove that

$\sin(6^{\circ})\sin(42^{\circ})\sin(66^{\circ})\sin(78^{\circ})=\frac{1}{16}$

Soln. We have,

$\sin(6^{\circ})\sin(42^{\circ})\sin(66^{\circ})\sin(78^{\circ})$

$$= \frac{1}{4\sin(54^\circ)}$$

$$(4\sin(6^\circ)\sin(60^\circ-6^\circ)\sin(60^\circ+6^\circ))$$

$$\times(\sin(78^\circ)\sin(42^\circ))$$

$$= \frac{1}{4\cos(36^\circ)}(\sin(18^\circ)\sin(72^\circ)\sin(42^\circ))$$

$$= \frac{1}{16\cos(36^\circ)}$$

$$(4\sin(18^\circ)\sin(60^\circ+18^\circ)\sin(60^\circ-18))$$

$$= \frac{1}{16\cos(36^\circ)}(\sin(54^\circ))$$

$$= \frac{1}{16\cos(36^\circ)}(\cos(36^\circ))$$

$$= \frac{1}{16}$$

Ex-5. Prove that

$$4(\sin(24^\circ)+\cos(6^\circ))=(1+\sqrt{5})$$

Soln. We have, $4(\sin(24^\circ)+\cos(6^\circ))$

$$= 4(\sin(24^\circ)+\sin(84^\circ))$$

$$= 4\left(2\sin\left(\frac{24^\circ+84^\circ}{2}\right)\cos\left(\frac{24^\circ-84^\circ}{2}\right)\right)$$

$$= 8(\sin(54^\circ)\cos(30^\circ))$$

$$= 8(\cos(36^\circ)\cos(30^\circ))$$

$$= 8\left(\frac{\sqrt{5}+1}{4}\times\frac{1}{2}\right)$$

$$= (\sqrt{5}+1)$$

Ex-6. Prove that

$$\tan(6^\circ)\tan(42^\circ)\tan(66^\circ)\tan(78^\circ)=1$$

Soln. We have,

$$\tan(6^\circ)\tan(42^\circ)\tan(66^\circ)\tan(78^\circ)$$

$$= (\tan(6^\circ)\tan(66^\circ))\times(\tan(42^\circ)\tan(78^\circ))$$

$$= \frac{1}{\tan(54^\circ)}$$

$$(\tan(6^\circ)\tan(54^\circ)\tan(66^\circ))$$

$$\times(\tan(42^\circ)\tan(78^\circ))$$

$$= \frac{1}{\tan(54^\circ)}$$

$$(\tan(6^\circ)\tan(60^\circ-6^\circ)\tan(60^\circ+6^\circ))$$

$$\times(\tan(42^\circ)\tan(78^\circ))$$

$$= \frac{1}{\tan(54^\circ)}(\tan(18^\circ))\times(\tan(42^\circ)\tan(78^\circ))$$

$$= \frac{1}{\tan(54^\circ)}(\tan(18^\circ)\tan(42^\circ)\tan(78^\circ))$$

$$= \frac{1}{\tan(54^\circ)}$$

$$(\tan(60^\circ-18^\circ)\tan(18^\circ)\tan(60^\circ+18^\circ))$$

$$= \frac{1}{\tan(54^\circ)}\times(.\tan(54^\circ))$$

$$= 1$$

Ex-8. Prove that

$$\left(1+\cos\left(\frac{\pi}{8}\right)\right).\left(1+\cos\left(\frac{3\pi}{8}\right)\right)$$

$$\left(1+\cos\left(\frac{5\pi}{8}\right)\right).\left(1+\cos\left(\frac{7\pi}{8}\right)\right)=\frac{1}{8}$$

Soln. We have, $\left(1+\cos\left(\frac{\pi}{8}\right)\right).\left(1+\cos\left(\frac{3\pi}{8}\right)\right)$

$$.\left(1+\cos\left(\frac{5\pi}{8}\right)\right).\left(1+\cos\left(\frac{7\pi}{8}\right)\right)$$

$$= \left(1+\cos\left(\frac{\pi}{8}\right)\right).\left(1+\cos\left(\frac{7\pi}{8}\right)\right)$$

$$.\left(1+\cos\left(\frac{3\pi}{8}\right)\right)\left(1+\cos\left(\frac{5\pi}{8}\right)\right)$$

$$= \left(1+\cos\left(\frac{\pi}{8}\right)\right)\left(1-\cos\left(\frac{\pi}{8}\right)\right)$$

$$.\left(1+\cos\left(\frac{3\pi}{8}\right)\right)\left(1-\cos\left(\frac{3\pi}{8}\right)\right)$$

$$= \left(1-\cos^2\left(\frac{\pi}{8}\right)\right)\left(1-\cos^2\left(\frac{3\pi}{8}\right)\right)$$

$$= \left(\sin^2\left(\frac{\pi}{8}\right)\right)\left(\sin^2\left(\frac{3\pi}{8}\right)\right)$$

$$= \frac{1}{4}\left(2\sin^2\left(\frac{\pi}{8}\right)\right)\left(2\sin^2\left(\frac{3\pi}{8}\right)\right)$$

$$= \frac{1}{4}\left(1-\cos\left(\frac{\pi}{4}\right)\right)\left(1-\cos\left(\frac{3\pi}{4}\right)\right)$$

$$= \frac{1}{4}\left(1-\frac{1}{\sqrt{2}}\right)\left(1+\frac{1}{\sqrt{2}}\right)$$

$$= \frac{1}{4}\left(1-\frac{1}{2}\right).$$

$$= \frac{1}{8}.$$

Ex-9. Prove that $\sin^4\left(\frac{\pi}{8}\right)+\sin^4\left(\frac{3\pi}{8}\right)$

$$+\sin^4\left(\frac{5\pi}{8}\right)+\sin^4\left(\frac{7\pi}{8}\right)=\frac{3}{2}$$

Soln. We have, $\sin^4\left(\frac{\pi}{8}\right)+\sin^4\left(\frac{3\pi}{8}\right)$

$$+\sin^4\left(\frac{5\pi}{8}\right)+\sin^4\left(\frac{7\pi}{8}\right)$$

$$= \sin^4\left(\frac{\pi}{8}\right)+\sin^4\left(\frac{3\pi}{8}\right)$$

$$+\sin^4\left(\frac{3\pi}{8}\right)+\sin^4\left(\frac{\pi}{8}\right)$$

$$= 2\left(\sin^4\left(\frac{\pi}{8}\right)+\sin^4\left(\frac{3\pi}{8}\right)\right)$$

$$= 2\left(\sin^4\left(\frac{\pi}{8}\right)+\cos^4\left(\frac{\pi}{8}\right)\right)$$

$$= 2\left(1-2\sin^2\left(\frac{\pi}{8}\right).\cos^2\left(\frac{\pi}{8}\right)\right)$$

$$= \left(2-2\sin^2\left(\frac{\pi}{8}\right).2\cos^2\left(\frac{\pi}{8}\right)\right)$$

$$= \left(2-\left(2\sin^2\left(\frac{\pi}{8}\right)\right)\left(2\cos^2\left(\frac{\pi}{8}\right)\right)\right)$$

$$= \left(2-\left(1-\cos\left(\frac{\pi}{4}\right)\right)\left(1+\cos\left(\frac{\pi}{4}\right)\right)\right)$$

$$= \left(2-\left(1-\frac{1}{\sqrt{2}}\right)\left(1+\frac{1}{\sqrt{2}}\right)\right)$$

$$= \left(2-\left(1-\frac{1}{2}\right)\right)$$

$$= \frac{3}{2}$$

Ex-10. Prove that $\cos^4\left(\frac{\pi}{8}\right)+\cos^4\left(\frac{3\pi}{8}\right)$

$$+\cos^4\left(\frac{5\pi}{8}\right)+\cos^4\left(\frac{7\pi}{8}\right)=\frac{3}{2}$$

Soln. We have, $\cos^4\left(\frac{\pi}{8}\right)+\cos^4\left(\frac{3\pi}{8}\right)$

$$+\cos^4\left(\frac{5\pi}{8}\right)+\cos^4\left(\frac{7\pi}{8}\right)$$

$$= \cos^4\left(\frac{\pi}{8}\right)+\cos^4\left(\frac{3\pi}{8}\right)$$

$$+\cos^4\left(\frac{3\pi}{8}\right)+\cos^4\left(\frac{\pi}{8}\right)$$

$$= 2\left(\cos^4\left(\frac{\pi}{8}\right)+\cos^4\left(\frac{3\pi}{8}\right)\right)$$

$$= 2\left(\cos^4\left(\frac{\pi}{8}\right)+\sin^4\left(\frac{\pi}{8}\right)\right)$$

$$= 2\left(1-2\cos^2\left(\frac{\pi}{8}\right).\sin^2\left(\frac{\pi}{8}\right)\right)$$

$$= \left(2-\left(2\cos^2\left(\frac{\pi}{8}\right)\right).\left(2\sin^2\left(\frac{\pi}{8}\right)\right)\right)$$

$$= \left(2-\left(1+\cos\left(\frac{\pi}{4}\right)\right).\left(1+\cos\left(\frac{\pi}{4}\right)\right)\right)$$

$$= \left(2-\left(1+\frac{1}{\sqrt{2}}\right).\left(1-\frac{1}{\sqrt{2}}\right)\right)$$

$$= \left(2-\left(1-\frac{1}{2}\right)\right)$$

$$= \frac{3}{2}$$

Exercise - 9

1. Prove that :

i) $\tan\frac{\theta}{2}+\cot\frac{\theta}{2}=2\operatorname{cosec}\theta$

ii) $\operatorname{cosec}\theta-\cot\theta=\tan\frac{\theta}{2}$

iii) $\frac{1+\sin\theta-\cos\theta}{1+\sin\theta+\cos\theta}=\tan\frac{\theta}{2}$

iv) $\sec\theta+\tan\theta=\tan\left(\frac{\pi}{4}+\frac{\theta}{2}\right)$

v) $(\cos A+\cos B)^2+(\sin A+\sin B)^2$

$=4\cos^2\left(\frac{A-B}{2}\right)$

2.. Prove that $\sin^2 24-\sin^2 6=\frac{\sqrt{5}-1}{8}$.

3. Prove that : $\sin^2 48^\circ-\cos^2 12^\circ=-\frac{\sqrt{5}+1}{8}$.

4. Prove that : $\left(1+\cos\frac{\pi}{8}\right)\left(1+\cos\frac{3\pi}{8}\right)\left(1+\cos\frac{5\pi}{8}\right)\left(1+\cos\frac{7\pi}{8}\right)=\frac{1}{8}$

5. Prove that : $\sin 12^\circ\sin 48^\circ\sin 54^\circ=\frac{1}{8}$

6. Prove that : $\sin 47^\circ+\sin 61^\circ-\sin 11^\circ-\sin 25^\circ=\cos 7^o$

7. Prove that : $\cos^4\frac{\pi}{8}+\cos^4\frac{3\pi}{8}+\cos^4\frac{5\pi}{8}+\cos^4\frac{7\pi}{8}=\frac{3}{2}$

8. Prove that $\sin^4\left(\frac{\pi}{8}\right)+\sin^4\left(\frac{3\pi}{8}\right)+\sin^4\left(\frac{5\pi}{8}\right)+\sin^4\left(\frac{7\pi}{8}\right)=\frac{3}{2}$

10. Prove that $\tan 20^o\tan 80^o=\sqrt{3}\tan 50^o$

11. Prove that $\tan 10^o\tan 70^o=\sqrt{3}\tan 40^o$

12. If $m=\tan 12^\circ\tan 48^\circ$ and $n=\tan 6^\circ\tan 66^\circ$, then find the value of $\frac{m}{n}+10$.

13. If $m=\sin 10^\circ\sin 50^\circ\sin 70^\circ$ and $m=\cos 20^\circ\cos 40^\circ\cos 80^\circ$, then find the value of $\frac{m}{n}+2012$.

14. If $m=\sin 6^o\sin 42^o$ and $n=\sin 66^\circ\sin 78^\circ$, then find the value of $16mn+10$.

15. If $m=\sin 8^\circ\sin 52^\circ\sin 66^\circ\sin 68^\circ$ and $8n=\cos 42^\circ$, then prove that $m=n$.

16. If $\sin 17^\circ\sin 43^\circ\sin 77^\circ=m\cos 49^\circ$, then find the value of m.

17. If $m=\sin\left(\frac{2\pi}{7}\right)+\sin\left(\frac{4\pi}{7}\right)+\sin\left(\frac{8\pi}{7}\right)$, then find the value of $\left(2m+2-\sqrt{7}\right)$.

18. If $m = \cos 13^o \cos 47^o \cos 73^o$ and $n = \frac{1}{4}\sin 51^o$, then prove that m = n.

19. Prove that $\sin 55^o - \sin 19^o + \sin 53^o - \sin 17^o = \cos 1^o$

20. Prove that : $\cos\frac{2\pi}{7} . \cos\frac{4\pi}{7} . \cos\frac{6\pi}{7} = \frac{1}{8}$.

21. Prove that $\cos\frac{2\pi}{7} + \cos\frac{4\pi}{7} + \cos\frac{6\pi}{7} = -\frac{1}{2}$

22. Find the value of $\tan 7\frac{1}{2}^\circ + \cot 7\frac{1}{2}^\circ$.

23. If $\cos\left(\frac{x}{2}\right) - \sqrt{3}\sin\left(\frac{x}{2}\right)$ takes its minimum value then find its x.

24. If α and β be two different roots $a\cos\theta + b\sin\theta = c$, then prove that $\sin(\alpha+\beta) = \frac{2ab}{a^2+b^2}$

Answers

12. 11, **13.** 2013, **14.** 10, 16. $\frac{1}{4}$, **17.** 2

23. $x = \frac{4\pi}{3}$.

1.19 CONDITIONAL TRIGONOMETRICAL IDENTITIES

Here, we shall deal with trigonometrical identities involving two or more angles. In establishing such identities we will be frequently using properties of supplementary and complementary angles and thus, students are advised to go through all the above formulae.

We have certain trigonometrical identities like, $\sin^2\theta + \cos^2\theta = 1$

and $\sec^2\theta = 1 + \tan^2\theta$ etc.

Such identities are identities in the sense that they hold for all values of the angles, which satisfy the given condition amongst them. Such identities are called **Conditional Identities.**

If A, B, C denote the angles of a triangle ABC, then the relation A + B+ C = π enables us to establish many important identities involving trigonometric raios of these angles.

i) If $A+B+C=\pi$, then $A+B = \pi - C$, $B+C = \pi - A$ and $C+A = \pi - B$

ii) If $A+B+C=\pi$, then $\sin(A+B) = \sin(\pi - C) = \sin C$

similarly, $\sin(B-C) = \sin(\pi - A) = \sin A$

and $\sin(C+A) = \sin(\pi - B) = \sin B$

iii) If $A+B+C=\pi$, then $\cos(A+B) = \cos(\pi - C) = -\cos C$

Similarly, $\cos(B+C) = \cos(\pi - A) = -\cos A$

and $\cos(C+A) = \cos(\pi - B) = -\cos B$

iv) If $A+B+C=\pi$, then $\tan(A+B) = \tan(\pi - C) = -\tan C$

Similarly, $\tan(B+C) = \tan(\pi - A) = -\tan A$

and $\tan(C+A) = \tan(\pi - B) = -\tan B$

v) If $A+B+C=\pi$, then $\frac{A+B}{2} = \frac{\pi}{2} - \frac{C}{2}$,

$\frac{B+C}{2} = \frac{\pi}{2} - \frac{A}{2}$ and $\frac{C+A}{2} = \frac{\pi}{2} - \frac{B}{2}$

Therefore, $\sin\left(\frac{A+B}{2}\right) = \sin\left(\frac{\pi}{2} - \frac{C}{2}\right) = \cos\left(\frac{C}{2}\right)$

$\cos\left(\frac{A+B}{2}\right) = \cos\left(\frac{\pi}{2} - \frac{C}{2}\right) = \sin\left(\frac{C}{2}\right)$

$\tan\left(\frac{A+B}{2}\right) = \tan\left(\frac{\pi}{2} - \frac{C}{2}\right) = \cot\left(\frac{C}{2}\right)$.

Note : Dear Students, please recollect the following formulae from basic trigonometry

Step - I :

(i) $\sin 2A = 2\sin A\cos A$

(ii) $\cos 2A = 2\cos^2 A - 1$.

(iii) $\cos 2A = 1 - 2\sin^2 A$

(iv) $\cos 2A = \cos^2 A - \sin^2 A$.

Step II:

(i) $1 + \cos 2A = 2\cos^2 A$

(ii) $1 - \cos 2A = 2\sin^2 A$

(iii) $1 + \cos A = 2\cos^2(A/2)$

(iv) $1 - \cos A = 2\sin^2(A/2)$.

Step III :

(i) $\cos(A+B) + \cos(A-B) = 2\cos A\cos B$

(ii) $\cos(A-B) - \cos(A+B) = 2\sin A\sin B$.

Step-IV :

(i) $\cos C + \cos D = 2\cos\left(\frac{C+D}{2}\right)\cos\left(\frac{C-D}{2}\right)$

(ii) $\cos C - \cos D = 2\sin\left(\frac{C+D}{2}\right)\sin\left(\frac{D-C}{2}\right)$

(iii) $\sin C + \sin D = 2\sin\left(\frac{C+D}{2}\right)\cos\left(\frac{C-D}{2}\right)$

(iv) $\sin C - \sin D = 2\cos\left(\frac{C+D}{2}\right)\sin\left(\frac{C-D}{2}\right)$.

1.20 SOME SOLVED EXAMPLES :

Ex-1. If $A+B+C=\pi$, then prove that
$\sin 2A + \sin 2B + \sin 2C = 4\sin A \sin B \sin C$

Soln. We have, $\sin 2A + \sin 2B + \sin 2C$

$= (\sin 2A + \sin 2B) + \sin 2C$

$= 2(\sin(A+B)\cos(A-B)) + \sin 2C$

$= 2(\sin(\pi - C).\cos(A-B)) + 2\sin C\cos C$

$= 2(\sin C.\cos(A-B)) + 2\sin C\cos C$

$= 2\sin C(\cos(A-B) + \cos C)$

$= 2\sin C(\cos(A-B) + \cos(\pi - (A+B)))$

$= 2\sin C(\cos(A-B) - \cos(A+B))$

$= 2\sin C(2\sin A\sin B)$

$= 4\sin A.\sin B.\sin C$

Ex-2. If $A+B+C=\pi$, then prove that,
$\cos 2A + \cos 2B + \cos 2C$
$= 1 - 4\cos A\cos B\cos C$

Soln. We have $\cos 2A + \cos 2B + \cos 2C$

$= (\cos 2A + \cos 2B) + \cos 2C$

$= 2\cos(A+B)\cos(A-B) + \cos 2C$

$= 2\cos\{\pi - C\}\cos(A-B) + \cos 2C$

$= -2\cos C\cos(A-B) + 2\cos^2 C - 1$

$= -1 - 2\cos C(\cos(A-B) - \cos C)$

$= -1 - 2\cos C(\cos(A-B) + \cos(A+B))$

$= -1 - 2\cos C(2\cos A.\cos B)$

$= -1 - 4\cos A.\cos B.\cos C$

Ex-3. If $A+B+C=\pi$, then prove that
$\sin^2 A + \sin^2 B - \sin^2 C = 4\sin A\sin B\cos C$

Soln. We have, $\sin^2 A + \sin^2 B - \sin^2 C$

$= \sin^2 A + (\sin^2 B - \sin^2 C)$

$= \sin^2 A + \sin(B+C)\sin(B-C)$

$= \sin^2 A + \sin(\pi - A)\sin(B-C)$

$= \sin^2 A + \sin A\sin(B-C)$

$= \sin A(\sin A + \sin(B-C))$

$= \sin A(\sin(B+C) + \sin(B-C))$

$= \sin A(2\sin B\cos C)$

$= 2\sin A.\sin B.\cos C$

Ex-4. If $A+B+C=\pi$, then prove that,
$\sin^2 A + \sin^2 B + \sin^2 C = 2 + 2\cos A\cos B\cos C$

Soln. We have, $\sin^2 A + \sin^2 B + \sin^2 C$

$= 1 - \cos^2 A + \sin^2 B + \sin^2 C$

$= 1 - (\cos^2 A - \sin^2 B) + (1 - \cos^2 C)$

$= 2 - (\cos^2 A - \sin^2 B) - \cos^2 C$

$= 2 - (\cos(A+B).\cos(A-B)) - \cos^2 C$

$= 2 - (\cos(\pi - C).\cos(A-B)) - \cos^2 C$

$= 2 + \cos C(.\cos(A-B) - \cos C)$

$= 2 + \cos C(.\cos(A-B) + \cos(A+B))$

$= 2 + \cos C(2\cos A.\cos B)$

$= 2 + 2\cos A.\cos B.\cos C$

Ex-5. In a triangle ΔABC, prove that,
$\tan A + \tan B + \tan C = \tan A.\tan B.\tan C$

Soln. We have, $A+B+C=\pi$

$\Rightarrow A+B = \pi - C$

$\Rightarrow \tan(A+B) = \tan(\pi - C)$

$\Rightarrow \frac{\tan A + \tan B}{1 - \tan A.\tan B} = -\tan C$

$\Rightarrow \quad \tan A + \tan B = -\tan C(1 - \tan A.\tan B)$

$\Rightarrow \quad \tan A + \tan B = -\tan C + \tan A.\tan B.\tan C$

$\Rightarrow \quad \tan A + \tan B + \tan C = \tan A.\tan B.\tan C$

Ex-6. In a triangle ΔABC, prove that,

$\cot A.\cot B + \cot B.\cot C + \cot C.\cot A = 1$

Soln. As we know that,

$\tan A + \tan B + \tan C = \tan A.\tan B.\tan C$

Dividing both the sides by 'tanA. tanB. tanC', we get,

$$= \frac{\tan A}{\tan A.\tan B.\tan C} + \frac{\tan B}{\tan A.\tan B.\tan C} + \frac{\tan C}{\tan A.\tan B.\tan C} = 1$$

$$\Rightarrow \quad \frac{1}{\tan B.\tan C} + \frac{1}{\tan A.\tan C} + \frac{1}{\tan A.\tan B} = 1$$

$\Rightarrow \quad \cot B.\cot C + \cot A.\cot C + \cot A.\cot B = 1$

Ex-7. If A, B, C and D be the angles of a quadri - lateral, then prove that

$$\frac{\tan A + \tan B + \tan C + \tan D}{\cot A + \cot B + \cot C + \cot D} = \tan A.\tan B.\tan C.\tan D$$

Soln. We have, $\quad A + B + C + D = 2\pi$

$\Rightarrow \quad A + B + C + D = 2\pi$

$\Rightarrow \quad A + B = 2\pi - (C + D)$

$\Rightarrow \quad \tan(A + B) = \tan\{2\pi - (C + D)\}$

$\Rightarrow \quad \tan(A + B) = -\tan(C + D)$

$$\Rightarrow \quad \frac{\tan A + \tan B}{1 - \tan A.\tan B} = -\frac{\tan C + \tan D}{1 - \tan C.\tan D}$$

$\Rightarrow \quad (\tan A + \tan B)(1 - \tan C.\tan D)$

$= -(\tan C + \tan D)(1 - \tan A.\tan B)$

$\Rightarrow \quad \tan A + \tan B - \tan A.\tan C.\tan D$

$- \tan B.\tan C.\tan D$

$= -\tan C - \tan D + \tan A. \tan B \tan C$

$+ \tan A. \tan B. \tan D$

$\Rightarrow \quad \tan A + \tan B + \tan C + \tan D$

$= \tan A. \tan C. \tan D + \tan B. \tan C. \tan D$

$+ \tan A. \tan B. \tan C + \tan A. \tan B. \tan D$

$$\Rightarrow \quad \frac{\tan A + \tan B + \tan C + \tan D}{\tan A.\tan B.\tan C.\tan D}$$

$$= \frac{\tan A.\tan B.\tan C}{\tan A.\tan B.\tan C.\tan D} + \frac{\tan A.\tan C.\tan D}{\tan A.\tan B.\tan C.\tan D} + \frac{\tan A.\tan B.\tan D}{\tan A.\tan B.\tan C.\tan D} + \frac{\tan B.\tan C.\tan D}{\tan A.\tan B.\tan C.\tan D}$$

$$\Rightarrow \quad \frac{\tan A + \tan B + \tan C + \tan D}{\tan A.\tan B.\tan C.\tan D}$$

$$= \frac{1}{\tan A} + \frac{1}{\tan B} + \frac{1}{\tan C} + \frac{1}{\tan D}$$

$$\Rightarrow \quad \frac{\tan A + \tan B + \tan C + \tan D}{\tan A.\tan B.\tan C.\tan D}$$

$= \cot A + \cot B + \cot C + \cot D$

$$\Rightarrow \quad \frac{\tan A + \tan B + \tan C + \tan D}{\cot A + \cot B + \cot C + \cot D}$$

$= \tan A.\tan B.\tan C.\tan D$

Ex-8. In a triangle ABC, prove that

$(\cot A + \cot B)(\cot B + \cot C)$

$(\cot C + \cot A)$=cosec A cosec B cosec C.

Soln. We have,

$$(\cot A + \cot B) = \frac{\cos A}{\sin A} + \frac{\cos B}{\sin B}$$

$$= \frac{\cos A \sin B + \sin A.\cos B}{\sin A \sin B}$$

$$= \frac{\sin(A+B)}{\sin A \sin B} = \frac{\sin C}{\sin A \sin B}.$$

Similarly, $(\cot B + \cot C) = \dfrac{\sin A}{\sin B \sin C}$

and $(\cot C + \cot A) = \dfrac{\sin B}{\sin A \sin C}$

Thus, $(\cot A + \cot B)(\cot B + \cot C)(\cot C + \cot A)$

$$= \frac{\sin C}{\sin A \sin B} \times \frac{\sin B}{\sin A \sin C} \times \frac{\sin A}{\sin B \sin C}$$

$$= \frac{1}{\sin A \sin B \sin C}$$

$$= \operatorname{cosec} A . \operatorname{cosec} B . \operatorname{cosec} C$$

Ex-9. If $xy + yz + zx = 1$, then prove that,

$$\frac{x}{1-x^2} + \frac{y}{1-y^2} + \frac{z}{1-z^2} = \frac{4xyz}{(1-x^2)(1-y^2)(1-z^2)}$$

Soln. Put $x = \tan A$, $y = \tan B$ and $z = \tan C$

Given, $xy + yz + zx = 1$

$\Rightarrow \tan A.\tan B + \tan B.\tan C + \tan C.\tan A = 1$

$\Rightarrow \tan B.\tan C + \tan C.\tan A = 1 - \tan A.\tan B$

$\Rightarrow \tan C(\tan B + \tan A) = 1 - \tan A.\tan B$

$\Rightarrow \dfrac{\tan A + \tan B}{1 - \tan A.\tan B} = \dfrac{1}{\tan C}$

$\Rightarrow \tan(A+B) = \cot C = \tan\left(\dfrac{\pi}{2} - C\right)$

$\Rightarrow (A+B) = \left(\dfrac{\pi}{2} - C\right)$

$\Rightarrow (A+B+C) = \dfrac{\pi}{2}$

Now, L.H.S $= \dfrac{x}{1-x^2} + \dfrac{y}{1-y^2} + \dfrac{z}{1-z^2}$

$$= \frac{\tan A}{1-\tan^2 A} + \frac{\tan B}{1-\tan^2 B} + \frac{\tan C}{1-\tan^2 C}$$

$$= \frac{1}{2}\left(\frac{2\tan A}{1-\tan^2 A} + \frac{2\tan B}{1-\tan^2 B} + \frac{2\tan C}{1-\tan^2 C}\right)$$

$$= \frac{1}{2}(\tan 2A + \tan 2B + \tan 2C)$$

$$= \frac{1}{2}(\tan 2A.\tan 2B.\tan 2C)$$

$$= \frac{1}{2}\left(\frac{2\tan A}{1-\tan^2 A}.\frac{2\tan B}{1-\tan^2 B}.\frac{2\tan C}{1-\tan^2 C}\right)$$

$$= \left(\frac{4\tan A.\tan B.\tan C}{(1-\tan^2 A)(1-\tan^2 B)(1-\tan^2 C)}\right)$$

$$= \frac{4xyz}{(1-x^2)(1-y^2)(1-z^2)}$$

Hence, the result.

Ex-10. If $xy + yz + zx = 1$, then prove that

$$\frac{x}{1+x^2} + \frac{y}{1+y^2} + \frac{z}{1+z^2} = \frac{2}{\sqrt{(1+x^2)(1+y^2)(1+z^2)}}$$

Soln. Put $x = \tan A$, $y = \tan B$ and $z = \tan C$

Given, $xy + yz + zx = 1$

$\Rightarrow \tan A.\tan B + \tan B.\tan C + \tan C.\tan A = 1$

$\Rightarrow A + B + C = \dfrac{\pi}{2}$

Now, L.H.S $= \dfrac{x}{1+x^2} + \dfrac{y}{1+y^2} + \dfrac{z}{1+z^2}$

$$= \frac{1}{2}\left(\frac{2x}{1+x^2} + \frac{2y}{1+y^2} + \frac{2z}{1+z^2}\right)$$

$$= \frac{1}{2}\left(\frac{2\tan A}{1+\tan^2 A} + \frac{2\tan B}{1+\tan^2 B} + \frac{2\tan C}{1+\tan^2 C}\right)$$

$$= \frac{1}{2}(\sin 2A + \sin 2B + \sin 2C)$$

$$= \frac{1}{2}(4\cos A.\cos B.\cos C)$$

$$= 2\cos A.\cos B.\cos C$$

$$= \frac{2}{\sec A.\sec B.\sec C}$$

$$= \frac{2}{\sqrt{(1+\tan^2 A)(1+\tan^2 B)(1+\tan^2 C)}}$$

$$= \frac{2}{\sqrt{(1+x^2)(1+y^2)(1+z^2)}}$$

Hence, the result.

Exercise - 10

1. If $A+B+C=\pi$, then prove that : sin 2A + sin 2 B + sin 2C = 4 sin A sin B sin C.
2. If $A+B+C=\pi$, then prove that : sin 2A + sin 2 B - sin 2C = 4 cos A cos B sin C.
3. If $A+B+C=\pi$, then prove that : cos 2A + cos 2 B + cos 2C = -1 - 4 cos A cos B cos C.
4. If $A+B+C=\pi$, then prove that : cos 2A + cos 2 B - cos 2C = 1 - 4 sin A sin B cos C.
5. If $A+B+C=\pi$, then prove that cos A + cos B + cos C = $1 + 4 \sin\frac{A}{2} \sin\frac{B}{2} \sin\frac{C}{2}$
6. If $A+B+C=\pi$, then prove that $\cos^2 A + \cos^2 B - \cos^2 C + 2 \cos A \cos B \cos C = 1$
7. If $A+B+C=\pi$, then prove that $\cos^2 A + \cos^2 B + \cos^2 C = 1 - 2 \sin A \sin B \cos C$.
8. If $A+B+C=\pi$, then prove that $\sin^2 A + \sin^2 B + \sin^2 C = 2(1 + \cos A \cos B \cos C)$
9. If $A + B + C = \pi$, then prove that $\frac{\sin 2A + \sin 2B + \sin 2C}{\cos A + \cos B + \cos C - 1} = 8 \cos\frac{A}{2} \cos\frac{B}{2} \cos\frac{C}{2}$
10. If $A+B+C=\pi$, then prove that $\sin^2\frac{A}{2} + \sin^2\frac{B}{2} - \sin^2\frac{C}{2} = 1 - 2\cos\frac{A}{2}\cos\frac{B}{2}\cos\frac{C}{2}$
11. If $A+B+C=\frac{\pi}{2}$, then prove that :
(i) $\sin^2 A + \sin^2 B + \sin^2 C = 1 - 2 \sin A \sin B \sin C$
(ii) $\cos^2 A + \cos^2 B + \cos^2 C = 2 + 2 \sin A \sin B \sin C$
12. If $A+B+C=\pi$, then prove that
(i) tan A + tan B + tan C = tan A tan B tan C.
(ii) tan 2A + tan2 B + tan 2C = tan 2A tan 2B tan 2C.
(iii) cot A cot B + cot B cot C + cot C cot A = 1
(iv) cot (A/2) cot (B/2)+cot (B/2) cot (C/2) +cot (C/2) cot (A/2) = 1.
13. If $A+B+C=\frac{\pi}{2}$, then prove that:
(i) cot A + cot B+ cot C = cot A cot B cot C
(ii) tan A tan B + tan B tan C + tan C tan A = 1
14. If xy + yz + zx = 1, then prove that $\frac{x}{1-x^2}+\frac{y}{1-y^2}+\frac{z}{1-z^2} = \frac{4xyz}{(1-x^2)(1-y^2)(1-z^2)}$
15. Prove that tan $(\alpha-\beta)+\tan(\beta-\gamma)+\tan(\gamma-\alpha) = \tan(\alpha-\beta)\tan(\beta-\gamma)\tan(\gamma-\alpha)$
16. In a ΔABC, if cot A + cot B + cot C = $\sqrt{3}$, then prove that the triangle is an equilateral.
17. If x + y + z = x y z, then prove that $\frac{3x-x^3}{1-3x^2}+\frac{3y-y^3}{1-3y^2}+\frac{3z-z^3}{1-3z^2} = \frac{3x-x^3}{1-3x^2}\cdot\frac{3y-y^3}{1-3y^2}\cdot\frac{3z-z^3}{1-3z^2}$
18. Prove that 1 + cos 56° + cos 58° – cos 66° = 4cos 28° cos 29° sin 33°.

1.21 TRIGONOMETRICAL SERIES

1.21.1 Introduction :

In this section, we are mainly connected with different procedures to find out the summation of trigonometrical series.

To find out the sum of different trigonometrical series, first we observe the nature of the angles in of the trigonometrical terms. We must observe whether the angles form any sequence or not ? If they form any sequences, then we must check, what kind of sequence it is ? we also observe the sequence formed (if any) by the coefficients of terms of the series.

So, our main attempt will be–

(i) to express each term as a difference between of the two terms directly or by manipulation and then add; or

(ii) to arrange the series in such a way that it follows some standard trigonometrical expansion.

1.22. DIFFRENT TYPES OF THE SUMMATION OF A TRIGONOMETRICAL SERIES :

(i) A trigonometrical series involved with the terms of sines or cosines.

Rule : Whenever angles are in A.P. and the trigonometrical terms involved with sines or cosines having power 1.

1. We must multiply each term by

$$2\sin\left(\frac{\text{common diffrence of angles}}{2}\right)$$

2. and then express each term as a difference between two terms.

3. Finally add them.

Ex.-1. Prove that

$$\sin\alpha+\sin(\alpha+\beta)+\sin(\alpha+2\beta)$$
$$+\sin(\alpha+3\beta)+.........+\sin(\alpha+(n-1)\beta)$$

$$=\frac{\sin\left(\frac{n\beta}{2}\right)}{\sin\left(\frac{\beta}{2}\right)}\times\sin\left[\alpha+(n-1)\frac{\beta}{2}\right]$$

Pf: Let $S=\sin\alpha+\sin(\alpha+\beta)+\sin(\alpha+2\beta)$
$$+\sin(\alpha+3\beta)+.........+\sin(\alpha+(n-1)\beta)$$

Now,

$$2\sin\alpha.\sin\left(\frac{\beta}{2}\right)=\cos\left(\alpha-\frac{\beta}{2}\right)-\cos\left(\alpha+\frac{\beta}{2}\right)$$

$$2\sin(\alpha+\beta).\sin\left(\frac{\beta}{2}\right)=\cos\left(\alpha+\frac{\beta}{2}\right)-\cos\left(\alpha+\frac{3\beta}{2}\right)$$

$$2\sin(\alpha+2\beta).\sin\left(\frac{\beta}{2}\right)=\cos\left(\alpha+\frac{3\beta}{2}\right)-\cos\left(\alpha+\frac{5\beta}{2}\right)$$

---------- ------------- --------- ---- ------

$$2\sin(\alpha+(n-1)\beta).\sin\left(\frac{\beta}{2}\right)$$

$$=\cos\left(\alpha+\frac{(2n-3)\beta}{2}\right)-\cos\left(\alpha+\frac{(2n-1)\beta}{2}\right)$$

Adding,we get,

$$=2\sin\left(\alpha+\frac{(n-1)\beta}{2}\right)\times\sin\left(\frac{\beta}{2}\right)$$

Thus
$$\frac{\sin\left(\frac{n\beta}{2}\right)}{\sin\left(\frac{\beta}{2}\right)}\times\sin\left(\alpha+\frac{(n-1)\beta}{2}\right)$$

Exercise - 11

Questions Based on Sine Series :

1. Prove that
$$\sin\alpha+\sin2\alpha+\sin3\alpha+..........+\sin n\alpha$$
$$=\frac{\sin\left(\frac{n\alpha}{2}\right)}{\sin\left(\frac{\alpha}{2}\right)}\times\sin(n+1)\frac{\alpha}{2}$$

2. Prove that
$$\sin\theta+\sin3\theta+\sin5\theta+......$$
$$....+\sin(2n-1)\theta=\frac{\sin^2 n\theta}{\sin\theta}$$

3. Prove that
$$\sin^2\alpha+\sin^2(\alpha+\beta)+\sin^2(\alpha+2\beta\beta+..$$
$$......+\sin^2(\alpha+(n-1)\beta).$$
$$=\frac{n}{2}-\frac{1}{2}\times\frac{\sin n\beta}{\sin\beta}\times\cos[2\alpha+(n-1)\beta]$$

4. Prove that $\sin^2\alpha+\sin^2\left(\alpha+\frac{2\pi}{n}\right)$
$$+\sin^2\left(\alpha+\frac{2\pi}{n}\right)+.....\text{to}-\text{n}-\text{terms}=\frac{n}{2}$$

5. If $4n\theta=\pi$, then prove that
$$\sin^2\theta+\sin^2 3\theta+\sin^2 5\theta+....\text{to}-\text{n}-\text{terms}$$
$$=n$$

6. Prove that

$\sin^3\theta + \sin^3 3\theta + \sin^3 5\theta + ...\text{to} - n - \text{terms}$

$= \frac{1}{4}\left[\frac{3\sin^2 n\theta}{\sin\theta} - \frac{3\sin^2 3n\theta}{\sin 3\theta}\right].$

7. Prove that $\sin\alpha + \cos(\alpha+\beta)$

$-\sin(\alpha+2)\beta - \cos(\alpha+2\beta\beta$

$+\sin(\alpha+4\beta\beta + ...\text{to} - n - \text{terms}$

$= \frac{\sin\left(\frac{n(2\beta+\pi)}{4}\right)}{\sin\left(\frac{2\beta+\pi}{4}\right)} \times \sin\left\{\alpha + \frac{(n-1)(2\beta+\pi)}{4}\right\}$

8. Prove that

$\sin^3\alpha + \sin^3\left(\alpha+\frac{2\pi}{n}\right) + \sin^3\left(\alpha+\frac{4\pi}{n}\right)$

$+.........\text{to} - n - \text{terms} = 0$

9. Prove that

$\sin\theta.\sin 2\theta + \sin 2\theta\theta.\text{si } 3\theta$

$+\sin 3\theta\ .\sin 4\theta + ...\text{to} - n - \text{terms}$

$= \frac{n}{2}\cos\theta - \frac{1}{2} \times \frac{\sin n\theta}{\sin\theta} \times \cos(n+2)\theta$

10. Prove that $\sin^4 x + \sin^4\left(x + \frac{2\pi}{n}\right)$

$+\sin^4\left(x+\frac{2\pi}{n}\right) + ...\text{to} - n - \text{terms} = \frac{3\pi}{8}$

2. A Trigonometrical Series involved with the terms of Cosines :

$\cos\alpha + \cos(\alpha+\beta) + \cos(\alpha+\beta)$

$+\cos(\alpha+\beta) + ... + \cos(\alpha+(n-1)\beta)$

$= \frac{\sin\left(\frac{n\beta}{2}\right)}{\sin\left(\frac{\beta}{2}\right)} \times \cos\left(\alpha + (n-1)\frac{\beta}{2}\right)$

Proof : Let $S = \cos\alpha + \cos(\alpha+\beta) + \cos(\alpha+\beta)$

$+\cos(\alpha+\beta) + ... + \cos(\alpha+(n-1)\beta)$

Now, $2\cos\alpha\sin\left(\frac{\beta}{2}\right)$

$= \sin\left(\alpha+\frac{\beta}{2}\right) - \sin\left(\alpha-\frac{\beta}{2}\right)$

$2\cos(\alpha+\beta)\sin\left(\frac{\beta}{2}\right)$

$= \sin\left(\alpha+\frac{3\beta}{2}\right) - \sin\left(\alpha+\frac{\beta}{2}\right)$

$2\cos(\alpha+3\beta)\sin\left(\frac{\beta}{2}\right)$

$= = \sin\left(\alpha+\frac{5\beta}{2}\right) - \sin\left(\alpha+\frac{3\beta}{2}\right)$

$2\cos(\alpha+(n-1)\beta)\sin\left(\frac{\beta}{2}\right)$

$= \sin\left(\alpha+\frac{(2n-1)\beta}{2}\right) - \sin\left(\alpha+\frac{(2n-3)\beta}{2}\right)$

Adding all, we get

$2\sin\left(\frac{\beta}{2}\right) \times S = \sin\left(\alpha+\frac{2n-1}{2}\beta\right) - \sin(\alpha-\beta)$

$\Rightarrow 2\sin\left(\frac{\beta}{2}\right) \times S = 2\cos\left(\alpha+\frac{n-1}{2}\beta\right) \times \sin\left(\frac{n\beta}{2}\right)$

$\Rightarrow \quad S = \frac{\sin\left(\frac{n\beta}{2}\right)}{\sin\left(\frac{\beta}{2}\right)} \times 2\cos\left(\alpha+\frac{n-1}{2}\beta\right)$

Exercise - 12

Questions Based on Cosine Series :

1. Prove that $\cos\alpha + \cos 2\alpha + \cos 3\alpha + ...$

$.... + \cos n\alpha = \frac{\sin\left(\frac{n\alpha}{2}\right)}{\sin\left(\frac{\alpha}{2}\right)} \times \cos\left(\frac{(n+1)\alpha}{2}\right)$

2. Prove that $\cos\alpha + \cos\left(\alpha+\frac{2\pi}{n}\right)$

$+\cos\left(\alpha+\frac{2\pi}{n}\right)+....to-n-terms$

$= \quad 0$

3. If B be the exist for angles of a regular polygon of n-sides and A is any constant, then prove that
$\cos A + \cos(A+B) + \cos(A+2B)$
+.............to n terms = 0.

4. Show that, if $n>2$,

$\cos^2\left(\frac{\pi}{n}\right)+\cos^2\left(\frac{3\pi}{n}\right)+\cos^2\left(\frac{5\pi}{n}\right)$

$+..........to-n-terms=\frac{n}{2}$

5. Prove that, $\sqrt{1-\sin 2\theta}+\sqrt{1-\sin 4\theta}$

$+\sqrt{1-\sin 6\theta}+......to-n-terms$

$=\frac{\sin\left(\frac{n\theta}{2}\right)}{\sin\left(\frac{\theta}{2}\right)}\times\left[\cos(n+1)\frac{\theta}{2}-\sin(n+1)\frac{\theta}{2}\right].$

6. Prove that, $\cos^2 x+\cos^2 3x+\cos^2 5x+...$

$........to-n-terms$

$=\frac{1}{2}\times\left[n+\frac{\sin 4nx}{2\sin 2x}\right].$

7. Prove that, cos x + sin 3 x + cos 5x
+ sin 7x +...........+sin (2n+1)x

$=\frac{\sin 2nx}{\sin 2x}\times[\cos(2n-1)x+\sin(2n+1)x].$

8. Prove that, $\cos^3\theta+\cos^3\left(\theta-\frac{2\pi}{n}\right)$

$+\cos^3\left(\theta-\frac{4\pi}{n}\right)+\cos^3\left(\theta-\frac{6\pi}{n}\right)$

$+....to-n-terms = 0$

9. Prove that, $\cos^2\alpha+\cos^2(\alpha+\beta)$

$+\cos^2(\alpha+2\beta)+.......to-n-terms$

$=\frac{n}{2}+\frac{1}{2}\times\frac{\sin n\beta}{\sin\beta}\times\cos[2\alpha+(n-1)\beta]$

10. Prove that, $\cos\theta\cos 2\theta+\cos 3\theta\cos 4\theta$

$+\cos 5\theta\cos 6\theta+......to-n-terms$

$=\frac{1}{2}\times\left[n\cos\theta+\frac{\sin 2n\theta}{\sin 2\theta}\times\cos(2n+1)\theta\right].$

3. A Trigonomeytrical series Based Method of difference :

Rule : 1. Express each term of the series as a diffreence of two expres sions.

2. Finally added them and we shall get the required result.

Ex-1. Find the sum of n-terms of the series

$\frac{\sin x}{\sin 2x.\sin 3x}+\frac{\sin x}{\sin 3x.\sin 4x}$

$+\frac{\sin x}{\sin 4x.\sin 5x}+....to-n-terms$

Soln. Let $t_n=\frac{\sin x}{\sin(n+1)x.\sin(n+2)x}$

$\Rightarrow \quad t_n=\frac{\sin[(n+2)x-(n+1)x]}{\sin(n+1)x.\sin(n+2)x}$

$=\frac{\sin(n+2)x\cos(n+1)x}{\sin(n+1)x.\sin(n+2)x}$

$-\frac{\cos(n+2)x\sin(n+1)x}{\sin(n+1)x.\sin(n+2)x}$

$=\cot(n+1)x-\cot(n+2)x$

Thus, $t_1 = \cot 2x - \cot 3x$

$t_2 = \cot 3x - \cot 4x$

$t_3 = \cot 4x - \cot 5x$

$t_n = \cot(n+1)x - \cot(n+2)x.$

Adding all,

we get, $S = \cot 2x - \cot(n+2)x$.

Exercise - 13

Questions Based on Method of Diffrences :

1. Prove that,

$\text{cosec}\,\theta.\text{cosec}\,2\theta+\text{cosec}\,2\theta.\text{cosec}\,3\theta$

$+\text{cosec}\,3\theta\theta.\text{cose}\,4\theta+....to-n-terms$

$=\text{cosec}\,\theta\,[\cot\theta-\cot(n+1)\theta]$

2. Prove that $\frac{1}{\cos\theta+\cos 2\theta}+\frac{1}{\cos\theta+\cos 5\theta}+\frac{1}{\cos\theta+\cos 7\theta}+....to-n-terms$

$=\operatorname{cosec}\theta\left[\tan(n+1)\theta-\tan\theta\right]$

3. Prove that

$\frac{\sin x}{\cos x+\cos 2x}+\frac{\sin x}{\cos x+\cos 4x}+\frac{\sin x}{\cos x+\cos 6x}+.....to-n-terms.$

$=\frac{1}{4}\times\operatorname{cosec}\left(\frac{n}{2}\right)\times\left[\sec(2n+1)\frac{x}{2}-\sec\left(\frac{x}{2}\right)\right].$

4. Prove that $\operatorname{cosec}\theta+\operatorname{cosec}2\theta+\operatorname{cosec}2^2\theta+\operatorname{cosec}2^4\theta+.....to-n-terms$

$=\cot\left(\frac{\theta}{2}\right)-\cot\left(2^{n-1}\theta\right).$

5. Prove that

$\operatorname{cosec}\theta+\operatorname{cosec}\left(\frac{\theta}{2}\right)+\operatorname{cosec}\left(\frac{\theta}{2^2}\right)+\operatorname{cosec}\left(\frac{\theta}{2^3}\right)+to-n-terms$

$=\cot\left(\frac{\theta}{2^n}\right)-\cot\theta.$

6. Prove that

tan x. tan 2x + tan 2x. tan 3x +.......+ tan nx. tan (n+1)x

= cot x [tan (n+1) x - tan x] - n.

7. Prove that,

$\tan^{-1}\left(\frac{x}{1+2x^2}\right)+\tan^{-1}\left(\frac{x}{1+6x^2}\right)+\tan^{-1}\left(\frac{x}{1+12x^2}\right)+..............+\tan^{-1}\left(\frac{x}{1+n(n+1)x^2}\right)$

$=\tan^{-1}(n+1)x-\tan^{-1}x$

8. Prove that,

$\tan^{-1}\left(\frac{1}{7}\right)+\tan^{-1}\left(\frac{1}{13}\right)+\tan^{-1}\left(\frac{1}{19}\right)+.....+\tan^{-1}\left(\frac{1}{n^2+3n+3}\right)$

$=\tan^{-1}(n+2)-\tan^{-1}2.$

9. Prove that, $\sin^{-1}\left(\frac{1}{\sqrt{2}}\right)+\sin^{-1}\left(\frac{\sqrt{2}-1}{\sqrt{2}}\right)+\sin^{-1}\left(\frac{\sqrt{3}-\sqrt{2}}{\sqrt{12}}\right)+......to\infty=\frac{\pi}{2}.$

10. Prove that, $\tan^{-1}\left(\frac{1}{3}\right)+\tan^{-1}\left(\frac{2}{9}\right)+\tan^{-1}\left(\frac{4}{23}\right)+......to\infty=\frac{\pi}{4}.$

Level– I QUESTIONS BASED ON FUNDAMENTALS

1. Find the values of the expression $3(\sin x-\cos x)^4+6(\sin x+\cos x)^2+4(\sin^6 x+\cos^6 x)$
2. If $m=\sin^6 x+\cos^6 x$, then find m.
3. If $\operatorname{cosec}\theta-\cot\theta=q$, then find $\operatorname{cosec}\theta$.
4. Find the value of $\tan\frac{\pi}{20}\cdot\tan\frac{3\pi}{20}\cdot\tan\frac{5\pi}{20}\cdot\tan\frac{7\pi}{20}\cdot\tan\frac{9\pi}{20}$
5. If $x=\frac{2\sin\theta}{1+\cos\theta+\sin\theta}$, then find the values of $\frac{1-\cos\theta+\sin\theta}{1+\sin\theta}$
6. If $\sin x+\sin^2 x=1$, then find the value of $\cos^8 x+2\cos^6 x+\cos^4 x$.
7. If $\sin\theta_1+\sin\theta_2+\sin\theta_3=3$, then find the value of $\cos\theta_1+\cos\theta_2+\cos\theta_3$.

8. If $\sin x + \sin^2 x + \sin^3 x = 1$, then find the value of $\cos^6 x - 4\cos^4 x + 8\cos^2 x$.

9. If $\cos x + \cos y + \cos\alpha = 0 = \sin x + \sin y + \sin\alpha$, then find the value of $\cot\left(\frac{x+y}{2}\right)$.

10. If $f(x) = \sin^2 x + \sin^2\left(x+\frac{\pi}{3}\right) + \cos x \cos\left(x+\frac{\pi}{3}\right)$, and $g\left(\frac{5}{4}\right) = 1$, then find the value of $(g \circ f)(x)$.

11. If $\sin x + \sin^2 x = 1$, then find the value of $\cos^2 x + \cos^4 x$.

12. If $A = \cos^2\theta + \sin^4\theta$, then find the range of A.

13. Find the maximum or minimum values of:
   i) $3\cos x + 4\sin x + 8$
   ii) $5\cos x + 3\cos\left(x+\frac{\pi}{3}\right) + 3$.

14. If $\cos 25° + \sin 25° = p$, then find $\cos 50°$.

15. Find the maximum value of $12\sin\theta - 9\sin^2\theta$

16. If $\operatorname{cosec}\theta - \sin\theta = a^3$ and $\sec\theta - \cos\theta = b^3$, then find the value of $a^2b^2(a^2+b^2)$

17. If $x = \sec\theta - \tan\theta$ and $y = \operatorname{cosec}\theta + \cot\theta$, then prove that $xy + x - y + 1 = 0$.

18. If $a\cos\theta - b\sin\theta = c$, then find the value of $a\sin\theta + b\cos\theta$.

19. If $\frac{x}{\cos\theta} = \frac{y}{\cos\left(\theta-\frac{2\pi}{3}\right)} = \frac{z}{\cos\left(\theta+\frac{2\pi}{3}\right)}$, then find the value of $x+y+z$.

20. Find the value of $\tan 70° - \tan 20°$.

21. If $\tan\theta = \frac{b}{a}$, then find $a\sin 2\theta + b\cos 2\theta$.

22. Find the value of $\tan 9° - \tan 27° - \tan 63° + \tan 81°$.

23. If $\tan\alpha + \tan\left(\alpha+\frac{\pi}{3}\right) + \tan\left(\alpha+\frac{2\pi}{3}\right) = \lambda\tan 3\alpha$, then find λ.

24. Find the value of $\frac{1}{\sin 10°} - \frac{\sqrt{3}}{\cos 10°}$

25. If $2\cos\theta = x + \frac{1}{x}$, $2\cos\phi = y + \frac{1}{y}$, then find the value of $\cos(\theta - \phi)$.

26. Prove that $\frac{\sec 8\theta - 1}{\sec 4\theta - 1} = \tan 8\theta \cdot \cot 2\theta$.

27. If α and β are the solutions of the equation $a\tan\theta + b\sec\theta = c$, then find $\tan(\alpha+\beta)$.

28. If $a\cos 2\theta + b\sin 2\theta = c$, has α and β as its solution, then find
   i) $\tan\alpha + \tan\beta$
   ii) $\tan\alpha \cdot \tan\beta$

29. If $\sin A + \sin B = a$ and $\cos A + \cos B = b$, then find $\cos(A+B)$

30. If $\left(1+\sqrt{1+y}\right)\tan y = 1 + \sqrt{1-y}$, then prove that $\sin 4y = y$.

ANSWER

1. 13	**2.** $\left[\frac{1}{4}, 1\right]$	**3.** $\frac{1}{2}\left(q+\frac{1}{q}\right)$	**4.** 1
5. x	**6.** 1	**7.** 0	**8.** 4
9. $\cot\alpha$	**10.** 1	**11.** 1	**12.** $\left[\frac{3}{4}, 1\right]$
13. (i) 3, 13 (ii) –4, 10	**14.** $P\sqrt{2-p^2}$	**15.** 4	**16.** 1
18. $\pm\sqrt{(a^2+b^2-c^2)}$	**19.** 0	**20.** 2tan 50°	**21.** b

22. 4 **23.** 3 **24.** 4 **25.** $\frac{1}{2}\left(\frac{x}{y}+\frac{y}{x}\right)$

27. $\frac{2ac}{a^2-c^2}$ **28.** (i) $\frac{2b}{c+a}$ (ii) $\frac{c-a}{c+a}$ **29.** $\frac{b^2-a^2}{b^2+a^2}$

Level– II FOR JEE-MAIN AND ADVANCED EXAMS ONLY

1: If $\sec x = p + \frac{1}{p}$, then sec x + tan x is

(a) p (b) 2p

(c) $\frac{1}{4p}$ (d) $\frac{4}{p}$.

2. If cosec x - sin x = a^3, sec x - cos x = b^3, then $a^2 b^2 (a^2 + b^2)$ is

(a) 0 (b) 1
(c) -1 (d) ab.

3. If sec x + cos x = 2, then the value of $\sec^3 x(1+\sec^3 x) + \cos^3 x(1+\cos^3 x)$. is

(a) 2 (b) 4
(c) 6 (d) 8.

4. The value of $\cos\frac{2\pi}{7} + \cos\frac{4\pi}{7} + \cos\frac{6\pi}{7}$ is

(a) 1/2 (b) -1/2
(c) 0 (d) None.

5. Which of the following is the smallest ?

(a) sin 1 (b) sin 2
(c) sin 3 (d) sin 4.

6. Which of the following is the greatest ?

(a) sin 1 (b) cos 1
(c) tan 1 (d) cot 1

7. If A= cos (cos x) + sin (cos x), then the least and the greatest value of A are

(a) 0, 2 (b) -1, 1
(c) $-\sqrt{2}, \sqrt{2}$ (d) $0, \sqrt{2}$.

8. If $A + B = \frac{\pi}{3}$, A, B > 0, then the maximum value of tan A. tan B is

(a) 1/3 (b) 1
(c) 1/2 (d) 2/3.

9. The maximum value of a sin 2 x + b cos 2 x for all real x is

(a) a+b (b) $\sqrt{a^2+b^2}$
(c) Max { |a|, |b|} (d) Max {a,b}.

10. Which of the following is / are true ?

(a) sin 1 > sin 1° (b) tan 1 > tan 1°
(c) sin 4 > sin 4° (d) tan 4 > tan 4°.

11. If $\cos 5x = a\cos^5 x + b\cos^3 x + c\cos x + d$, then

(a) a = 16 (b) b - 20
(c) c = 5 (d) d = 2.

12. If $\sin^3 x \sin 3x = c_0 + c_1 \cos x + c_2 \cos 2x + c_3 \cos 3x + \dots + c_n \cos nx$, then

(a) the highest value of n is 6
(b) $c_0 = 1/8$ (c) $c_2 = -c_4$
(d) $c_1 = c_3 = c_5$.

13. If f(x) = cos [π]x + cos [πx], where [,] is the greatest integer function, then f(π/2) is

(a) 0 (b) cos 3
(c) cos 4 (d) None.

14. Let f (x) =

$$\begin{vmatrix} 1+\sin^2 x & \cos^2 x & 4\sin 2x \\ \sin^2 x & 1+\cos^2 x & 4\sin 2x \\ \sin^2 x & \cos^2 x & 1+4\sin 2x \end{vmatrix},$$

then the maximum value of f(x) is

(a) 0 (b) 2
(c) 6 (d) None.

15. For any real x, the maximum value of $\cos^2(\cos x) + \sin^2(\sin x)$ is

(a) 1 (b) $1 + \sin^2 1$
(c) $1+\cos^2 1$ (d) None.

16. If $a = \sin\frac{\pi}{18}\sin\frac{5\pi}{18}\sin\frac{7\pi}{18}$ and x is the solution of the equation $y = 2[x] + 2$ and $y = 3[x - 2]$, where [,] = G.I.F, then a is

(a) $[x]$ (b) $1/[x]$
(c) $2[x]$ (d) $[x]^2$.

17. The minimum value of $\sin^8 x + \cos^8 x$ is

(a) 0 (b) 1
(c) 1/8 (d) 2.

18 If $\frac{\sin^4\theta}{a} + \frac{\cos^4\theta}{b} = \frac{1}{a+b}$, then $\frac{\sin^8\theta}{a^3} + \frac{\cos^8\theta}{b^3}$ is

(a) $\frac{1}{a^3+b^3}$ (b) $\frac{1}{(a+b)^3}$
(c) $\frac{1}{(a-b)^3}$ (d) None.

19. The value of $\tan\frac{\pi}{7}\tan\frac{2\pi}{7}\tan\frac{3\pi}{7}$ is

(a) 1 (b) $\frac{1}{\sqrt{7}}$
(c) $\sqrt{7}$ (d) None.

20. If α and β are the solutions of $\sin^2 x + a\sin x + b = 0$ as well as that of $\cos^2 x + c\cos x + d = 0$ then $\sin(\alpha + \beta)$ is

(a) $\frac{2bd}{b^2+d^2}$ (b) $\frac{a^2+c^2}{2ac}$
(c) $\frac{b^2+d^2}{2bd}$ (d) $\frac{2ac}{a^2+c^2}$.

21. If $\sec\theta + \tan\theta = 1$, then one of the roots of the equation $a(b-c)x^2 + b(c-a)x + c(a-b) = 0$ is

(a) $\tan\theta$ (b) $\sec\theta$
(c) $\cos\theta$ (d) $\sin\theta$.

22. If α is the common +ve root of the equation $x^2 - ax + 12 = 0$, $x^2 - bx + 15 = 0$ and $x^2 - (a+b)x + 36 = 0$ and $\cos x + \cos 2x + \cos 3x = 0$, then $\sin x + \sin 2x + \sin 3x$ is

(a) 3 (b) -3
(c) 0 (d) 2

23. For any real θ, the maximum value of $\cos^2(\cos\theta) + \sin^2(\sin\theta)$ is

(a) 1 (b) $1+\sin^2 1$
(c) $1+\cos^2 1$ (d) $1-\cos^2 1$

24. If $\alpha + \beta = \pi/2$ and $\beta + \gamma = \alpha$, then $\tan\alpha$ is

(a) $2(\tan\beta + \tan\gamma)$
(b) $(\tan\beta + \tan\gamma)$
(c) $(\tan\beta + 2\tan\gamma)$
(d) $(2\tan\beta + \tan\gamma)$.

25. The maximum value of $\cos\alpha_1.\cos\alpha_2.\cos\alpha_3.............\cos\alpha_n$ under the restriction $0 \le \alpha_1, \alpha_2, \alpha_3, ..., \alpha_n \le \frac{\pi}{2}$ and $\cot\alpha_1.\cot\alpha_2.\cot\alpha_3..............\cot\alpha_n = 1$, is

(a) $\frac{1}{2^{n/2}}$ (b) $\frac{1}{2^n}$
(c) $\frac{1}{2n}$ (d) 1

26. If A> 0 and B > 0 and $A + B = \frac{\pi}{3}$, then the maximum value of tan A. tan B is

(a) $\frac{1}{2}$ (b) $\frac{1}{3}$
(c) $\frac{1}{\sqrt{3}}$ (d) $\frac{1}{\sqrt{2}}$

27. If $\tan\beta = 2\sin\alpha.\sin\gamma.\text{cosec}(\alpha+\gamma)$,

then $\cot\alpha$, $\cot\beta$, $\cot\gamma$ are in

(a) A.P (b) G.P
(c) H.P (d) A.G..P

28. The minimum value of the expression $\sin\alpha + \sin\beta + \sin\gamma$, where α, β, γ are real +ve angles satisfying $\alpha+\beta+\gamma = \pi$, is

(a) +ve (b) - ve
(c) 0 (d) - 3

29. The value of $4\cos 20^o - \sqrt{3}\cot 20^o$ is

(a) 1 (b) -1
(c) - 1/2 (d) 1/4

30. The maximum value of

$4\sin^2 x + 3\cos^2 x + \sin\left(\frac{x}{2}\right) + \cos\left(\frac{x}{2}\right)$ is

(a) $4+\sqrt{2}$ (b) $3+\sqrt{2}$
(c) $4-\sqrt{2}$ (d) 4

ANSWER

1.(b)	**2.**(b)	**3.**(b)	**4.**(b)	**5.**(d)	**6.**(c)	**7.**(c)	**8.**(a,b)	**9.**(d)	**10.**(a,b,d)
11.(a,c)	**12.**(a,c,d)	**13.**(c)	**14.**(c)	**15.**(b)	**16.**(b)	**17.**(a)	**18.**(b)	**19.**(c)	**20.**(d).
21. (b, c)	**22.** (c)	**23.**(b)	**24.** (c)	**25.** (a)	**26.** (b)	**27.** (c)	**28.** (a)	**29.** (b)	**30.** (a)

Level– II FOR JEE - ADVANCED EXAM ONLY

1. Prove that the sum of tan x tan 2 x + tan 2x tan 3x +........... + tan n x tan (n+1) x. = cot x tan (n+1) x - (n+1).

2. Prove that cosec x + cosec 2x + cosec 4x +.........+n - terms = $\cot\frac{\theta}{2} - \cot 2^{n-1}\theta$.

3. Prove that sin x + sin 3 x + sin 5 x +...........+ sin (2n-1)x = $\frac{\sin^2 nx}{\sin x}$.

5. If $\theta = \frac{\pi}{2^n - 1}$, prove that $2^n \cos\theta$.

$\cos 2\theta.\cos 4\theta.\cos 8\theta$..........$\cos 2^{n-1}\theta$.

$= -1$.

6. Prove that $\sin\frac{2\pi}{7} + \sin\frac{4\pi}{7} + \sin\frac{8\pi}{7} = \frac{\sqrt{7}}{2}$.

7. Prove that $\tan^2\frac{\pi}{16} + \tan^2\frac{2\pi}{16} + \tan^2\frac{7\pi}{16}$

$= 35$.

8. Prove that $\left(\tan^2\frac{\pi}{7} + \tan^2\frac{2\pi}{7} + \tan^2\frac{3\pi}{7}\right)$

$\times\left(\cot^2\frac{\pi}{7} + \cot^2\frac{2\pi}{7} + \cot^2\frac{3\pi}{7}\right) = 105$.

9. Prove that $\frac{\sin x}{\cos 3x} + \frac{\sin 3x}{\cos 9x} + \frac{\sin 9x}{\cos 27x}$

10. Prove that $\sin\frac{\pi}{14}$ is a root of the equation $8x^3 - 4x^2 - 4x + 1 = 0$.

11. Prove that

$\frac{3}{4} \le \sin^2\theta + \cos^4\theta \le 1$ for all θ.

12. If cos A + cos B + cos C = 0 = sin A + sin B + sin C, then prove that $\sin^2 A + \sin^2 B + \sin^2 C = \cos^2 A + \cos^2 B + \cos^2 C = 3/2$.

13. Let A, B, C be three angles such that

$A = \frac{\pi}{4}$ and tan B. tan C = p. Find all possible values of p such that A, B, C are three angles of a triangle.

14. If $\frac{\tan 3A}{\tan A} = k$, show that $\frac{\sin 3A}{\sin A} = \frac{2k}{k-1}$ and k cannot lie between 1/3 and 3.

15. If A + B + C = π, then prove that cot A + cot B + cot C - cosec A cosec B cosec C = cot A. cot B. Cot C

16. If $A + B + C = \pi$, sin A. sin B. sin C = p and cos A. cos B. cos C = q, then obtain the cubic equation whose roots are tan A, tan B and tan C.

17. If α and β are acute angles and

$\cos 2\alpha = \dfrac{3\cos 2\beta - 1}{3 - \cos 2\beta}$, prove that

$\tan\alpha : \tan\beta = \sqrt{2} : 1$.

18. If $\tan^3\left(\dfrac{\alpha}{2}+\dfrac{\pi}{4}\right) = \tan\left(\dfrac{\beta}{2}+\dfrac{\pi}{4}\right)$, then

prove that $\sin\beta = \dfrac{(3+\sin^2\alpha)\sin\alpha}{1+3\sin^2\alpha}$

19. If $\sin\beta = \dfrac{1}{5}\sin(2\alpha+\beta)$, then prove that

$\tan(\alpha+\beta) = \dfrac{3}{2}\tan\alpha$.

20. Prove that

sin 55 - sin 19 + sin 53 - sin 17 = cos 1.

21. Prove that

$\tan 9^o - \tan 27^o - \tan 63^o + \tan 81^o = 4$.

22. Prove that $\sin^2 12 + \sin^2 21 + \sin^2 39 + \sin^2 48 = 1 + \sin^2 9 + \sin^2 18$.

23. Prove that $\dfrac{\sec^2\theta - \tan\theta}{\sec^2\theta + \tan\theta}$ lies between 1/3 and 3 for all real θ.

24. Prove that the maximum and minimum values of

$\sqrt{a^2\cos^2\theta + b^2\sin^2\theta} + \sqrt{a^2\sin^2\theta + b^2\cos^2\theta}$

are $\sqrt{2(a^2+b^2)}$ and a + b respectively..

26. If $\dfrac{\tan(\alpha+\beta-\gamma)}{\tan(\alpha-\beta+\gamma)} = \dfrac{\tan\gamma}{\tan\beta}$, then prove that $\sin(\beta-\gamma) = 0$ or

$\sin 2\alpha + \sin 2\beta + \sin 2\gamma = 0$.

27. If $A + B + C = \pi$, then prove that

$\cot A + \dfrac{\sin A}{\sin B \sin C} = \cot B + \dfrac{\sin B}{\sin A \sin C}$

$= \cot C + \dfrac{\sin C}{\sin A \sin B}$.

28. If cos A = tan B, cos B = tan C and cos C = tan A, then prove that

sinA = sin B = sinC = 2 sin 18.

29. If cos (x - y) = -1, then prove that cos x + cos y = 0 and sin x + sin y = 0.

30. Prove that

cosec x + cosec 2x + cosec 4x + cosec 8 x = cot (x/2) - cot 8 x..

ANSWERS

13. $p \in (-\infty, 3-2\sqrt{2}] \cup [3+2\sqrt{2}, \infty)$.

16. $qx^3 - px^2 + (1+q)x - p = 0$

COMPREHENSIVE LINK PASSAGE

PASSAGE – I

Increasing product with angles are in G. P

$$\cos\alpha.\cos 2\alpha.\cos 2^2\alpha..........\cos 2^{n-1}\alpha = \begin{cases} \dfrac{\sin 2^n\alpha}{2^n\sin\alpha}, & if\ \alpha \neq n\pi \\ \dfrac{1}{2^n}, & if\ \alpha = \dfrac{\pi}{2^n+1} \\ -\dfrac{1}{2^n}, & if\ \alpha = \dfrac{\pi}{2^n-1} \end{cases}$$

, where n is an integer.

On the basis of above information, answer the following questions:

1. The value of $\cos\left(\frac{2\pi}{7}\right).\cos\left(\frac{4\pi}{7}\right).\cos\left(\frac{6\pi}{7}\right)$ is

(a) -1/2 (b) 1/2 (c) ¼ (d) 1/8

2. If $\alpha = \frac{\pi}{13}$, then the value of $\prod_{r=1}^{6} \cos(r\alpha)$ is

(a) 1/64 (b) -1/64 (c) 1/32 (d) -1/8

3. The value of $\sin\left(\frac{\pi}{14}\right).\sin\left(\frac{3\pi}{14}\right).\sin\left(\frac{5\pi}{14}\right).\sin\left(\frac{7\pi}{14}\right).\sin\left(\frac{9\pi}{14}\right).\sin\left(\frac{11\pi}{14}\right).\sin\left(\frac{13\pi}{14}\right)$ is

(a) 1 (b) 1/8 (c) 1/32 (d) 1/64.

4. The value of $\sin\left(\frac{\pi}{18}\right).\sin\left(\frac{5\pi}{18}\right).\sin\left(\frac{7\pi}{18}\right)$ is

(a) 1/16 (b) 1/8 (c) - 1/8 (d) -1

5. The value of $64\sqrt{3} \times \sin\left(\frac{\pi}{48}\right).\cos\left(\frac{\pi}{48}\right).\cos\left(\frac{\pi}{24}\right).\cos\left(\frac{\pi}{12}\right).\cos\left(\frac{\pi}{6}\right)$ is

(a) 8 (b) 6 (c) 4 (d) -1.

PASSAGE – II

If $\cos\left(\frac{\pi}{7}\right)$, $\cos\left(\frac{3\pi}{7}\right)$, $\cos\left(\frac{5\pi}{7}\right)$ are the roots of the equation $8x^3 - 4x^2 - 4x + 1 = 0$.

On the basis of the above information, answer the following questions.

1. The value of $\sec\left(\frac{\pi}{7}\right) + \sec\left(\frac{3\pi}{7}\right) + \sec\left(\frac{5\pi}{7}\right)$ is

(a) 2 (b) 4 (c) 8 (d) None.

2. The value of $\sin\left(\frac{\pi}{14}\right).\sin\left(\frac{3\pi}{14}\right).\sin\left(\frac{5\pi}{14}\right)$ is

(a) ¼ (b) 1/8 (c) $\frac{\sqrt{7}}{4}$ (d) $\frac{\sqrt{7}}{8}$.

3. The value of $\cos\left(\frac{\pi}{14}\right).\cos\left(\frac{3\pi}{14}\right).\cos\left(\frac{5\pi}{14}\right)$ is

(a) ¼ (b) 1/8 (c) $\frac{\sqrt{7}}{4}$ (d) $\frac{\sqrt{7}}{8}$

4. The equation whose roots are $\tan^2\left(\frac{\pi}{7}\right), \tan^2\left(\frac{3\pi}{7}\right), \tan^2\left(\frac{5\pi}{7}\right)$ is

(a) $x^3-35x^2+7x-21=0$ (b) $x^3-35x^2+21x-7=0$

(c) $x^3-21x^2+35x-7=0$ (d) $x^3-21x^2+7x-35=0$

5. The value of $\sum_{r=1}^{3}\tan^2\left(\frac{2r-1}{7}\right)\pi \times \sum_{r=1}^{3}\cot^2\left(\frac{2r-1}{7}\right)\pi$ is

(a) 15 (b) 105 (c) 21 (d) 147

ANSWERS

Passage-I

1.(d) **2.**(a) **3.**(d) **4.**(b) **5.**(b)

Passage-II

1.(b) **2.**(b) **3.**(d) **4.**(c) **5.**(b)

MATCH MATRIX

1. Match the following columns :

Column – I		Column – II	
(A)	If $\theta+\phi=\frac{\pi}{2}$, where θ and ϕ are positive, then $(\sin\theta+\sin\phi)\sin\left(\frac{\pi}{4}\right)$ is always less than	(P)	1
		(Q)	2
(B)	If $\sin\theta-\sin\phi=a$ and $\cos\theta+\cos\phi=b$, then a^2+b^2 can not exceed	(R)	3
		(S)	4
(C)	If $3\sin\theta+5\cos\theta=5$, $(\theta\neq 0)$, then the value Of $5\sin\theta-3\cos\theta$ is	(T)	5

2. **Match the following columns :**

	Column - I		Column - II
(A)	The value of $\cos(20^o)\cos(40^o)\cos(80^o)$ is	(P)	$\sqrt{3}/8$
(B)	The value of $\cos(20^o)\cos(40^o)\cos(60^o)\cos(80^o)$ is	(Q)	$\sqrt{3}/16$
(C)	The value of $\sin(20^o)\sin(40^o)\sin(80^o)$ is	(R)	$\sqrt{3}/32$
(D)	The value of $\sin(20^o)\sin(40^o)\sin(60^o)\sin(80^o)$ is	(S)	1/16
		(T)	1/8

3. **Match the following columns :**

	Column – I		Column – II
(A)	If maximum and minimum values of $\dfrac{7+6\tan\theta-\tan^2\theta}{1+\tan^2\theta}$	(P)	$\lambda+\mu=2$
	For all real values of $\theta-\dfrac{\pi}{2}$ are λ and μ respectively, then	(Q)	$\lambda-\mu=6$
(B)	If maximum and minimum values of $5\cos\theta+3\cos\left(\theta+\dfrac{\pi}{3}\right)+3$	(R)	$\lambda+\mu=6$
	For all real values of θ are λ and μ resp., then	(S)	$\lambda-\mu=10$
(C)	If maximum and minimum values of	(T)	$\lambda-\mu=14$
	$1+\sin\left(\dfrac{\pi}{4}+\theta\right)+2\cos\left(\dfrac{\pi}{4}-\theta\right)$		
	For all real values of θ are λ and μ resp., then		

4. **Match the following columns :**

	Column - I		Column-II
	In a triangle ABC		
(A)	$\sin 2A+\sin 2B+\sin 2C$ is	(P)	4 sin A. sin B. sin C
(B)	$\cos 2A+\cos 2B+\cos 2C$ is	(Q)	-1 - 4 cos A. cos B. cos C
(C)	$\sin^2 A+\sin^2 B+\sin^2 C$ is	(R)	2 + 2 cos A. cos B. cos C
(D)	$\cos^2 A+\cos^2 B+\cos^2 C$ is	(S)	1 - 2 cos A. cos B. cos C

5. **Match the following columns :**

	Column - I		Column-II
(A)	The value of $\cos^4\left(\frac{\pi}{8}\right)+\cos^4\left(\frac{3\pi}{8}\right)+\cos^4\left(\frac{5\pi}{8}\right)+\cos^4\left(\frac{7\pi}{8}\right)$ is	(P)	1/8
		(Q)	- 3/2
(B)	The value of $\sin(12^o)\sin(48^o)\sin(54^o)$ is	(R)	3/2
(C)	The value of $\sin(6^o)\sin(42^o)\sin(66^o)\sin(78^o)$ is	(S)	1/16
(D)	The value of $\tan(6^o)\tan(42^o)\tan(66^o)\tan(78^o)$ is	(T)	1

6. **Match the following columns :**

If $A+B=\frac{\pi}{4}$, then the value of $(1+\tan A)(1+\tan B)$ is 2

	Column - I		Column - II
(A)	The value of $(1+\tan 21^o)(1+\tan 22^o)(1+\tan 23^o)(1+\tan 24^o)$ is	(P)	2
		(Q)	4
(B)	The value of $(1+\tan 2058^o)(1-\tan 2013^o)$ is	(R)	8
(C)	The value of $\left(1+\tan\left(\frac{\pi}{8}-x\right)\right)\left(1+\tan\left(x+\frac{\pi}{8}\right)\right)$ is	(S)	- 8
(D)	The value of $(1+\tan 235^o)(1-\tan 190^o)$ is	(T)	- 4

7. **Match the following columns :**

	Column - I		Column-II
(A)	The value of $2\tan(50^o)+\tan(20^o)$ is	(P)	3
(B)	The value of $\tan(40^o)+2\tan(10^o)$ is	(Q)	5
(C)	The value of $\tan(20^o)\tan(40^o)\tan(60^o)\tan(80^o)$ is	(R)	$\tan(70^o)$
(D)	If $3\sin x+4\cos x=5$, then the value of $2\sin x+\cos x+4\tan x$ is	(S)	$\tan(50^o)$

8. Match the following columns :

	Column - I		Column - II
(A)	The minimum value of $2\sin^2\theta+3\cos^2\theta$ is	(P)	1
(B)	The maximum value of $\sin^2\theta+\cos^4\theta$ is	(Q)	3/4
(C)	The least value of $\sin^4\theta+\cos^2\theta$ is	(R)	2
(D)	The greatest value of $\sin^{2014}\theta+\cos^{2010}\theta$ is	(S)	4

9. Match the following columns :

If α and β are the solutions of $a\cos\theta+b\sin\theta=c$, then

	Column-I		Column-II
(A)	the value of $\sin\alpha+\sin\beta$ is	(P)	$\dfrac{c^2-b^2}{a^2+b^2}$
(B)	the value of $\sin\alpha.\sin\beta$ is	(Q)	$\dfrac{2ac}{a^2+b^2}$
(C)	the value of $\cos\alpha+\cos\beta$ is	(R)	$\dfrac{c^2-a^2}{a^2+b^2}$
(D)	the value of $\cos\alpha.\cos\beta$ is	(S)	$\dfrac{2bc}{a^2+b^2}$

10. Match the following columns :

	Column - I		Column-II
(A)	The value of $\cos(12^\circ)+\cos(84^\circ)+\cos(156^\circ)+\cos(132^\circ)$ is	(P)	0
		(Q)	1
(B)	The value of $2\tan\left(\dfrac{\pi}{10}\right)+3\sec\left(\dfrac{\pi}{10}\right)-4\cos\left(\dfrac{\pi}{10}\right)$ is	(R)	2
(C)	The value of $\sqrt{3}\cot(20^\circ)-4\cos(20^\circ)$ is	(S)	-1/2
(D)	The value of $\tan(20^\circ)+2\tan(50^\circ)-\tan(70^\circ)$ is	(T)	- 1

ANSWERS

1. $A \to (P,Q,R,S,T); B \to (S,T); C \to (R)$
2. $A \to (T), B \to (S), C \to (P), D \to (Q)$.
3. $A \to (R,S); B \to (R,T); C \to (P,Q)$
4. $A \to (P), B \to (Q), C \to (R), D \to (S)$.
5. $(A) \to (R); (B) \to (P); (C) \to (S), (D) \to (T)$
6. $(A) \to (Q); (B) \to (P); (C) \to (P), (D) \to (P)$
7. $(A) \to (R); (B) \to (S); (C) \to (P), (D) \to (Q)$
8. $(A) \to (R); (B) \to (P); (C) \to (Q); (D) \to (P)$
9. $(A) \to (S); (B) \to (R); (C) \to (Q), (D) \to (P)$
10. $(A) \to (S); (B) \to (P); (C) \to (Q), (D) \to (P)$

ASSERTION & REASON

Codes

(A) **Both A and R are individually true and R is the correct explanation of A.**

(B) **Both A and R are individually true and R is not the correct explanation of A.**

(C) **A is true but R is false.**

(D) **A is false but R is true.**

1. **Assertion (A) :** $\sin\theta = x + \frac{1}{x}$ is impossible if $x \in R - \{0\}$.

Reason (R) : A.M $\geq$ G.M

(a) A (b) B
(c) C (d) D

2. **Assertion (A)** :A is an obtuse angle in ΔABC, then tan B. tan C > 1

Reason (R): In ΔABC,

$$\tan A = \frac{\tan B + \tan C}{\tan B \tan C - 1}$$

(a) A (b) B
(c) C (d) D

3. **Assertion (A) :**

$$\sin\left(\frac{2\pi}{7}\right) + \sin\left(\frac{4\pi}{7}\right) + \sin\left(\frac{6\pi}{7}\right) = -\frac{1}{2}.$$

Reason (R) :

$\cos\left(\frac{2\pi}{7}\right) + i\sin\left(\frac{2\pi}{7}\right)$ is complex 7th root of unity.

(a) A (b) B
(c) C (d) D

4. **Assertion (A) :** $\tan\alpha + 2\tan(2\alpha)$

$4\tan(4\alpha) + 8\tan(8\alpha)\ 16\cot(16\alpha) = \cot\alpha$

Reason (R): $\cot\alpha - \tan\alpha = 2\cot 2\alpha$

(a) A (b) B
(c) C (d) D

5. **Assertion (A) :** $\cos^2\alpha + \cos^2\left(\alpha + \frac{2\pi}{3}\right)$

$\cos^2\left(\alpha + \frac{4\pi}{3}\right) = = 3\cos\alpha$

$\cos\left(\alpha + \frac{2\pi}{3}\right).\cos\left(\alpha + \frac{4\pi}{3}\right)$

Reason (R) :If a + b + c = 0, then $a^3 + b^3 + c^3 = 3abc$

(a) A (b) B
(c) C (d) D

6. **Assertion (A) :** $\tan(5\theta) - \tan(3\theta) - \tan(\theta)$

$= \tan(5\theta).\tan(3\theta).\tan(\theta)$

Reason (R) :If x = y + z, then

tan x – tan y – tan z = tan x. tan y. tan z

(a) A (b) B
(c) C (d) D

7. **Assertion (A) :The maximum value of** $\sin\theta + \cos\theta$ is 2

Reason (R): The maximum value of $\sin\theta$ is 1 and that of $\cos\theta$ is also 1.

(a) A (b) B
(c) C (d) D.

8. **Assertion (A) :The maximum value of**

$\prod_{i=1}^{n} \cos\alpha_i$ under the restriction

$0 \le \alpha_1, \alpha_2, \alpha_3, ..., \alpha_n \le \frac{\pi}{2}$ is $\frac{1}{2^{n/2}}$

Reason (R) : $\prod_{i=1}^{n} \cot\alpha_i = 1$

(a) A (b) B
(c) C (d) D

9. **Assertion (A) :If A + B + C =** π, then the maximum value of

tan A. tan B. tan C is $3\sqrt{3}$

Reason (R) : A.M $\ge$ G.M

(a) A (b) B
(c) C (d) D

10. **Assertion (A) :** $\sec^2\theta = \frac{4xy}{(x+y)^2}$ is positive for all real values of x and y only when x = y

Reason (R) : $\sec^2\theta \ge 1$.

(a) A (b) B
(c) C (d) D

ANSWERS

1.(a) **2.**(d) **3.**(d) **4.**(a) **5.**(a) **6.**(a) **7.**(d) **8.**(a) **9.**(a) **10.**(a)

QUESTIONS ASKED IN ROORKEE EXAM

1. If $x+y+z = xyz$, prove that

$$\frac{2x}{1-x^2}+\frac{2y}{1-y^2}+\frac{2z}{1-z^2} = \frac{2x}{1-x^2}.\frac{2y}{1-y^2}.\frac{2z}{1-z^2}$$

[Roorkee – 1983]

2. No questions asked in 1984.

3. If $\cos\alpha+\cos\beta+\cos\gamma = 0$

and $\sin\alpha+\sin\beta+\sin\gamma = 0$

, then prove that

$\cos(3\alpha)+\cos(3\beta)+\cos(3\gamma)-0$

and $\sin(3\alpha)+\sin(3\beta)+\sin(3\gamma)-0$

[Roorkee – 1985]

4. No questions asked in 1986, 1987, 1988.

5. Show that (without using tables)

$\tan 9^o - \tan 27^o - \tan 63^o + \tan 81^o = 4$

[Roorkee – 1989]

6. Find 'a' and 'b' such that the inequality

$a \le 3\cos x + 5\sin\left(x - \frac{\pi}{6}\right) \le b$ holds good

for all x.

[Roorkee – 1989]

7. No questions asked in 1990, 1991.

8. If $A = \cos^2\theta + \sin^4\theta$, then for all values of θ, find range of A.

[Roorkee – 1992]

9. Given the product p of sines of the angles of a triangle and the product q of their cosines, find the cubic equation, whose co-efficients are functions of p and q and whose roots are the tangents of the angles of the triangle.

[Roorkee Main – 1992]

10. No questions asked in 1993, 1994.

11. If $x = \cos 10^o \; \cos 20^o \; \cos 40^o$,

then find the value of x

[Roorkee – 1995]

12. No questions asked in 1996, 1997, 1998, 1999.

13. Find the real values of x for which

$27^{\cos 2x}.\,81^{\sin 2x}$ is minimum.

Also find the minimum value.

[Roorkee Main – 2000]

14. If $e^{i\theta - \log\cos(x-iy)} = 1$, then find the values of x and y in terms of θ.

[Roorkee Main – 2001]

ANSWERS

6. $a = -\sqrt{19}$, $b = \sqrt{19}$

8. $\frac{3}{4} \le A \le 1$

9. $q\,x^3 - px^2 + (1+q)x - p = 0$

11. $x = \frac{1}{8} \times \cot 10^o$

13. 1/243.

14. $x = 2n\pi \pm \sin^{-1}\sqrt{\sin\theta}$,

$y = \frac{\sin^{-1}\left(\pm\sqrt{\sin\theta}\right)}{h}$

QUESTIONS ASKED IN IIT-JEE

1. Prove that sin x sin (x – y) + sin y sin z sin (y – z) + sin z sin x sin (z – x) + sin (x – y)+ sin (y – z) + sin (z – x) = 0

[IIT-JEE – 1978]

2. If $\cos(\alpha+\beta) = \frac{4}{5}$, $\sin(\alpha-\beta) = \frac{5}{13}$ and α, β lie between 0 and π/4, find $\tan 2\alpha$.

[IIT-JEE – 1979]

3. Given $A = \sin^2\theta + \cos^4\theta$ for all values of θ, then

(a) $1 \le A \le 2$ (b)*$3/4 \le A \le 1$
(c) $13/6 \le A \le 1$ (d)$3/4 \le A \le 13/6$

[IIT-JEE – 1980]

4. If $\tan A = \frac{1-\cos B}{\sin B}$, then Tan 2A = tan B. Is it true / false ?

[IIT-JEE – 1980]

5. Suppose $\sin^3 x \sin 3x = \sum_{m=0}^{n} C_m \cos n x$ is an identity in x, when $C_0, C_1, \ldots\ldots\ldots, C_n$ are constants and $C_n \ne 0$ is the value of n = ...

[IIT-JEE – 1981]

6. If $\alpha+\beta+\gamma=\pi$, then prove that,

$\sin^2\alpha + \sin^2\beta + \sin^2\gamma$

$= 2\sin\alpha.\sin\beta.\sin\gamma$

[IIT-JEE – 1983]

7. Without using the tables, prove that

$\sin 12^o \sin 54^o \sin 48^o = \frac{1}{8}$

[IIT-JEE – 1983]

8. Prove that

$16\cos\left(\frac{2\pi}{15}\right)\cos\left(\frac{4\pi}{15}\right)\cos\left(\frac{8\pi}{15}\right)\cos\left(\frac{16\pi}{15}\right) = 1$

[IIT-JEE – 1983]

9. The value of $\left(1+\cos\frac{\pi}{8}\right)\left(1+\cos\frac{3\pi}{8}\right)\left(1+\cos\frac{5\pi}{8}\right)\left(1+\cos\frac{7\pi}{8}\right)$ is equal to

(a) 1/2 (b) $\cos\frac{\pi}{8}$
(c)* 1/8 (d) $\frac{1+\sqrt{2}}{2\sqrt{2}}$.

[IIT-JEE – 1984]

10. The expression

$3\left[\sin^2\left(\frac{3\pi}{2}-\alpha\right) + \sin^4(3\pi+\alpha)\right]$ -

$2\left[\sin^6\left(\frac{\pi}{2}+\alpha\right) + \sin^6(5\pi-\alpha)\right]$ is equal to

(a) 0 (b)* 1
(c) 3 (d) $\sin 4\alpha + \cos\alpha$.

[IIT-JEE – 1986]

11. The set of all x in the interval $(0, \pi)$ for which $2\sin^2 x - 3\sin x + 1 \geq 0$ is ....

[IIT-JEE – 1987]

12. The value of the expression $\sqrt{3}\ \operatorname{cosec} 20^o - \sec 20^o$ is equal to

(a) 2 (b) $\dfrac{2\sin 20^o}{\sin 40^o}$

(c)* 4 (d) $\dfrac{4\sin 20^o}{\sin 40^o}$.

[IIT-JEE – 1988]

13. Prove that $\tan\alpha + 2\tan 2\alpha + 4\tan 4\alpha$ $8\cot 8\alpha = \cot\alpha$

[IIT-JEE – 1988]

14. Find the value of

$\sin\left(\dfrac{\pi}{14}\right).\sin\left(\dfrac{3\pi}{14}\right).\sin\left(\dfrac{5\pi}{14}\right)\sin\left(\dfrac{7\pi}{14}\right)$

$\sin\left(\dfrac{9\pi}{14}\right).\sin\left(\dfrac{11\pi}{14}\right).\sin\left(\dfrac{13\pi}{14}\right)$

[IIT-JEE – 1991]

15. Match the following columns :

Column – I	Column – II
(i) Positive	(A) $\left(\dfrac{13\pi}{48}, \dfrac{14\pi}{48}\right)$
(ii) Negative	(B) $\left(\dfrac{14\pi}{48}, \dfrac{18\pi}{48}\right)$
	(C) $\left(\dfrac{18\pi}{48}, \dfrac{23\pi}{48}\right)$
	(D) $\left(0, \dfrac{\pi}{2}\right)$

[IIT-JEE – 1992]

16. Number of solution of the equation t an x + sec x = 2 cos x, lying in the interval $[0, 2\pi]$, is

(a) 0 (b) 1

*(c) 2 (d) 3

[IIT-JEE – 1993]

17. If $k = \sin\left(\dfrac{\pi}{18}\right)\sin\left(\dfrac{5\pi}{18}\right)\sin\left(\dfrac{7\pi}{18}\right)$, then the numerical value of k is.........

[IIT-JEE – 1993]

18. If A > 0, B > 0 and $A + B = \dfrac{\pi}{3}$, then the maximum value of tan A tan B is ......

[IIT-JEE – 1993]

19. Let $0 < x < \dfrac{\pi}{4}$, then $(\sec 2x - \tan 2x)$ equals

(a) $\tan\left(x - \dfrac{\pi}{4}\right)$ *(b) $\tan\left(\dfrac{\pi}{4} - x\right)$

(c) $\tan\left(x + \dfrac{\pi}{4}\right)$ (d) $\tan^2\left(x + \dfrac{\pi}{4}\right)$

[IIT-JEE – 1994]

20. If ω is an imaginary cube root of unity, then the value of $\sin\left\{\left(\omega^{10} + \omega^{23}\right)\pi - \dfrac{\pi}{4}\right\}$ is

(a) $-\sqrt{3}/2$ (b) $-1/\sqrt{2}$

(c)* $1/\sqrt{2}$ (d) $\dfrac{\sqrt{3}}{2}$

[IIT-JEE – 1994]

21. If the lengths of the sides of triangle are 3, 5, 7 , then the largest angle of the triangle is

(a) $\pi/2$ (b) $5\pi/6$

(c)* $2\pi/3$ (d) $3\pi/4$

[IIT-JEE – 1994]

22. The value of the expression

$$3(\sin^4 x+\cos^4 x)+6(\sin x+\cos x)^2$$

$$+ \quad 4(\sin^6 x+\cos^6 x) \text{ is}$$

(a) 11 (b) 12
(c)* 13 (d) 14

[IIT-JEE – 1995]

23. The minimum value of the expression $\sin\alpha+\sin\beta+\sin\gamma$, where α, β, γ are real numbers satisfying $\alpha+\beta+\gamma=\pi$ is

(a) positive (b) zero
(c)* negative (d) -3.

[IIT-JEE – 1995]

24. $\sec^2\theta=\dfrac{4xy}{(x+y)^2}$ is true if and only if

(a) $x+y=0$ (b)* $x=y, x\neq 0$
(c) $x=y$ (d) $x\neq 0, y\neq 0$.

[IIT-JEE – 1996]

25. The length of a longest interval in which the function $3\sin x-4\sin^3 x$ is increasing in

(a)* $\pi/3$ (b) $\pi/2$
(c) $3\pi/2$ (d) π.

[IIT-JEE – 2002]

26. Using the relation $2(1-\cos x)<x^2,\ x\neq 0$ or otherwise prove that $\sin(\tan x)\geq x$, $\forall\, x\in\left[0,\dfrac{\pi}{4}\right]$.

[IIT-JEE – 2003]

27. Given that both θ and ϕ are acute angles and $\sin\theta=\dfrac{1}{2},\ \cos\phi=\dfrac{1}{3}$, then the value of $(\theta+\phi)$ belongs to the interval

(a) $\left(\dfrac{\pi}{3},\dfrac{\pi}{2}\right)$ (b)* $\left(\dfrac{\pi}{2},\dfrac{2\pi}{3}\right)$

(c) $\left(\dfrac{2\pi}{3},\dfrac{5\pi}{6}\right)$ (d) $\left(\dfrac{5\pi}{6},\pi\right)$

[IIT-JEE – 2004]

28. Prove that

$$\sin x+2x\geq\frac{3x(x+1)}{\pi},\ \forall\, x\in[0,\pi/2]$$

[IIT-JEE – 2004]

29. Find the range of values of t for which

$$2\sin t=\frac{1-2x+5x^2}{3x^2-2x-1},\quad t\in\left(-\frac{\pi}{2},\frac{\pi}{2}\right)$$

[IIT-JEE – 2005]

30. $\cos(\alpha-\beta)=1$ and $\cos(\alpha+\beta)=1/c$, where $\alpha, \beta\in[-\pi,\pi]$. Values of α, β which satisfy the
equations is/are

(a) 0 (b) 1
(c) 2 *(d) 4.

[IIT-JEE – 2005]

31. The area of the equilateral triangle which containing three coins of unit radius is

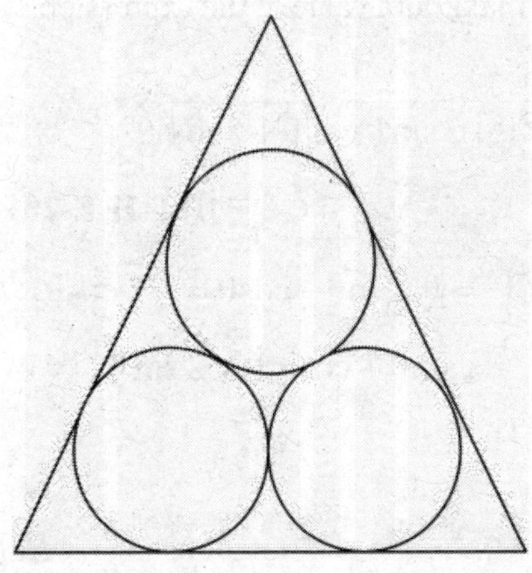

[IIT-JEE – 2005]

(a) $12+2\sqrt{3}$ sq. units

(b)* $6+4\sqrt{3}$ sq. units

(c) $8+\sqrt{3}$ sq. units

(d) $4+\dfrac{7\sqrt{3}}{2}$ sq. units

32.. Let $\theta\in(0,\pi/4)$ and $t_1=(\tan\theta)^{\tan\theta}$, $t_2=(\tan\theta)^{\cot\theta}$, $t_3=(\cot\theta)^{\tan\theta}$ and $t_4=(\cot\theta)^{\cot\theta}$, then

(a) $t_1 > t_2 > t_3 > t_4$

(b)* $t_4 > t_3 > t_2 > t_1$

(c) $t_3 > t_1 > t_2 > t_4$

(d) $t_2 > t_3 > t_1 > t_4$. [IIT-JEE – 2006]

33. No questions asked in 2007, 2008

34. If $\frac{\sin^4 x}{2} + \frac{\cos^4 x}{3} = \frac{1}{5}$, then

(a)* $\tan^2 x = \frac{2}{3}$

(b)* $\frac{\sin^8 x}{8} + \frac{\cos^8 x}{27} = \frac{1}{125}$

(c) $\frac{\sin^8 x}{8} + \frac{\cos^8 x}{27} = \frac{2}{125}$

(d) $\tan^2 x = \frac{1}{3}$ **[IIT-JEE - 2009]**

35. The maximum value of the expression

$\frac{1}{\sin^2\theta + 3\sin\theta\cos\theta + 5\cos^2\theta}$ is.........

[IIT-JEE-2010]

36. Let $P = \{\theta : \sin\theta - \cos\theta = \sqrt{2}\cos\theta\}$ and $Q = \{\theta : \sin\theta + \cos\theta = \sqrt{2}\sin\theta\}$ be two sets. Then

(a) $P \subset Q$ and $Q - P \neq \varphi$

(b) $Q \not\subset P$

(c) $P \not\subset Q$

(d)* $P = Q$ **[IIT-JEE - 2011]**

37. The positive integral value of $n > 3$ satisfying the equation

$$\frac{1}{\sin\left(\frac{\pi}{n}\right)} = \frac{1}{\sin\left(\frac{2\pi}{n}\right)} + \frac{1}{\sin\left(\frac{3\pi}{n}\right)}$$

is............ **[IIT-JEE - 2011]**

38. Let $f : (-1,1) \to R$ be such that $f(\cos 4\theta) = \frac{2}{2-\sec^2\theta}$ for $\theta \in \left(0, \frac{\pi}{4}\right) \cup \left(\frac{\pi}{4}, \frac{\pi}{2}\right)$. Then the values of $f\left(\frac{1}{3}\right)$ is /are

(a)* $1-\sqrt{\frac{3}{2}}$ (b)* $1+\sqrt{\frac{3}{2}}$

(c) $1-\sqrt{\frac{2}{3}}$ (d) $1+\sqrt{\frac{2}{3}}$

[IIT-JEE - 2012]

ANSWERS

2. 56/33
4. True
5. $n = 6$
11. $x \in \left[0, \frac{\pi}{6}\right] \cup \left\{\frac{\pi}{2}\right\} \cup \left[\frac{5\pi}{6}, \pi\right]$
14. 1/64
15. (i) $\to$ C ; (ii) $\to$ A.
17. 1 / 3
18. 1 / 3
29. $t \in \left[-\frac{\pi}{2}, -\frac{\pi}{10}\right] \cup \left[\frac{3\pi}{2}, \frac{\pi}{2}\right]$
35. 2
37. $n = 7$

SELF ASSESMENT TEST-I

Time : 3 Hrs. Ch: Trigometric Ratios and Identities M.M. : 100

Give answers of the following questions by yourself only.

1. Find the minimum values of $27^{\cos x} + 81^{\sin x}$

2. If cos 3 x = 4 cos x. cos (x + p). cos (x - p), 0 < p < 90, then find the value of p.

3. If $\frac{x}{\cos\theta} = \frac{y}{\cos(\theta - \frac{2\pi}{3})} = \frac{z}{\cos(\theta + \frac{2\pi}{3})}$, then find the value of x + y + z.

4. Find the numerical value of $4 \sin 50 - \sqrt{3} \tan 50$.

5. If sin a + cos a = k, then find the value of $|\sin^6 a - \cos^6 a|$ in terms of k.

6. Prove that $\sin(\frac{\pi}{14})$ is a root of the equation $8x^3 - 4x^2 - 4x + 1 = 0$.

7. If tan $\theta = 2$ is satisfied by α, β then find the value of $\tan\left(\frac{\alpha}{2}\right) + \tan\left(\frac{\beta}{2}\right)$.

8. Find the numerical value of $\cos\left(\frac{2\pi}{7}\right) + \cos\left(\frac{4\pi}{7}\right) + \cos\left(\frac{6\pi}{7}\right)$.

9. Find the numerical value of $\cos\left(\frac{\pi}{7}\right) + \cos\left(\frac{3\pi}{7}\right) + \cos\left(\frac{5\pi}{7}\right)$.

10. If A, B, C are in A.P and $B = \frac{\pi}{4}$, then find the value of tan A. tan B. tan C.

SELF ASSESMENT TEST-II

Time : 3 Hrs. Ch: Trigometric Ratios and Identities M.M. : 100

Give answers of the following questions by yourself only.

1. Prove that $\sin 9 = \frac{1}{4}[\sqrt{3+\sqrt{5}} - \sqrt{5-\sqrt{5}}]$.

2. Find the value of $\cos^2 x + \cos^2\left(\frac{\pi}{3} + x\right) - \cos x . \cos\left(\frac{\pi}{3} + x\right)$.

3. If $\tan x + \tan\left(\frac{\pi}{3} + x\right) + \tan\left(\frac{2\pi}{3} + x\right) = k \tan 3x$, then find k.

4. Find the range of $\sin\sqrt{x^2 - \frac{\pi}{36}}$.

5. Find the value of $\sum_{k=1}^{6}\left(\sin\frac{2\pi k}{7} - i\cos\frac{2\pi k}{7}\right)$, where $i = \sqrt{-1}$.

6. If $\cos\alpha = \frac{2\cos\beta - 1}{2 - \cos\beta}$, $0 < \alpha < \beta < \pi$, then $\frac{\tan\frac{\alpha}{2}}{\tan\frac{\beta}{2}}$ is

   (a) 1 (b) $\sqrt{2}$

   (c) $\sqrt{3}$ (d) $\frac{1}{\sqrt{3}}$.

7. If $0 < \theta < \pi$, then

   (a) $1 + \cot\theta \le \cot\frac{\theta}{2}$

   (b) $1 + \cot\theta \ge \cot\frac{\theta}{2}$

   (c) $1 + \cot\frac{\theta}{2} \ge \cot\theta$

   (d) $1 + \cot\frac{\theta}{2} \le \cot\theta$.

8. If sinx + cosy = a and cos x + sin y = b, then $\tan\frac{x-y}{2}$ is equal to

   (a) a+b (b) a - b

   (c) $\frac{a+b}{a-b}$ (d) $\frac{a-b}{a+b}$.

9. Find the value of $\sin\frac{2\pi}{7} + \sin\frac{4\pi}{7} + \sin\frac{8\pi}{7}$.

10. Find the number of ordered pair (a, x) satisfying the equation $\sec^2(a+2)x + a^2 - 1 = 0$, where $-\pi < x < \pi$.

CHAPTER 2

GRAPH OF TRIGONOMETRIC FUNCTIONS

INTRODUCTION

There are six trigonometric ratios. In this section, we shall describe each trigonometric function and their characteristics and graph.

1. Sine function: For each real number x, there is *a* sine function defined as f(x) = sin x.

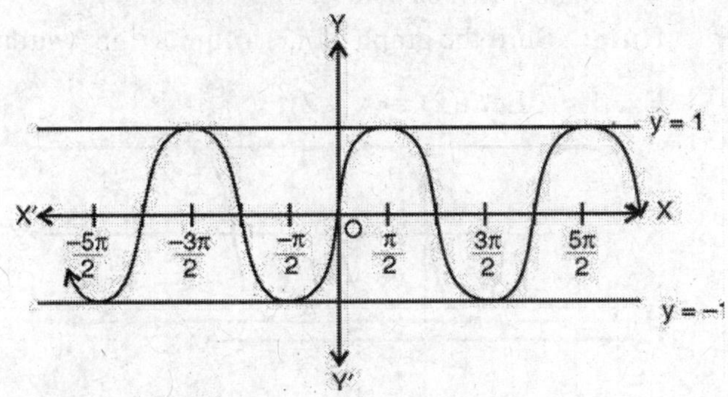

Characteristics of sine function :

1. It is an odd function, since $\sin(-x) = -\sin x$
2. It is symmetrical about the origin.
3. It is a periodic function with period 2π.
4. Let $f(x) = \sin^n x$, $n \in I - \{0\}$. Then the periods of

$$f(x) \text{ is } \begin{cases} = 2\pi, n = \text{odd} \\ = \pi, n = \text{even} \end{cases}$$

5. $\sin x = 1 \Rightarrow \quad x = (4n+1)\frac{\pi}{2},\ n \in I$
6. $\sin x = 0 \Rightarrow \quad x = n\pi,\ n \in I$
7. $\sin x = -1 \Rightarrow \quad x = (2n-1)\frac{\pi}{2},\ n \in I$
8. $\sin x > 0 \Rightarrow \quad x \in (2n\pi, (2n+1)\pi),\ n \in I.$
9. $\sin x < 0 \Rightarrow \quad x \in ((2n-1)\pi, 2n\pi),\ n \in I.$
10. $\sin x \le x,\ \forall\, x \in R.$
11. If $x > y \Rightarrow \sin x > \sin y,\ \forall\, x, y \in \left(\frac{-\pi}{2}, \frac{\pi}{2}\right)$
12. If $x > y \Rightarrow \sin x < \sin y,\ \forall\, x, y \in \left(\frac{\pi}{2}, \frac{3\pi}{2}\right)$
13. $D_f = R$ and $R_f = [-1, 1]$
14. It is non monotonic function .

2. Rule to draw the graphs of different types of functions:

Type-I: when $f(x) \to f(x) + a$

Rule: Lift the graph of f(x), *a*-units upwards.

Ex. 1. Let $f(x) = x + 2$

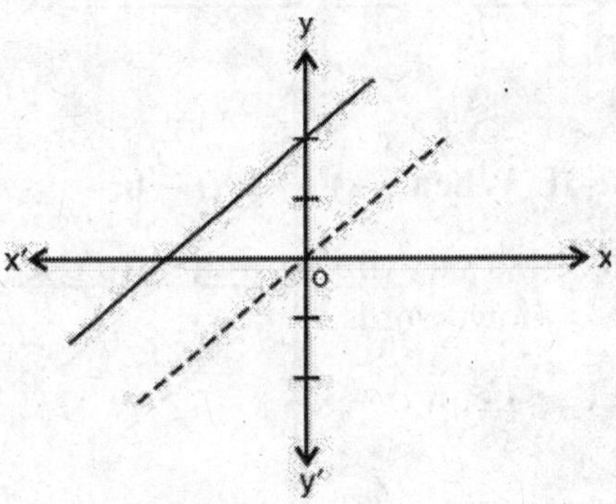

Ex. 2. Let $f(x) = x^2 + 1$

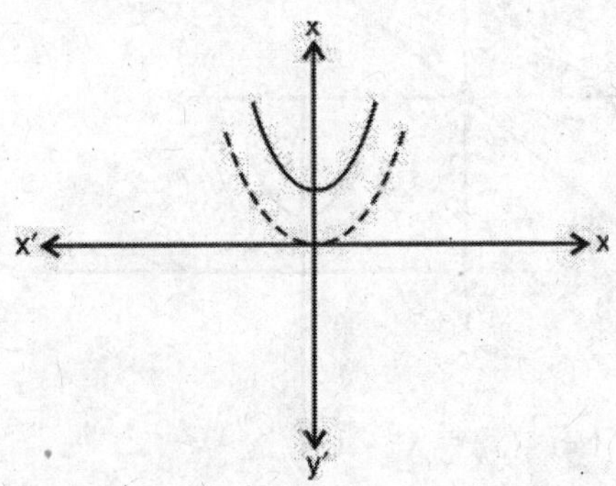

Ex.3. Let $f(x) = x^3 + 1$

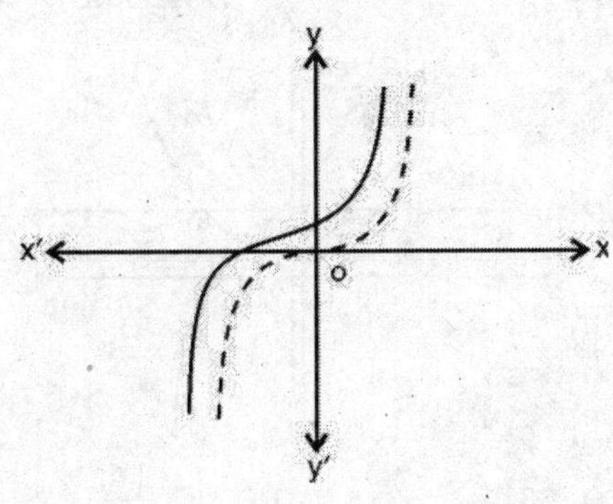

Ex-4. Let f(x) = sin x + 1

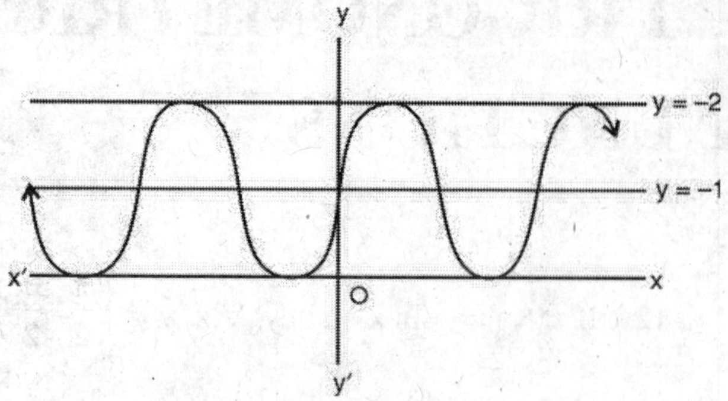

Ex. 5. Let $f(x) = e^x + 1$

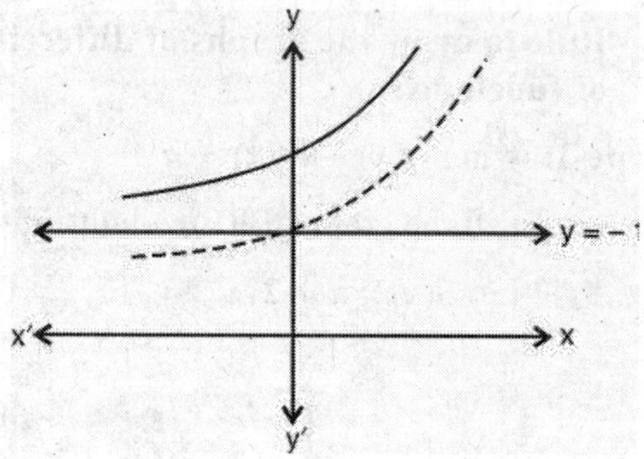

Type-II. When f(x) → f(x) – b.

Rule: Down the graph of f(x), *b*-units downwords.

Ex. 1. Let f(x) = x – 2

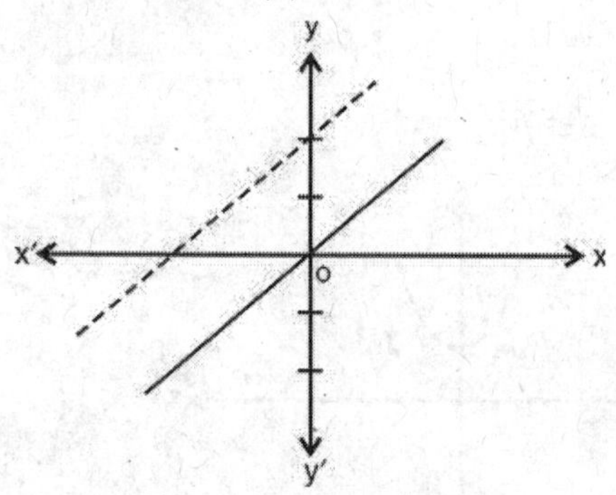

Ex-2. Let f(x) = x^2 - 1

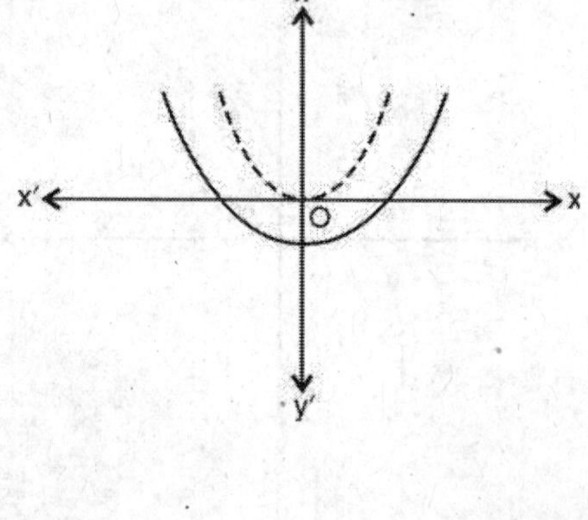

Ex-3. Let f(x) = 2^x - 1

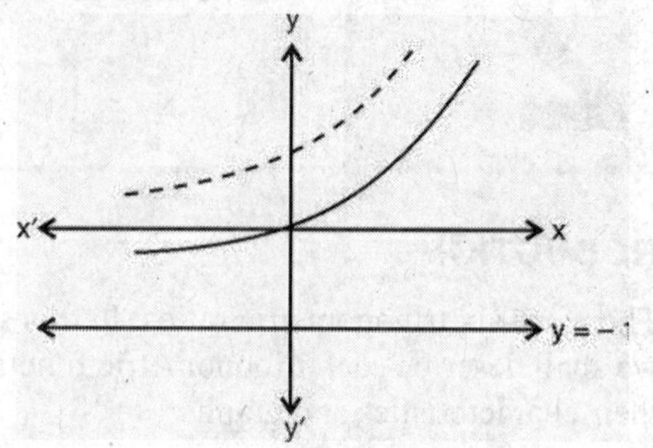

Type-III. When f(x) → f(x – a).

Rule: Shift the graph of *f*(*x*), *a*-units right wards.

Ex. 1. Let f(x) = $(x – 2)^2$

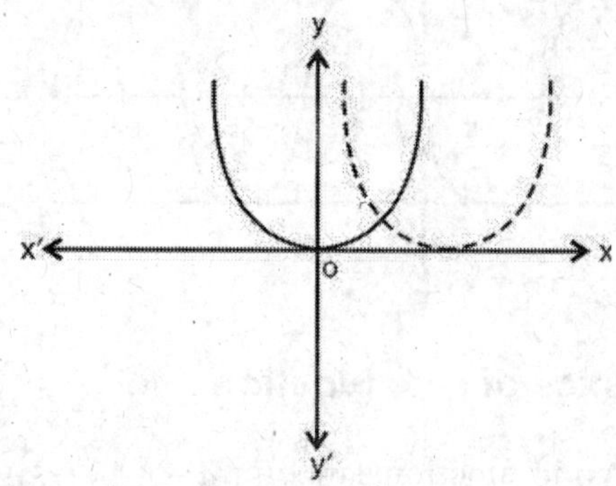

Ex. 2. Let $f(x) = |x - 1|$

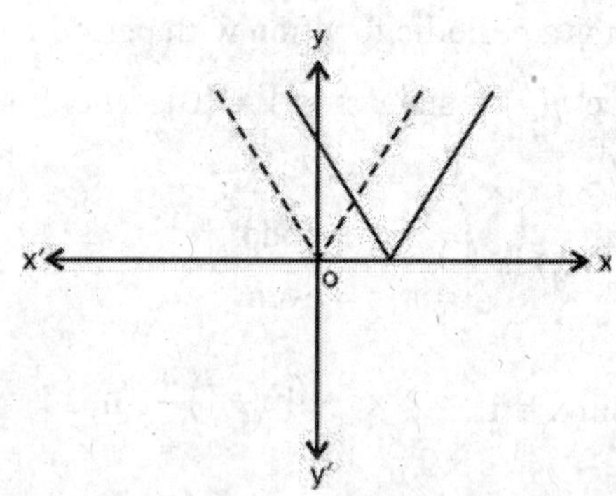

Ex. 3. Let $f(x) = (x – 1)^3$

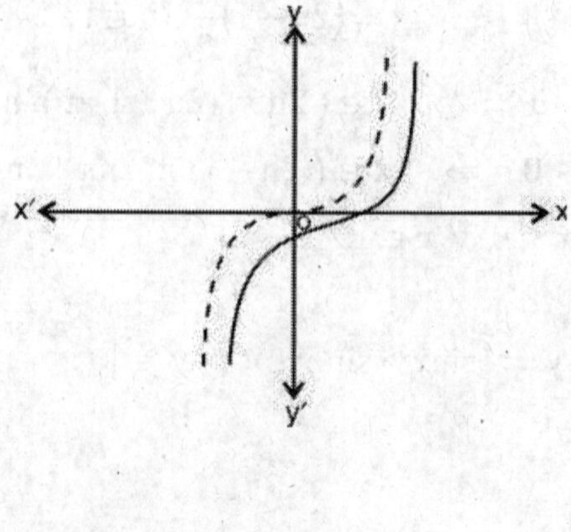

Ex. 4. Let $f(x) = \sin\left(x - \frac{\pi}{4}\right)$.

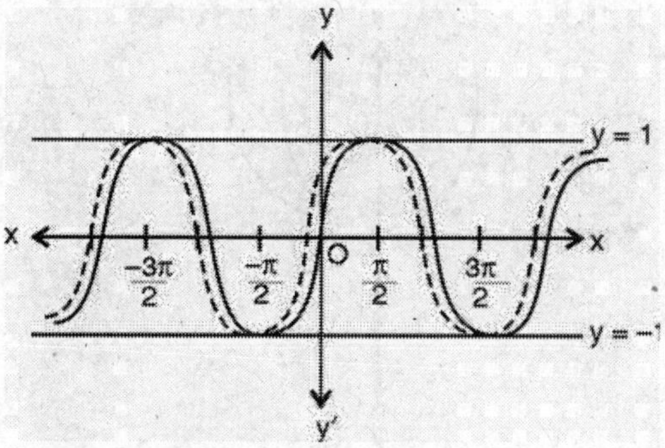

Type-IV. When f(x) → f(x + b).

Rule: Shift the graph of f(x), b-units left wards.

Ex. 1. Let $f(x) = (x + 2)^2$

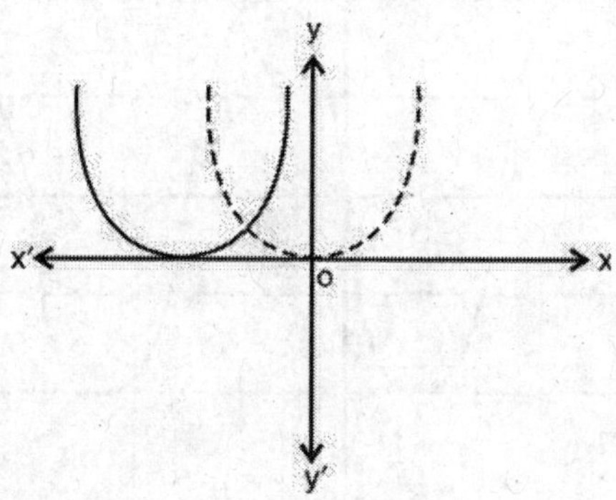

Ex.2. Let $f(x) = |x + 2|$

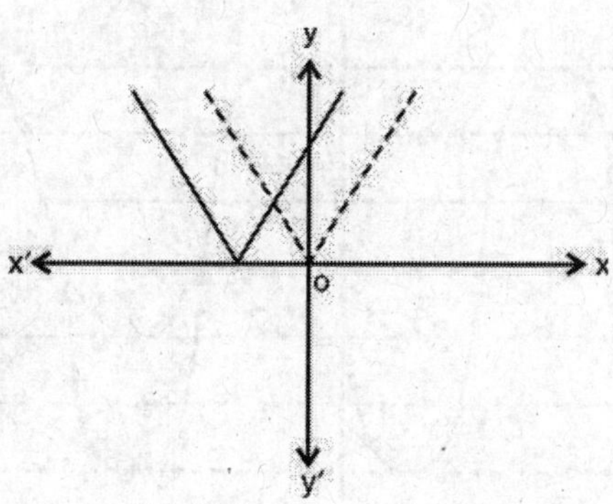

Ex. 3. Let $f(x) = 2^{x+1}$

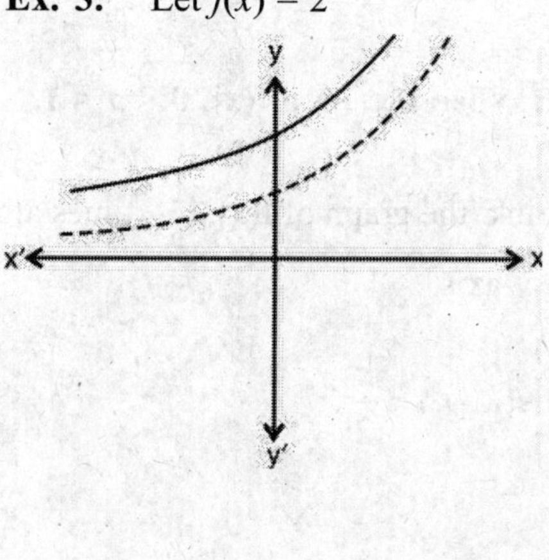

Ex. 4. Let $f(x) = \sin\left(x + \frac{\pi}{4}\right)$

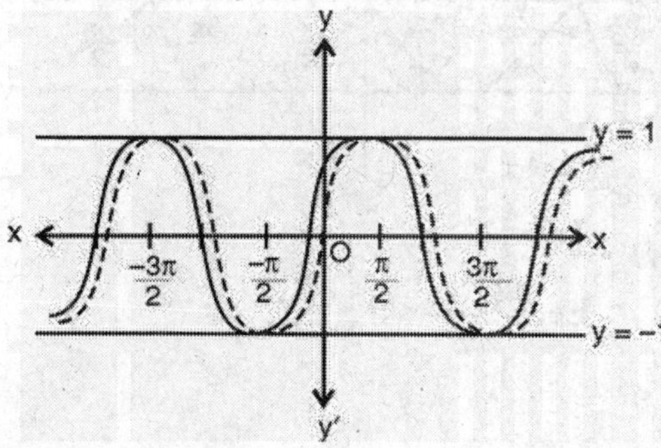

Type - V : When f(x) → f(ax), where a > 1.

Rule : Shrink the graph of $f(x)$, 'a'-times along the x-axis.

Ex. 1. Let y = sin 2x.

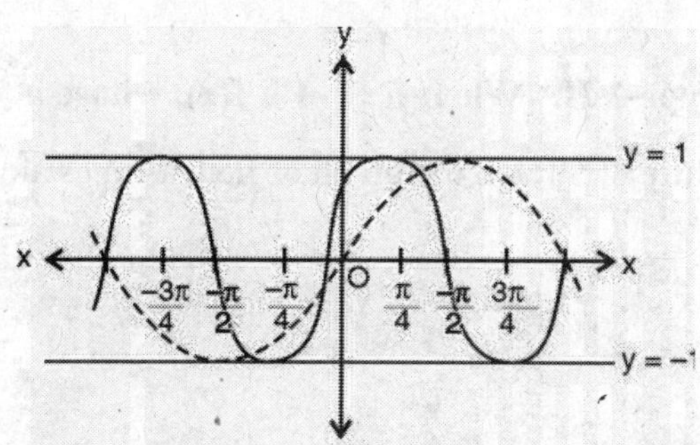

Ex. 2. Let y = sin 3x.

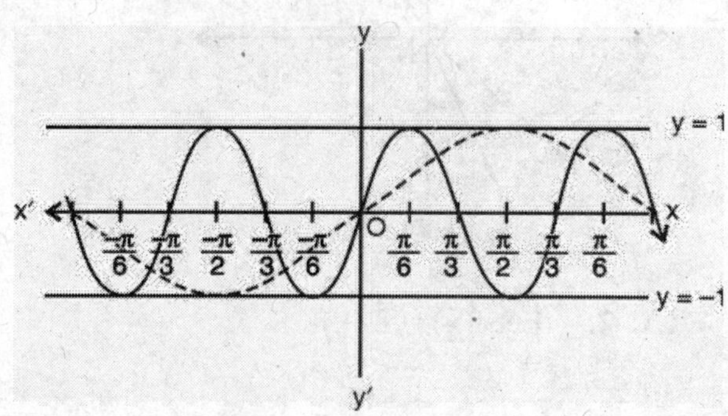

Type-VI. When f(x) → f(a x), where 0 < a < 1.

Rule: Stretch the graph of $f(x)$, $\frac{1}{a}$, times along the x-axis.

Ex. 1. Let $f(x) = \sin\left(\frac{x}{2}\right)$

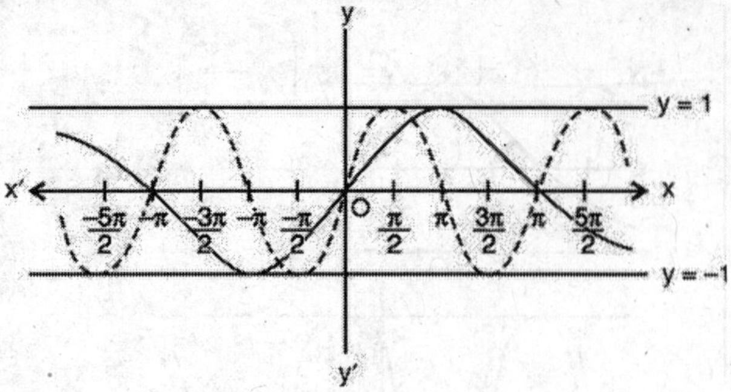

Ex. 2. Let f (x) = sin $\left(\dfrac{x}{3}\right)$.

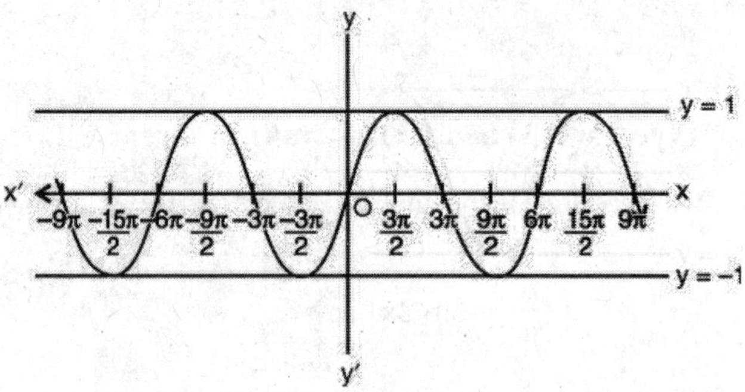

Type-VII. When f(x) → a f(x), whose a > 1.

Rule: Stretch the graph of f(x), a-times along *y*-axis.

Ex. 1. Let f(x) = 2x.

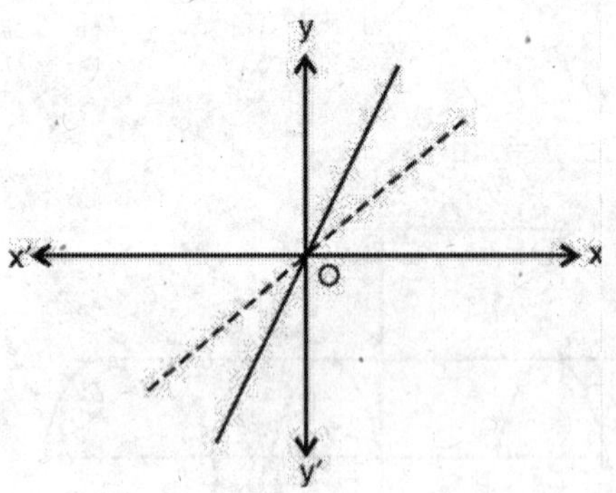

Ex. 2. Let f(x) = 2 (x).

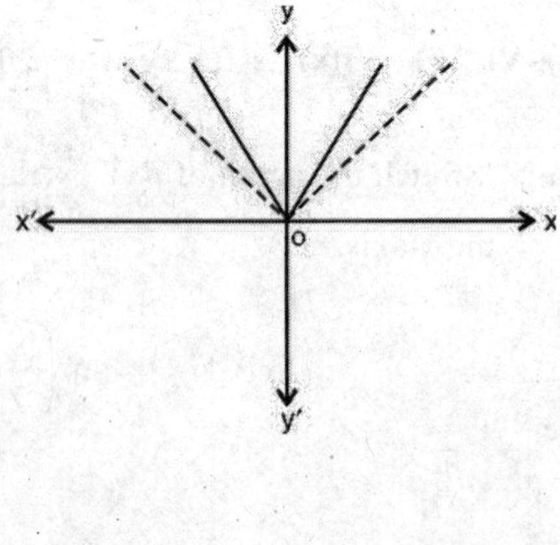

Ex. 3. Let *f*(*x*) = $2x^2$.

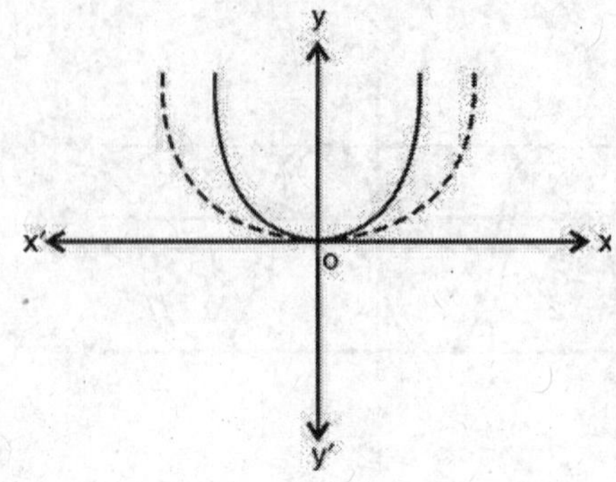

Ex. 3. Let f(x) = 2 sinx.

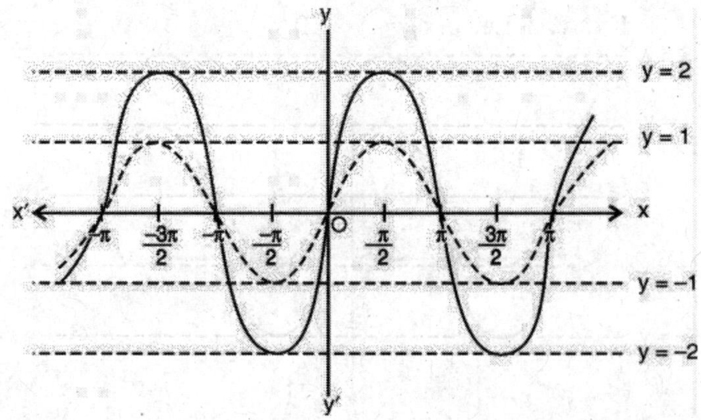

Ex. 4. Let *f*(*x*) = 3 sin*x*.

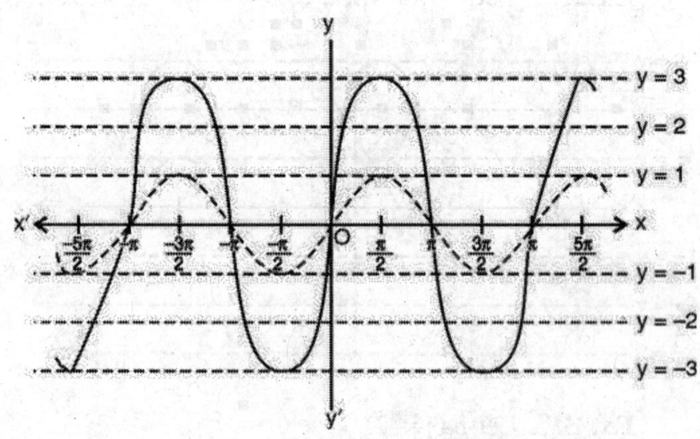

Type-VIII. When f(x) → a f(x), 0 < *a* < 1.

Rule: Shrink the graph of f(x), $\dfrac{1}{a}$ – times along the y-axis.

Ex. 1. Let $f(x) = \dfrac{x}{2}$

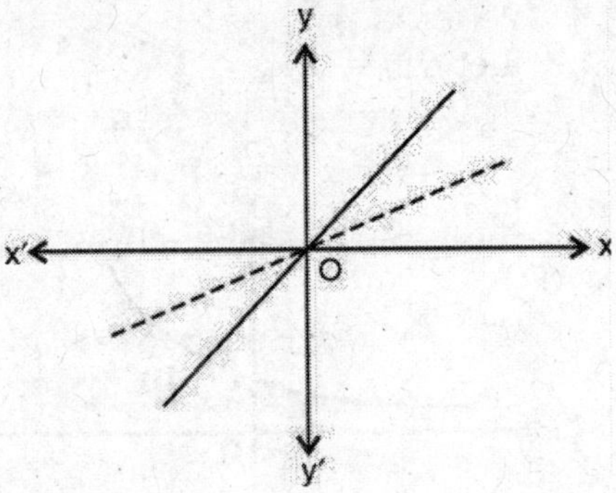

Ex. 2. Let $f(x)=\frac{|x|}{2}$

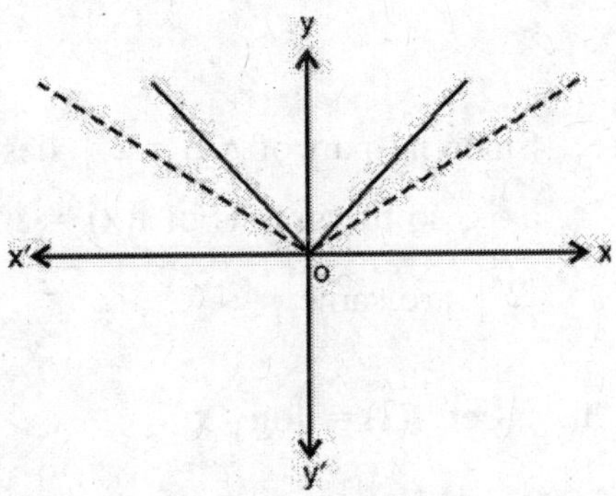

Ex. 3. Let $f(x) = \frac{1}{2}x^2$.

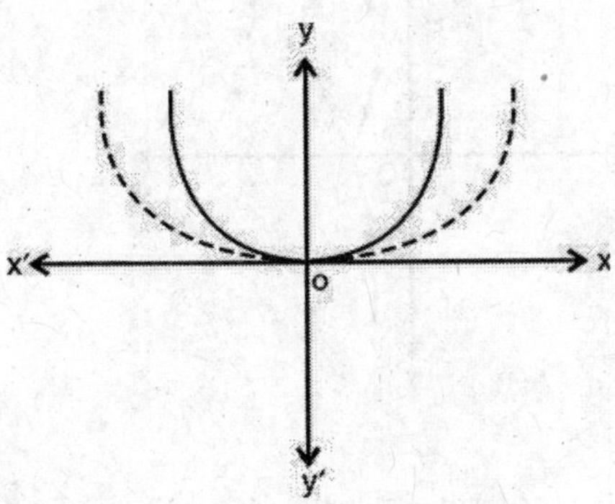

Ex. 4. Let $f(x) = \frac{1}{2}$ sin x.

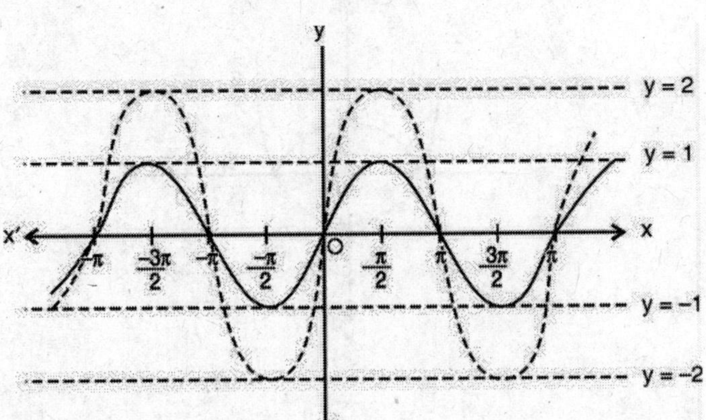

Type-IX. When $f(x) \to -f(x)$

Rule: Take the image of the graph of $f(x)$ w.r.t. x-axis..

Ex. 1. Let f(x) = –x.

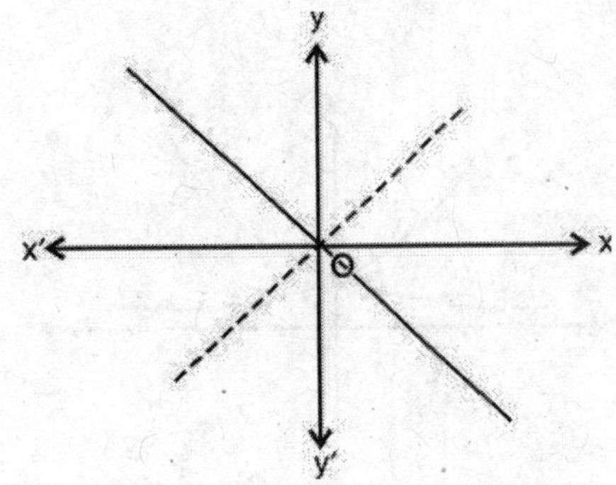

Ex. 2. Let $f(x) = -e^x$.

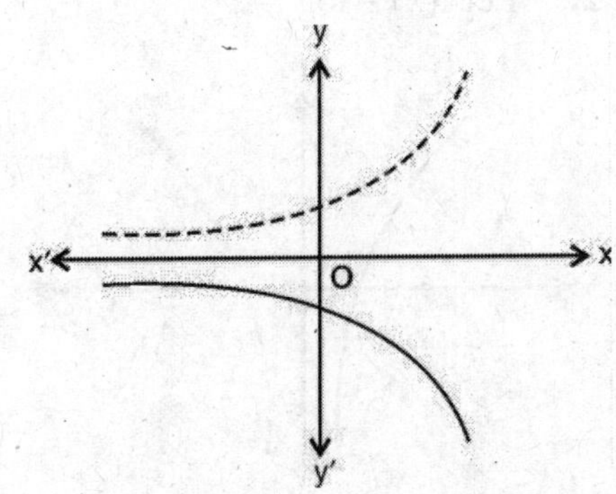

Ex. 3. Let $f(x) = -\log_e x$.

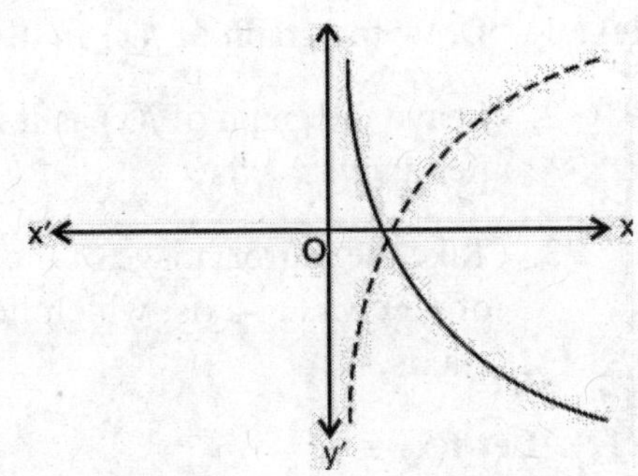

Ex. 4. Let f(x) = –sin x.

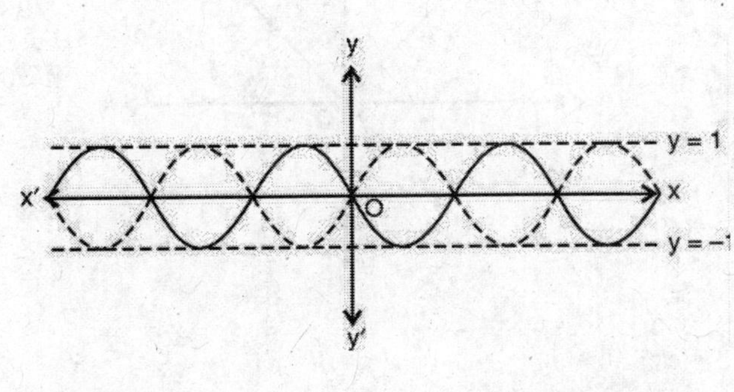

Type X. When $f(x) \to f(-x)$

Rule: Take the image of the graph of $f(x)$ w.r.t. y-axis..

Ex. 1. Let $f(x) = 2^{-x}$.

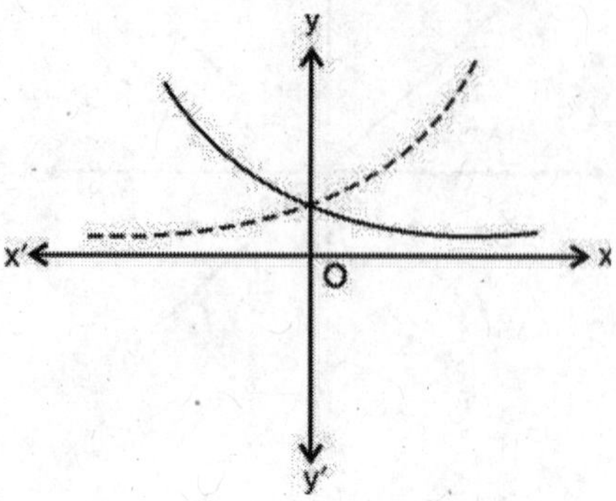

Ex. 2. Let $f(x) = \log_e(-x)$.

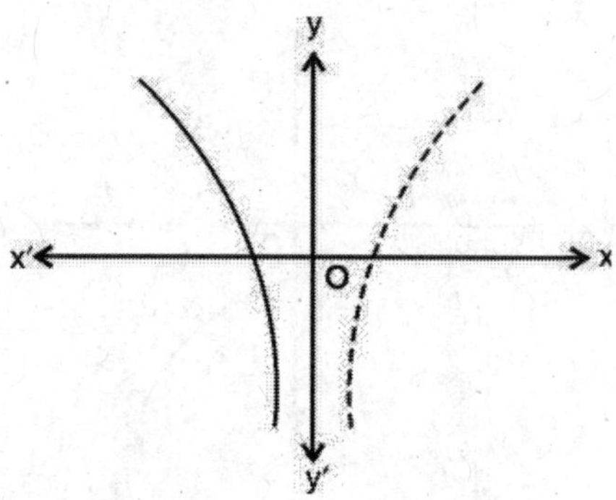

Type XI. When $f(x) \to |f(x)|$

Rules: 1. Draw the graph of $f(x)$ first.

2. Leave the graph of $f(x)$ as it is, which lies above x-axis.

3. Take the mirror image of the graph of $f(x)$ w.r.t. x-axis, which lies below x-axis.

Ex. 1. Let $f(x) = |x|$

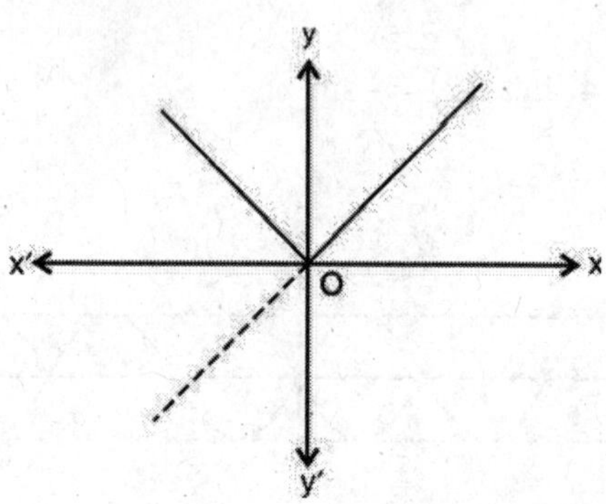

Ex. 2. Let $f(x) = |2^x|$.

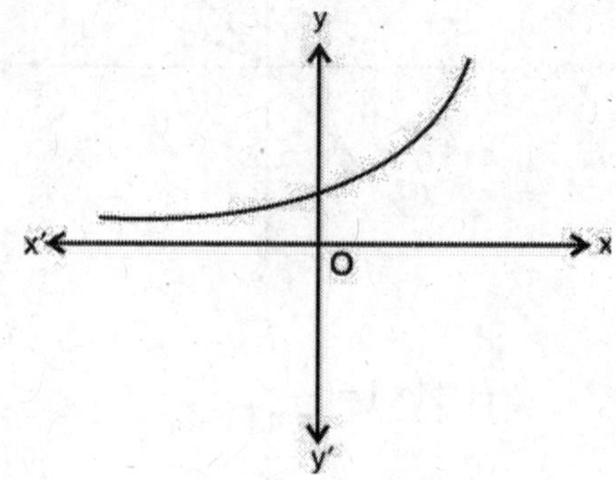

Note: Since no part of $f(x) = |2^x|$ lies below x-axis, so the graphs of $f(x) = 2^x$ & $f(x) = |2^x|$ are same.

Ex. 3. Let $f(x) = \log_e x$.

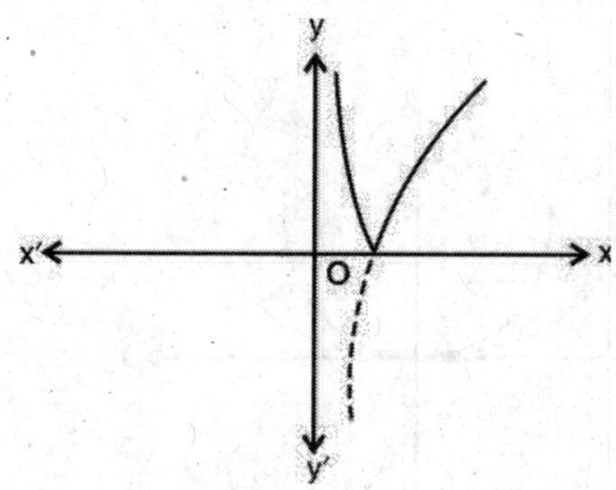

Ex. 4. Let $f(x) = |x^2 - 1|$

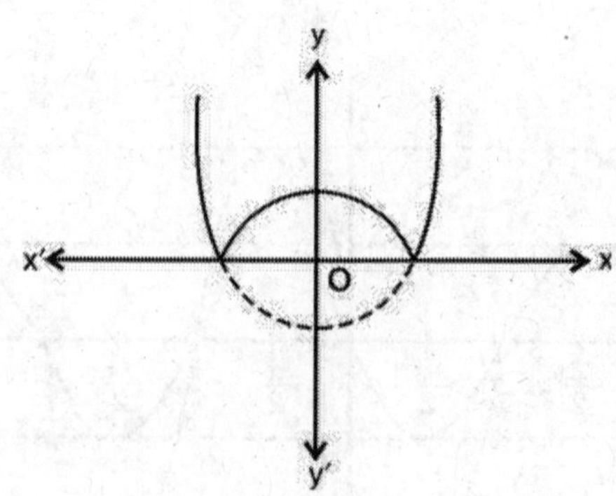

Ex. 5. Let $f(x) = \left|\frac{1}{x}\right|$

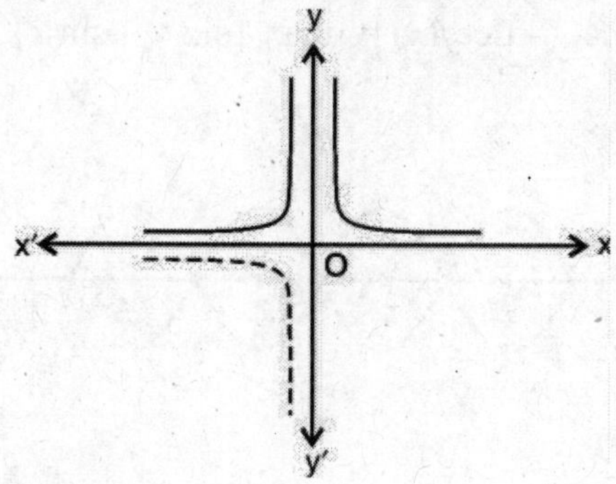

Ex. 6. Let $f(x) = |\sin x|$

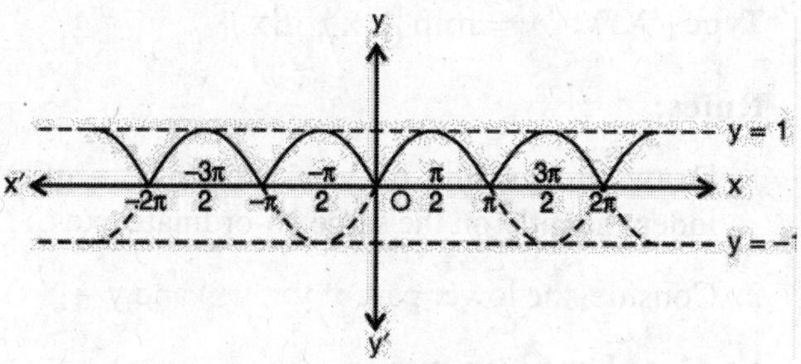

Ex. 7. Let $f(x) = |\sin x| + 1$

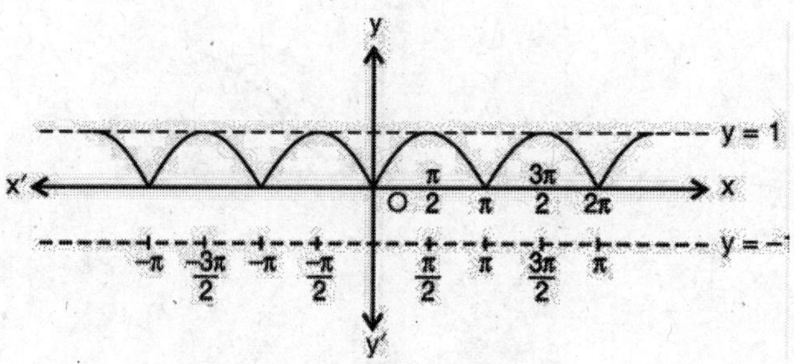

Type XII. When $f(x) \to f(|x|)$

Rules:

1. Draw the graph of f(x) first.
2. Remove the part of f(x), which lies left of *x*-axis.
3. Take the image of the part of f(x) w.r.t. y-axis, which lies right of x-axis.

Ex. 1. Let $f(x) = e^{|x|}$

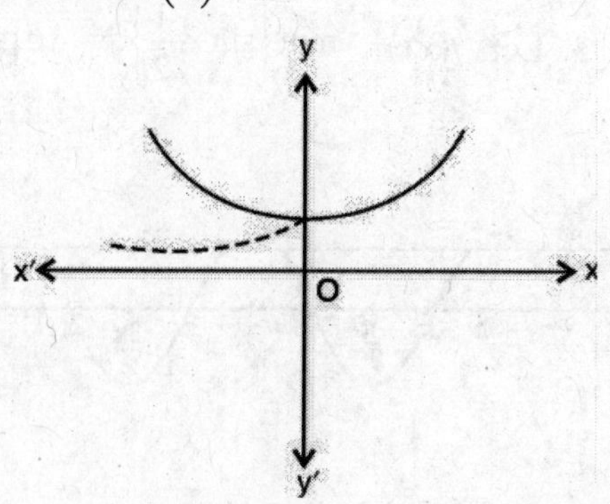

Ex. 2. Let $f(x) = 2^{-|x|}$.

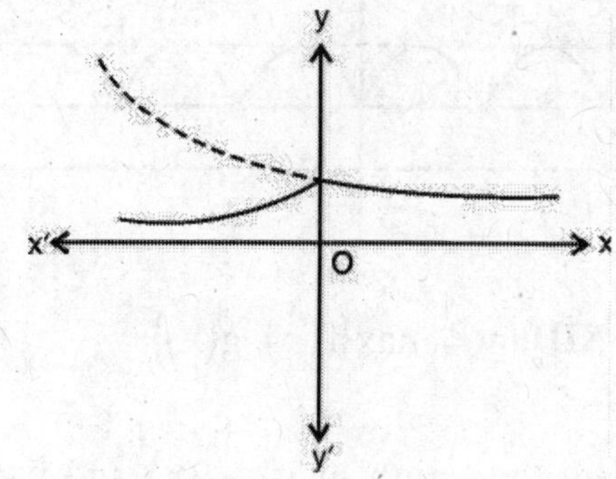

Ex. 3. Let f(x) = $\sin|x|$.

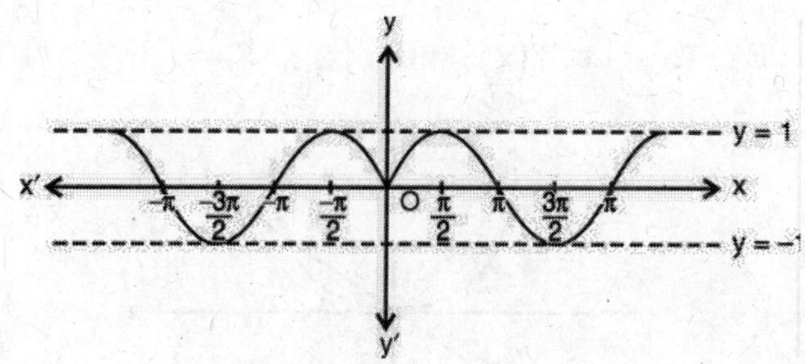

Ex. 4. Let $f(x) = \log_e|x|$

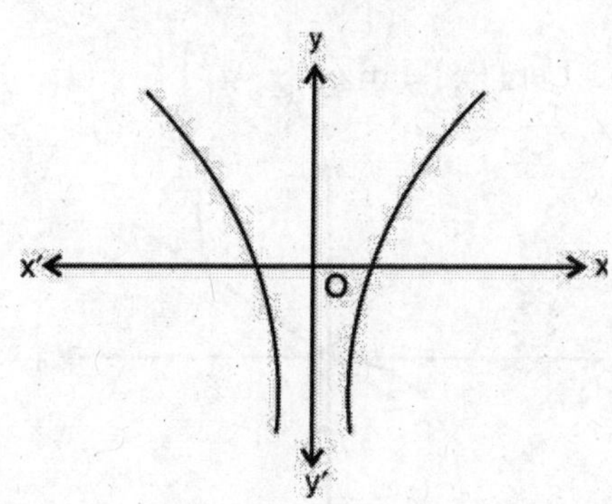

Ex. 5. Let $f(x) = x^2 - 3|x| + 2 = |x|^2 - 3|x| + 2$

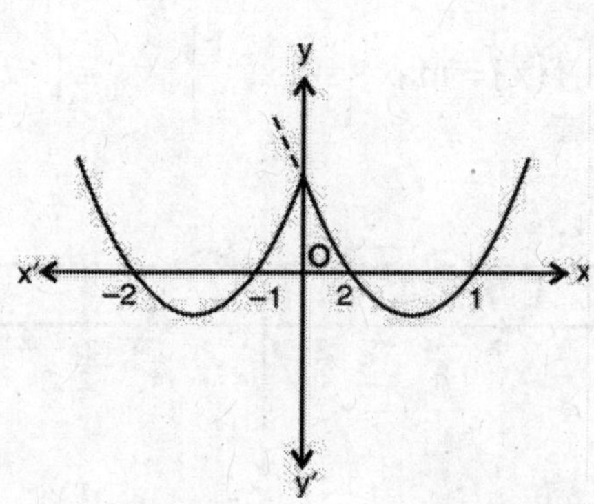

Ex. 6. Let $f(x) = |\sin|x|| = |\sin x|$

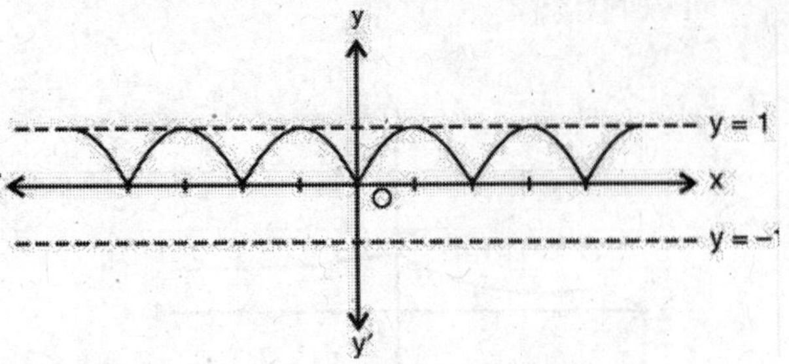

Type XIII. $y = \max\{f(x), g(x)\}$

Rules:

1. Draw the graph of y = f(x) and y = g(x) independently on the same co-ordinate axes.
2. Consider the upper part between the graphs of y = f(x) and y = g(x).

Ex. 1. Let $f(x) = \max\{x, x^2\}$

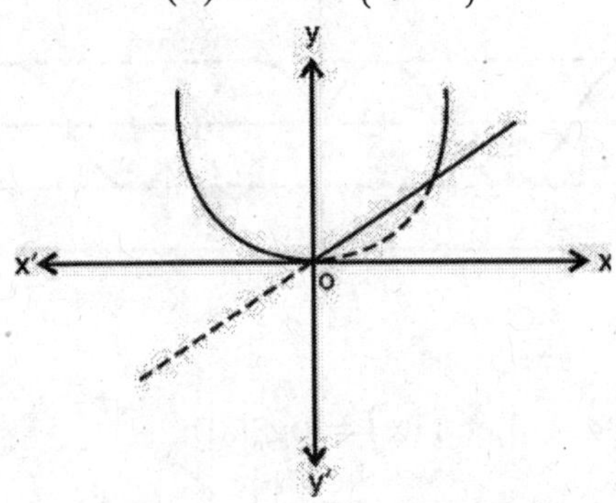

Ex. 2. Let $f(x) = \max\{x, x^3\}$

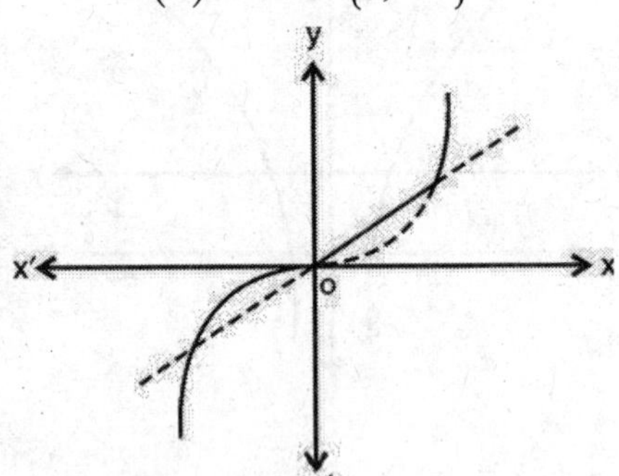

Ex. 3. Let

$$f(x) = \max\left\{\sin x, \frac{1}{2}\right\}, \forall x \in [-2\pi, 2\pi]$$

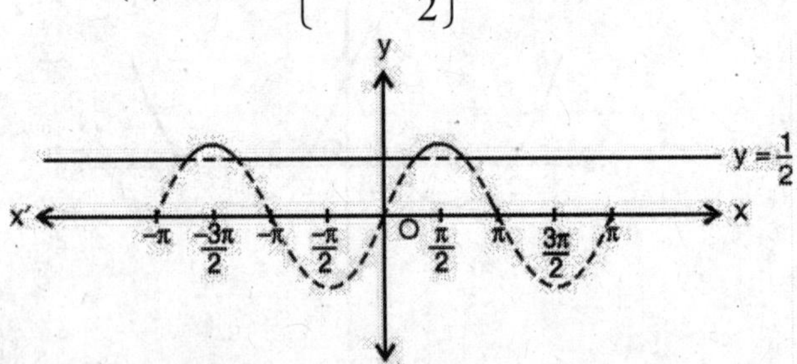

Ex .4. Let $f(x) = \max\{\sin x, -\sin x\}$.

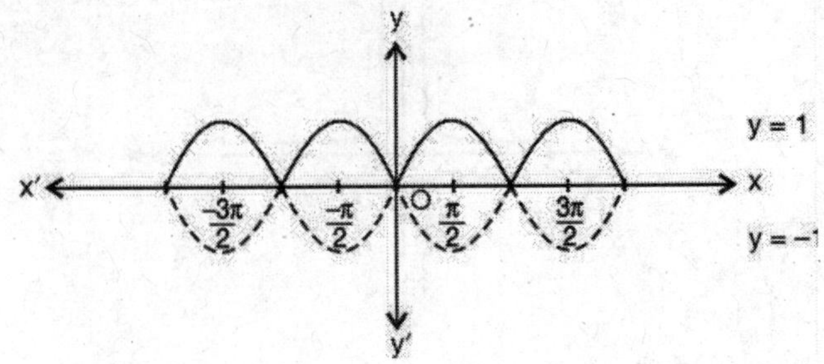

Note:

Type - XIV: $y = \min\{f(x), g(x)\}$

Rules:

1. Draw the graphs of y = f(x) & y = g(x) independently on the same co-ordinate axes.
2. Conside, the lower part of y = f(x) and y = g(x)

Ex. 1. Let $f(x) = \min\{x, x^2\}$.

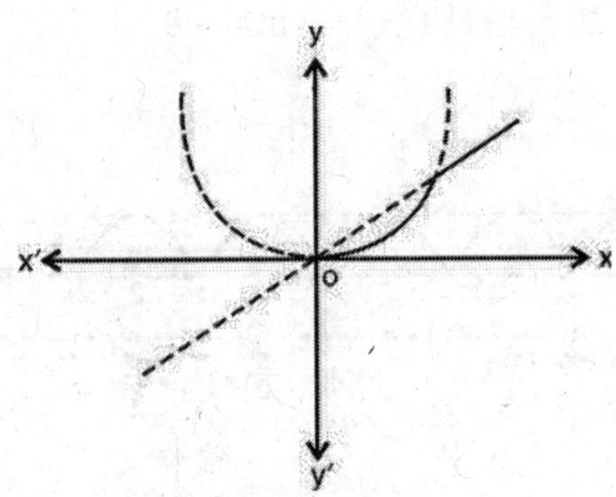

Ex. 2. Let $f(x) = \min\{x, x^3\}$.

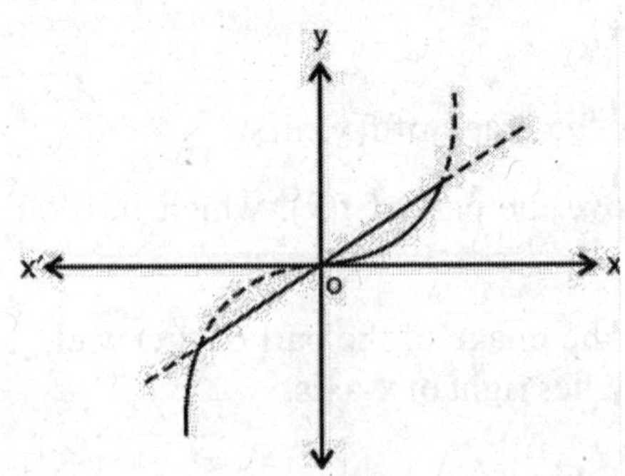

Ex. 3. Let $f(x) = \min\left\{\sin x, \frac{1}{2}\right\}, \forall x \in [-2\pi, 2\pi]$

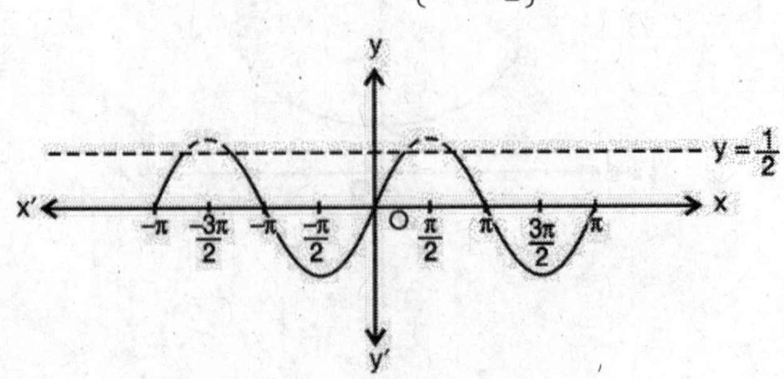

Ex. 4. Let $f(x$

$$= \min\left\{\sin x, \sin\left(x - \frac{\pi}{4}\right)\right\}, \forall\, x \in [-2\pi, 2\pi]$$

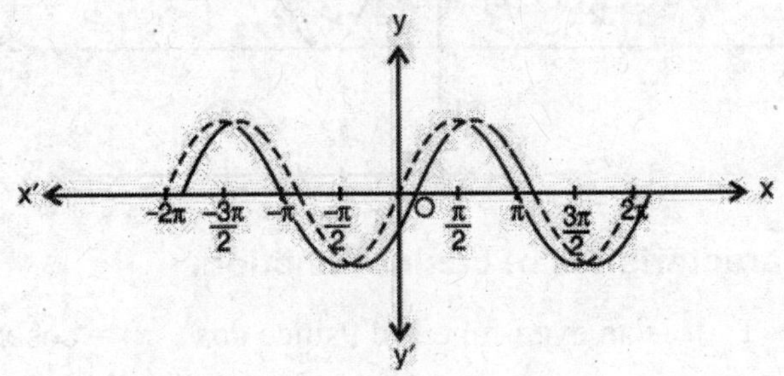

3. Some solved examples:

Ex. 1. Draw the graph of $y = \sin^2 x$.

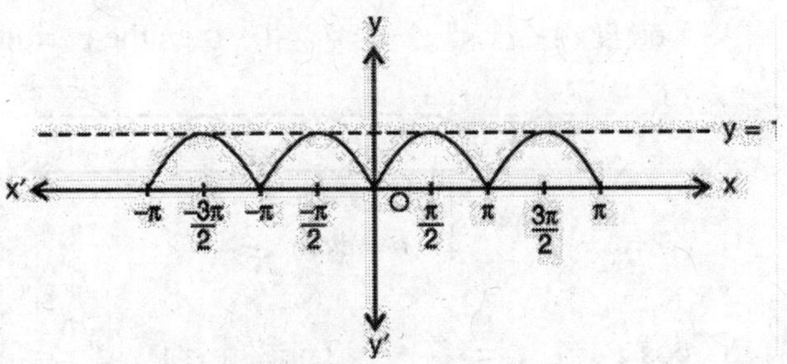

Ex. 2. Draw the graph of $y = \sin^3 x$.

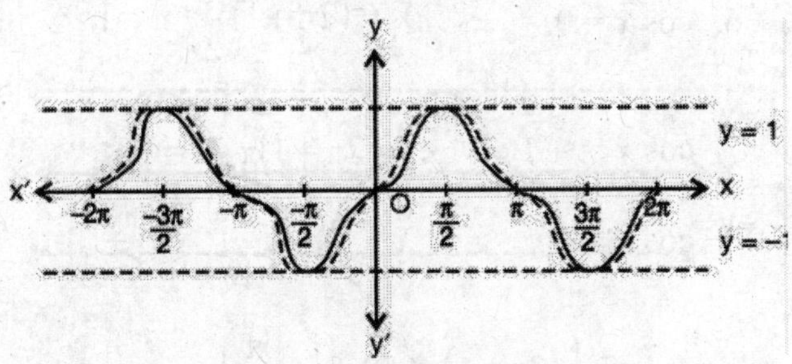

Ex. 3. Draw the graph of $y = x\sin x$.

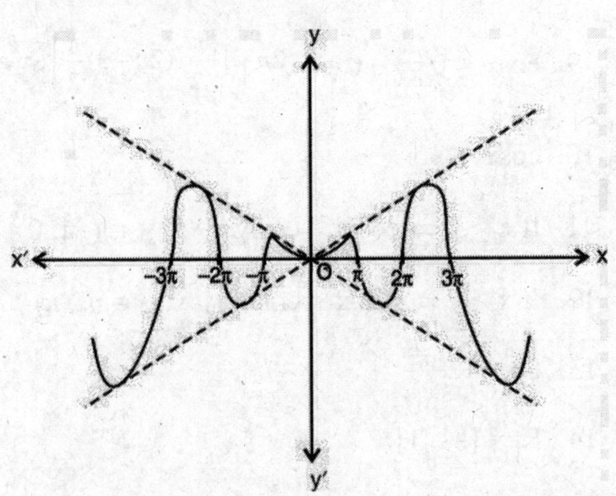

Ex. 4. Draw the graph of $y = x + \sin x$.

$$-1 \le \sin x \le 1$$

$$\Rightarrow \quad x - 1 \le x + \sin x \le x + 1$$

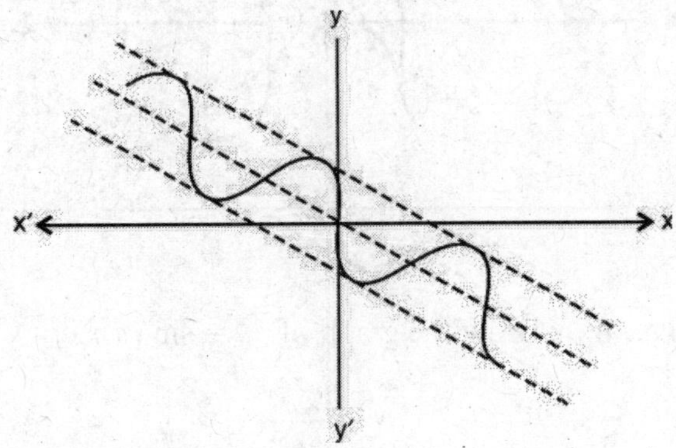

Ex. 5. Draw the graph of $y = 2^{\sin x}$, $\forall\, x \in [-2\pi, 2\pi]$

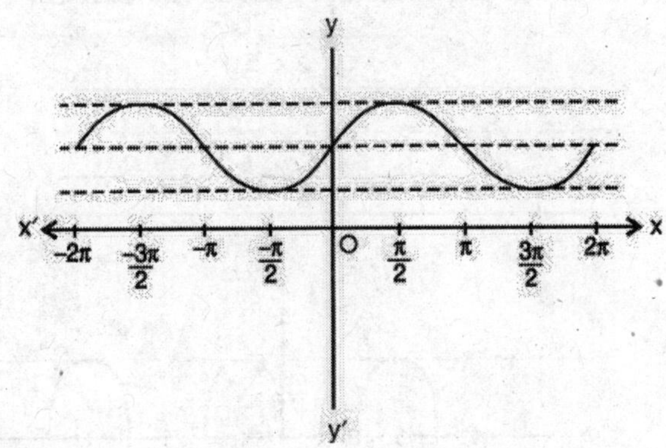

Ex. 6. Draw the graph of $y = \sin|x| + |x|$

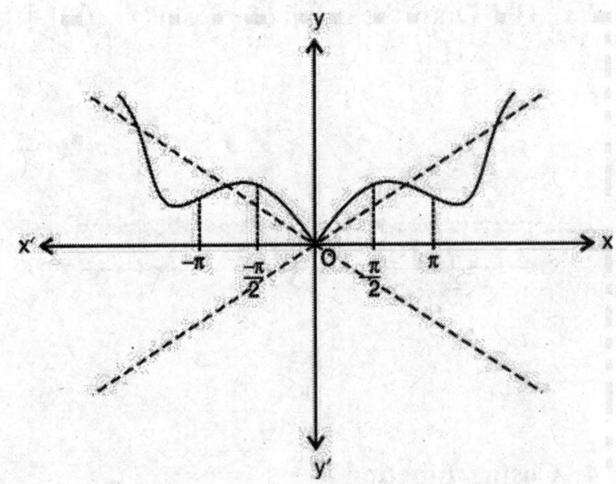

Ex. 7. Draw the graph of $y = \sin|x| - |x|$

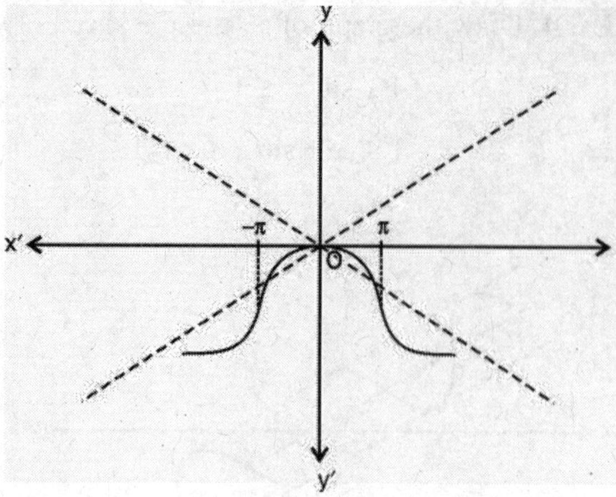

Ex. 8. Draw the graph of $y = \sin(\pi x)$

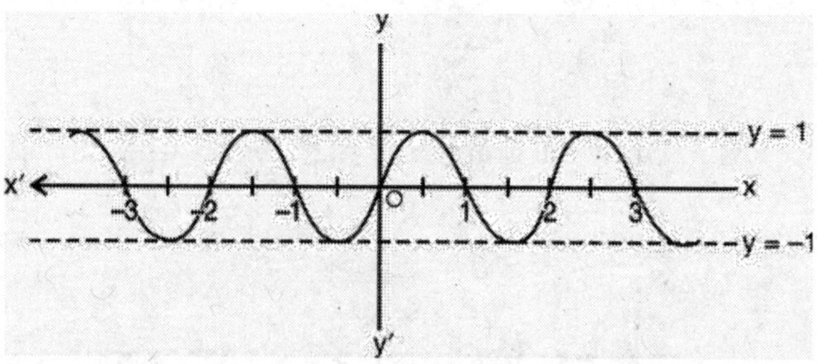

Ex. 9. Draw the graph of $y = \sqrt{\sin x}$.

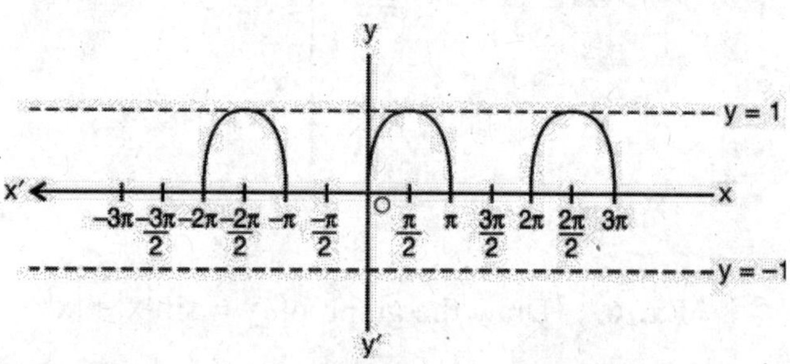

Ex. 10. Draw the graph of $y = \sin(x - (x))$, [,] = G.I.F.

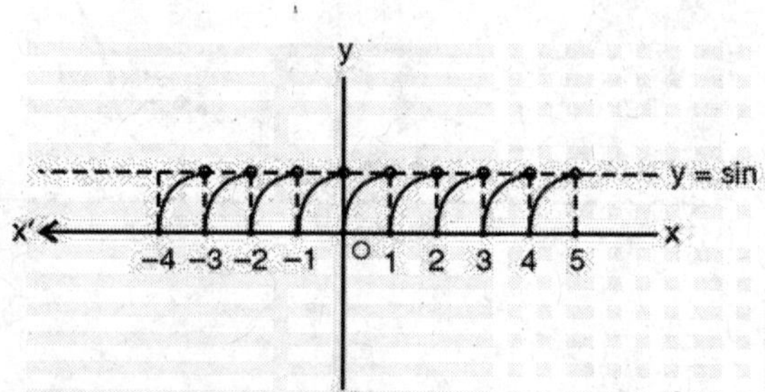

4. Cosine function:

For each real number x, there is a cosine function defined as $f(x) = \cos x$.

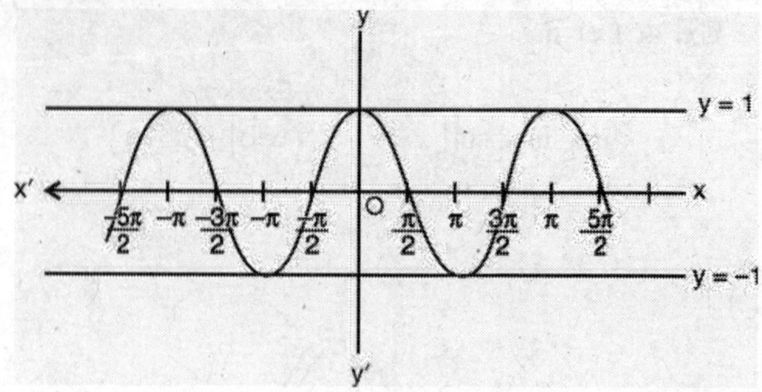

Characteristics of cosine function:

1. It is an even function , since $\cos(-x) = \cos x$.
2. It is symmetrical about y -axis.
3. It is a periodic function with period 2π.
4. Let $f(x) = \cos^n x$, $n \in I - \{0\}$ then the period of

$$f(x) \text{ is } \begin{cases} 2\pi, n = \text{odd} \\ \pi, n = \text{even} \end{cases}$$

5. $\cos x = 1 \Rightarrow x = 2n\pi, n \in I$.
6. $\cos x = 0 \Rightarrow x = (2n+1)\frac{\pi}{2}, n \in I$
7. $\cos x = -1 \Rightarrow x = (2n+1)\pi, n \in I$
8. $\cos x > 0$

$$\Rightarrow x \in \left((2n-1)\frac{\pi}{2}, (2n+1)\frac{\pi}{2}\right) n \in I$$

9. $\cos x < 0 \Rightarrow x \in \left((2n+1)\frac{\pi}{2}, (2n+3)\frac{\pi}{2}\right), \pi \in I.$
10. $\cos x \leq x$.
11. If $x > y \Rightarrow \cos x > \cos y$, $\forall x, y \in (-\pi, 0)$
12. If $x > y \Rightarrow \cos x < \cos y$, $\forall x, y \in (0, \pi)$
13. $D_f = R$.
14. $R_f = [-1, 1]$.

4. Some solved examples

Ex. 1. Draw the graph of y = cos x + 1.

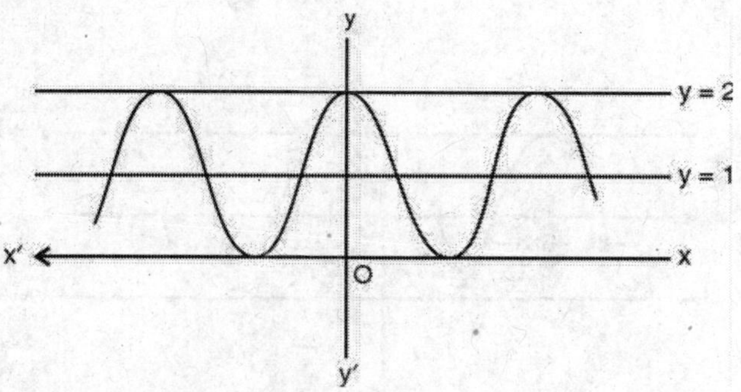

Ex. 2. Draw the graph of y = cos x – 1.

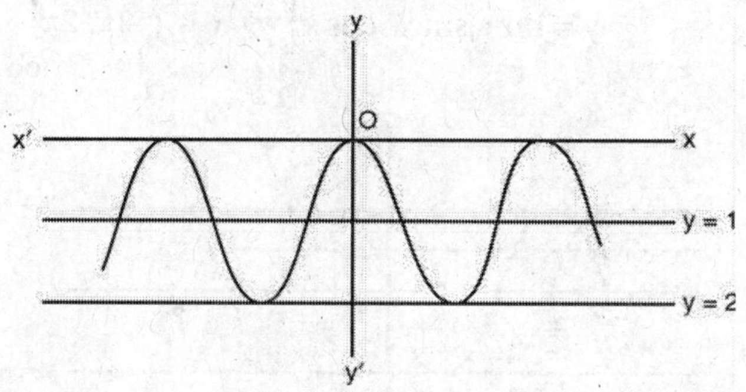

Ex. 3. Draw the graph of y = –cos x.

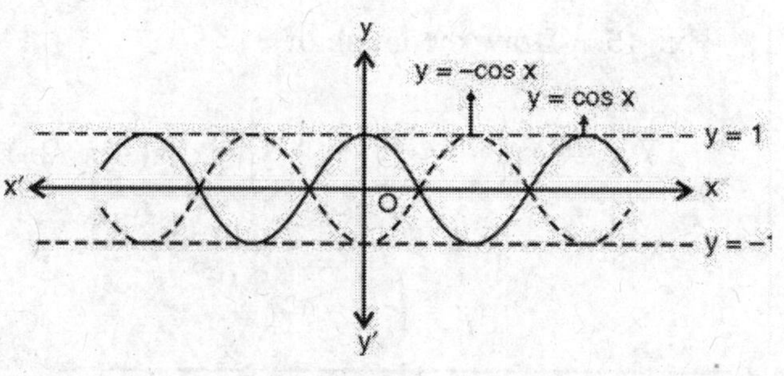

Ex. 4. Draw the graph of y = 1 – cos x.

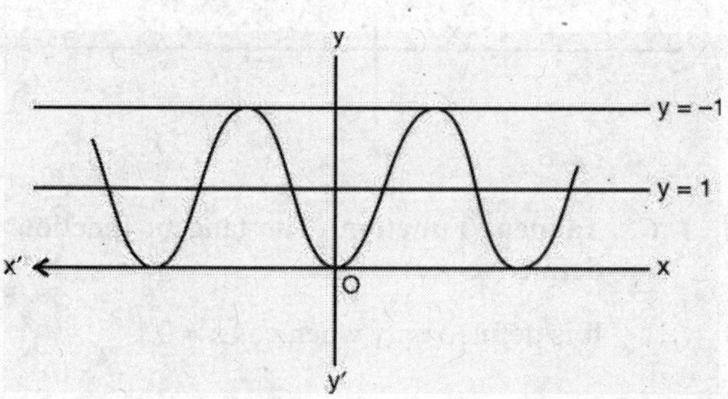

Ex. 5. Draw the graph of y = cos 2x.

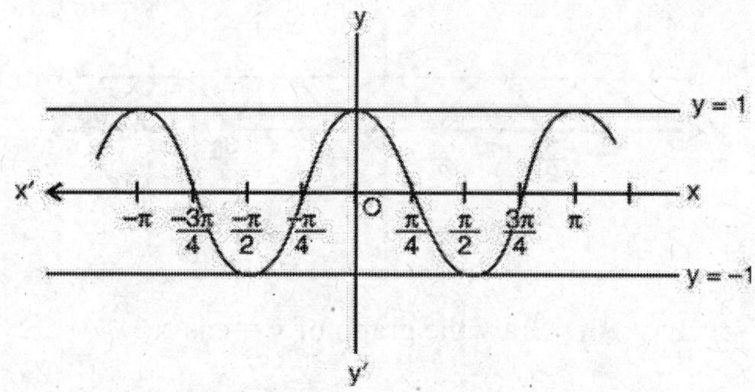

Ex. 6. Draw the graph of y = 2cosx.

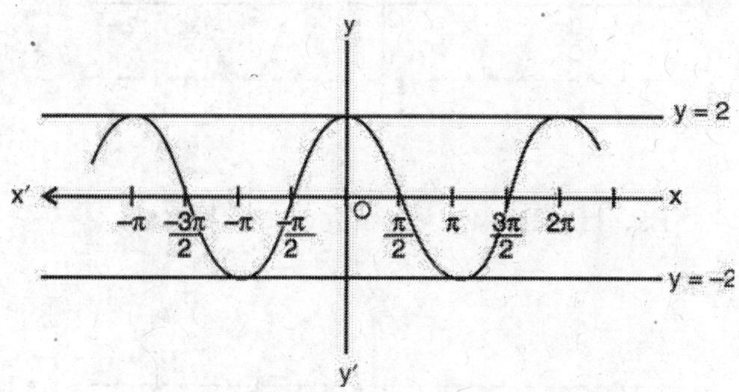

Ex. 7. Draw the graph of y = cos $\left(\frac{x}{2}\right)$.

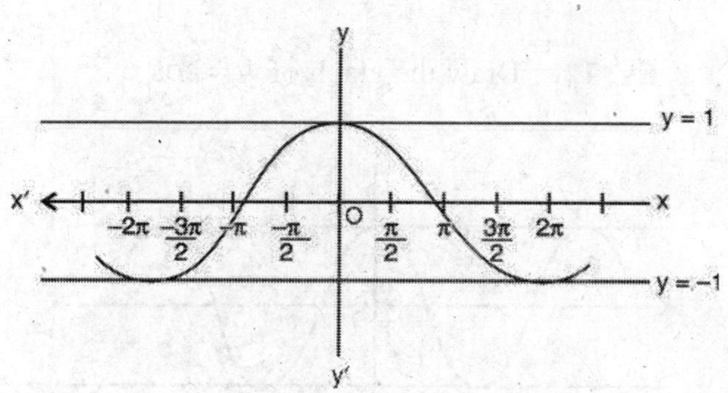

Ex. 8. Draw the graph of y = $\frac{1}{2}$ cos x.

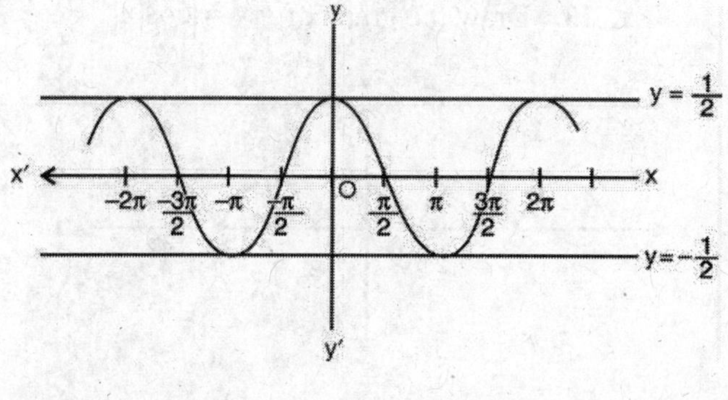

Ex. 9. Draw the graph of $y = \cos^2 x$.

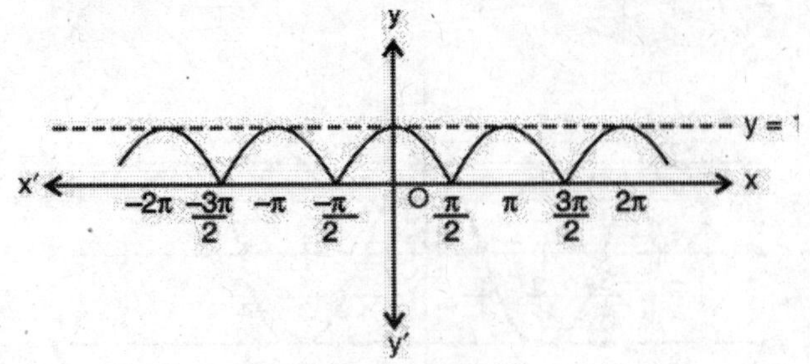

Ex. 10. Draw the graph of $y = \cos^3 x$.

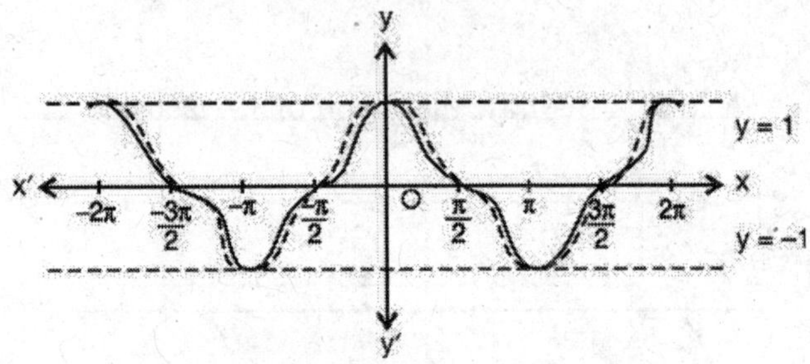

Ex. 11. Draw the graph of $y = |\cos x|$.

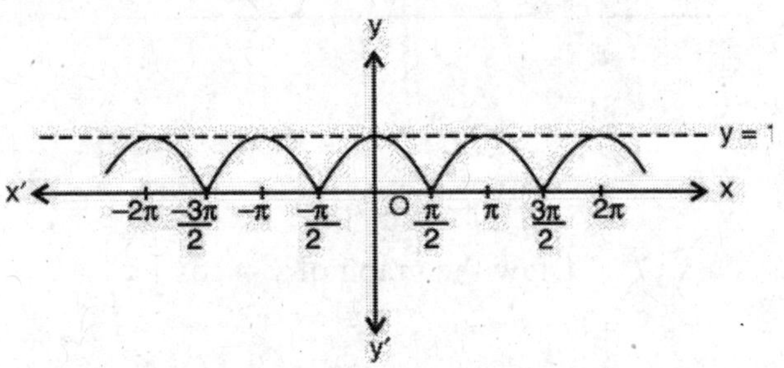

Ex. 12. Draw the graph of $y = \cos|x|$.

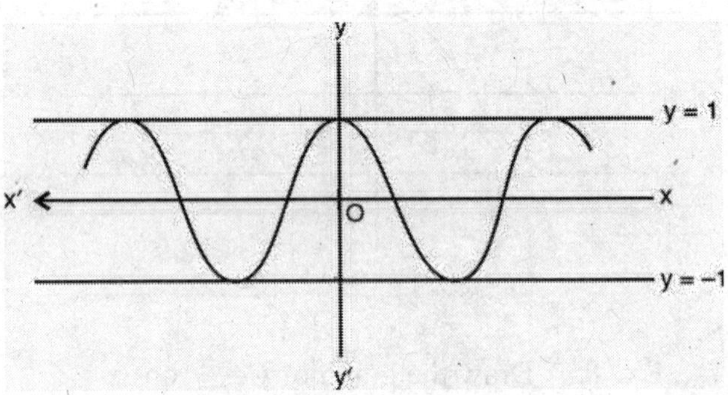

Ex. 13. Draw the graph of $y = |\cos|x||$

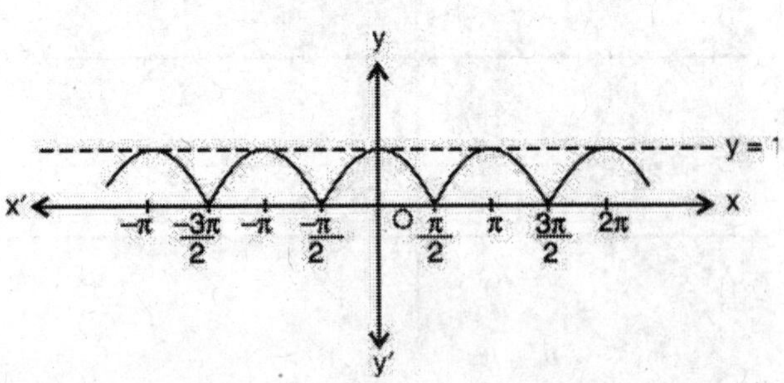

Ex. 14. Draw the graph of

$$y = \max\{\sin x, \cos x\} \ \forall x \in (-2\pi, 2\pi).$$

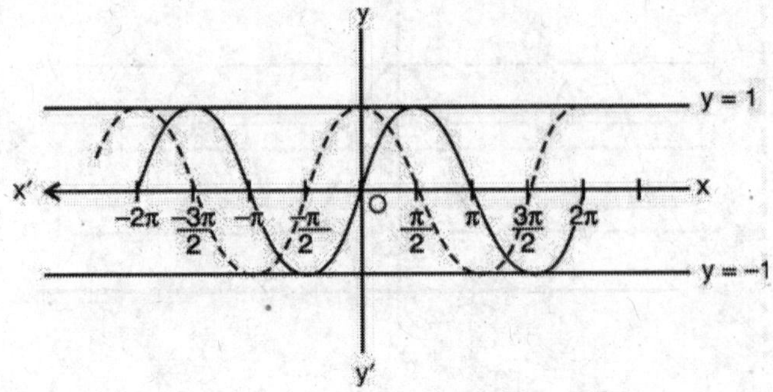

Ex. 15. Draw the graph of

$$y = \min\{\sin x, \cos x\} \ \forall x \in (-2\pi, 2\pi).$$

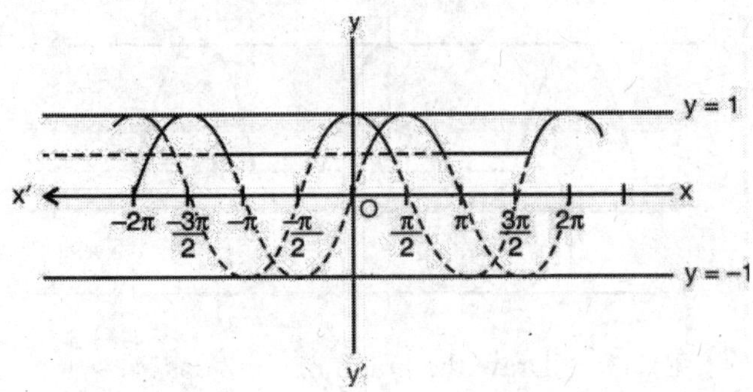

Ex. 15. Draw the graph of

$$y = \max\left\{\sin x, \frac{1}{2}, \cos x\right\} \forall x \in (-2\pi, 2\pi).$$

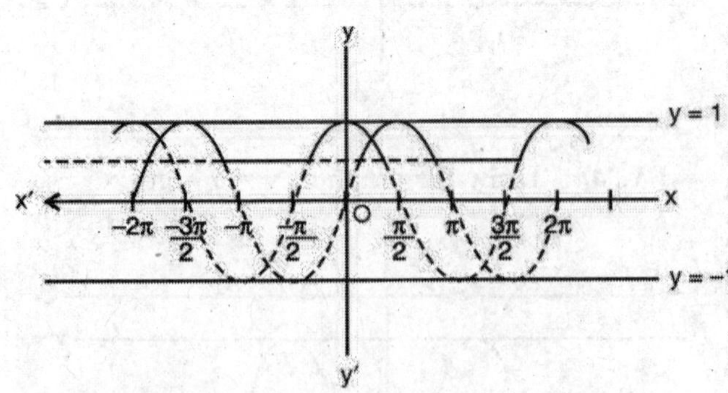

6. **Tangent Function:** The tangent function is defined as $f(x) = \tan x$.

It is defined only when $\cos x \neq 0$.

$$\Rightarrow \quad x \neq (2n + 1)\frac{\pi}{2}, n \in I.$$

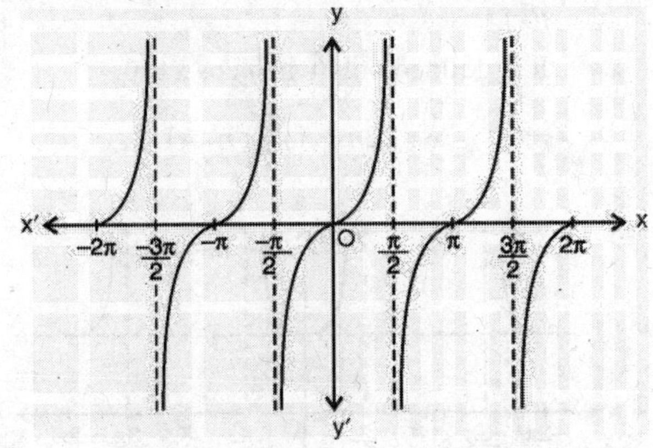

Characteristics of tangent function :

1. It is an odd function, $\tan(-x) = -\tan x$.
2. It is symmetrical about the origin.
3. It is a periodic function with period π.
4. Let $f(x) = \tan^n x$. Then the periods of $f(x)$ is π, whatever n may be.
5. $\tan x = 1 \Rightarrow x = (4n+1)\dfrac{\pi}{4}, n \in I$.
6. $\tan x = 0 \Rightarrow x = n\pi,\ n \in I$.
7. $\tan x = -1 \Rightarrow x = (4n-1)\dfrac{\pi}{4},\ n \in I$.
8. $\tan x > 0 \Rightarrow x \in (n\pi, (2n+1)\dfrac{\pi}{2}),\ n \in I$.
9. $\tan x < 0 \Rightarrow x \in ((2n-1)\dfrac{\pi}{2}, n\pi),\ n \in I$.
10. $\tan x \geq x$.
11. If $x > y \Rightarrow \tan x > \tan y,\ \forall x, y \in R - (2n+1)\dfrac{\pi}{2}, n \in I$
12. $D_f = R - (2n+1)\dfrac{\pi}{2},\ n \in I$.
13. $R_f = R$.

Ex. 1. Draw the graph of $f(x) = |\tan x|$

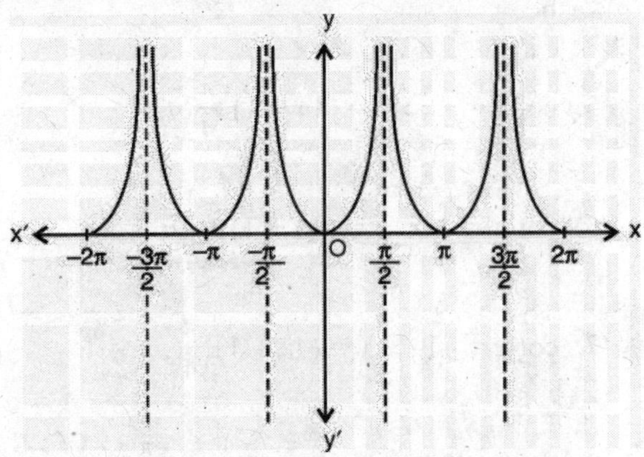

Ex. 2. Draw the graph of $f(x) = \tan^2 x$

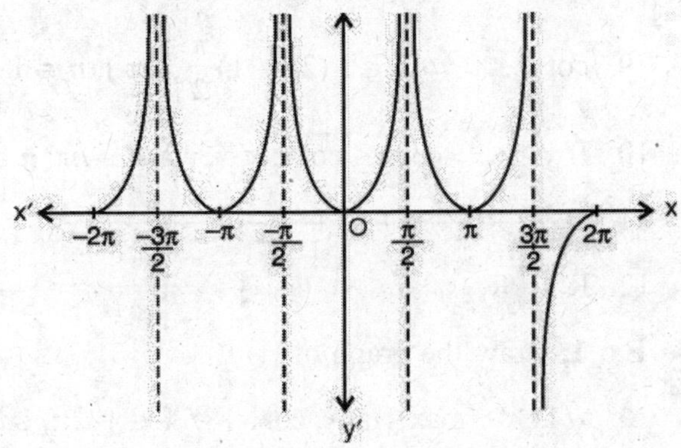

7. **Co-tangent function**: The cotangent function is defined as $f(x) = \cot x = \dfrac{\cos x}{\sin x}$.

It is defined only when $\sin x \neq 0 \Rightarrow x = n\pi$, $n \in I$.

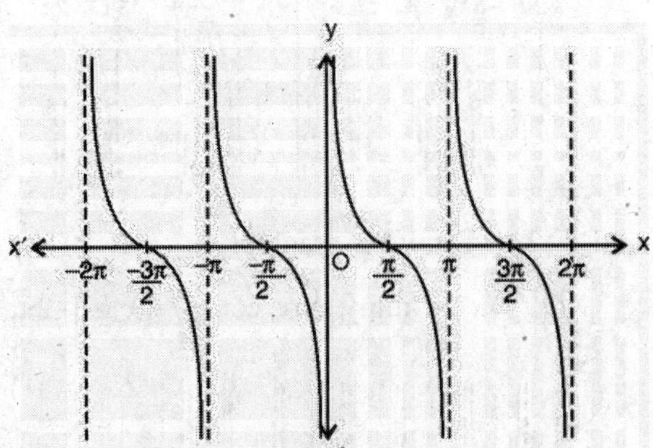

Characteristics of co-tangent function

1. It is an odd function, since $\cot(-x) = -\cot x$.
2. It is symmetrical about the origin.
3. It is a periodic function with period π.
4. Let $f(x) = \cot^n x,\ n \in I - \{0\}$.

Then the periods of f(x) is π, whatever n may be.

5. $\cot x = 1 \Rightarrow x = (4n + 1)\ \frac{\pi}{4}, n \in I.$

6. $\cot x = 0 \Rightarrow x = (2n + 1)\frac{\pi}{2},\ n \in I.$

7. $\cot x = -1 \Rightarrow x = (4n - 1)\ \frac{\pi}{4},\ n \in I.$

8. $\cot x > 0 \Rightarrow x \in (n\pi, (2n + 1)\frac{\pi}{2}),\ n \in I.$

9. $\cot x < 0 \Rightarrow x \in (\ (2n - 1)\frac{\pi}{2},\ n\pi\ ),\ n \in I.$

10. If $x > y \Rightarrow \cot x < \cot y,\ \forall\ x, y \in R - n\pi,\ n \in I.$

11. $D_f = R - n\pi,\ n \in I.$

12. $R_f = R.$

Ex. 1. Draw the graph of

$f(x) = \max\ \{\tan x, \cot x\}\ \forall\ x \in [-2\pi, 2\pi]$

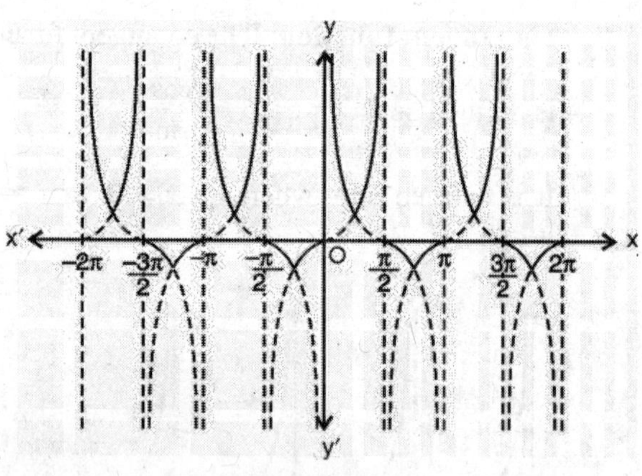

Ex. 2. Draw the graph of

$f(x) = \min\ \{\tan x, \cot x\}\ \forall\ x \in [-2\pi, 2\pi]$

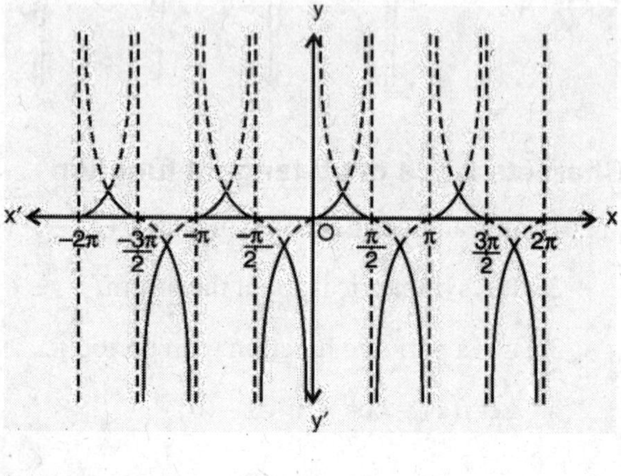

7. Secant function: Secant function is defined as $f(x) = \sec x$.

It is defined only when $\cos x \neq 0$

$\Rightarrow \quad x \neq (2n + 1)\frac{\pi}{2}$, n $\in$ I.

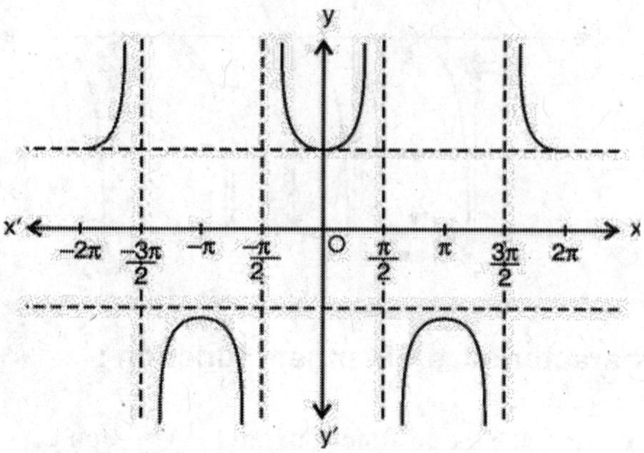

Characteristics of secant function

1. It is an even function, $\sec(-x) = \sec x$.
2. It is symmetrical about y-axis.
3. It is a periodic function with period 2π.
4. Let $f(x) = \sec^n x,\ n \in I - \{0\}$. Then the periods of $f(x)$ is $\begin{cases} 2\pi, n = \text{odd} \\ \pi, n = \text{even} \end{cases}$.
5. $\sec x = 1 \Rightarrow x = 2n\pi,\ n \in I.$
6. $\sec x$ can never be zero.
7. $\sec x = -1 \Rightarrow x = (2n + 1)\pi,\ n \in I.$
8. $\sec x > 0 \Rightarrow x \in ((4n + 1)\frac{\pi}{2},\ (4n + 1)\frac{\pi}{2}),\ \forall$ $n \in$ I.
9. $\sec x < 0 \Rightarrow x \in (\ (4n + 1)\frac{\pi}{2},\ (4n + 1)\frac{\pi}{2}),\ \forall$ $n \in$ I.
10. If $x > y \Rightarrow \sec x > \sec y,\ \forall\ x, y \in (0, \frac{\pi}{2})$.
11. $D_f = R - (2n + 1)\frac{\pi}{2},\ n \in I.$
12. $R_f = (-\infty, -1] \cup [1, \infty)$.

8. Co-secant function: Cosecant function is defined as $f(x) = \text{cosec}\ x$.

It is defined only when $\sin x \neq 0 \Rightarrow x \neq n\pi$, n $\in$ I.

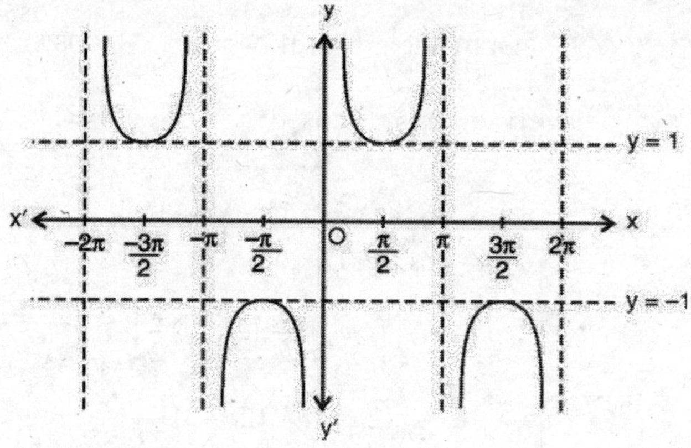

Characteristics of co-secant function:

1. It is an odd function, since cosec $(-x) = -$cosec x.
2. It is symmetrical about the origin.
3. It is a periodic function with period 2π.
4. Let $f(x) = \text{cosec}^n x$, $n \in \text{I} - \{0\}$.

   Then the periods of $f(x)$ is $\begin{cases} 2\pi, n = \text{odd} \\ \pi, n = \text{even} \end{cases}$.
5. cosec $x = 1 \Rightarrow x = (4n + 1)\frac{\pi}{2}$, $n \in \text{I}$.
6. cosec $x = -1 \Rightarrow n = (4n - 1)\frac{\pi}{2}$, $n \in \text{I}$..
7. cosec x can never be zero.
8. cosec $x > 0 \Rightarrow x \in (2n\pi, (2n + 1)\pi)$, $n \in \text{I}$.
9. cosec $x < 0 \Rightarrow x \in ((2n - 1)\pi, 2n\pi)$, $n \in \text{I}$.
10. If $x > y \Rightarrow$ cosec $x <$ cosec y, $\forall\, x, y \in (0, \frac{\pi}{2})$.
11. $D_f = R - n\pi$, $n \in \text{I}$.
12. $R_f = (-\infty, -1) \cup (1, \infty)$.

9. Some solved examples:

Ex. 1. Draw the graph of $y = 2^{\cos x}$

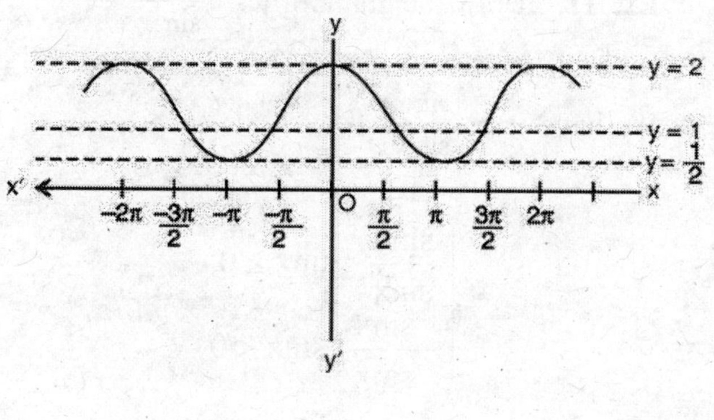

Ex. 2. Draw the graph of $y = 2^{-\cos x}$

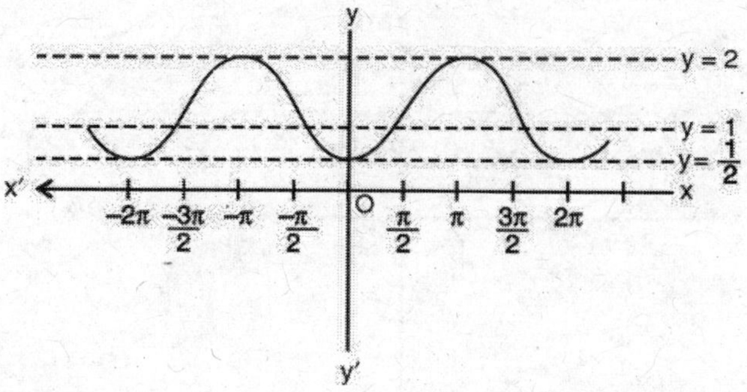

Ex. 3. Draw the graph of $y = 2^{\sin x}$

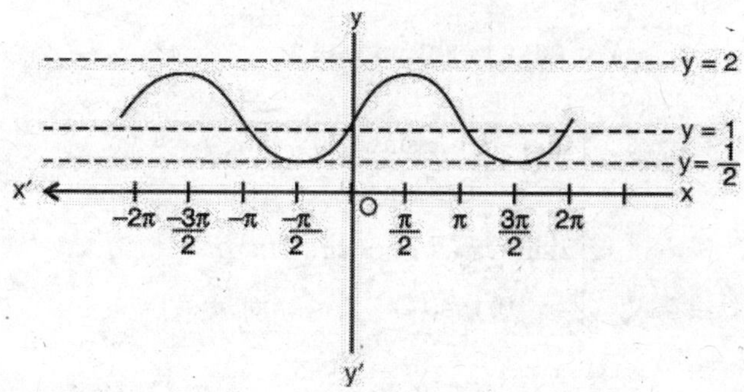

Ex. 4. Draw the graph of $y = 2^{-\sin x}$

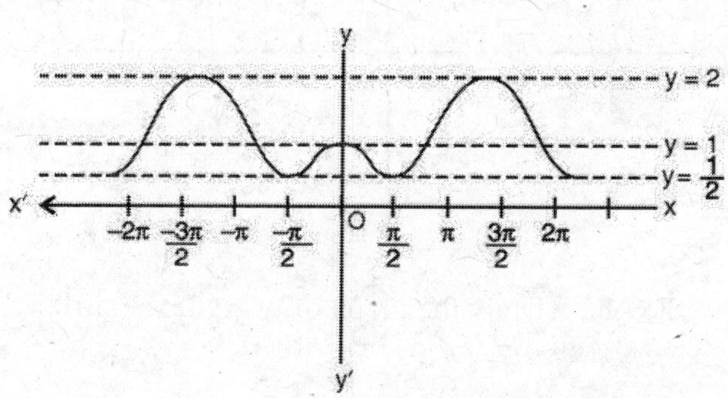

Ex. 5. Draw the graph of $y = 2^{|\cos x|}$

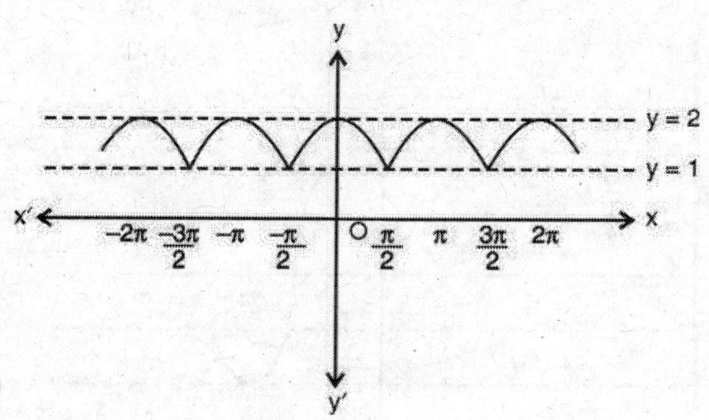

Ex. 6. Draw the graph of $y = 2^{|\sin x|}$

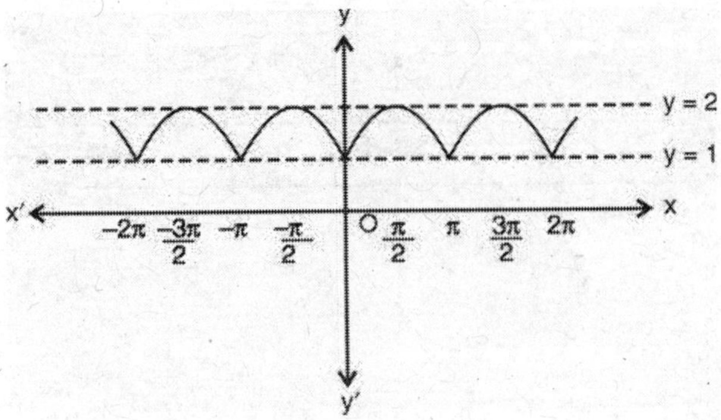

Ex. 7. Draw the graph of $y = \sin x + |\sin x|$

We have $y = \sin x + |\sin x|$

$$= \begin{cases} \sin x + \sin x, \sin x \geq 0 \\ \sin x - \sin x, \sin x < 0 \end{cases}$$

$$= \begin{cases} 2\sin x, x \in (2x\pi, (2n+1)\pi), n \in I \\ 0, x \in ((2n-1)\pi, 2n\pi), n \in II \end{cases}$$

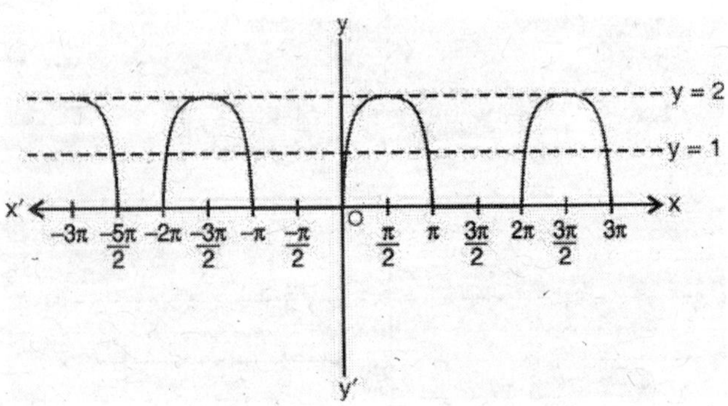

Ex. 8. Draw the graph of $y = \sin x + \sin|x|$

We have $y = \sin x + \sin|x|$

$$= \begin{cases} 2\sin x, x \geq 0 \\ 0, x < 0 \end{cases}$$

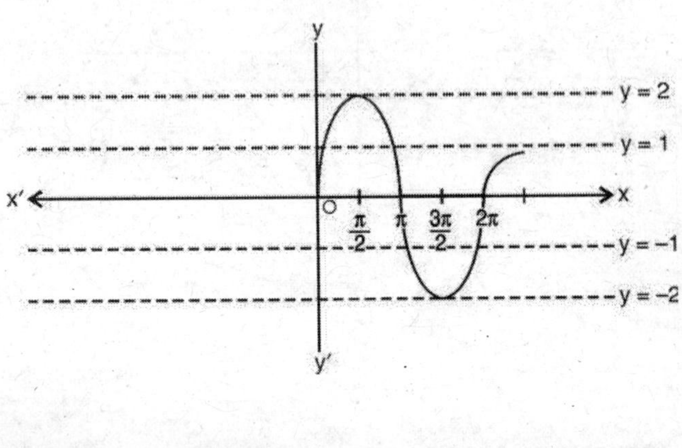

Ex. 9. Draw the graph of $y = \cos x + |\cos x|$

We have $y = \cos x + |\cos x|$

$$= \begin{cases} \cos x + \cos x, \cos x \geq 0 \\ \cos x - \cos x, \cos x < 0 \end{cases}$$

$$= \begin{cases} 2\cos x, x \in \left((2n-1)\frac{\pi}{2}, (2n+1)\frac{\pi}{2}\right) \\ 0, x \in \left((2n+1)\frac{\pi}{2}, (2n+3)\frac{\pi}{2}\right) \end{cases}$$

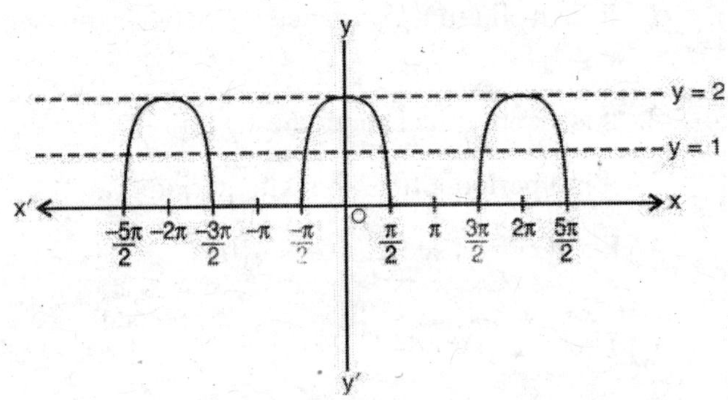

Ex. 10. Draw the graph of $y = \tan x \cdot \cot x$

We have $y = \tan x \cdot \cot x$

$$= 1, x \neq n\pi, (2n+1)\frac{\pi}{2}, \; n \in I$$

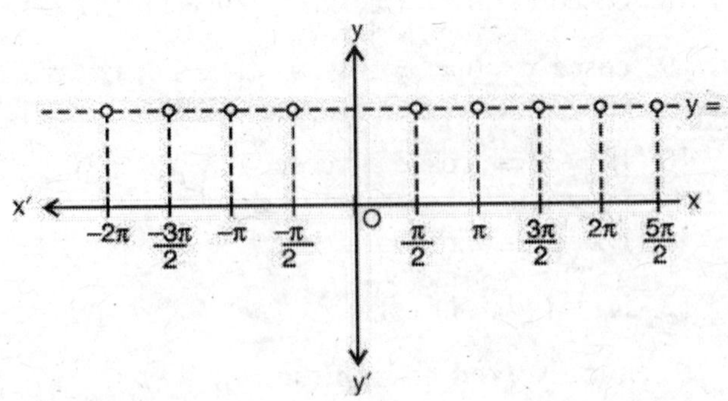

Ex. 11. Draw the graph of $y = \dfrac{|\sin x|}{\sin x}$.

We have $y = \dfrac{|\sin x|}{\sin x}$.

$$= \begin{cases} \dfrac{\sin x}{\sin x} : \sin x \geq 0 \\ -\dfrac{\sin x}{\sin x} : \sin x < 0 \end{cases}$$

$$= \begin{cases} 1 & : x \in (2n\pi, (2n+1)\pi),\ n \in I \\ -1 & : x \in ((2n-1)\pi, 2n\pi),\ n \in I \end{cases}$$

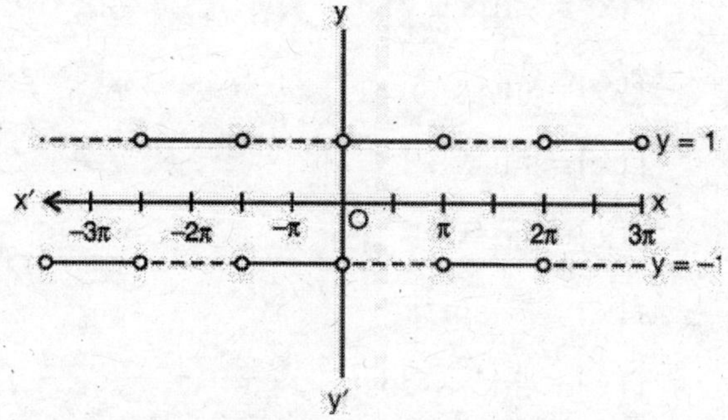

Ex. 12. Draw the graph of $y = 2^x \sin x$.

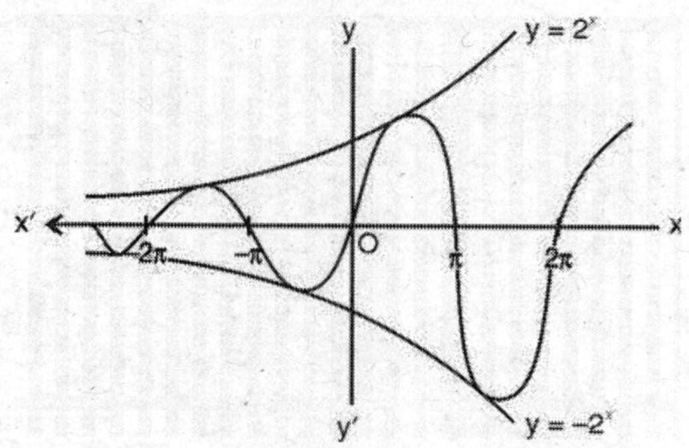

Ex. 13. Draw the graph of $y = \dfrac{\sin x}{x}$.

We have $-1 \le \sin x \le 1$.

$$\Rightarrow \quad -\frac{1}{x} \le \frac{\sin x}{x} \le \frac{1}{x}$$

Also $\quad \lim\limits_{x \to 0} \dfrac{\sin x}{x} = 1.$

Hence, $y = \dfrac{\sin x}{x}$ lies between $y = -\dfrac{1}{x}$ & $y = \dfrac{1}{x}$.

It is now periodic curve & cuts the X-axis at $x = n\pi$,

$n \in I - \{0\}$, and does not cut the y-axis.

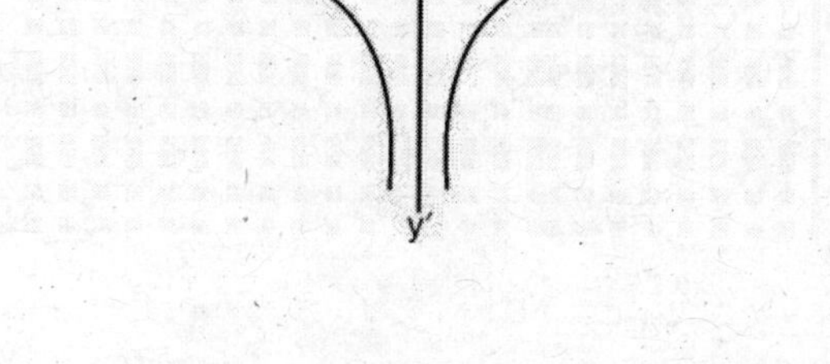

Ex. 14. Draw the graph of $y = \sin\left(\dfrac{1}{x}\right)$.

We have $-1 \le \sin\left(\dfrac{1}{x}\right) \le 1.$

Its cut the n-axis at $x = \dfrac{1}{n\pi}, n \in I - \{0\}$.

But it does not cut y-axis.

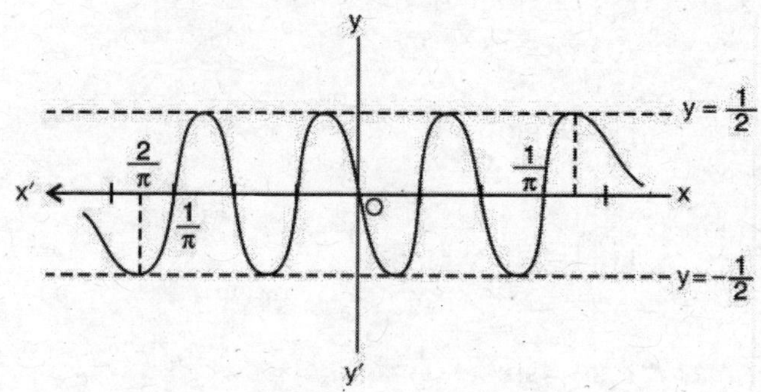

Ex. 15. Draw the graph of $y = x \sin\left(\dfrac{1}{x}\right)$

We have $-1 \le \sin\left(\dfrac{1}{x}\right) \le 1.$

$$\Rightarrow \qquad -x \le x \sin\left(\frac{1}{x}\right) \le x.$$

Hence, the curve $y = x \sin\left(\dfrac{1}{x}\right)$ lies between

the curves $y = -x$ *and* $y = x$.

Also, $\underset{x \to 0^+}{\text{Lt}} \left(x \sin\left(\dfrac{1}{x}\right)\right) = 0 = \underset{x \to 0^-}{\text{Lt}} \left(x \sin\left(\dfrac{1}{x}\right)\right)$.

It cuts the x-axis at $x = \dfrac{1}{n\pi}, n \in I - \{0\}$.

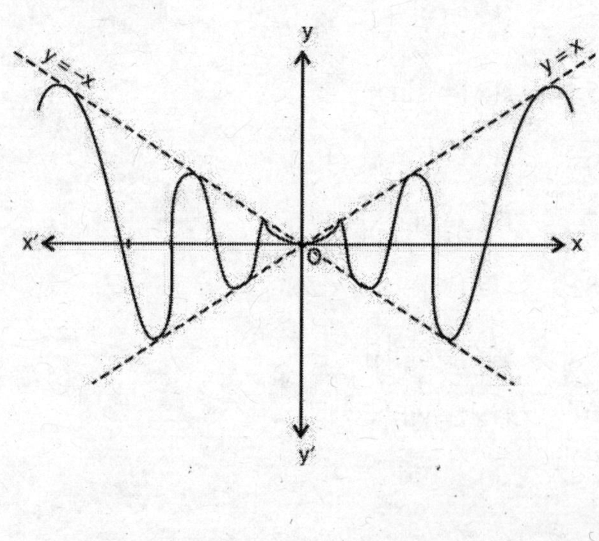

Level– I QUESTIONS BASED ON FUNDAMENTALS

Q. Draw the graphs of

1. $f(x) = \sin 2x$
2. $f(x) = \sin 3x$
3. $f(x) = \sin\left(\frac{x}{2}\right)$
4. $f(x) = \sin\left(\frac{x}{3}\right)$
5. $f(x) = \sin x + \frac{1}{2}$
6. $f(x) = \sin x - \frac{1}{2}$
7. $f(x) = \sin\left(x - \frac{\pi}{6}\right)$
8. $f(x) = \sin\left(x + \frac{\pi}{6}\right)$
9. $f(x) = \sin(x-1)$
10. $f(x) = \sin(x+1)$
11. $f(x) = -\sin x$
12. $f(x) = \frac{1}{2} - \sin x$
13. $f(x) = \sin\left(x - \frac{\pi}{4}\right) + 1$
14. $f(x) = \sin\left(x - \frac{\pi}{4}\right) - 1$
15. $f(x) = |\sin x|$
16. $f(x) = |\sin x| + 1$
17. $f(x) = |\sin x| - 1$
18. $f(x) = 1 - |\sin x|$
19. $f(x) = \sin|x|$
20. $f(x) = \sin|x-1|$
21. $f(x) = \sin|x+1|$
22. $f(x) = \sin^2 x$
23. $f(x) = \sin^2 x + 1$
24. $f(x) = \sin^2 x - 1$
25. $f(x) = 1 - \sin^2 x.$
26. $f(x) = \sin^2 x + \cos^2 x$
27. $f(x) = |\sin|x||$
28. $f(x) = |\sin|x|| + 1$
29. $f(x) = |\sin|x|| - 1$
30. $f(x) = \cos 2x$
31. $f(x) = \cos 4x$
32. $f(x) = \cos\left(\frac{x}{2}\right)$
33. $f(x) = \cos\left(\frac{x}{3}\right)$
34. $f(x) = -\cos x$
35. $f(x) = 1 - \cos x$
36. $f(x) = \cos\left(x - \frac{\pi}{4}\right)$
37. $f(x) = \cos\left(x + \frac{\pi}{4}\right)$
38. $f(x) = \cos(x-1)$
39. $f(x) = \cos(x+1)$
40. $f(x) = \cos\left(x - \frac{\pi}{4}\right) + 1$
41. $f(x) = \cos\left(x - \frac{\pi}{4}\right) - 1$
42. $f(x) = |\cos x|$
43. $f(x) = -|\cos x|$
44. $f(x) = 1 - |\cos x|$

45. $f(x)=|\cos|x||$

46. $f(x)=|\cos(x-1)|$

47. $f(x)=\cos|x|$

48. $f(x)=\cos|x|+1$

49. $f(x)=\cos^2 x$

50. $f(x)=\cos^3 x$

51. $f(x)=\cos^2 x-\sin^2 x$

52. $f(x)=\frac{1}{2}\cos^2 x$

53. $f(x)=\sin x+\cos x$

54. $f(x)=\sin x-\cos x$

55. $f(x)=\tan 2x$

56. $f(x)=\tan^2 x$

57. $f(x)=\tan^3 x$

58. $f(x)=|\tan x|$

59. $f(x)=\cot 2x$

60. $f(x)=|\cot x|$

61. $f(x)=\tan x.\cot x$

62. $f(x)=\frac{1-\cos 2x}{\sin 2x}$

63. $f(x)=\frac{1+\cos 2x}{\sin 2x}$

64. $f(x)=\frac{\sin 2x}{1+\cos 2x}$

65. $f(x)=\frac{\sin 2x}{1-\cos 2x}$

66. $f(x)=\operatorname{cosec} 2x$

67. $f(x)=\sec 2x$

68. $f(x)=\tan x+\cot x$

69. $f(x)=\cot x-\tan x$

70. $f(x)=\sqrt{\sin^2 x}$

71. $f(x)=\sqrt{\cos^2 x}$

72. $f(x)=\sqrt{\sin x}$

73. $f(x)=\sqrt{-\sin x}$

74. $f(x)=\sqrt{-\cos x}$

75. $f(x)=\sqrt{\sin\left(x-\frac{\pi}{4}\right)}$

Level– II FOR JEE-MAIN AND ADVANCED EXAMS ONLY

Q. Draw the graphs of

1. $f(x)=x\sin x$
2. $f(x)=x+\sin x$
3. $f(x)=x-\sin x$
4. $f(x)=2^x.\sin x$
5. $f(x)=e^x.\sin x$
6. $f(x)=\sin x-\cos x$
7. $f(x)=\sin x+\cos x$
8. $f(x)=x\cos x$
9. $f(x)=x+\cos x$
10. $f(x)=x-\cos x$
11. $f(x)=\sqrt{\sin x}$
12. $f(x)=\sqrt{-\sin x}$
13. $f(x)=\sqrt{\sin x}+\sqrt{-\sin x}$
14. $f(x)=\sqrt{\sin x}-\sqrt{-\sin x}$
15. $f(x)=\sin x+|\sin x|$
16. $f(x)=\sin x-|\sin x|$
17. $f(x)=\sin x+\sin|x|$
18. $f(x)=\sin x-\sin|x|$
19. $f(x)=\cos x+|\cos x|$

20. $f(x)=\cos x-|\cos x|$

21. $f(x)=\cos x+\cos|x|$

22. $f(x)=\cos x-\cos|x|$

23. $f(x)=\max\left\{\sin x,\frac{1}{2}\right\}$

24. $f(x)=\max\left\{\sin x,-\frac{1}{2}\right\}$

25. $f(x)=\max\{\sin(x-1),\sin(x+1)\}$

26. $f(x)=\max\left\{\sin\left(x-\frac{\pi}{4}\right),\sin\left(x+\frac{\pi}{4}\right)\right\}$

27. $f(x)=\tan x+\tan|x|$

28. $f(x)=\tan x-\tan|x|$

29. $f(x)=\tan x.\cot x$

30. $f(x)=\sin x.\text{cosec}\, x$

31. $f(x)=\cos x.\sec x$

32. Find the number of solutions of

(i) $\sin x=x$

(ii) $\sin x=|x|$

(iii) $\sin x=2|x|$

(iv) $\sin x=\frac{1}{2}|x|$

(v) $\sin x=x+\frac{1}{x},\ x>0$

(vi) $\sin x=x+\frac{1}{x},\ x<0$

(vii) $\sin x=x^2+x+1$

(viii) $\sin x=-x^2+x-1$

(ix) $\sin x=2^x+2^{-x}$

(x) $\sin x=2^x-2^{-x}$

(xi) $2^{\sin x}=\frac{3}{4},\ \forall\ x\in[-2\pi,2\pi]$

(xii) $2^{\cos x}=\frac{3}{4},\ \forall\ x\in[-2\pi,2\pi]$

(xiii) $2^{|\sin x|}=\frac{5}{6},\ \forall\ x\in[-2\pi,2\pi]$

(xiv) $2^{|\cos x|}=1,\ \forall\ x\in[-2\pi,2\pi]$

(xv) $3^{\sin x}=\frac{1}{2},\ \forall\ x\in[-2\pi,2\pi]$

(xvi) $2^{\sin x}=|\sin x|,\ \forall\ x\in[-2\pi,2\pi]$

(xvii) $2^{\cos x}=|\sin x|,\ \forall\ x\in[-2\pi,2\pi]$

(xviii) $2^{\cos x}=|\cos x|,\ \forall\ x\in[-2\pi,2\pi]$

33. Find the max and minimum values of

(i) $f(x)=\sin x+\cos x$

(ii) $f(x)=\sin x-\cos x$

(iii) $f(x)=3\sin x+4\cos x+10$

(iv) $f(x)=5\sin x+12\cos x+12$

(v) $f(x)=\sin(x-1)+\cos x$

(vi) $f(x)=\sin x+\cos(x-1)$

(vii) $f(x)=3\sin x+4$

(viii) $f(x)=2-4\sin x$

(ix) $f(x)=-3-2\sin x$

(x) $f(x)=3\cos x+5$

(xi) $f(x)=-2\cos x+3$

(xii) $f(x)=3\sin^2 x+4$

(xiii) $f(x)=2-4\sin^2 x$

(xiv) $f(x)=3\sin^2 x+4\cos^2 x$

(xv) $f(x)=2\sin^2 x-3\cos^2 x$

(xvi) $f(x)=3\sin^2 x+4\cos^2 x+10\sin x\cos x$

(xvii) $f(x)=\frac{1}{3\sin^2 x+2\cos^2 x}$

(xviii) $f(x)=\frac{1}{3\sin^2 x-2\cos^2 x}$

(xix) $f(x)=2^{5\sin^2 x+6\cos^2 x}$

(xx) $f(x)=\log_2(3\sin^2 x+1)$.

34. Find the domains and ranges of

(i) $f(x)=2\sin x+4$

(ii) $f(x)=\sin x+\cos x$

(iii) $f(x)=\sin x+\cos x+1$

(iv) $f(x)=3\sin x+4\cos x+10$

(v) $f(x)=2\sin\left(x-\frac{\pi}{4}\right)+\cos x$

(vi) $f(x)=\sin\left(x-\frac{\pi}{4}\right)+\cos\left(x-\frac{\pi}{6}\right)$

(vii) $f(x)=\sin\left(x-\frac{\pi}{4}\right)+\cos\left(x-\frac{\pi}{4}\right)$

(viii) $f(x)=3\sin^2 x+2$

(ix) $f(x)=3\sin^2 x+2\cos^2 x$

(x) $f(x)=3\sin^2 x-4\cos^2 x$.

ANSWER

32. (i) 1 (ii) 1 (iii) 1 (iv) 3 (v) 0
(vi) 0 (vii) 0 (viii) 0 (ix) 0 (x) 3
(xi) 4 (xii) 4 (xiii) 0 (xiv) 4 (xv) 4
(xvi) 4 (xvii) 8 (xviii) 4

33. (i) Max $v=\sqrt{2}$, Min $v=-\sqrt{2}$

(ii) Max $v=\sqrt{2}$, Min $v=-\sqrt{2}$

(iii) Max $v=15$, Min $v=5$

(iv) Max $v=25$, Min $v=-1$

(v) Max $v=\sqrt{2}\left(\cos\frac{1}{2}-\sin\frac{1}{2}\right)$

Max $v=-\sqrt{2}\left(\cos\frac{1}{2}-\sin\frac{1}{2}\right)$

(vi) Max $v=\sqrt{2}\left(\cos\frac{1}{2}+\sin\frac{1}{2}\right)$

Max $v=-\sqrt{2}\left(\cos\frac{1}{2}+\sin\frac{1}{2}\right)$

(vii) Max $v=7$, Min $v=1$

(viii) Max $v=6$, Min $v=-2$

(ix) Max $v=-1$, Min $v=-5$

(x) Max $v=8$, Min $v=2$

(xi) Max $v=5$, Min $v=+1$

(xii) Max $v=7$, Min $v=4$

(xiii) Max $v=2$, Min $v=-2$

(xiv) Max $v=4$, Min $v=3$

(xv) Max $v=2$, Min $v=-3$

(xvi) Max $v=\frac{13}{2}+\frac{\sqrt{101}}{2}$,

Min $v=\frac{13}{2}-\frac{\sqrt{101}}{2}$.

(xvii) Max $v=\frac{1}{2}$, Min $v=\frac{1}{3}$

(xviii) Max $v=\frac{1}{3}$, Min $v=-\frac{1}{2}$

(xix) Max $v=2^6$, Min $v=2^5$

(xx) Max $v=2$, Min $v=0$.

34. (i) $_f=R$; $R_f=[2, 6]$

(ii) $D_f=R$; $R_f=\left[-\sqrt{2},\sqrt{2}\right]$

(iii) $D_f=R$; $R_f=\left[1-\sqrt{2},\sqrt{2}+1\right]$

(iv) $D_f=R$; $R_f=[5, 15]$

(v) $D_f=R$; $R_f=\left[-\sqrt{5-2\sqrt{2}},\sqrt{5+2\sqrt{2}}\right]$

(vi) $D_f=R$;

$R_f=\left[-\sqrt{5+\sqrt{2}+\sqrt{6}},\sqrt{5+\sqrt{2}+\sqrt{6}}\right]$

(vii) $D_f=R$; $R_f=\left[-\sqrt{2},\sqrt{2}\right]$

(viii) $D_f=R$; $R_f=[2, 5]$

(ix) $D_f=R$; $R_f=[2, 3]$

(x) $D_f=R$; $R_f=[-4, 3]$

Level– III FOR JEE - ADVANCED EXAM ONLY

Q. Draw the graphs of

1. $f(x)=\sin\left(\frac{1}{x}\right)$

2. $f(x)=x\sin\left(\frac{1}{x}\right)$

3. $f(x)=\sin x+|x|$

4. $f(x)=\sin x-|x|$

5. $f(x)=|x|+\sin|x|$

6. $f(x)=|x|-\sin|x|$

7. $f(x)=|x|+\cos|x|$

8. $f(x)=|x|-\cos|x|$

9. $f(x)=\log_2(\sin x)$

10. $f(x)=\log_2(\cos x)$

11. $f(x)=[\sin x]$, where [,] = G.I.F

12. $f(x)=[2\sin x]$, [,]=G.I.F

13. $f(x)=[\sin x+\cos x]$, [,] = G.I.F

14. $f(x)=\max\{\sin x, \cos x\}$ $,\forall\, x\in[-2\pi, 2\pi]$

15. $f(x)=\max\left\{\sin x, \frac{1}{2}, \cos x\right\}$ $,\forall\, x\in[-2\pi, 2\pi]$

16. $f(x)=\frac{\sin x+\cos x}{2}-\left|\frac{\sin x-\cos x}{2}\right|$ $,\forall\, x\in[-2\pi, 2\pi]$

17. $f(x)=\frac{\sin x+\cos x}{2}+\left|\frac{\sin x-\cos x}{2}\right|$ $,\forall\, x\in[-2\pi, 2\pi]$

18. $f(x)=x\sin\left(\frac{1}{x}\right)$

19. $f(x)=\operatorname{sgn}(\sin x)$

20. $f(x)=\operatorname{sgn}(\cos x)$

21. Find the max and minimum values of

(i) $f(x)=\sin^2 x+\cos^4 x$

(ii) $f(x)=\sin^4 x+\cos^2 x$

(iii) $f(x)=\sin^4 x+\cos^4 x$

(iv) $f(x)=\sin^6 x+\cos^6 x$

22. Find the number of solutions of

(i) $2^{\sin x}=\sin x$, $\forall\, x\in[0,4\pi]$

(ii) $2^{\sin x}=|\sin x|$ $\forall\, x\in[0,4\pi]$

(iii) $2^{\cos x}=|\sin x|$ $\forall\, x\in[0,8\pi]$

(iv) $2^{\cos x}=|\cos x|$ $\forall\, x\in[0,10\pi]$

(v) $2^{\cos x}=\cos x$, $\forall\, x\in[0,20\pi]$

(vi) $x\sin x-1=0$, $\forall\, x\in[0,2\pi]$

(vii) $x^2\sin x-1=0$, $\forall\, x\in[0,2\pi]$

(viii) $\sin x=2^x+2^{-x}$, $\forall\, x\in[0,2\pi]$

(ix) $\cos x=2^x-2^{-x}$, $\forall\, x\in[0,2\pi]$

(x) $x^2\cos x-2=0, \forall\, x\in[0,2\pi]$

ANSWER

21. (i) Max $v = 1$, Min $v = \frac{3}{4}$

(ii) Max $v = 1$, Min $v = \frac{3}{4}$

(iii) Max $v = 1$, Min $v = \frac{1}{2}$

(iv) Max $v = 1$, Min $v = \frac{1}{4}$

22. (i) 0 (ii) 4 (iii) 16

(iv) 10 (v) 0

(vi) 2 (vii) 2 (viii) 0

(ix) 1 (x) 3

CHAPTER

3 TRIGONOMETRIC EQUATIONS

2.1 DEFINITION :

An equation involving one or more trigonometrical ratios of unknown angle is called trigonometrical equation..

For examples, $\sin^2 x - \cos x - 2$, $\tan x = 1$, $\cos^2 x + \cos x - 2 = 0$ etc. are trigonomteric equations.

2.2 SOLUTION OF A TRIGONOMETRIC EQUATION :

A value of the unknown angle which satisfies the given trigonometrical equation is called a solution or root of the equation

For example, $\sin\theta = \frac{1}{2} \Rightarrow \theta = \frac{\pi}{6}, \frac{5\pi}{6}$.

Types of Solutions :

(i) Principal Solution

(ii) General Solution.

Principal Solution :

The smallest numerical value of the angle which satisfies the given equation is called the principal solutions.

General Solutions :

Since trigonometric functions are periodic function, solution of trigonometric equations can be generalised with the help of periodicity of a trigonometrical function. The solution consisting of all possible solution of a trigonometrical equation is called the general solution.

2.3 GENERAL SOLUTION OF TRIGONOMETRIC EQUATIONS :

The general solution of the equations are

Step - I

1. $\sin\theta = 0 \quad \Rightarrow \quad \theta = n\pi, n \in Z.$

2. $\cos\theta = 0 \quad \Rightarrow \quad \theta = (2n+1)\frac{\pi}{2}, n \in Z.$

3. $\tan\theta = 0 \quad \Rightarrow \quad \theta = n\pi, n \in Z.$

Step- II

1. $\sin\theta = \sin\alpha \quad \Rightarrow \quad \theta = n\pi + (-1)^n\alpha, n \in Z$

2. $\cos\theta = \cos\alpha \quad \Rightarrow \quad \theta = 2n\pi \pm \alpha, n \in Z.$

3. $\tan\theta = \tan\alpha \quad \Rightarrow \quad \theta = n\pi + \alpha,, n \in Z.$

Step-III

1. $\sin^2\theta = \sin^2\alpha \quad \Rightarrow \quad \theta = n\pi \pm \alpha,, n \in Z.$

2. $\cos^2\theta = \cos^2\alpha \quad \Rightarrow \quad \theta = n\pi \pm \alpha,, n \in Z.$

3. $\tan^2\theta = \tan^2\alpha \quad \Rightarrow \quad \theta = n\pi \pm \alpha, n \in Z.$

2. 4. SOME SOLVED EXAMPLES :

Ex-1. Solve for θ : $\sin 3\theta = 0$

Soln. We have $\sin 3\theta = 0 = 0$

$\Rightarrow \quad 3\theta = n\pi$

$\Rightarrow \quad \theta = \frac{n\pi}{3}$, where $n \in I$

Ex-2. Solve for θ : $\cos^2(5\theta) = 0$

Soln. We have $\cos^2(5\theta) = 0$

$\Rightarrow \quad \cos^2(5\theta) = \cos^2\left(\frac{\pi}{2}\right)$

$\Rightarrow \quad (5\theta) = n\pi \pm \left(\frac{\pi}{2}\right)$

$\Rightarrow \quad \theta = \frac{1}{5}\left(n\pi \pm \left(\frac{\pi}{2}\right)\right)$, where $n \in I$

Ex-3. Solve for θ : $\tan\theta = \sqrt{3}$

Soln. We have , $\tan\theta = \sqrt{3}$

$$\Rightarrow \tan\theta = \tan\left(\frac{\pi}{3}\right)$$

$$\Rightarrow \theta = n\pi + \left(\frac{\pi}{3}\right), \text{ where } n \in I$$

Ex-4. Solve for θ : $\sin 2\theta = \sin\theta$

Soln. We have, $\sin 2\theta = \sin\theta$

$$\Rightarrow 2\sin\theta\cos\theta = \sin\theta$$

$$\Rightarrow \sin\theta(2\cos\theta - 1) = 0$$

$$\Rightarrow \sin\theta = 0 \ \& \ (2\cos\theta - 1) = 0$$

$$\Rightarrow \sin\theta = 0 \ \& \ \cos\theta = \frac{1}{2}$$

$$\Rightarrow \theta = n\pi \ \& \ \theta = 2n\pi \pm \frac{\pi}{3}, \text{where } n \in I$$

Ex-5. Solve for θ : $\sin(9\theta) = \sin\theta$

Soln. We have, $\sin(9\theta) = \sin\theta$

$$\Rightarrow \sin(9\theta) - \sin\theta = 0$$

$$\Rightarrow 2\cos\left(\frac{9\theta + \theta}{2}\right)\sin\left(\frac{9\theta - \theta}{2}\right) = 0$$

$$\Rightarrow 2\cos(5\theta)\sin(4\theta) = 0$$

$$\Rightarrow \cos(5\theta) = 0 \ \& \ \sin(4\theta) = 0$$

$$\Rightarrow (5\theta) = (2n+1)\frac{\pi}{2} \ \& \ (4\theta) = n\pi$$

$$\Rightarrow \theta = (2n+1)\frac{\pi}{10} \ \& \ \theta = \left(\frac{n\pi}{4}\right)$$

where $n \in I$

Ex-6. Solve for θ : $5\sin^2\theta + 3\cos^2\theta = 4$

Soln. We have, $5\sin^2\theta + 3\cos^2\theta = 4$

$$\Rightarrow 2\sin^2\theta + 3(\sin^2\theta + \cos^2\theta) = 4$$

$$\Rightarrow 2\sin^2\theta + 3 = 4$$

$$\Rightarrow 2\sin^2\theta = 1$$

$$\Rightarrow \sin^2\theta = \frac{1}{2}$$

$$\Rightarrow \sin^2\theta = \left(\frac{1}{\sqrt{2}}\right)^2 = \sin^2\left(\frac{\pi}{4}\right)$$

$$\Rightarrow \theta = n\pi \pm \left(\frac{\pi}{4}\right), \text{ where } n \in I$$

Ex-7. Solve for θ :

$$3\tan(\theta - 15^\circ) = \tan(\theta + 15^\circ)$$

Soln. We have, $3\tan(\theta - 15^\circ) = \tan(\theta + 15^\circ)$

$$\Rightarrow \frac{\tan(\theta - 15^\circ)}{\tan(\theta + 15^\circ)} = \frac{3}{1}$$

$$\Rightarrow \frac{\tan(\theta - 15^\circ) + \tan(\theta + 15^\circ)}{\tan(\theta - 15^\circ) - \tan(\theta + 15^\circ)} = \frac{3+1}{3-1}$$

$$\Rightarrow \frac{\sin(\theta + 15^\circ + \theta - 15^\circ)}{\sin(\theta + 15^\circ - \theta + 15^\circ)} = \frac{3+1}{3-1}$$

$$\Rightarrow 2\sin(2\theta) = 2$$

$$\Rightarrow \sin(2\theta) = 1$$

$$\Rightarrow 2\theta = n\pi + (-1)^n\left(\frac{\pi}{2}\right)$$

$$\Rightarrow \theta = \frac{1}{2}\left(n\pi + (-1)^n\left(\frac{\pi}{2}\right)\right), \text{ where}$$

$n \in I$

Ex-8. Solve for θ : $\tan^2(\theta) + \cot^2(\theta) = 2$

Soln. We have, $\tan^2(\theta) + \cot^2(\theta) = 2$

$$\Rightarrow \tan^2(\theta) + \frac{1}{\tan^2(\theta)} = 2$$

$$\Rightarrow \quad \tan^4(\theta) - 2\tan^2(\theta) + 1 = 0$$

$$\Rightarrow \quad \left(\tan^2(\theta) - 1\right)^2 = 0$$

$$\Rightarrow \quad \left(\tan^2(\theta) - 1\right) = 0$$

$$\Rightarrow \quad \theta = n\pi \pm \left(\frac{\pi}{4}\right), \; n \in I$$

Ex-9. Solve for θ :

$$\cos(\theta) + \cos(2\theta) + \cos(3\theta) = 0$$

Soln. We have

$$(\cos(3\theta) + \cos(\theta)) + \cos(2\theta) = 0$$

$$\Rightarrow \quad (\cos(3\theta) + \cos(\theta)) + \cos(2\theta) = 0$$

$$\Rightarrow \quad 2\cos(2\theta)\cos(\theta) + \cos(2\theta) = 0$$

$$\Rightarrow \quad \cos(2\theta)(2\cos(\theta) + 1) = 0$$

$$\Rightarrow \quad \cos(2\theta) = 0 \;\&\; (2\cos(\theta) + 1) = 0$$

$$\Rightarrow \quad \cos(2\theta) = 0 \;\&\; \cos(\theta) = -\frac{1}{2}$$

$$\Rightarrow \quad \theta = (2n+1)\left(\frac{\pi}{4}\right) \;\&\; \theta = n\pi \pm \left(\frac{2\pi}{3}\right)$$

Ex-10. Solve for θ :

$$\sin(2\theta) + \sin(4\theta) + \sin(6\theta) = 0$$

Soln. We have $\sin(2\theta) + \sin(4\theta) + \sin(6\theta) = 0$

$$\Rightarrow \quad \sin(6\theta) + \sin(2\theta) + \sin(4\theta) = 0$$

$$\Rightarrow \quad 2\sin(4\theta).\cos(2\theta) + \sin(4\theta) = 0$$

$$\Rightarrow \quad \sin(4\theta)(2\cos(2\theta) + 1) = 0$$

$$\Rightarrow \quad \sin(4\theta) = 0 \;\&\; (2\cos(2\theta) + 1) = 0$$

$$\Rightarrow \quad (4\theta) = n\pi \;\&\; \cos(2\theta) = -\frac{1}{2}$$

$$\Rightarrow \quad \theta = \left(\frac{n\pi}{4}\right) \;\&\; (2\theta) = n\pi \pm \left(\frac{2\pi}{3}\right)$$

$$\Rightarrow \quad \theta = \left(\frac{n\pi}{4}\right) \;\&\; \theta = \left(\frac{n\pi}{2}\right) \pm \left(\frac{\pi}{3}\right),$$

where $n \in I$

Ex-11. Solve for θ :

$$\tan(\theta) + \tan(2\theta) + \tan(\theta)\tan(2\theta) = 1$$

Soln. We have

$$\tan(\theta) + \tan(2\theta) + \tan(\theta)\tan(2\theta) = 1$$

$$\Rightarrow \quad \tan(2\theta) + \tan(\theta) = 1 - \tan(\theta)\tan(2\theta)$$

$$\Rightarrow \quad \frac{\tan(2\theta) + \tan(\theta)}{1 - \tan(\theta)\tan(2\theta)} = 1$$

$$\Rightarrow \quad \tan(3\theta) = 1$$

$$\Rightarrow \quad \tan(3\theta) = \tan\left(\frac{\pi}{4}\right)$$

$$\Rightarrow \quad (3\theta) = n\pi + \left(\frac{\pi}{4}\right)$$

$$\Rightarrow \quad \theta = \left(\frac{n\pi}{3}\right) + \left(\frac{\pi}{12}\right), \; n \in I$$

Ex-12. Solve for θ : $\tan(\theta) + \tan(2\theta) + \tan(3\theta) = \tan(\theta).\tan(2\theta).\tan(3\theta)$

Soln. We have $\tan(\theta) + \tan(2\theta) + \tan(3\theta) = \tan(\theta).\tan(2\theta).\tan(3\theta)$

$$\Rightarrow \quad \tan(\theta) + \tan(2\theta) = -\tan(3\theta) + \tan(\theta).\tan(2\theta).\tan(3\theta)$$

$$\Rightarrow \quad \tan(\theta) + \tan(2\theta) = -\tan(3\theta)(1 - \tan(\theta).\tan(2\theta))$$

$$\Rightarrow \quad \left(\frac{\tan(\theta) + \tan(2\theta)}{(1 - \tan(\theta).\tan(2\theta))}\right) = -\tan(3\theta)$$

$$\Rightarrow \quad \tan(3\theta) = -\tan(3\theta)$$

$$\Rightarrow \quad 2\tan(3\theta) = 0$$

$\Rightarrow \quad (3\theta) = n\pi$

$\Rightarrow \quad \theta = \left(\frac{n\pi}{3}\right),\ n \in I$

Exercise - 1

Solve the general values of θ :

1. (i) $\sin 2\theta = 0$

(ii) $\cos 3\theta = 0$

(iii) $\tan 5\theta = 0$

(iv) $\sin(5\theta/2) = 0$

(v) $\cos(7\theta/2) = 0$

(vi) $\tan(7\theta/3) = 0$

(vii) ... $^2 7\theta = 0$

(viii) $\cos^2 3\theta = 0.$

2. (i) $\sin\theta = \frac{\sqrt{3}}{2}$

(ii) $\cos\theta = \frac{1}{2}$

(iii) $\tan\theta = \sqrt{2}$

(iv) $2\sin\theta + 1 = 0$

(v)` $\cos 3\theta = -\frac{1}{\sqrt{2}}$

(vi) $\tan 3\theta + 1 = 0$

(vii) $2\sin\theta - 1 = 0$

(viii) $2\cos\theta - \sqrt{3} = 0$

(ix) $\sqrt{3}\tan\theta - 1 = 0$

(x) $\sec(\theta) + 2 = 0$

3. (i) $\sin 3\theta = \sin\theta$

(ii) $\sin 5\theta = \cos 2\theta$

(iii) $7\cos^2\theta + 3\sin^2\theta = 4.$

4. (i) $\sin\theta + \sin 3\theta + \sin 5\theta = 0.$

(ii) $\cos\theta + \cos 3\theta - 2\cos 2\theta = 0$

(iii) $\sin 2\theta + \sin 4\theta + \sin 6\theta = 0.$

5. (i) $\cot^2\theta + \frac{3}{\sin\theta} + 3 = 0$

(ii) $2\tan\theta - \cot\theta = -1$

(iii) $\tan^2\theta + (1-\sqrt{3})\tan\theta - \sqrt{3} = 0$

6. (i) $\tan\theta + \tan 2\theta + \tan\theta\tan 2\theta = 1$

(ii) $\tan\theta + \tan 2\theta + \tan 3\theta$

$= \tan\theta\tan 2\theta\tan 3\theta$

(iii) $\tan\theta + \tan\left(\theta + \frac{\pi}{3}\right)$

$+ \tan\left(\theta + \frac{2\pi}{3}\right) = 3$

(iv) $3\tan(\theta - 15^o) = \tan(\theta + 15^o)$

7. (i) $\tan\theta + \tan 2\theta + \tan 3\theta = 0$

(ii) $\cos 2\theta . \cos 4\theta = 1/2$

(iii) $\cot\theta - \tan\theta = \cos\theta - \sin\theta$

(iv) $(1 - \tan\theta)(1 + \sin 2\theta) = 1 + \tan\theta.$

8. (i) $2\sin^2\theta + \sin^2 2\theta = 2$

(ii) $\sin 3\theta$

$= 4\sin\theta in\theta.si + \alpha).\sin(\theta - \alpha)$

$, \alpha \neq n\pi\pi n \in Z$

(iii) $4\sin\theta\sin 2\theta\sin 4\theta = \sin 3\theta.$

Answers

3 .(i) $\theta = n\pi,\ n\pi \pm \frac{\pi}{3},\ n \in I$

(ii) $\theta = \frac{2n\pi}{7} \pm \frac{\pi}{14},\ n \in I$

(iii) $\theta = n\pi \pm \frac{\pi}{3},\ n \in I$

4. (i) $\theta = \frac{n\pi}{3},\ 2n\pi \pm \frac{2\pi}{3},\ n \in I$

(ii) $\theta = (2n+1)\frac{\pi}{4},\ 2n\pi\ n \in I$

(iii) $\theta = \frac{n\pi}{4},\ n\pi \pm \frac{\pi}{3},\ n \in I$

5. (i) $\theta = (4n-1)\frac{\pi}{2},\ n\pi + (-1)^n \frac{\pi}{6},\ n \in I$

(ii) $\theta = n\pi + \tan^{-1}\left(\frac{1}{2}\right),\ n\pi + \frac{3\pi}{4},\ n \in I$

(iii) $\theta = n\pi + \frac{\pi}{3},\ n\pi + \frac{3\pi}{4},\ n \in I$

6. (i) $\theta = \frac{n\pi}{3} + \frac{\pi}{12},\ n \in I$

(ii) $\theta = \frac{n\pi}{3},\ n \in I$

(iii) $\theta = \frac{n\pi}{3} + \frac{\pi}{12},\ n \in I$

(iv) $\theta = \frac{n\pi}{2} + (-1)^n\left(\frac{\pi}{4}\right),\ n \in I$

7. (i) $\theta = n\pi,\ n\pi \pm \frac{\pi}{3}$

$,n\pi \pm \tan^{-1}\left(\frac{1}{\sqrt{2}}\right),\ n \in I$

(ii) $\theta = \frac{n\pi}{6},\ (2n+1)\frac{\pi}{4}$

$(2n+1)\frac{\pi}{12},\ n \in I$

(iii) $\theta = 2n\pi + \frac{\pi}{2},\ 2n\pi,\ n \in I$

(iv) $\theta = n\pi - \frac{\pi}{4},\ n\pi,\ n \in I$

8. (i) $\theta = (2n+1)\frac{\pi}{2},\ n \in I$

(ii) $\theta = n\pi + \frac{\pi}{3},\ n \in I$

(iii) $\theta = n\pi + \frac{\pi}{3},\ n \in I$

2.5 EQUATIONS OF THE FORM $a\cos\theta \pm b\sin\theta = c$

Rule : 1. Divide by $\sqrt{a^2+b^2}$ on both the sides

2. Reduce the given equation into either $\sin(\theta \pm \alpha)$ or $\cos(\theta \pm \alpha)$

3. simplify the given equation.

Ex-1. Solve for θ : $\sin(\theta)+\cos(\theta)=1$

Soln. We have $\sin(\theta)+\cos(\theta)=1$

$$\Rightarrow \sqrt{2}\left(\frac{1}{\sqrt{2}}\sin(\theta)+\frac{1}{\sqrt{2}}\cos(\theta)\right)=1$$

$$\Rightarrow \left(\frac{1}{\sqrt{2}}\sin(\theta)+\frac{1}{\sqrt{2}}\cos(\theta)\right)=\frac{1}{\sqrt{2}}$$

$$\Rightarrow \sin\left(\theta+\frac{\pi}{4}\right)=\frac{1}{\sqrt{2}}$$

$$\Rightarrow \sin\left(\theta+\frac{\pi}{4}\right)=\left(\sin\left(\frac{\pi}{4}\right)\right)$$

$$\Rightarrow \left(\theta+\frac{\pi}{4}\right)=\left(n\pi+(-1)^n\left(\frac{\pi}{4}\right)\right)$$

$$\Rightarrow \theta=\left(n\pi+(-1)^n\left(\frac{\pi}{4}\right)-\frac{\pi}{4}\right),\ n \in I$$

Ex-2. Solve for θ : $\sqrt{3}\sin(\theta)+\cos(\theta)=2$

Soln. We have $\sqrt{3}\sin(\theta)+\cos(\theta)=2$

$$\Rightarrow \frac{\sqrt{3}}{2}\sin\theta+\frac{1}{2}\cos\theta=1$$

$$\Rightarrow \sin\left(\theta+\frac{\pi}{6}\right)=1$$

$\Rightarrow \quad \sin\left(\theta+\frac{\pi}{6}\right)=\sin\left(\frac{\pi}{2}\right)$

$\Rightarrow \quad \left(\theta+\frac{\pi}{6}\right)=n\pi+(-1)^n\left(\frac{\pi}{2}\right)$

$\Rightarrow \quad \theta=n\pi+(-1)^n\left(\frac{\pi}{2}\right)-\frac{\pi}{6}$

Ex-3. Solve for θ :

$\sin(2\theta)+\cos(2\theta)+\sin(\theta)+\cos(\theta)+1=0$

Soln. We have

$\sin(2\theta)+\cos(2\theta)+\sin(\theta)+\cos(\theta)+1=0$

$\Rightarrow \quad (\sin(\theta)+\cos(\theta))+(1+\sin(2\theta))$

$+\cos(2\theta)=0$

$\Rightarrow \quad (\sin(\theta)+\cos(\theta))+(\sin(\theta)+\cos(\theta))^2$

$+\left(\cos^2\theta-\sin^2\theta\right)=0.$

$\Rightarrow \quad (\sin(\theta)+\cos(\theta))+(\sin(\theta)+\cos(\theta))^2$

$+(\cos\theta+\sin\theta)(\cos\theta-\sin\theta)=0$

$\Rightarrow \quad (\sin(\theta)+\cos(\theta))$

$(1+(\sin(\theta)+\cos(\theta))+(\cos\theta-\sin\theta))=0$

$\Rightarrow \quad (\sin(\theta)+\cos(\theta))(1+2\cos\theta)=0$

$\Rightarrow \quad (\sin(\theta)+\cos(\theta))=0 \ \& \ (1+2\cos\theta)=0$

$\Rightarrow \quad \left(\sin\left(\frac{\pi}{4}+\theta\right)\right)=0 \ \& \ \cos\theta=-\frac{1}{2}$

$\Rightarrow \quad \left(\frac{\pi}{4}+\theta\right)=n\pi \ \& \ \theta=2n\pi\pm\left(\frac{2\pi}{3}\right)$

$\Rightarrow \quad \theta=n\pi-\frac{\pi}{4} \ \& \ \theta=2n\pi\pm\left(\frac{2\pi}{3}\right)$, where $n\in I$

Ex-4. Solve for $\theta:\sin^3\theta+\sin\theta\cos\theta+\cos^3\theta=1$

Soln. We have $\sin^3\theta+\sin\theta\cos\theta+\cos^3\theta=1$

$\Rightarrow \quad \left(\sin^3\theta+\cos^3\theta\right)+\sin\theta\cos\theta=1$

$\Rightarrow \quad (\sin\theta+\cos\theta)(1-\sin\theta\cos\theta)+\sin\theta\cos\theta=1$

$\Rightarrow \quad (\sin\theta+\cos\theta)(1-\sin\theta\cos\theta)=(1-\sin\theta\cos\theta)$

$\Rightarrow \quad (\sin\theta+\cos\theta-1)(1-\sin\theta\cos\theta)=0$

$\Rightarrow \quad (\sin\theta+\cos\theta-1)=0 \ \& \ (1-\sin\theta\cos\theta)=0$

$\Rightarrow \quad (\sin\theta+\cos\theta)=1 \ \& \ \sin(2\theta)=\frac{1}{2}$

$\Rightarrow \quad \left(\sin\left(\theta+\frac{\pi}{4}\right)\right)=\sin\left(\frac{\pi}{2}\right)$

$\& \ \sin(2\theta)=\sin\left(\frac{\pi}{6}\right)$

$\Rightarrow \quad \theta=n\pi+(-1)^n\left(\frac{\pi}{2}\right)-\left(\frac{\pi}{4}\right)$

$\& \ \theta=\frac{1}{2}\left(n\pi+(-1)^n\left(\frac{\pi}{6}\right)\right)$

where $n\in I$

Exercise - 2

1. $\sin\theta+\sqrt{3}\cos\theta=\sqrt{2}$
2. $\sqrt{2}\sec\theta+\tan\theta=1$
3. $\cot\theta+\operatorname{cosec}\theta=\sqrt{3}$
4. $\sin\theta+\cos\theta=\sqrt{2}$
5. $\sqrt{3}\cos\theta+\sin\theta=1$
6. $\sin\theta+\cos\theta=1$
7. $\operatorname{cosec}\theta=1+\cot\theta$
8. $\tan\theta+\sec\theta=\sqrt{3}$
9. $\cos\theta+\sqrt{3}\sin\theta=2\cos 2\theta$
10. $\sqrt{3}(\cos\theta-\sqrt{3}\sin\theta)=4\sin 2\theta.\cos 3\theta$.

Answers

1. $\theta=n\pi+(-1)^n.\frac{\pi}{4}-\frac{\pi}{3}, n\in I.$
2. $\theta=2n\pi-\frac{\pi}{4}, n\in I.$
3. $\theta=(4n+1)\frac{\pi}{2}-\frac{\pi}{6}, n\in I.$
4. $\theta=(4n+1)\frac{\pi}{2}-\frac{\pi}{4}, n\in I.$
5. $\theta=2n\pi\pm\frac{\pi}{3}+\frac{\pi}{6}, n\in I.$

6. $\theta = n\pi + (-1)^n.\frac{\pi}{4} - \frac{\pi}{4}, n \in I.$

7. $\theta = n\pi + (-1)^n.\frac{\pi}{4} - \frac{\pi}{4}, n \in I.$

8. $\theta = 2n\pi \pm \frac{\pi}{3} - \frac{\pi}{6}, n \in I.$

9. $\theta = -\left(2n\pi + \frac{\pi}{3}\right), \frac{2n\pi}{3} + \frac{\pi}{9}, n \in I.$

10. $\theta = \frac{n\pi}{3} \pm \frac{\pi}{18}, n \in I.$

2.68. PRINCIPAL VALUE

The numerically least angle is called the principal value.

For example, $\sin\theta = \frac{1}{2}$

Then ,

$$\theta = \frac{\pi}{6}, \frac{5\pi}{6}, \frac{13\pi}{6}, \frac{17\pi}{6} \ldots - \frac{11\pi}{6} \ldots - \frac{7\pi}{6}$$

Among all these values of θ, $\frac{\pi}{6}$ is the numerically smallest.

So principal value of $\sin\theta = \frac{1}{2}$ is $\theta = \frac{\pi}{6}$.

2.7. METHOD TO FIND OUT THE PRINCIPAL VALUE :

(i) First draw a trigonometric circle and mark the quadrant in which the angle may lie.

(ii) Select anti-clockwise direction for 1st and 2nd quadrant and select clockwise direction for 3rd and 4th quadrants.

(iii) Find the angle in the first rotation.

(iv) Select the numerically least angle among these two values. The angle thus formed will be the principal value.

(v) In case, two angles, one with +ve sign and the other with -ve sign qualify for the numerically least angle, then it is the conventional of mathematics, to consider the angle with +ve signs as principal value.

Ex-1. Find the principal value of $\sin(\theta) = -\frac{1}{2}$

Soln. We have $\sin(\theta) = -\frac{1}{2}$

$\Rightarrow \quad \theta = -\frac{\pi}{6}$

Hence, the principal value of θ is $\left(-\frac{\pi}{6}\right)$

Ex-2. Find the proncipal value of $\sin(\theta) = \frac{1}{\sqrt{2}}$

$\Rightarrow \quad \theta = \frac{\pi}{4}, \frac{3\pi}{4}$

Hence, the proincipal value of θ is $\frac{\pi}{4}$

Ex-3. Find the principal value of $\tan(\theta) = -\sqrt{3}$

Soln. We have $\tan(\theta) = -\sqrt{3}$

$\Rightarrow \quad (\theta) = -\frac{\pi}{3}$

Hence, the principal value of θ is $-\frac{\pi}{3}$

Exercise - 3

Find the principal values of

1. $\tan\theta = -1$
2. $\cos\theta = \frac{1}{2}$
3. $\cos\theta = -\frac{1}{2}$
4. $\tan\theta = -\sqrt{3}$
5. $\sec\theta = \sqrt{2}$ ·

Answers

1. $\theta = \frac{3\pi}{4}$
2. $\theta = \frac{\pi}{3}$.
3. $\theta = \frac{2\pi}{3}$.
4. $\theta = \frac{2\pi}{3}$.
5. $\theta = \frac{\pi}{4}$.

2.8. SOLUTIONS IN CASE OF TWO EQUATIONS ARE GIVEN :

Two equations are given and we have to find

the value of θ which may satishfy both the given equations like $\cos\theta = \cos a$, $\sin\theta = \sin\alpha$ and $\tan\theta = \tan\alpha$.

The common solution is $\theta = 2n\pi + \alpha$, where $n \in Z$

Similarly, $\sin\theta = \sin\alpha$, $\tan\theta = \tan\alpha$.

The common solution is

$\theta = 2n\pi + \alpha, n \in Z$

Rule : (i) Find the common value of θ between 0 and 2π

(ii) Add $2n\pi$ to this common value

(iii) then we shall get the general value of the given two equations.

Ex-1. If $\sin(\theta) = \frac{1}{\sqrt{2}}$ and $\cos(\theta) = -\frac{1}{\sqrt{2}}$, then find the general values of θ

Soln. Now, $\sin(\theta) = \frac{1}{\sqrt{2}} \Rightarrow \theta = \frac{\pi}{4}, \frac{3\pi}{4}$

and $\cos(\theta) = -\frac{1}{\sqrt{2}} \Rightarrow \theta = \frac{3\pi}{4}$

Thus, the common value of θ is $\frac{3\pi}{4}$

Hence, the genral values of θ is $\left(2n\pi + \frac{3\pi}{4}\right)$.

Ex-2. If $\sin(\theta) = \frac{1}{\sqrt{2}}$ and $\tan(\theta) = -1$, then find the general values of θ

Soln. We have $\sin(\theta) = \frac{1}{\sqrt{2}}$

$\Rightarrow \theta = \frac{\pi}{4}, \frac{3\pi}{4}$

Also, $\tan(\theta) = -1 \Rightarrow \theta = \frac{3\pi}{4}, \frac{7\pi}{4}$

Thus, the common value of θ is $\frac{3\pi}{4}$

Hence, the general values of θ is $\left(2n\pi + \frac{3\pi}{4}\right)$, where $n \in I$.

Ex-3. If $(1+\tan A)(1+\tan B) = 2$, then find all the values of A + B.

Soln. We have $(1+\tan A)(1+\tan B) = 2$

$\Rightarrow 1 + \tan A + \tan B + \tan A.\tan B = 2$

$\Rightarrow \tan A + \tan B = 1 - \tan A.\tan B$

$\Rightarrow \left(\frac{\tan A + \tan B}{1 - \tan A.\tan B}\right) = 1$

$\Rightarrow \tan(A+B) = 1$

$\Rightarrow \tan(A+B) = \tan\left(\frac{\pi}{4}\right)$

$\Rightarrow (A+B) = n\pi + \left(\frac{\pi}{4}\right)$, where $n \in I$

Ex-4. If $\sin(\pi\cos\theta) = \cos(\pi\sin\theta)$, then prove that, $\cos\left(\theta \pm \frac{\pi}{4}\right) = \frac{1}{2\sqrt{2}}$

Soln. We have $\sin(\pi\cos\theta) = \cos(\pi\sin\theta)$

$\Rightarrow \sin(\pi\cos\theta) = \sin\left(\frac{\pi}{2} - \pi\sin\theta\right)$

$\Rightarrow (\pi\cos\theta) = \left(\frac{\pi}{2} - \pi\sin\theta\right)$

$\Rightarrow \cos\theta = \left(\frac{1}{2} - \sin\theta\right)$

$\Rightarrow \cos\theta + \sin\theta = \frac{1}{2}$

$\Rightarrow \frac{1}{\sqrt{2}}\cos\theta + \frac{1}{\sqrt{2}}\sin\theta = \frac{1}{2\sqrt{2}}$

$\Rightarrow \cos\left(\theta + \frac{\pi}{4}\right) = \frac{1}{2\sqrt{2}}$

Similarly, we can prove that

$$\cos\left(\theta-\frac{\pi}{4}\right)=\frac{1}{2\sqrt{2}}$$

Ex-5. If $\tan(\pi\cos\theta)=\cot(\pi\sin\theta)$, then prove that $\cos\left(\theta-\frac{\pi}{4}\right)=\frac{1}{2\sqrt{2}}$

Soln. We have, $\tan(\pi\cos\theta)=\cot(\pi\sin\theta)$

$$\Rightarrow \tan(\pi\cos\theta)=\tan\left(\frac{\pi}{2}-\pi\sin\theta\right)$$

$$\Rightarrow (\pi\cos\theta)=\left(\frac{\pi}{2}-\pi\sin\theta\right)$$

$$\Rightarrow \cos(\theta)+\sin(\theta)=\frac{1}{2}$$

$$\Rightarrow \frac{1}{\sqrt{2}}\cos(\theta)+\frac{1}{\sqrt{2}}\sin(\theta)=\frac{1}{2\sqrt{2}}$$

$$\Rightarrow \cos\left(\theta-\frac{\pi}{4}\right)=\frac{1}{2\sqrt{2}}.$$

Ex-6. If sin A = sin B and cos A = cos B, then find the values of A in terms of B.

Soln. Given sin A = sin B................(i)

and cos A = cos B..............(ii)

Dividing (i) and (ii), we get,

$$\frac{\sin A}{\cos A}=\frac{\sin B}{\cos B}$$

$$\Rightarrow \tan A=\tan B$$

$$\Rightarrow A=n\pi+B,\text{ where } n\in I$$

Ex-7. If A and B are acute +ve angles satisfying the equations $3\sin^2 A+2\sin^2 B=1$ and $3\sin 2A-2\sin 2B=0$, then find A + 2B.

Soln. Given equations are

$3\sin^2 A+2\sin^2 B=1$ ....(i) and

$3\sin 2A-2\sin 2B=0$...............(ii)

From (ii), we get, $3\sin 2A=2\sin 2B$

$$\Rightarrow \frac{\sin 2A}{2}=\frac{\sin 2B}{3}$$

$$\Rightarrow \frac{\sin 2B}{\sin 2A}=\frac{3}{2}$$

From (i), we get,

$$\frac{3}{2}(2\sin^2 A)+(2\sin^2 B)=1$$

$$\Rightarrow \frac{3}{2}(1-\cos 2A)+(1-\cos 2B)=1$$

$$\Rightarrow \frac{3}{2}\cos 2A+\cos 2B=\frac{3}{2}$$

$$\Rightarrow \frac{\sin 2B}{\sin 2A}\cos 2A+\cos 2B=\frac{\sin 2B}{\sin 2A}$$

$$\Rightarrow \sin 2B\cos 2A+\sin 2A\cos 2B=\sin 2B$$

$$\Rightarrow \sin(2A+2B)=\sin 2B$$

$$\Rightarrow \sin(2A+2B)=\sin(\pi-2B)$$

$$\Rightarrow (2A+2B)=(\pi-2B)$$

$$\Rightarrow (2A+4B)=\pi$$

$$\Rightarrow (A+2B)=\frac{\pi}{2}.$$

Ex-8. Solve: $x+y=\frac{\pi}{4}$ and $\tan x+\tan y=1$.

Soln. Given $x+y=\frac{\pi}{4}$ and $\tan x+\tan y=1$

$$\Rightarrow \tan(x+y)=\tan\left(\frac{\pi}{4}\right)$$

$$\Rightarrow \frac{\tan x+\tan y}{1-\tan x.\tan y}=1$$

$$\Rightarrow 1-\tan x.\tan y=1$$

$\Rightarrow \quad \tan x . \tan y = 0$

$\Rightarrow \quad \tan x = 0 \ \& \ \tan y = 0$

$\Rightarrow \quad x = n\pi = y$

Thus, no values of x and y satisfying the given equations.

Therefore, the given equations have no solutions.

Ex-9. Solve: $\sin x + \sin y = 1$,

$\cos 2x - \cos 2y = 1$.

Soln. Given $\sin x + \sin y = 1$..........(i) and

$\cos 2x - \cos 2y = 1$............(ii)

From (ii), we get, $\cos 2x - \cos 2y = 1$

$\Rightarrow \quad 1 - 2\sin^2 x - 1 + \sin^2 y = 1$

$\Rightarrow \quad \sin^2 x - \sin^2 y = -\frac{1}{2}$

$\Rightarrow \quad y = n\pi + (-1)^n \sin^{-1}\left(\frac{3}{4}\right)$, where $n \in I$

$\Rightarrow \quad \sin x - \sin y = -\frac{1}{2}$.......................(iii)

Adding (i) and (iii), we get,

$2 \sin x = \frac{1}{2}$

$\Rightarrow \quad \sin x = \frac{1}{4}$

$\Rightarrow \quad x = n\pi + (-1)^n \sin^{-1}\left(\frac{1}{4}\right)$, where $n \in I$

Subtracting (i) and (iii), we get,

$2 \sin y = \frac{3}{2}$

$\Rightarrow \quad \sin y = \frac{3}{4}$

Ex-10. If $r\sin\theta = 3$ and $r = 4(1 + \sin\theta)$, where $0 \le \theta \le 2\pi$, then find the value of θ

Soln. Given equations are

$r\sin\theta = 3$.................(i) and

$r = 4(1 + \sin\theta)$..........(ii)

Eliminating (i) and (ii), we get,

$4(1 + \sin\theta)\sin\theta = 3$

$\Rightarrow \quad 4\sin^2\theta + 4\sin\theta - 3 = 0$

$\Rightarrow \quad 4\sin^2\theta + 6\sin\theta - 2\sin\theta - 3 = 0$

$\Rightarrow \quad 2\sin\theta(2\sin\theta + 3) - 1(2\sin\theta + 3) = 0$

$\Rightarrow \quad (2\sin\theta + 3)(2\sin\theta - 1) = 0$

$\Rightarrow \quad \sin\theta = -\frac{3}{2}, \frac{1}{2}$

$\Rightarrow \quad \sin\theta = \frac{1}{2}$

$\Rightarrow \quad \theta = \frac{\pi}{6}, \frac{5\pi}{6}$

Ex-11. Find the set of values of x for which

$$\frac{\tan 3x - \tan 2x}{1 + \tan 3x . \tan 2x} = 1$$

Soln. We have, $\frac{\tan 3x - \tan 2x}{1 + \tan 3x . \tan 2x} = 1$

$\Rightarrow \quad \tan(3x - 2x) = 1$

$\Rightarrow \quad \tan x = 1$

$\Rightarrow \quad x = n\pi + \frac{\pi}{4}$, where $n \in I$

But the values of x do not satisfy the given equation.

Hence, the set of values of x is ϕ

Ex-12. Find the number of solutions of the equation $\tan x + \sec x = 2\cos x$ lying in the interval $[0, 2\pi]$

Soln. Given equation is $\tan x + \sec x = 2\cos x$

$\Rightarrow \quad (1 + \sin x) = 2\cos^2 x$

$\Rightarrow \quad (1+\sin x)=2\,(1-\sin^2 x)$

$\Rightarrow \quad (1+\sin x)=2\,(1+\sin x).(1-\sin x)$

$\Rightarrow \quad (1+\sin x)(1-2+2\sin x)=0$

$\Rightarrow \quad (1+\sin x)(2\sin x - 1)=0$

$\Rightarrow \quad (1+\sin x)=0 \ \& \ (2\sin x - 1)=0$

$\Rightarrow \quad \sin x=-1 \ \& \ \sin x=\frac{1}{2}$

$\Rightarrow \quad x=\frac{\pi}{2},\frac{\pi}{6},\frac{5\pi}{6}$

But $x=\frac{\pi}{2}$ doest not satisfy the given equation.

Thus, the values of x are $\frac{\pi}{6}$ and $\frac{5\pi}{6}$

Hence, the number of solutions is 2.

Ex-13. Find the number of values of x in the interval $[0, 3\pi]$ satisfying the equation

$2\sin^2 x+5\sin x-3=0$

Soln. Given equation is $2\sin^2 x+5\sin x-3=0$

$\Rightarrow \quad 2\sin^2 x+6\sin x-\sin x-3=0$

$\Rightarrow \quad 2\sin x(\sin x+3)-1(\sin x+3)=0$

$\Rightarrow \quad (\sin x+3)\,(2\sin x-1)=0$

$\Rightarrow \quad \sin x=-3,\frac{1}{2}$

$\Rightarrow \quad \sin x=\frac{1}{2}$

$\Rightarrow \quad x=\frac{\pi}{6},\frac{5\pi}{6},\frac{13\pi}{6},\frac{17\pi}{6}$

Hence, the number of values of x is 4.

Ex-14. Find the smallest positive value of x such that

$\tan(x+20^\circ)=\tan(x-10^\circ)\tan x.\tan(x+10^\circ)$

Soln. Given,

$\tan(x+20^\circ)=\tan(x-10^\circ)\tan x.\tan(x+10^\circ)$

$\Rightarrow \quad \frac{\tan(x+20^\circ)}{\tan x}=\tan(x-10^\circ).\tan(x+10^\circ)$

$\Rightarrow \quad \frac{\sin(x+20^\circ)\cos x}{\cos(x+20^\circ)\sin x}=\frac{\sin(x-10^\circ)\sin(x+10^\circ)}{\cos(x-10^\circ)\cos(x+10^\circ)}$

$\Rightarrow \quad \frac{\sin(x+20^\circ)\cos x+\cos(x+20^\circ)\sin x}{\sin(x+20^\circ)\cos x-\cos(x+20^\circ)\sin x}$

$=\frac{\sin(x-10^\circ)\sin(x+10^\circ)+\cos(x-10^\circ)\cos(x+10^\circ)}{\sin(x-10^\circ)\sin(x+10^\circ)-\cos(x-10^\circ)\cos(x+10^\circ)}$

$\Rightarrow \quad \frac{\sin(x+20^\circ+x)}{\sin(x+20^\circ-x)}=-\frac{\cos(x+10^\circ-x+10^\circ)}{\cos(x+10^\circ+x-10^\circ)}$

$\Rightarrow \quad \frac{\sin(2x+20^\circ)}{\sin(20^\circ)}=-\frac{\cos(20^\circ)}{\cos(2x)}$

$\Rightarrow \quad \sin(2x+20^\circ)\cos(2x)=-\sin(20^\circ)\cos(20^\circ)$

$\Rightarrow \quad 2\sin(2x+20^\circ)\cos(2x)=-2\sin(20^\circ)\cos(20^\circ)$

$\Rightarrow \quad \sin(4x+20^\circ)+\sin(20^\circ)=-\sin(40^\circ)$

$\Rightarrow \quad \sin(4x+20^\circ)=-\sin(40^\circ)-\sin(20^\circ)$

$\Rightarrow \quad \sin(4x+20^\circ)=-2\sin(30^\circ)\cos(10^\circ)$

$\Rightarrow \quad \sin(4x+20^\circ)=-\cos(10^\circ)$

$\Rightarrow \quad \sin(4x+20^\circ)=-\sin(80^\circ)$

$\Rightarrow \quad \sin(4x+20^\circ)=\sin(-80^\circ)$

$\Rightarrow \quad \sin(4x+20^\circ)=\sin(\pi-(-80^\circ))$

$\Rightarrow \quad (4x+20^\circ)=(\pi-(-80^\circ))$

$\Rightarrow \quad (4x+20^\circ)=260^\circ$

$\Rightarrow \quad 4x=260^\circ-20^\circ=240^\circ$

$\Rightarrow \quad x=60^\circ$

Hence, the smallest positive value of x is 60°.

Exercise - 4

1. If $\cos\theta = \frac{1}{\sqrt{2}}$ and $\tan\theta = -1$, then find the genral value of θ.

2. Find the most genral value of θ which satisfy the equations $\sin\theta = \frac{1}{2}$ and $\tan\theta = \frac{1}{\sqrt{3}}$

3. If A and B are acute +ve angles satishfying the equations $3\sin^2 A + 2\sin^2 B = 1$ and $3\sin 2A - 2\sin 2B = 0$, then find $A + 2B$.

4. If $\tan(A-B) = 1$ and $\sec(A+B) = \frac{2}{\sqrt{3}}$, then find the smallest +ve values of A and B and their most genral values.

5. Solve: $x + y = \frac{2\pi}{3}$ and $\sin x = 2\sin y$.

6. Solve: $x + y = \frac{2\pi}{3}$ and $\cos x + \cos y = \frac{3}{2}$

7. Solve: $x + y = \frac{\pi}{4}$ and $\tan x + \tan y = 1$.

8. Solve: $r\sin\theta = 3$ and $r = 4(1+\sin\theta), 0 \le \theta \le 2\pi$

9. If $\sin A = \sin B$ and $\cos A = \cos B$, then find the values of A in terms of B.

10. Solve: $\sin x + \sin y = 1$, $\cos 2x - \cos 2y = 1$.

11. Find the co-ordinates of the point of inter section of the curves $y = \cos x$ and $y = \sin 2x$

12. Find all points of x, y that satisfying the equations $\cos x + \cos y + \cos(x+y) = -\frac{3}{2}$

13. If $0 < \theta, \phi < \pi$ and $8\cos\theta\cos\varphi\cos(\theta+\varphi) + 1 = 0$, then find θ & ϕ.

14. Solve: $4^{\sin x} + 3^{\frac{1}{\cos y}} = 11$,
$5.16^{\sin x} + 2.3^{\frac{1}{\cos y}} = 2$.
if $-\frac{\pi}{2} \le x \le \frac{\pi}{2}$

Answers

1. $\theta = 2n\pi + \frac{7\pi}{4}, n \in Z$

2. $\theta = 2n\pi + \frac{7\pi}{6}, n \in Z$

3. $\theta = 2n\pi + \frac{\pi}{6}, n \in Z$

4. $A = \frac{1}{2} \times \left[(2m+n)\pi + \frac{\pi}{4} \pm \frac{\pi}{6}\right]$,
$B = \frac{1}{2} \times \left[(2m-n)\pi - \frac{\pi}{4} \pm \frac{\pi}{6}\right], m, n \in Z$

5. $x = \frac{\pi}{2} + n\pi, y = \frac{\pi}{6} - n\pi, n \in Z$

6. No. Solution., $x = \phi = y$

7. $x = n\pi$, $y = \frac{\pi}{4} - n\pi$ or, $x = \frac{\pi}{4} - n\pi$, $y = n\pi$

8. $r = 6, \theta = \frac{\pi}{6} or \frac{5\pi}{6}$

9. A-B $= 2n\pi, n \in Z$

10. $x = n\pi + (-1)^n \frac{\pi}{6}, y = 2m\pi \pm \frac{\pi}{2}, m, n \in Z$

11. $\left(-\frac{3\pi}{8}, \cos\frac{3\pi}{8}\right), \left(\frac{\pi}{8}, \cos\frac{\pi}{8}\right), \left(\frac{\pi}{4}, \frac{1}{\sqrt{2}}\right)$

12. $x = 2m\pi \pm \frac{2\pi}{3}, y = 2(m-n)\pi \pm \frac{2\pi}{3}$, $m, n \in Z$

13. $x = \frac{\pi}{3} or \frac{2\pi}{3} = \phi$

14. $x = n\pi + (-1)^n \frac{\pi}{6}$

$\& \ y = 2m\pi \pm \frac{\pi}{3}, m, n \in Z$

2.9. SOME IMPORTANT REMARKS TO KEEP IN MIND WHILE SOLVING A TRIGONOMETRIC EQUATION.

1. Squaring the equation at any stage should be avoided as far as possible. If squaring is necessary, check the solutions for extraneous roots.

2. Never cancel terms containing unknown terms on both the sides, which are in product form. It may cause roots getting loose.

3. Domain should not be changed. If it is changed, necessary correction must be made.

4. Check whether the denominator is non zero at any stage, while solving equations.

5. The answer should not contain such values of angles as where any of the terms undefined.

6. Some-times, we may find our answers differ from those in the package in their notations.This leads us to the different methods of solving the same problem.

   Whenever we come across such situation, we must check authencity. This will ensure that our answer is correct.

7. Some-times the two soltuion set consists partly of common values. In all such cases, the common part must be presented only once.

2.10 TYPES OF TRIGONOMETRIC EQUATIONS :

1. **A Trigonometric equation reduces to Quadratic / Higher degree Equations.**

Rules : 1. Transform the terms to be a only one trigonomet ratio involving angles in same form.

2. Factorize the equation and express it in $f(x) \times g(x) = 0$

   $\Rightarrow$ $\mathbf{f(x) = 0}$ or $g(x) = 0$

3. Solve both the equations one by one to get general value of the variables.

Ex- 1. Solve: $5\cos 2x + 2\cos^2\left(\frac{x}{2}\right) + 1 = 0$.

Soln. The given equation can be expressed as

$5(2\cos^2 x - 1) + (1 + \cos x) + 1 = 0$

$\Rightarrow$ $10\cos^2 x + \cos x - 3 = 0$

$\Rightarrow$ $(5\cos x + 3)(2\cos x - 1) = 0$

$\Rightarrow$ $5\cos x + 3 = 0$ or $2\cos x - 1 = 0$.

$\Rightarrow$ $\cos x = -3/5 = \cos \alpha$,

$\cos x = 1/2 = \cos(\pi/3)$

$\Rightarrow$ $x = 2n\pi \pm \alpha = 2n\pi \pm \cos^{-1}\left(\frac{\pi}{3}\right)$;

$x = 2n\pi \pm \frac{\pi}{3}, n \in Z$.

Exercise-5

1. $4\sin^4 x + \cos^4 x = 1$

2. $4\cos^2 x \sin x - 2\sin^2 x = 2\sin x$

3. $\sin x - \cos x - 4\cos^2 x \sin x = 4\sin^2 x$.

4. $2\cos 2x + \sqrt{2\sin x} = 2$

5. $1 + \sin^2 x + \cos^2 x = \frac{3}{2}\sin 2x$

6. $\sin^6 x + \cos^6 x = \frac{7}{16}$

7. $\sin^8 x + \cos^8 x = \frac{17}{16}\cos^2 2x$

8. $2\sin^3 x = \cos^2 3x$

9. $\cos 4x = \cos^2 3x$

10. $\cos 2x = 6\tan^2 x - 2\cos^2 x$

Answers

1. $\theta = n\pi + \alpha$, where $\sin\alpha = \sqrt{\frac{2}{5}}, n \in Z$

2. $x = n\pi\pi x = n\pi + (-1)^n \frac{\pi}{16}$

$, x = n\pi - (-1)^n \frac{3\pi}{16}, n \in Z$

3. $x = n\pi + \alpha$, where $\alpha = \tan^{-1}\left(-\frac{1}{3}\right), n \in Z$

4. $x = n\pi, x = n\pi + (-1)^n \frac{\pi}{16}, n \in Z$

5. $x = 2n\pi + \frac{\pi}{4} \pm \frac{3\pi}{4}, n \in Z.$

6. $x = n\pi + (-1)^n \frac{\pi}{6}, n \in Z$

7. $.x = (2n+1)\frac{\pi}{8}, n \in Z$

8. $x = n\pi + \frac{\pi}{4}, n \in Z$

9. $x = n\pi, x = \frac{n\pi}{2} \pm \frac{\pi}{2}$

10. $x = \frac{n\pi}{2} \pm \frac{\pi}{6}, n \in Z.$

2. **A Trigonometric Equation is Solved by Factorization Method.**

Rule : Simply reduce to a single trigonometric ratio of the unknown angles and factorize by basic algebraic method.

Ex.-1 Solve : $(2\sin x - \cos x)(1 + \cos x) = \sin^2 x.$

Soln. The given equation can be written as

$(2\sin x - \cos x)(1 + \cos x)$

$= (1 - \cos x)(1 + \cos x)$

$\Rightarrow (1 + \cos x)(2\sin x - \cos x - 1 + \cos x) = 0$

$\Rightarrow (1 + \cos x)(2\sin x - 1) = 0.$

$\Rightarrow \cos x = -1, \sin x = 1/2$

$\Rightarrow \cos x = -1 = \cos\pi, \ \sin x = \frac{1}{2} = \sin\frac{\pi}{6}$

$\Rightarrow x = 2n\pi \pm \pi, \ x = n\pi + (-1)^n \frac{\pi}{6}, n \in Z$

Exercise - 6

1. $2\sin^2 x + \sin x - 1 = 0$, where $0 \le x \le 2\pi$

2. $5\sin^2 x + 7\sin x - 6 = 0$, where $0 \le x \le 2\pi$

3. $\sin^2 x - \cos x = \frac{1}{4}$, where $0 \le x \le 2\pi$

4. $\tan^2 x - 2\tan x - 3 = 0$

5. $2\cos^2 x - \sqrt{3}\sin x + 1 = 0$

Answers

1. $x = \frac{\pi}{6}, \frac{5\pi}{6}, \frac{7\pi}{12}$

2. $x = \sin^{-1}(0.6), \ x = \pi - \sin^{-1}(0.6)$

3. $x = \frac{\pi}{2}, \frac{5\pi}{3}$

4. $x = n\pi \pm \frac{\pi}{3}, \ n \in Z$

5. $x = n\pi + (-1)^n \frac{\pi}{3}, n \in Z$

3. **A Trigonometric equation is solved by transformation as a sum or difference into a product.**

Rule : 1. The given equation is reducible to $f(x) \times g(x) = 0$

$\Rightarrow f(x) = 0, g(x) = 0$

2. Solve both the equations one by one to get the genral value of the variable x

Ex.-1 Solve : $\sin x + \sin 3x + \sin 5x = 0,$

$$0 \le x \le \frac{\pi}{2}$$

Soln. The given equation can be written as

$(\sin x + \sin 5x) + \sin 3x = 0$

$\Rightarrow$ $2 \sin 3x . \cos 2x + \sin 3x = 0$

$\Rightarrow$ $\sin 3x (2 \cos 2x + 1) = 0$

$\Rightarrow$ $\sin 3x = 0, \cos 2x = -1/2$

$\Rightarrow$ $\sin 3x = 0, \ \cos 2x = \cos\frac{2\pi}{3}$

$\Rightarrow$ $3x = n\pi, \quad 2x = 2n\pi \pm \frac{2\pi}{3}, \ n \in Z$

$\Rightarrow$ $x = \frac{n\pi}{3}, x = n\pi \pm \frac{\pi}{3}, n \in Z$

$\Rightarrow$ $x = 0, \ \frac{\pi}{3}$

Exercise - 7

1. Solve : $\cos x - \cos 2x = \sin 3x$
2. Solve : $\sin 7x + \sin 4x + \sin x = 0,$ $0 \le x \le \frac{\pi}{2}$
3. Solve : $\cos 3x + \cos 2x = \sin\left(\frac{3x}{2}\right) + \sin\left(\frac{x}{2}\right), \ 0 \le x \le 2\pi$
4. Solve : $\sin x + \sin 2x + \sin 3x = \cos x + \cos 2x + \cos 3x, \ -\pi \le x \le \pi$
5. Solve : $\cos 2x + \cos 4x = 2 \cos x$
6. Solve : $\sin 2x + \cos 2x + \sin x + \cos x + = 0.$
7. Solve : $\tan x + \tan 2x + \tan 3x = 0.$
8. Solve : $\tan 3x + \tan x = 2 \tan 2x.$
9. Solve : $(1 - \tan x)(1 + \sin 2x) = 1 + \tan 4x.$
10. Solve : $\sin x - 3 \sin 2x + \sin 3x = \cos x - 3 \cos 2x + \cos 3x.$

Answers

1. $x = (4n+1)\frac{\pi}{2}, (4n-1)\frac{\pi}{2}, n \in Z$
2. $x = 0, \frac{\pi}{4}, \frac{\pi}{2}. \frac{2\pi}{9}, \frac{4\pi}{9}$
3. $x = \frac{\pi}{7}, \frac{5\pi}{7}. \pi, \frac{9\pi}{7}, \frac{13\pi}{7}$
4. $x = \frac{\pi}{8}, \frac{5\pi}{8}, \frac{2\pi}{3}, \frac{4\pi}{3}, \frac{9\pi}{8}, \frac{13\pi}{8}.$
5. $x = (2n+1)\frac{\pi}{2}, \frac{2n\pi}{3}, n \in Z$
6. $x = n\pi - \frac{\pi}{4}, 2n\pi \pm \frac{2\pi}{3}, n \in Z$
7. $x = n\pi \pm \alpha, n \in Z$
8. $x = n\pi, \frac{n\pi}{2}, n \in Z$
9. $x = n\pi - \frac{\pi}{4}, n \in Z$
10. $x = (4n+1)\frac{\pi}{8}, n \in Z$

4. **A Trigonometric Equation is solved by transformation as a product into a sum or difference :**

Rule : 1. The given equation can be reduced to $f(x) \times g(x) = 0$

$\Rightarrow f(x) = 0, g(x) = 0$

2. Solve both the equation, one -by-one to get the general value of the variable x.

Ex.-1. Solve: $4 \sin x. \sin 2x. \sin 4x = \sin 3x$.

Soln. The given equation can be written as

$(2 \sin 2x. \sin x). 2 \sin 4x - \sin 3x = 0$

$\Rightarrow$ $2(\cos x - \cos 3x) \sin 4x - \sin 3x = 0$

$\Rightarrow$ $2 \sin 4x \cos x - 2\sin 4x \cos 3x - \sin 3x = 0$

$\Rightarrow$ $(\sin 5x + \sin 3x) - (\sin 7x + \sin x) - \sin 3x = 0$

$\Rightarrow$ $(\sin 7x - \sin 5x) + \sin x = 0$

$\Rightarrow$ $\sin x(2 \cos 6x + 1) = 0$

$\Rightarrow$ $\sin x = 0, \cos 6x = -1/2,$

$\Rightarrow$ $\sin x = 0, \cos 6x = \cos \frac{2\pi}{3}$

$\Rightarrow$ $x = n\pi, 6x = 2n\pi \pm \frac{2\pi}{3}, n \in Z$

$\Rightarrow$ $\Rightarrow x = n\pi, x = (3n \pm 1)\frac{\pi}{9}, n \in Z$.

Exercise - 8

1. $\cos x. \cos 2x. \cos 3x = 1/4,\ 0 \le x \le \pi$
2. $\sin 3\alpha = 4 \sin \alpha. \sin(x+\alpha). \sin(x-\alpha)$
3. $\sin 2x. \sin 4x. + \cos 2x = \cos 6x$.
4. $\sec x. \cos 5x + 1 = 0.,\ 0 \le x \le \pi$
5. $\cos x. \cos 6x = -1$.

Answers

1. $x = \frac{\pi}{6}, \frac{\pi}{3}, \frac{3\pi}{8}, \frac{5\pi}{8}, \frac{2\pi}{3}, \frac{7\pi}{8}$
2. $x = n\pi \pm \frac{\pi}{3}, n \in Z$
3. $x = \frac{n\pi}{2}, \frac{n\pi}{8} + (-1)^n \frac{3\pi}{16}, n \in Z$
4. $x = \frac{\pi}{6}, \frac{\pi}{4}, \frac{3\pi}{4}, \frac{5\pi}{6}, \frac{7\pi}{4}, \frac{11\pi}{6}$
5. $x = (2n+1)\pi, n \in Z$.

5. **A Trigonometric equation is of the form**

$$b_0 \sin^n x + b_1 \sin^{n-1} x.\cos x + b_2 \sin^{n-2} x.\cos^2 x + ... + b_n \cos^n x = 0$$

where, $b_0, b_1, b_2, ..., b_n \in R$ is a homogenous equation of sinx and cos x , where cos x is nonzero.

Rule: 1. Divide both the sides by highest power of cos x.

2. The given equation can be reduced to $b_0 \tan^n x + b_1 \tan^{n-1} x + ... + b_n = 0$

3. and then use factorization method.

Ex.-1 Solve:

$2\sin^2 x - 5\sin x \cos x - 8\cos^2 x = -2$

Soln. The given equation can be written as

$2\sin^2 x - 5\sin x \cos x - 8\cos^2 x$

$= -2(\sin^2 x + \cos^2 x)$

$\Rightarrow$ $2\tan^2 x - 5\tan x - 8 = -2(\tan^2 x + 1)$

$\Rightarrow$ $4\tan^2 x - 5\tan x - 6 = 0$

$\Rightarrow$ $(\tan x - 2)(4\tan x + 3) = 0$

$\Rightarrow$ $\tan x = -2, \tan x = -\frac{3}{4}$

$\Rightarrow$ $x = n\pi + \alpha, x = n\pi + \beta$, where

$\alpha = \tan^{-1}(2), \beta = \tan^{-1}\left(-\frac{3}{4}\right), n \in Z$.

Exercise - 9

1. Solve: $5\sin^2 x - 7\sin x \cos x + 6\cos^2 x = 4$
2. Solve: $2\sin^2 x - 5\sin x \cos x - 8\cos^2 x = -3$
3. Solve: $\sin 3x.\cos x + \sin^2 x \cos^2 x$ $, 0 \le x \le 2\pi$

Answers

1. $x = n\pi + \tan^{-1}(3), n\pi + \tan^{-1}(4), n \in Z$

2. No. Solution

3. No. solution.

6. **A Trigonometric equation is of the form R (sin m x, cos n x, tan p x, cot q x) = 0 , Where R is a rational function and m, n, p, q$\in N$, can be reduced to a rational function with respect to sin x, cos x, tanx and cot x.**

Rule : 1. Use half angle formulae of tangents :

$$\sin x = \frac{2\tan\left(\frac{x}{2}\right)}{1+\tan^2\left(\frac{x}{2}\right)},$$

$$\cos x = \frac{1-\tan^2\left(\frac{x}{2}\right)}{1+\tan^2\left(\frac{x}{2}\right)}$$

$$\tan x = \frac{2\tan\left(\frac{x}{2}\right)}{1-\tan^2\left(\frac{x}{2}\right)},$$

$$\cot x = \frac{1-\tan^2\left(\frac{x}{2}\right)}{2\tan\left(\frac{x}{2}\right)}.$$

2. Substitute $\tan\left(\frac{x}{2}\right)$ = t, and solve it.

Ex.-1 Solve : $1 + 2\operatorname{cosec} x = -\frac{\sec^2\left(\frac{x}{2}\right)}{2}$.

Soln. The given equation can be written as

$$1 + \frac{2}{\sin x} = -\frac{1}{2}\left(1+\tan^2\left(\frac{x}{2}\right)\right)$$

$$\Rightarrow \quad 2(\sin x + 2) = -\left(1+\tan^2\left(\frac{x}{2}\right)\right)\sin x$$

$$\Rightarrow \quad 2\left(\frac{2\tan\frac{x}{2}}{1+\tan^2\frac{x}{2}} + 2\right)$$

$$= -\left(1+\tan^2\frac{x}{2}\right)\cdot\left(\frac{2\tan\frac{x}{2}}{1+\tan^2\frac{x}{2}}\right)$$

$$\Rightarrow \quad 2\left(\frac{2t}{1+t^2} + 2\right) = -\left(1+t^2\right)\times\left(\frac{2t}{1+t^2}\right),$$

where t = tan (x/2)

$$\Rightarrow \quad t^3 + 2t^2 + 3t + 2 = 0$$

$$\Rightarrow \quad t^3 + t^2 + t^2 + t + 2t + 2 = 0$$

$$\Rightarrow \quad (t+1)(t^2+t+2) = 0$$

$$\Rightarrow \quad t+1=0,\ t^2+t+2 \neq 0$$

$$\Rightarrow \quad \tan\left(\frac{x}{2}\right) = -1 = \tan\left(\frac{-\pi}{4}\right)$$

$$\Rightarrow \quad \frac{x}{2} = n\pi - \frac{\pi}{4}, n \in Z$$

$$\Rightarrow \quad x = 2n\pi - \frac{\pi}{2}, n \in Z.$$

Exercise - 10

1. Solve: $(\cos x - \sin x)(2\tan x + \sec x) + 2 = 0$

2. Solve: $\frac{\sin^3\frac{x}{2} - \cos^3\frac{x}{2}}{2+\sin x} = \frac{\cos x}{3}$

3. Solve: $\cot\frac{x}{2} - \operatorname{cosec}\frac{x}{2} = \cot x$

4. If $\theta_1, \theta_2, \theta_3, \theta_4$ be the four roots of the equation $\sin(\theta+\alpha) = k\sin 2\theta$, no two of which differ by a multiple of 2π, then prove that

$$\theta_1+\theta_2+\theta_3+\theta_4 = (2n+1)\frac{\pi}{4}, n \in Z.$$

Answers

1. $x = 2n\pi \pm \frac{\pi}{3}, n \in Z$

2. $x = (4n+1)\frac{\pi}{2}, n \in Z$

3. $x = 4n\pi \pm \frac{2\pi}{3}, n \in Z$.

7. **A Trigonometric equation is of the form R(sinx +cosx, sinx. cos x) = 0, where R is a rational function of the argument of sinx and cosx.**

Rule : 1. Put sin x + cos x = t

2. Use the identity

$(\sin x + \cos x)^2 = 1 + 2\sin x \cos x.$

$\Rightarrow \quad \sin x \cos x = \frac{t^2-1}{2}$

3. So, the given equation reduces to

$R\left(t, \frac{t^2-1}{2}\right)$ and then solve it.

Ex.-1 Solve: sin x + cos x - $2\sqrt{2}$ sin x cos x = 0.

Soln. Let sin x + cos x = t

$\Rightarrow \quad \sin x.\cos x = \frac{t^2-1}{2}$

So, the given equation can be reduced to

$$t - 2\sqrt{2}\left(\frac{t^2-1}{2}\right) = 0$$

$$\Rightarrow \quad \sqrt{2}t^2 - t - \sqrt{2} = 0$$

$$\Rightarrow \quad (\sqrt{2}t+1)(t-\sqrt{2}) = 0$$

$$\Rightarrow \quad t = \sqrt{2}, -\frac{1}{\sqrt{2}}$$

When sin x + cos x = $\sqrt{2}$

$$\Rightarrow \quad \frac{1}{\sqrt{2}}\sin x + \frac{1}{\sqrt{2}}\cos x = 1$$

$$\Rightarrow \quad \sin\left(x+\frac{\pi}{4}\right) = 1 = \sin\left(\frac{\pi}{2}\right)$$

$$\Rightarrow \quad x + \frac{n\pi}{4} = n\pi + (-1)^n\frac{\pi}{2}, n \in Z$$

When sin x + cos x = $-\frac{1}{\sqrt{2}}$

$$\Rightarrow \quad \frac{1}{\sqrt{2}}\sin x + \frac{1}{\sqrt{2}}\cos x = -\frac{1}{2}$$

$$\Rightarrow \quad \sin\left(x+\frac{\pi}{4}\right) = \sin\left(-\frac{\pi}{6}\right)$$

$$\Rightarrow \quad \left(x+\frac{\pi}{4}\right) = n\pi + (-1)^n\left(-\frac{\pi}{6}\right)$$

$$\Rightarrow \quad x = n\pi - (-1)^n\frac{\pi}{6} - \frac{\pi}{4}, n \in Z$$

Exercise - 11

1. Solve: $\sin^3 x + \sin x \cos x + \cos^3 x = 1$

2. Solve: sin x + cos x = 1 - sin x cos x.

3. Solve: $1 + \sin^3 x + \cos^3 x = \frac{3}{2}\sin 2x$

4. Solve: sin 2 x - 12 (sin x - cos x) + 12 = 0 $, 0 \le x \le 2\pi$.

Answers Key :

1. $x = 2n\pi, 2n\pi + \frac{\pi}{3}, n \in Z$

2. $x = 2n\pi, (4n+1)\frac{\pi}{2}, n \in Z$

3. $x = 2n\pi \pm \frac{\pi}{4} \pm \frac{3\pi}{4}, n \in Z$

4. $x = \frac{\pi}{2}, \pi$.

8. **A Trigonometrical equation is based on extreme values of sinx and cos x.**

Rule : 1. Whenever terms are sin, cos in

power 1 and all terms connected with plus sign and number of terms in L.H.S (with +ve or -ve sign) then each term must have in ex treme value.

2. In such problems, each term will be +1 when the value of R.H.S is +ve and each term will be (-1) when the value of R.H.S is -ve.

Ex.-1 Solve: $\sin 6x + \cos 4x + 2 = 0$.

Soln. The given equation can be rewritten as

$\sin 6x + \cos 4x = -2$

$\Rightarrow$ $\sin 6x = -1$ and $\cos 4x = -1$

$$\Rightarrow \sin 6x = \sin\frac{3\pi}{2},\ \cos 4x = \cos\pi$$

$$\Rightarrow 6x = 2n\pi + \frac{3\pi}{2}, 4x = 2n\pi + \pi, n \in Z$$

$$\Rightarrow x = \frac{n\pi}{3} + \frac{\pi}{4}, x = \frac{n\pi}{2} + \frac{\pi}{4}, n \in Z$$

$$\Rightarrow x = \frac{\pi}{4}, \frac{5\pi}{4}, \frac{7\pi}{12}, \frac{11\pi}{12}, \ldots$$

$$\ldots x = \frac{\pi}{4}, \frac{3\pi}{4}, \frac{5\pi}{4}, \frac{7\pi}{4}, \ldots\ldots$$

$$\Rightarrow x = \frac{\pi}{4}, \frac{5\pi}{4}$$

Hence, the genral solution will be,

$$x = 2n\pi + \frac{\pi}{4}, 2n\pi + \frac{5\pi}{4}, n \in Z$$

$$\Rightarrow x = 2n\pi + \frac{\pi}{4}, (2n+1)\pi + \frac{\pi}{4}, n \in Z$$

$$\Rightarrow x = m\pi + \frac{\pi}{4}, m \in Z.$$

<u>Exercise - 12</u>

1. Solve: $\sin^6 x = 1 + \cos^4 3x$
2. Solve: $\sin^4 x = 1 + \tan^8 x$
3. Solve: $\sin^2 x + \cos^2 y = 2\sec^2 z$
4. Solve: $\sin 3x + \cos 2x + 2 = 0$
5. Solve: $\cos 4x + \sin 5x = 2$.

Answers

1. No Solution
2. No. Solution
3. No. Solution
4. $x = 2n\pi + \frac{\pi}{2}, n \in Z$
5 $x = 2n\pi + \frac{\pi}{2}, n \in Z$.

9. **A Trigonometrical equation involving with exponential, logarithmic and modulii terms :**

Rule: Whenever equation contains power term, then we should use the following method.

1. Equate the base if possible
2. If, it is not possible to equate the base, take log of both the sides and make its R.H.S. a is zero, then we proceed further.

Ex.-1 Find the values of x in $(-\pi, \pi)$ which satisfy the equation

$$8^{1+|\cos x|+\cos^2 x+|\cos x|^3+\cos^4 x+|\cos x|^5+\ldots\ldots\ldots to\infty} = 64$$

Soln. The given equation can be rewritten as

$$8^{1+|\cos x|+\cos^2 x+|\cos x|^3+\cos^4 x+|\cos x|^5+\ldots\ldots\ldots to\infty} = 8^2$$

$$\Rightarrow 1 + |\cos x| + \cos^2 x + |\cos x|^3 + \cos^4 x + ..to\infty$$

$$= 2$$

$$\Rightarrow \frac{1}{1-|\cos x|} = 2$$

$$\Rightarrow \cos x = \pm\frac{1}{2}$$

When $\cos x = \frac{1}{2}$

$\Rightarrow \quad \cos x = \cos\left(\frac{\pi}{3}\right)$

$\Rightarrow \quad x = 2n\pi \pm \frac{\pi}{3}, n \in Z$

When $\cos x = -\frac{1}{2} = \cos\left(\frac{2\pi}{3}\right)$

$\Rightarrow \quad x = 2n\pi \pm \frac{2\pi}{3}, n \in Z$.

Hence, the values of x are $\pm\frac{\pi}{3}, \pm\frac{2\pi}{3}$..

Exercise - 13

1.Solve : $2^{1+|\cos x|+\cos^2 x+|\cos x|^3+\cos^4 x+|\cos x|^5+.............to\infty} = 4$

2. Solve : $1+\sin\theta+\sin^2\theta+\sin^3\theta+..........to\infty$

$= 4+2\sqrt{3}$

3. Solve: $|\cos x|^{\sin^2 x-\frac{3}{2}\sin x+\frac{1}{2}} = 1$

4. Solve: $e^{\sin x} - e^{-\sin x} - 4 = 0$

5. If $e^{[\sin^2 x+\sin^4 x+\sin^6 x+..............to\infty]\log_e 2}$ satisfies the equations $x^2 - 9x + 8 = 0$, then find the value of $\frac{\cos x}{\cos x + \sin x}, 0 < x < \frac{\pi}{2}$.

6. Solve : $\log_{\cos x} \tan x + \log_{\sin x} \cot x = 0$.

7. Solve : $3^{\sin 2x+2\cos^2 x} + 3^{1-\sin 2x+2\sin^2 x} = 28$.

8. Solve: $\log_{\cos x} \sin x + \log_{\sin x} \cos x = 2$, where $x > 0$.

Answers

1. $\pm\frac{\pi}{3}, \pm\frac{2\pi}{3}$

2. $\theta = \frac{\pi}{3}, \frac{2\pi}{3}$

3. $x = n\pi, n\pi + (-1)^n \frac{\pi}{6}, n \in Z$.

4. $\frac{1}{2}(\sqrt{3}-1)$

5. No solution

6. $x = 2n\pi + \frac{\pi}{4}, n \in Z$

7. $x = n\pi - \frac{\pi}{4}, (2n+1)\frac{\pi}{2}, n \in Z$

8. $x = \frac{\pi}{4}$.

10. A Trigonometrical Equation involving the terms of two sides are of different nature :

Rule :

1. Let y = each side of the equation and break the equation in two parts.
2. Find the inequality for y taking L.H.S of the equation and also for the R.H.S of the equation.
3. Use the A.M $\geq$ G.M. on the right hand side of the equation.
4. If there is any value of y satisfying both the inequalities, then the equation will have real solution, otherwise no solution..

Exercise - 14

1. Solve: $2\cos^2\left(\frac{x}{2}\right)\sin^2 x = x^2 + \frac{1}{x^2}$ $, 0 < x < \frac{\pi}{2}$

2. Solve: $2\cos^2\left(\frac{x^2+x}{6}\right) = 2^x + 2^{-x}$.

Answers

1. No Solution 2. $x = 0$.

Level– I **QUESTIONS BASED ON FUNDAMENTALS**

1. Solve : $7\cos^2\theta + 3\sin^2\theta = 4$

2. Solve : $\sin 2\theta + \sin 4\theta + \sin 6\theta = 0$.

3. Solve : $\tan^2\theta + (1-\sqrt{3})\tan\theta - \sqrt{3} = 0$

4. Solve : $\tan\theta + \tan 2\theta + \tan\theta\tan 2\theta = 1$

5. Solve : $3\tan(\theta - 15^o) = \tan(\theta + 15^o)$

6. Solve : $\tan\theta + \tan 2\theta + \tan 3\theta = 0$

7. Solve : $4\sin\theta\sin 2\theta\sin 4\theta = \sin 3\theta$.

8. Solve: $\sqrt{2}\sec\theta + \tan\theta = 1$

9. Solve : $\sin(2013\theta) + \cos(2013\theta) = 2$.

10. Solve : $\cos\theta + \sqrt{3}\sin\theta = 2\cos 2\theta$

11 Solve: $x+y = \frac{2\pi}{3}$ and $\cos x + \cos y = \frac{3}{2}$

12. Solve: $x + y = \frac{\pi}{4}$ and $\tan x + \tan y = 1$.

13. Solve: $r\sin\theta = 3$ and $r = 4(1+\sin\theta)$, $0 \le \theta \le 2\pi$

14. Solve: $\sin x + \sin y = 1$, $\cos 2x - \cos 2y = 1$.

15. If A and B are acute +ve angles satisfying the equations $3\sin^2 A + 2\sin^2 B = 1$ and $3\sin 2A - 2\sin 2B = 0$, then find A+2B.

16. If tan (A-B) = 1 and $\sec(A+B) = \frac{2}{\sqrt{3}}$, then find the smallest +ve values of A and B and their most general values.

17. If sin A = sin B and cos A = cos B, then find the values of A in terms of B.

18. Solve: $4\sin^4 x + \cos^4 x = 1$

19. Solve: $4\cos^2 x\sin x - 2\sin^2 x = 2\sin x$

20. Solve: $\sin^6 x + \cos^6 x = \frac{7}{16}$

21. Solve: $\sin 7x + \sin 4x + \sin x = 0$, $0 \le x \le \frac{\pi}{2}$

22. Solve: $\cos 3x + \cos 2x = \sin\left(\frac{3x}{2}\right) + \sin\left(\frac{x}{2}\right)$, $0 \le x \le 2\pi$

23. Solve: $\sin x + \sin 2x + \sin 3x = \cos x + \cos 2x + \cos 3x$, $-\pi \le x \le \pi$

24. Solve: $\cos 2x + \cos 4x = 2\cos x$

25. Solve : $\sin 2x + \cos 2x + \sin x + \cos x + 1 = 0$.

26 Solve: $\cos x . \cos 2x . \cos 3x = 1/4$, $0 \le x \le \pi$

27 Solve $\sin 3\alpha = 4\sin\alpha . \sin(x+\alpha) . \sin(x-\alpha)$

28. Solve :. $\sin 2x . \sin 4x . + \cos 2x = \cos 6x$.

29. Solve : $\sin 3x . \cos x + \sin^2 x\cos^2 x + \sin x . \cos^3 x = 1, 0 \le x \le 2\pi$

30. If $\theta_1, \theta_2, \theta_3, \theta_4$ be the four roots of the equation $\sin(\theta + \alpha) = k\sin 2\theta$, no two of which differ by a multiple of 2π, then prove that $\theta_1 + \theta_2 + \theta_3 + \theta_4 = (2n+1)\frac{\pi}{4}, n \in Z$..

ANSWER

18. $\theta = n\pi + \alpha$, where $\sin\alpha = \sqrt{\frac{2}{5}}, n \in Z$

19. $x = n\pi\pi x = n\pi + (-1)^n \frac{\pi}{16}$,

$x = n\pi - (-1)^n \frac{3\pi}{16}, n \in Z$

20. $x = n\pi + (-1)^n \frac{\pi}{6}, n \in Z$

21. $x = 0, \frac{\pi}{4}, \frac{\pi}{2}. \frac{2\pi}{9}, \frac{4\pi}{9}$

22. $x = \frac{\pi}{7}, \frac{5\pi}{7}. \pi, \frac{9\pi}{7}, \frac{13\pi}{7}$

23. $x = \frac{\pi}{8}, \frac{5\pi}{8}, \frac{2\pi}{3}, \frac{4\pi}{3}, \frac{9\pi}{8}, \frac{13\pi}{8}$

24. $x = (2n+1)\frac{\pi}{2}, \frac{2n\pi}{3}, n \in Z$

25. $x = n\pi - \frac{\pi}{4}, 2n\pi \pm \frac{2\pi}{3}, n \in Z$

26. $x = \frac{\pi}{6}, \frac{\pi}{3}, \frac{3\pi}{8}, \frac{5\pi}{8}, \frac{2\pi}{3}, \frac{7\pi}{8}$

27. $x = n\pi \pm \frac{\pi}{3}, n \in Z$

28. $x = \frac{n\pi}{2}, \frac{n\pi}{8} + (-1)^n \frac{3\pi}{16}, n \in Z$

29. No solution

Level– II FOR JEE-MAIN AND ADVANCED EXAM ONLY

Solve the following equations and tick the correct one.

1. $\sin^2\theta - \cos\theta = \frac{1}{2}, 0 \le \theta \le 2\pi$

(a) $\frac{2\pi}{3}. \frac{\pi}{3}$ (b) $\frac{\pi}{3}. \frac{5\pi}{3}$

(c) $-\frac{\pi}{3}. \frac{2\pi}{3}$ (d) $\frac{2\pi}{3}. \frac{5\pi}{3}$

2. If $3\tan^2\theta - 2\sin\theta = 0$, then θ is

(a) $n\pi$ (b) $n\pi + (-1)^n \frac{\pi}{6}$

(c) $n\pi - (-1)^n \frac{\pi}{6}$ (d) $n\pi + \frac{\pi}{3}$.

3. If $\tan^2 x + (1-\sqrt{3})\tan x - \sqrt{3} = 0$, then x is

(a) $n\pi + \frac{\pi}{3}$ (b) $n\pi - \frac{\pi}{3}$

(c) $n\pi + \frac{\pi}{4}$ (d) $n\pi - \frac{\pi}{4}$

4. If $\tan^2\theta + \cot^2\theta = 2$, then θ is

(a) $n\pi + \frac{\pi}{6}$ (b) $n\pi - \frac{\pi}{6}$

(c) $n\pi + \frac{\pi}{4}$ (d) $n\pi - \frac{\pi}{4}$.

5. If $\tan\theta + \cot\theta = 2$, then θ is

(a) $n\pi + \frac{\pi}{4}$ (b) $n\pi - \frac{\pi}{4}$

(c) $n\pi + \frac{\pi}{3}$ (d) $n\pi - \frac{\pi}{3}$.

6. The set of values of x for which $\frac{\tan 3x - \tan 2x}{1 + \tan 3x \tan 2x} = 1$ is

(a) ϕ (b) $\frac{\pi}{4}$

(c) $n\pi+\frac{\pi}{3}$ (d) $2n\pi+\frac{\pi}{4}$

7. If sin 5x + sin 3 x + sin x = 0, then the value of x other than zero, lying between $0 \le x \le \frac{\pi}{2}$

(a) $\frac{\pi}{6}$ (b) $\frac{\pi}{12}$

(c) $\frac{\pi}{3}$ (d) $\frac{\pi}{4}$.

8. If α and β are acute positive angles satisfying the equations $3\sin^2\alpha+2\sin^2\beta=1$ and $3\sin 2\alpha-2\sin 2\beta=0$, then $\alpha+2\beta$ is

(a) 0 (b) $\frac{\pi}{4}$

(c) $\frac{\pi}{3}$ (d) $\frac{\pi}{2}$.

9. If $2\sin^2 x+\sin^2 2x=2$, $-\pi<x<\pi$, then x is

(a) $\pm\frac{\pi}{2}$ (b) $\pm\frac{\pi}{4}$

(c) $\pm\frac{3\pi}{4}$ (d) None.

10. The real roots of the equation $\cos^7 x+\sin^4 x=1$ in $(-\pi,\pi)$ are

(a) $-\frac{\pi}{2},0$ (b) $-\frac{\pi}{2},0,\frac{\pi}{2}$

(c) $0,\frac{\pi}{2}$ (d) $0,\frac{\pi}{4},\frac{\pi}{2}$.

11. The general solution of $\cos^5 x-\sin^5 x-1=0$ is

(a) $n\pi$ (b) $2n\pi$

(c) $n\pi+\frac{\pi}{2}$ (d) $2n\pi+\frac{\pi}{2}$.

12. If $4\sin^4 x+\cos^4 x=1$, then x is

(a) $n\pi$ (b) $n\pi\pm\sin^{-1}\sqrt{\frac{2}{5}}$

(c) $\frac{2n\pi}{3}$ (d) $2n\pi\pm\frac{\pi}{4}$.

13. The number of points of intersection of 2y = 1 and y = cos x in $-\frac{\pi}{2}\le x\le\frac{\pi}{2}$, is

(a) 1 (b) 2

(c) 3 (d) 4

14. The number of values of x in the interval $[0,3\pi]$ satisfying the equation $2\sin^2 x+5\sin x-3=0$ is

(a) 6 (b) 1

(c) 2 (d) 4 .

15. The number of values of x in $[0,5\pi]$ satisfying the equation $3\sin^2 x-7\sin x+2=0$ is

(a) 0 (b) 5

(c) 6 (d) 10 .

16. The number of solution of the equation $|\cot x|=\cot x+\frac{1}{\sin x}$, $0<x<2\pi$, is

(a) 1 (b) 2

(c) 3 (d) 4.

17. The number of solution of $|\cos x|=\sin x$ such that $0<x<2\pi$, is

(a) 2 (b) 4

(c) 8 (d) None.

18. The number of solution of the equation $\tan x.\tan 4x=1$, $0<x<\pi$, is

(a) 1 (b) 2

(c) 5 (d) 8.

19. The number of solution of the equation $12\cos^3 x-7\cos^2 x+4\cos x-9=0$, is

(a) 0 (b) 2
(c) infinity (d) None.

20. The sum of all solution of the equation $\cos\theta.\cos\left(\frac{\pi}{3}+\theta\right).\cos\left(\frac{\pi}{3}-\theta\right)=\frac{1}{4}$ is

(a) 15π (b) 30π
(c) $\frac{100\pi}{3}$ (d) None.

21. The number of solution of $16^{\sin^2 x}+16^{\cos^2 x}=10$, $0\le x\le 2\pi$, is

(a) 2 (b) 4
(c) 6 (d) 8

22. The smallest positive value of x such that $\tan\left(x+20^{\circ}\right)=\tan\left(x+10^{\circ}\right)\tan x.\tan\left(x-10^{\circ}\right)$, is

(a) 30° (b) 45°
(c) 60° (d) 75°.

23. The maximum value of $\sin\left(x+\frac{\pi}{6}\right)+\cos\left(x+\frac{\pi}{6}\right)$ in $\left(0,\frac{\pi}{2}\right)$ is attained at

(a) $\frac{\pi}{12}$ (b) $\frac{\pi}{6}$
(c) $\frac{\pi}{3}$ (d) $\frac{\pi}{2}$.

24. The minimum value of $2^{\sin x}+2^{\cos x}$ is

(a) 1 (b) $2^{1-\frac{1}{\sqrt{2}}}$
(c) $2^{-\frac{1}{\sqrt{2}}}$ (d) $2-\frac{1}{\sqrt{2}}$.

25. If $\cos p\theta+\cos q\theta=0$, then the different values of θ are in A.P., whose common difference is

(a) $\frac{\pi}{p+q}$ (b) $\frac{\pi}{p-q}$
(c) $\frac{2\pi}{p\pm q}$ (d) $\frac{3\pi}{p\pm q}$.

26. If tan2x. tan x = 1, then x is

(a) $\frac{\pi}{3}$ (b) $(6n\pm1)\frac{\pi}{6}$
(c) $(4n\pm1)\frac{\pi}{6}$ (d) $(2n\pm1)\frac{\pi}{6}$

27. The maximum value of $5\sin\theta+3\sin(\theta-\alpha)$ is 7, then the set of all possible values of α is

(a) $2n\pi\pm\frac{\pi}{3}$ (b) $2n\pi\pm\frac{2\pi}{3}$
(c) $\left[\frac{\pi}{3},\frac{2\pi}{3}\right]$ (d) None.

28. If $\tan\left(\frac{\pi}{2}\sin\theta\right)=\cot\left(\frac{\pi}{2}\sin\theta\right)$, then $\sin\theta+\cos\theta$ is

(a) 2n-1 (b) 2n+1
(c) 2n (d) n

29. If $\sin\left(\frac{\pi}{4}\cot\theta\right)=\cos\left(\frac{\pi}{4}\tan\theta\right)$, then θ is

(a) $n\pi+\frac{\pi}{4}$ (b) $2n\pi\pm\frac{\pi}{4}$
(c) $n\pi-\frac{\pi}{4}$ (d) $2n\pi\pm\frac{\pi}{6}$.

30. If $\tan(\pi\cos\theta)=\cot(\pi\sin\theta)$, then the value (s) of $\cos\left(\theta-\frac{\pi}{4}\right)$ is (are)

(a) $\frac{1}{2}$ (b) $\frac{1}{\sqrt{2}}$
(c) $\pm\frac{1}{2\sqrt{2}}$ (d) None.

31. If $3\tan\left(\theta-15^{\circ}\right)=\tan\left(\theta+15^{\circ}\right)$, then θ is

(a) $n\pi+\frac{\pi}{4}$ (b) $n\pi+\frac{\pi}{8}$

(c) $n\pi+\frac{\pi}{3}$ (d) None.

32. If $\tan\theta+\tan\left(\theta+\frac{\pi}{3}\right)+\tan\left(\theta+\frac{2\pi}{3}\right)=3$, then θ is

(a) $(2n+1)\frac{\pi}{12}$ (b) $n\pi\pm\frac{\pi}{3}$

(c) $(4n+1)\frac{\pi}{12}$ (d) None.

33. The equation a sin2x + cos 2x = 2a - 7 posses a solution if

(a) $a>6$ (b) $2\le a\le 6$
(c) $a>2$ (d) None.

34. The number of all possible triplets (a_1,a_2,a_3) such that $a_1+a_2\cos 2x+a_3\sin^2 x=0$ for all x, the number of possible 5-tuplets is

(a) 0 (b) 1
(c) 2 (d) None.

35. If $a_1+a_2\sin x+a_3\cos x$

$+a_4\sin 2x+a_5\cos 2x=0$

holds for all x, then the number of possible 5-tuplets is

(a) 0 (b) 1
(c) 2 (d) infinity.

36. The number of solution of the equation $1+\sin x.\sin^2\frac{x}{2}=0$ in $[-\pi,\pi]$ is

(a) 0 (b) 1
(c) 2 (d) 3

37. The solution of $\sin^4 x+\cos^4 x+\sin 2x+\alpha=0$ is solvable for

(a) $-\frac{1}{2}\le\alpha\le\frac{1}{2}$ (b) $-3\le\alpha\le 1$

(c) $-\frac{3}{2}\le\alpha\le\frac{1}{2}$ (d) $-1\le\alpha\le 1$.

38. The equation $\sin^4 x-2\cos^2 x+a^2=0$ is solvable for

(a) $-\sqrt{3}\le a\le\sqrt{3}$ (b) $-\sqrt{2}\le a\le\sqrt{2}$

(c) $-1\le a\le 1$ (d) None.

39. The number of pairs (x,y) satisfying the equations $\sin x+\sin y=\sin(x+y)$ and $|x|+|y|=1$, is

(a) 2 (b) 4
(c) 6 (d) infinity.

40. The value of 'a' for which the equation $4\operatorname{cosec}^2[\pi(a+x)]+a^2-4a=0$, has a real solution, if

(a) a = 1 (b) a = 2
(c) a = 3 (d) None.

41. If $\sin x+\cos x=\sqrt{y+\frac{1}{y}}$, $x\in[0,\ \pi]$, then

(a) $x=\frac{\pi}{4}$, $y=1$ (b) $y=0$

(c) $y=2$ (d) $x=\frac{3\pi}{4}$

42. $|\tan x+\sec x|=|\tan x|+|\sec x|$; $x\in[0,\ 2\pi]$, iff x belongs to that interval

(a) $[0,\ \pi]$ (b) $\left[0,\frac{\pi}{2}\right)\cup\left(\frac{\pi}{2},\pi\right]$

(c) $\left[0,\frac{3\pi}{2}\right)\cup\left(\frac{3\pi}{2},2\pi\right]$

(d) $(\pi,2\pi]$

43. The number of solutions of $\sum_{r=1}^{5}\cos rx=5$ in the interval $[0,2\pi]$ is

(a) 0 (b) 1
(c) 5 (d) 10

44. If $f(x)=\max\{\tan x,\cot x\}$. The number of roots of the equation $f(x)=\frac{1}{2+\sqrt{3}}$ in $(0,2\pi)$ is

(a) 0 (b) 2
(c) 4 (d) ∞

45. If $\sin x + \cos x + \tan x + \cot x + \sec x + \operatorname{cosec} x = 7$ and $\sin 2x = a - b\sqrt{c}$, then $a - b + 2c$ is

(a) 0 (b) 14
(c) 2 (d) $\frac{3}{2}$

46. If $\sin^4 x + \cos^4 x + 2 = 4 \sin x \cos y$ and $0 \le x, y \le \frac{\pi}{2}$, then $\sin x + \cos y$ is

(a) –2 (b) 0
(c) 2 (d) $\frac{3}{2}$

47. The equation $\cos^4 x - (\lambda + 2)\cos^2 x - (\lambda + 3) = 0$ possesses a solution if

(a) $\lambda > -3$ (b) $\lambda < -2$ (c) $-3 \le \lambda \le -2$ (d) $\lambda \in z^+$

48. If $0 < \theta < 2\pi$ and $2\sin^2\theta - 5\sin\theta + 2 > 0$, then the range of θ is

(a) $\left(0, \frac{\pi}{6}\right) \cup \left(\frac{5\pi}{6}, 2\pi\right)$

(b) $\left(0, \frac{5\pi}{6}\right) \cup (\pi, 2\pi)$

(c) $\left(0, \frac{\pi}{6}\right) \cup (\pi, 2\pi)$

(d) None

49. The number of values of x for which $\sin 2x + \cos 4x = 2$ is

(a) 0 (b) 1
(c) 2 (d) ∞

50. The number of solutions of the equation $x^3 + x^2 + 4x + 2 \sin x = 0$ in $0 \le x \le 2\pi$ is

(a) 0 (b) 1
(c) 2 (d) 4

51. The number of solutions of the equation $\tan x + \sec x = 2\cos x$ lying in the interval $[0, 2\pi]$ is

(a) 0 (b) 1
(c) 2 (d) 3

52. The number of solutions of the equation $2(\sin^4 2x + \cos^4 2x) + 3 \sin^2 x \cos^2 x = 0$ is

(a) 0 (b) 1
(c) 2 (d) 3

53. $\cos 2x + a \sin x = 2a - 7$ possesses a solution for

(a) all a (b) $a > 6$
(c) $a < 2$ (d) $a \in [2, 6]$

54. If $0 < x < 2\pi$ and $81^{\sin^2 x} + 81^{\cos^2 x} = 30$, then x is

(a) $\frac{\pi}{6}$ (b) $\frac{\pi}{2}$
(c) π (d) $\frac{\pi}{4}$

55. If $1 + \sin\theta + \sin^2\theta + \ldots\ldots$ to $\infty = 4 + 2\sqrt{3}$, $0 < \theta < \pi$, $\theta \ne \frac{\pi}{2}$, then

(a) $\theta = \frac{\pi}{6}$ (b) $\theta = \frac{\pi}{3}$
(c) $\theta = \frac{\pi}{3}$ or $\frac{\pi}{6}$ (d) $\frac{\pi}{3}$ or $\frac{2\pi}{3}$

56. If $\tan(\pi \cos\theta) = \cot(\pi \sin\theta)$, then the value of $\cos\left(\theta - \frac{\pi}{4}\right)$ is

(a) $\frac{1}{2}$ (b) $\frac{1}{\sqrt{2}}$
(c) $\frac{1}{2\sqrt{2}}$ (d) None

57. The most general values of x for which $\sin x + \cos x = \min\limits_{a \in R} \{1, a^2 - 4a + 6\}$ are given by

(a) $2n\pi, n \in N$

(b) $2n\pi + \frac{\pi}{2}, n \in N$

(c) $n\pi + (-1)^n \frac{\pi}{4} - \frac{\pi}{4}, n \in N$

(d) None

58. If $x \in [0, 2\pi]$ and $\sin x + \sin y = 2$ then the valuee of $x + y$ is

(a) π (b) $\frac{\pi}{2}$

(c) 3π (d) None

59. The number of roots of the equation

$x + 2\tan x = \frac{\pi}{2}$ in the interval $[0, 2\pi]$ is

(a) 1 (b) 2

(c) 3 (d) ∞

60. The number of solutions of the equation

$\cos(\pi\sqrt{x-4})\cos(\pi\sqrt{x}) = 1$ is

(a) None (b) 1

(c) 2 (d) >2

61. The number of solutions of the equation $\sin\left(\frac{\pi x}{2\sqrt{3}}\right) = x^2 - 2\sqrt{3}\,x + 4$

(a) forms an empty set

(b) only one

(c) is only two

(d) is greater than two

62. Number of real roots of the equation $\sec\theta + \operatorname{cosec}\theta = \sqrt{15}$ lying between 0 and 2π is

(a) 8 (b) 4

(c) 2 (d) 0

63. The general solution of the equation $\sin^{100}x - \cos^{100}x = 1$, is

(a) $2n\pi + \frac{\pi}{3}, n \in z$

(b) $n\pi + \frac{\pi}{2}, n \in z$

(c) $n\pi + \frac{\pi}{4}\pi, n \in z$

(d) $2n\pi\ \frac{\pi}{3}, n \in z$

64. The number of solution of the equation $2^{\cos x} = |\sin x|$ in $[-2\pi, 2\pi]$ is

(a) 1 (b) 2

(c) 3 (d) 4

65. The general solution of the equation $2^{\cos 2x} + 1 = 3.2^{-\sin^2 x}$ is

(a) $n\pi.\ n \in z$ (b) $(n+1)\pi, n \in z$

(c) $(n-1)\pi, n \in z$ (d) None

66. If $x \in (0, 1)$, the greatest root of the equation $\sin 2\pi x = \sqrt{2}\cos \pi x$ is

(a) $\frac{1}{4}$ (b) $\frac{1}{2}$

(c) $\frac{3}{4}$ (d) None

67. The number of solutions of $\tan(5\pi\cos\alpha) = \cot(5\pi\sin\alpha)$ for $\alpha \in (0, 2\pi)$ is

(a) 7 (b) 14

(c) 21 (d) 3

68. The number of solution of the equation $1 + \sin x.\ \sin^2\left(\frac{x}{2}\right) = 0$ in $[-\pi, \pi]$ is

(a) 0 (b) 1

(c) 2 (d) 3

69. The number of solution of the equation $|\cot x| = \cot x + \frac{1}{\sin x}$, $\forall x \in [0, 2\pi]$ is

(a) 0 (b) 1

(c) 2 (d) 3

70. The real roots of the equation $\cos^7 x + \sin^4 x = 1$ in $(-\pi, \pi)$ are

(a) $-\frac{\pi}{2}, 0$ (b) $-\frac{\pi}{2}, 0, \frac{\pi}{2}$

(c) $\frac{\pi}{2}, 0$ (d) $0, \frac{\pi}{4}, \frac{\pi}{2}$

ANSWER

1. (b)	2. (a,b)	3. (a,d)	4. (c)	5. (a)	6. (a)	7. (c)	8. (d)	9 (a,b,c)	10 (b)
11 (a)	12 (a,b)	13 (b)	14 (d)	15 (c)	16 (b)	17 (b)	18 (c)	19 (c)	20 (b)
21 (b)	22 (c)	23 (a)	24 (b)	25 (c)	26 (b)	27 (a)	28 (b)	29 (a)	30 (c)
31 (a)	32 (c)	33 (b)	34 (d)	35 (b)	36 (a)	37 (c)	38 (b)	39 (c)	40. (b)
41. (a)	42. (b)	43. (b)	44. (a)	45. (c)	46 (c)	47. (c)	48. (a)	49. (a)	50. (b)
51. (c)	52. (a)	53. (d)	54. (a)	55. (d)	56. (c)	57. (c)	58. (a)	59. (c)	60. (b)
61. (b)	62. (b)	63. (b)	64. (d)	65. (a)	66. (c)	67. (b)	68. (a)	69. (c)	70. (b)

Level– III FOR JEE - ADVANCED EXAM ONLY

Solve the following trigonometric equations :

1. $\cot\left(\frac{x}{2}\right)-\cos ec\left(\frac{x}{2}\right)=\cot x$.

2. $8\cos x.\cos 2x.\cos 4x=\frac{\sin 6x}{\sin x}$.

3. $\frac{\tan x}{\tan 2x}+\frac{\tan 2x}{\tan x}+2=0$.

4, $2(\cos x+\cos 2x)+\sin 2x(1+2\cos x)=2\sin x\ ,-\pi\le x\le\pi$

5. $2(\sin x-\cos 2x)-\sin 2x(1+2\sin x)+2\cos x\ ,-\pi\le x\le\pi$

6. $\sin 2x+5\cos x+5\sin x+1=0$.

7. $\sin x+\sin 2x+\sin 3x=\cos x+\cos 2x+\cos 3x$ in the interval $0\le x\le\pi$.

8. $\sin^2 x\tan x+\cos^2 x\cot x-\sin 2x=1+\tan x\cot x$

9. $\sin^2 4x+\cos^2 x=2\sin 4x\cos^4 x$.

10. $\sin^4 x+\cos^4 x=\frac{7}{2}\sin x\cos x$

11. $\sin^4 x+\cos^4 x=2\cos\left(2x+\frac{\pi}{6}\right)\cos\left(2x-\frac{\pi}{6}\right)$

12. $\sin^4 x+\sin^4\left(x+\frac{\pi}{4}\right)=\frac{1}{4}$.

13. If $\cos\left(x+\frac{\pi}{3}\right)+\cos x=a$, then find all values of a so that the equation has a real solution.

14. Find real values of x for which $27^{\cos 2x}.81^{\sin 2x}$ is minimum. Also find the minimum value.

15. If $32^{\tan^8\theta}=2\cos^2\alpha-3\cos\alpha$ and $3\cos 2\theta=1$ then find the general value of α.

16. Find the general values of x and y satisfying the equations 5 sinx cos y = 1, 4 tanx = tan y.

17. Find the values of α and β, $0<\alpha,\beta<\frac{\pi}{2}$ satisfying the following equations $\cos\alpha.\cos\beta.\cos(\alpha+\beta)=-\frac{1}{8}$.

18. Find the number of solution of the equation sin 5x. cos 3 x = sin 6 x. cos 2 x in $[0,\pi]$.

19. Find the number of solution of the equation cos 3 x. tan 5 x = sin 7 x lying in $\left[0,\frac{\pi}{2}\right]$.

20. The angles B and C (B > C) of a triangle satisfying the equation

$2\tan x - \lambda\left(1+\tan^2 x\right) = 0$, then

find the angle A, if $0 < \lambda < 1$.

21. Determine all values of 'a' for which the equation

$\cos^4 x - (a+2)\cos^2 x - (a+3) = 0$ has a solution and find those.

22. Find all the solution of the equation

$$\sin\alpha + \sin\frac{\pi}{6}\left(\sqrt{(1-\cos x)^2 + \sin^2 x}\right) = 0$$

in $\left[\frac{5\pi}{2}, \frac{7\pi}{2}\right]$.

ANSWERS

1. $x = 4n\pi \pm \frac{2\pi}{3}, n \in Z$

2. $x = \left(n + \frac{1}{2}\right)\frac{\pi}{7}, n \in Z$

3. $x = n\pi \pm \frac{\pi}{3}, n \in Z$

4. $x = -\pi, -\frac{\pi}{2}, -\frac{\pi}{3}, \frac{\pi}{3}, \pi$

5. $x = 2n\pi + \frac{3\pi}{2}, n \in Z$

6. $x = n\pi - \frac{\pi}{4}, n \in Z$

7. $x = \frac{\pi}{8}, \frac{5\pi}{8}, \frac{7\pi}{8}, \frac{13\pi}{8}, \frac{2\pi}{3}, \frac{4\pi}{3}$

8. $x = \frac{n\pi}{2} - (-1)^n \frac{\pi}{12}, n \in Z$

9. $x = n\pi + \frac{\pi}{2}, n \in Z$

10. $x = \frac{n\pi}{2} + (-1)^n \frac{\pi}{12}, n \in Z$

11. $x = \frac{n\pi}{2} + \cos^{-1}\left(\frac{1}{3}\right), n \in Z$

12. $x = n\pi, n\pi - \frac{\pi}{4}, n \in Z$

13. $-\sqrt{3} \le a \le \sqrt{3}$

14. $x = (2n \pm 1)\frac{\pi}{2} + \frac{1}{2}\tan^{-1}\left(\frac{4}{3}\right), \frac{1}{243}$.

15. $\alpha = 2n\pi \pm \frac{2\pi}{3}, n \in Z$

16. $x = 2n\pi - \sin^{-1}\left(\frac{1}{\sqrt{5}}\right)$

$, y = 2m\pi + \left(\pi - \cos^{-1}\left(\frac{1}{\sqrt{5}}\right)\right)$

17. $\alpha = \frac{\pi}{3} = \beta$

18. 5

19. 1

20. 90°

21. $x = n\pi + \alpha$, where $\alpha = \cos^{-1}\sqrt{(a+3)}$

22. $x = \frac{13\pi}{4}$.

COMPREHENSIVE LINK PASSAGE

PASSAGE – I

An equation is of the form

$f(\sin x \pm \cos x, \pm \sin x \cos x) = 0$ can be solved by changing the variable. Let $\sin x \pm \cos x = t$

$\Rightarrow \sin^2 x + \cos^2 x \pm 2\sin x \cos x = t^2$

$\Rightarrow 1 \pm 2\sin x \cos x = t^2$

Thus, the given equation is reducible to

$f\left(t, \frac{t^2-1}{2}\right) = 0$

On the basis of above information, answer the following questions.

1. If $1 - \sin 2x = \cos x - \sin x$, then x is

(a) $2n\pi$, $2n\pi - \frac{\pi}{2}$, $n \in I$

(b) $2n\pi$, $n\pi + \frac{\pi}{4}$, $n \in I$

(c) $2n\pi - \frac{\pi}{2}$, $n\pi + \frac{\pi}{4}$, $n \in I$

(d) None.

2. If $\sin x + \cos x = 1 + \sin x \cos x$, then x is

(a) $2n\pi$, $2n\pi + \frac{\pi}{2}$, $n \in I$

(b) $2n\pi$, $n\pi + \frac{\pi}{4}$, $n \in I$

(c) $2n\pi - \frac{\pi}{2}$, $n\pi + \frac{\pi}{4}$, $n \in I$

(d) None

3. If $\sin^4 x + \cos^4 x = \sin x \cos x$, then x is

(a) $n\pi$,, $n \in I$

(b) $(6n+1)\frac{\pi}{6}$, $n \in I$

(c) $(4n+1)\frac{\pi}{4}$, $n \in I$

(d) None.

4. If $1 + \sin^3 x + \cos^3 x = \frac{3}{2}\sin 2x$, then x is

(a) $(2n+1)\pi$, $2n\pi + \frac{\pi}{4}$, $n \in I$

(b) $(2n+1)\pi$, $2n\pi - \frac{\pi}{2}$, $n \in I$

(c) $2n\pi + \frac{\pi}{4}$, $2n\pi + \frac{\pi}{2}$, $n \in I$

(d) None.

5. If $(\sin x + \cos x) - 2\sqrt{2}\sin x \cos x = 0$, then x is

(a) $2n\pi + \frac{\pi}{4}$, $n \in I$

(b) $2n\pi - \frac{\pi}{4}$, $n \in I$

(c) $n\pi + \frac{\pi}{4}$, $n \in I$

(d) $n\pi - \frac{\pi}{4}$, $n \in I$

PASSAGE- II

α is a root of the equation
(2sinx - cos x) (1 + cos x) = $\sin^2$ x
β is a root of the equation
$3\cos^3 x - 10\cos x + 3 = 0$
and γ is a root of the equation
$1 - \sin 2x = \cos x - \sin x$,
$0 \le \alpha, \beta, \gamma \le \pi/2$. Then

On the basis of above information, answer the following questions.

1. $\cos\alpha + \cos\beta + \cos\gamma$ is equal to

(a) $\frac{3\sqrt{6} + 2\sqrt{2} + 6}{6\sqrt{2}}$ (b) $\frac{3\sqrt{3} + 8}{6}$

(c) $\frac{3\sqrt{3}+2}{6}$ (d) None.

2. $\sin\alpha+\sin\beta+\sin\gamma$ is equal to

(a) $\frac{14+3\sqrt{2}}{6\sqrt{2}}$ (b) $\frac{5}{6}$

(c) $\frac{3+4\sqrt{2}}{6}$ (d) $\frac{1+\sqrt{2}}{2}$.

3. $\sin(\alpha-\beta)$ is equal to

(a) 1 (b) 0

(c) $\frac{1-2\sqrt{6}}{6}$ (d) $\frac{\sqrt{3}-2\sqrt{2}}{6}$

PASSAGE – III

Solutions of equations $a\sin x \pm b\cos x = c$. General value satisfying two equations.

$a\cos\theta \pm b\sin\theta = c$, where θ satisfying two equations.

(a) The equations be first converted to, where $a = r\cos\theta$, $b = r\sin\theta$.

(b) Satisfying two equations, find the common value of lying between 0 and 2π and then add $2n\pi$.

On the basis of above information, answer the following questions:

1. The number of integral values of k for which the equation 7 cos x + 5 sin x = 2 k + 1 has a solution is

(a) 4 (b) 8

(c) 10 (d) 12.

2. If $\cos 3x+\sin\left(2x-\frac{7\pi}{6}\right)=-2$, then x is

(a) $(6n-1)\frac{\pi}{3}$ (b) $(6n+1)\frac{\pi}{3}$

(c) $(2n+1)\frac{\pi}{3}$ (d) None.

3. The value of x such that $-\pi<x<\pi$ and satisfying the equation

$$8^{\left[1+|\cos x|+|\cos x|^2+|\cos x|^3+\dots\dots\dots to\ \infty\right]}=4^3,$$

then x is

(a) $\pi/3$ (b) $2\pi/3$

(c) $-\pi/3$ (d) $-2\pi/3$

4. The number of solutions of the equation $e^{\sin x}-e^{-\sin x}-4=0$ is

(a) 1 (b) 2

(c) 4 (d) 0.

PASSAGE – IV

Suppose equation is $f(x)-g(x)=0$ or $f(x)=g(x)=y$, say, then draw the graphs of y = f(x) and y = g(x).

If graphs of y = f(x) and y = g(x) cuts at one, two, three,………, no points, then number of solutions are one, two, three, ……, zero respectively.

On the basis of above information, answer the following questions:

1. The number of solutions of $\sin x=\frac{|x|}{10}$ is

(a) 4 (b) 6

(c) 5 (d) None.

2. Total number of solutions of the equation $3x+2\tan x=\frac{5\pi}{2}$, where $x\in[0, 2\pi]$ is

(a) 1 (b) 2

(c) 3 (d) 4

3. Total number of solutions of $\sin\{x\}=\cos\{x\}$, where $\{,\}$ = F.P.F, in is

(a) 3 (b) 5

(c) 7 (d) None.

4. If it has no solution, when , then

(a) (b)

(c) (d)

5. Total number of solutions of, where is

(a) 3 (b) 4
(c) 5 (d) 6

PASSAGE – V

Whenever the terms of two sides of the equation are of different nature, then equations are known as Non standard form, some of them are in the form of an ordinary equation but cannot be solved by standard procedures.

Non standard problems require high degree of logic, they also require the use of graphs, inverse properties of functions, inequalities.

On the basis of above information, answer the following questions:

1. The number of solutions of the equation

$2\cos\left(\frac{x}{2}\right)=3^x+3^{-x}$ is

(a) 1 (b) 2
(c) 3 (d) None.

2. The number of solution of the equation

$2\cos^2\left(\frac{x}{2}\right)\sin^2 x=x^2+x^{-2},\ 0<x\le\frac{\pi}{2}$ is

(a) 1 (b) >1
(c) 0 (d) None.

3. The number of real solutions of the equation $\sin\left(e^x\right)=5^x+5^{-x}$ is

(a) 0 (b) 1
(c) 2 (d) infinitely many.

4. If $0\le x\le 2\pi$ and

$2^{\operatorname{cosec}^2 x}\sqrt{\frac{1}{2}y^2-y+1}\le\sqrt{2}$, then the number of ordered pairs of (x, y) is

(a) 1 (b) 2
(c) 3 (d) infinitely many.

5. The number of solutions of the equation

$\sin x=x^2+x+1$ is

(a) 0 (b) 1
(c) 2 (d) None.

ANSWERS

Passage-I

1.(d) 2.(a) 3.(c) 4.(b) 5.(a)

Passage-II

1. (a, b) 2. (a, c) 3.(c)

Passage-III

1. (b) 2. (b) 3. (a, b, c, d) 4. (d)

Passage-IV

1.(b) 2.(c) 3.(b) 4.(b) 5.(b)

Passage-V

1.(a) 2.(c) 3.(a) 4.(b) 5.(a)

MATCH MATRIX

1. **Match the following columns :**

	Column - I		Column - II
(A)	The equation sin x + cos x = 2 has	(P)	One solution
(B)	The equation $\sqrt{3}\sin x+\cos x=4$ has	(Q)	Two solution
(C)	The equation $3\sin x+4\cos x=6$ has	(R)	Three solution.
(D)	The equation $\sin x\cos x=2$ has	(S)	No solution

2. Match the dollowing columns :

	Column - I		Column - II
(A)	If $\begin{vmatrix} \cos(A+B) & -\sin(A+B) & \cos(2B) \\ \sin A & \cos A & \sin B \\ -\cos A & \sin A & \cos B \end{vmatrix} = 0$ then B is	(P)	$n\pi$
(B)	If $\begin{vmatrix} \cos\theta & \sin\theta & \cos\theta \\ -\sin\theta & \cos\theta & \sin\theta \\ -\cos\theta & -\sin\theta & \cos\theta \end{vmatrix} = 0$, then θ is	(Q)	$(2n+1)\frac{\pi}{2}$
(C)	If $\begin{vmatrix} 1+\sin^2\theta & \cos^2\theta & 4\sin 4\theta \\ \sin^2\theta & 1+\cos^2\theta & 4\sin 4\theta \\ \sin^2\theta & \sin^2\theta & 1+4\sin 4\theta \end{vmatrix} = 0,$	(R)	$(2n-1)\frac{\pi}{2}$
	then θ is	(S)	$\frac{7\pi}{24}$

3. Match the following columns ;

	Column - I		Column -II
(A)	If $4\sin^4 x + \cos^4 x = 1$, then x is	(P)	$\frac{\pi}{4}$
(B)	If $\sec x.\cos(5x) + 1 = 0$, where $0 < x < 2\pi$, then x is	(Q)	$\frac{\pi}{6}$
(C)	If $81^{\sin^2 x} + 81^{\cos^2 x} = 30$, where $0 < x < 2\pi$, then x is	(R)	$-\frac{\pi}{4}$
(D)	If $2\sin^2 x + \sin^2 2x = 2$ where $0 < x < 2\pi$, then x is	(S)	$n\pi,\ n \in I$

4. Match the following columns ;

	Column - I		Column -II
(A)	If $\cos\theta + \cos 3\theta + \cos 5\theta + \cos 7\theta = 0$, then θ is	(P)	$\frac{n\pi}{2} + \frac{\pi}{6},\ n \in I$
(B)	If $\sin x - 3\sin 2x + \sin 3x = \cos x - 3\cos 2x + \cos 3x$ then x is	(Q)	$n\pi + \frac{\pi}{2},\ n \in I$
(C)	If $\sin 4\theta - \sec 2\theta = 2$, then θ is	(R)	$\frac{n\pi}{5} + \frac{\pi}{10},\ n \in I$

(D) If $\tan(x+100^\circ)=\tan(x+50^\circ)\tan x.\tan(x-50^\circ)$ then x is — (S) $\frac{n\pi}{5}$, $n \in I$

(T) $\pm\left(\frac{\pi}{3}\right)$

5. **Match the following columns :**

Column - I	Column - II
(A) The number of real roots of $\cos^7 x+\sin^4 x=1$ in $(-\pi, \pi)$, is	(P) 8
(B) The number of real roots of cosec $x = 1+\cot x$ in $(-2\pi, 2\pi)$, is	(Q) 4
(C) The number of integral values of k for which, the equation $7\cos x+5\sin x=2k+1$ has a solution is	(R) 3
(D) The number of solutions of the pair of equations $2\sin^2\theta-\cos 2\theta=0$ and $2\cos^2\theta-3\sin\theta=0$ in $[0, 2\pi]$ is	(S) 2
	(T) 7

6. **Match the following columns :**

Column - I	Column - II
(A) If $\cos 3x.\cos^3 x+\sin 3x.\sin^3 x=0$, then x is	(P) $\left(n\pi\pm\frac{\pi}{3}\right)$, $n \in I$
(B) If $\sin 3\alpha=4\sin\alpha\sin(x+\alpha)\sin(x-\alpha)$, then α is, where $\alpha \neq n\pi$	(Q) $\left(n\pi+\frac{\pi}{4}\right)$, $n \in I$
(C) If $\lvert 2\tan x-1\rvert+\lvert 2\cot x-1\rvert=2$, then x is	(R) $\left(\frac{n\pi}{4}+\frac{\pi}{8}\right)$, $n \in I$
(D) If $\sin^{10}x+\cos^{10}x=\frac{29}{16}\cos^4(2x)$, then x is	(S) $\left(\frac{n\pi}{2}\pm\frac{\pi}{4}\right)$, $n \in I$

7. **Match the following columns :**

If α & β are the roots of $a\cos\theta+b\sin\theta=c$, then

Column - I	Column - II
(A) $\sin\alpha+\sin\beta$ is	(P) $\frac{2b}{a+c}$
(B) $\sin\alpha.\sin\beta$ is	(Q) $\frac{c-a}{c+a}$

(C) $\tan\left(\frac{\alpha}{2}\right)+\tan\left(\frac{\beta}{2}\right)$ is (R) $\frac{2bc}{a^3+b^3}$

(D) $\tan\left(\frac{\alpha}{2}\right).\tan\left(\frac{\beta}{2}\right)$ is (S) $\frac{c^2-a^2}{a^2+b^2}$

8. **Match the following columns :**

Column - I		**Column - II**	
(A)	If $\sin 5x = 16\sin^5 x$, then x is	(P)	$(2n+1)\frac{\pi}{4}$, $n\in I$
(B)	If $4\cos^2 x.\sin x - 2\sin^2 x = 3\sin x$, then x is	(Q)	$n\pi$, $n\in I$
(C)	If $\tan^2(2x)+\cot^2(2x)+2\tan(2x)+2\cot(2x)=0$	(R)	$n\pi+\frac{\pi}{6}$, $n\in I$
(D)	If $\tan^2 x.\tan^2 3x.\tan 4x = \tan^2 x - \tan^2 3x + \tan 4x$ then x is	(S)	$n\pi+(-1)^n\frac{\pi}{8}$, $n\in I$

9. **Observe the following columns :**

Column - I		**Column - II**	
(A)	If $\cos(6\theta)+\cos(4\theta)+\cos(2\theta)+1=0$, then θ is	(P)	$2n\pi$, $n\in I$
(B)	If $3-2\cos\theta-4\sin\theta-\cos 2\theta+\sin 2\theta=0$, then θ is	(Q)	$\frac{n\pi}{3}$, $n\in I$
(C)	If $\cos\theta.\cos 2\theta.\cos 3\theta=\frac{1}{4}$, then θ is	(R)	$(4n+1)\frac{\pi}{2}$, $n\in I$
(D)	If $\sin(5\theta)+\sin(\theta)=\sin(3\theta)$, then θ is	(S)	$(2n+1)\frac{\pi}{12}$, $n\in I$

ANSWERS

1. (A) → (S), (B) → (S), (C) → (S), (D) → (S)
2. (A) → (Q), (B) → (R), (C) → (S).
3. (A) → (S), (B) → (P), (C) → (Q), (D) → (P, Q)
4. (A) → (S), (B) → (P), (C) → (Q, R), (D) → (T)
5. (A) → (R), (B) → (S), (C) → (P), (D) → (S)
6. (A) → (S), (B) → (P), (C) → (Q), (D) → (R)
7. (A) → (R), (B) → (S), (C) → (P), (D) → (Q)
8. (A) → (Q), (B) → (R), (C) → (S), (D) → (P)
9. (A) → (S), (B) → (R), (C) → (Q), (D) → (P).

ASSERTION & REASON

Codes :

(A) Both A and R are individually true and R is the correct explanation of A

(B) Both A and R are individually true and R is not the correct explanantion of A.

(C) A is true but R is false.

(D) A is false but R is true.

1. Assertion (A): The number of real solutions of $\sin x = x^2 + x + 1$ is 1

   Reson (R): since $|\sin x| \le 1$

   (a) A (b) B
   (c) C (d) D

2. Assertion (A): The number of real solutions of $\cos x = 3^x + 3^{-x}$

   Reason (R): since $|\cos x| \le 1$

   (a) A (b) B
   (c) C (d) D

3. Assertion (A): The maximum value of $3 \sin x + 4 \cos x + 10$ is 15

   Reason (R): The least value of $2\sin^2 x + 4$ is 4

   (a) A (b) B
   (c) C (d) D

4. Assertion (A): The greatest value of $\sin^4 x + \cos^2 x$ is 1

   Reason (R): The range of the function $f(x) = \sin^2 x + \cos^2 x$ is 1

   (a) A (b) B
   (c) C (d) D

5. Assertion (A): $a \cos x + b \cos 3x \le 1$ for every x in R

   Reason (R): since $|b| \le 1$

   (a) A (b) B
   (c) C (d) D

6. Assertion (A): The set of values of x for which $\frac{\tan 3x - \tan 2x}{1 + \tan 3x . \tan 2x} = 1$ is ϕ

   Reason (R): since tanx is not defined at $x = (2n+1)\frac{\pi}{2}, n \in I$

   (a) A (b) B
   (c) C (d) D

7. Assertion (A): The number of solutions of the equation $\tan x + \sec x = 2 \cos x$ lying in the interval $[0, 2\pi]$ is 2

   Reason (R): The number of solutions of the equation $3\sin^2 x - 7\sin x + 2 = 0$ in $[0, 5\pi]$ is 6

   (a) A (b) B
   (c) C (d) D

8. Assertion (A): The number of solutions of $\tan x . \tan 4x = 1$ in $(0, \pi)$ is 5

   Reason (R): The number of solutions of $|\cos x| = \sin x$ in $[0, 4\pi]$ is 4

   (a) A (b) B
   (c) C (d) D

9. Assertion (A): If $\tan\left(\frac{\pi}{2}\sin\theta\right) = \cot\left(\frac{\pi}{2}\cos\theta\right)$, then $\sin\theta + \cos\theta = \pm\sqrt{2}$

   Reason (R): $-\sqrt{2} \le \sin\theta + \cos\theta \le \sqrt{2}$

   (a) A (b) B
   (c) C (d) D

10. Assertion (A): $\sin A = \sin B = \sin C = 2\sin(18^\circ)$

    Reason (R): If $\cos A = \tan B$, $\cos B = \tan C$, $\cos C = \tan A$

    (a) A (b) B
    (c) C (d) D

ANSWERS

1.(a) **2.**(a) **3.**(b) **4.**(b) **5.**(a) **6.**(a) **7.**(b) **8.**(b) **9.**(d) **10.**(a)

QUESTIONS ASKED IN ROORKEE EXAM

1. Find all the angles θ between π and $-\pi$ that satisfy the equation

$$5\cos(2\theta)+2\cos^2\left(\frac{\theta}{2}\right)+1=0.$$

Roorkee – 1984]

2. No questions asked in **1985, 1986**.

3. Find the general solution to the following equation

$2(\sin x - \cos 2x) - \sin 2x(1+2\sin x)$
$+2\cos x = 0$

[Roorkee – 1987]

4. Solve for x and y ;

$x\cos^3 y+3x\cos y\sin^2 y=14,$

$x\sin^3 y+3x\cos^2 y\sin y=13$

[Roorkee – 1988]

5. Solve for x ; $4\sin^4 x+\cos^4 x=1$

[Roorkee – 1989]

6. Find all the values of 'a' for which the equation $\sin^4 x+\cos^4 x+\sin 2x+a=0$ is valid.
Also find the general solution of the equation.

[Roorkee Main – 1990]

7. No questions asked in **1991.**

8. Find the general solution of the equation
$(\sqrt{3}-1)\sin\theta+(\sqrt{3}+1)\cos\theta=2.$

[Roorkee – 1992]

9. No questions in **1993.**

10. Solve for θ ; $\sec\theta-\operatorname{cosec}\theta=\frac{4}{3}$

[Roorkee Main – 1994]

11. No questions asked in **1995.**

12. If $32\tan^8\theta=2\cos^2\alpha-3\cos\alpha$ and $3\cos 2\theta=1$, then find the general values of α.

[Roorkee Main – 1996]

13. No questions asked in **1997.**

14. Find the general values of x and y and satisfying the equations $5\sin x\cos y=1$,
$4\tan x=\tan y.$

[Roorkee Main – 1998]

15. No questions asked in **1999.**

16. Find the smallest positive value of x and y satisfying $x-y=\frac{\pi}{4}$, $\cot x+\cot y=2.$

[Roorkee Main – 2000]

17. Solve the following equations for x and y ;

$5^{\left(\operatorname{cosec}^2 x-3\sec^2 y\right)}=1$

$2^{\left(2\operatorname{cosec} x+\sqrt{3}|\sec y|\right)}=64$

[Roorkee Main – 2001].

ANSWERS

1. $\theta=-\frac{\pi}{3},\frac{\pi}{3},\pi-\cos^{-1}\left(\frac{3}{5}\right),-\pi+\cos^{-1}\left(\frac{3}{5}\right)$

3. $x=n\pi+(-1)^n\left(-\frac{\pi}{2}\right),\ n\in I$

4. $x=5\sqrt{5},\ y=\tan^{-1}\left(\frac{1}{2}\right)$

5. $x=2n\pi,\ (2m-1)\pi,\ 2p\pi\pm\cos^{-1}\left(\sqrt{\frac{3}{5}}\right),\ n,m,p\in I$

6. $a \ge -\frac{3}{2}$, $x = \frac{n\pi}{2} + \frac{(-1)^n}{2}\beta$, $n \in I$

8. $\theta = 2n\pi \pm \frac{\pi}{4} + \frac{\pi}{12}$, $n \in I$

10. $\theta = n\pi + \frac{1}{2}\sin^{-1}\left(\frac{3}{4}\right)$, $n \in I$

12. $\alpha = 2n\pi \pm \frac{2\pi}{3}$, $n \in I$

14. $x = \frac{1}{2}\left[(m+n)\pi + (-1)^n\left(\frac{\pi}{2}\right) + (-)^{m+1}\sin^{-1}\left(\frac{3}{5}\right)\right]$

$y = \frac{1}{2}\left[(m-n)\pi + (-1)^n\left(\frac{\pi}{2}\right) - (-)^{m+1}\sin^{-1}\left(\frac{3}{5}\right)\right]$

16. $x = \tan^{-1}(2+\sqrt{3}), y = \frac{\pi}{6}$,

$x = \tan^{-1}(2-\sqrt{3}), y = \frac{5\pi}{6}$

17. $x = \frac{\pi}{6}, \frac{5\pi}{6}, \frac{13\pi}{6}$ and $y = 2n\pi + \frac{\pi}{3}, 2n\pi + \frac{\pi}{6}$

QUESTIONS ASKED IN IIT-JEE

1. If $\tan(\theta) = -\frac{4}{3}$, then $\sin(\theta)$ is

(a) -4/5 but not 4/5 (b)* - 4/5 or 4/5

(c) 4/5 but not – 4/5 (d) None.

[IIT-JEE – 1978]

2. Solve for x : $2(\cos x + \cos 2x) + (1 + 2\cos x)\sin 2x = 2\sin x$, $-\pi \le x \le \pi$

[IIT-JEE – 1978]

3. No question asked in 1979.

4. There exists a solution of θ between 0 and 2π that satisfies the equation

$\sin^4\theta - 2\sin^2\theta - 1 = 0$

Is it true or false ?

[IIT-JEE – 1980]

5. The equation

$2\cos^2\left(\frac{x}{2}\right)\sin^2 x = x^2 + \frac{1}{x}$, $x \le \frac{\pi}{9}$ has

(a)* no solution

(b) one solution

(c) more than one real solution

(d) Cannot be said. [IIT-JEE – 1981]

6. The solution of $\sin x + \cos x = 1$ is

(a) $x = 2n\pi$ (b) $x = 2n\pi + \frac{\pi}{2}$

(c)* $x = n\pi + (-1)^n\frac{\pi}{4} - \frac{\pi}{4}$

(d) None. [IIT-JEE – 1981]

7 For all $\theta \in \left[0, \frac{\pi}{2}\right]$, show that

$\cos(\sin\theta) \ge \sin(\cos\theta)$

[IIT-JEE – 1981]

7. Find the co-ordinates of the point of intersection of the curves y = cos x and y = sin 3 x, if $-\frac{\pi}{2} \le x \le \frac{\pi}{2}$

[IIT-JEE – 1982]

8. Show that $e^{\sin x} - e^{-\sin x} = 4$ has no solution

[IIT-JEE – 1982]

9. Find all solutions of

$4\sin^2 x\cos^2 x - 2\sin^2 x = 3\sin x$

[IIT-JEE – 1983]

10. The larger of $\cos(\log\theta)$ and $\log(\cos\theta)$, $e^{-\pi/2} < \theta < \pi/2$, is ……..

[IIT-JEE – 1983]

11. Find all the solutions of

$2^{1+|\cos x|+|\cos x|^2+|\cos x|^3+|\cos x|^4+\ldots\ldots\text{to } \infty} = 4$

in $(-\pi, \pi)$

[IIT-JEE – 1984]

12. The solution set of the equations $x+y=\frac{2\pi}{3}$, $\cos x+\cos y=\frac{3}{2}$, where x and y are real is................

[IIT-JEE – 1986]

13. The general solution of $\sin x-3\sin 2x+\sin 3x = \cos x-3\cos 2x+\cos 3x$ is

(a) $n\pi+\frac{\pi}{8}$ (b)* $\frac{n\pi}{2}+\frac{\pi}{8}$

(c) $(-1)^n\frac{n\pi}{2}+\frac{\pi}{8}$ (d) $2n\pi+\cos^{-1}\left(\frac{3}{2}\right)$

[IIT-JEE – 1989]

14. The equation $(\cos P-1)x^2+(\cos P)x+\sin P=0$ in the variable x, has real roots. Then P can take any value in the interval

(a) $(0,\ 2\pi)$ (b) $(-\pi,\ 0)$

(c) $\left(-\frac{\pi}{2},\frac{\pi}{2}\right)$ (d)* $(0,\ \pi)$.

[IIT-JEE – 1990]

15. In a triangle ABC, angle A is greater than angle B. If the measure of angles A and B satisfy the equation

3 sin x – 4 sin ^{3} x – k = 0, o < k < 1, then the measure of the angle C is

(a) $\pi/3$ (b) $\pi/2$

(c)* $2\pi/3$ (d) $5\pi/6$.

[IIT-JEE – 1990]

16. The number of solutions of the equation $\sin(e^x)=5^x+5^{-x}$ is

(a)* 0 (b) 1

(c) 2 (d) infinite

IIT-JEE – 1990]

17. Show that 2 sin x + tan x > 3 x, where $0\le x\le\pi/2$ [IIT-JEE – 1990]

18. Prove that for any positive integer k,

$$\frac{\sin 2kx}{\sin x} = 2[\cos x+\cos 3x+\cos 5x+....+\cos(2k-1)x]$$

Hence, prove that $\int_0^{\frac{\pi}{2}}(\sin 2kx\cot x)\,dx=\frac{\pi}{2}$

[IIT-JEE – 1990]

19. Determine the smallest positive value of x (in degrees) for which $\tan(x+100^\circ) = \tan(x+50^\circ)\tan x.\tan(x-50^\circ)$

[IIT-JEE – 1993]

20. Let n be a positive integer such that $\sin\left(\frac{\pi}{2n}\right)+\cos\left(\frac{\pi}{2n}\right)=\frac{\sqrt{n}}{2}$. Then

(a) $6\le n\le 8$ (b) $4<n\le 8$

(c) $4\le n\le 8$ (d)* $4<n<8$

[IIT-JEE – 1994]

21. Let $2\sin^2 x+3\sin x-2>0$ and $x^2-x-2<0$ (x is measured in radians).

Then x lies in the interval

(a) $\left(\frac{\pi}{6},\frac{5\pi}{6}\right)$ (b) $\left(-1,\frac{5\pi}{6}\right)$

(c) $(-1,2)$ (d)* $\left(\frac{\pi}{6},2\right)$

[IIT-JEE – 1994]

22. The genral values of θ satisfying the equation $2\sin^2\theta-3\sin\theta-2=0$ is

(a) $n\pi + (-1)^n \frac{\pi}{6}$

(b) $n\pi + (-1)^n \frac{\pi}{2}$

(c) $n\pi + (-1)^n \frac{5\pi}{6}$

(d)* $n\pi + (-1)^n \frac{7\pi}{6}$

[IIT-JEE – 1995]

23. Find the smallest positive value of p for which the equation

$\cos(p \sin x) = \sin(p \cos x)$

[IIT-JEE – 1995]

24. Find all values of θ in the interval $\left(-\frac{\pi}{2}, \frac{\pi}{2}\right)$ satisfying the equation

$(1 - \tan\theta)(1 + \tan\theta)\sec^2\theta + 2^{\tan^2\theta} = 0$

[IIT-JEE – 1996]

25. General values of θ satisfying the equation $\tan^2\theta + \sec^2\theta = 1$ is..........................

[IIT-JEE 1996]

26. The real roots of the equation $\cos^7 x + \sin^4 x = 1$ in the interval $(-\pi, \pi)$ are……..and…………..

[IIT-JEE – 1997]

27. If cos (x – y), cos x, cos (x + y) are in H.P, then the value of $\cos x \sec\left(\frac{y}{2}\right)$ is………

[IIT-JEE – 1997]

28. Prove that the value of the function $\frac{\sin x \cos 3x}{\cos x \sin 3x}$ does not lie between 1/3 and 3 for any real x.

IIT-JEE – 1997]

29. Le n be an odd integer. If $\sin n\theta = \sum_{r=0}^{n} b_r \sin^r\theta$, for every value of θ, then

(a) $b_0 = 1, b_1 = 3$

(b)* $b_0 = 0, b_1 = n$

(c) $b_0 = -1, b_1 = n$

(d) $b_0 = 0, b_1 = n^2 - 3n + 3$

[IIT-JEE – 1998]

30. The number of values of x in the interval $(0, 5\pi)$ satisfying equation $3\sin^2 x - 7\sin x + 2 = 0$ is

(a)* 0 (b) 5

(c) 6 (d) 10.

[IIT-JEE – 1998]

31. Which of the following numbers are rational ?

(a) $\sin 15^o$ (b) $\cos 15^o$

(c)* $\sin 15^o \cos 15^o$ (d) $\sin 15^o \cos 75^o$.

[IIT-JEE – 1998]

32. In a triangle PQR, $\angle R = \pi/2$. If $\tan\left(\frac{P}{2}\right)$ and $\tan\left(\frac{Q}{2}\right)$ are the roots of the equation $ax^2 + bx + c = 0$, $(a \neq 0)$, then

[IIT-JEE – 1999]

(a)* a + b = c (b) b + c = a

(c) a + c = b (d) b = c.

33. For a positive integer n, let

$f(\theta) = \tan\left(\frac{\theta}{2}\right)(1 + \sec\theta)$

$(1 + \sec 2\theta)(1 + \sec 2^2\theta)..(1 + \sec 2^n\theta)$, then

(a) $f_2\left(\frac{\pi}{16}\right)=1$ (b) $f_3\left(\frac{\pi}{32}\right)=1$

(c) $f_4\left(\frac{\pi}{64}\right)=1$ (d) $f_5\left(\frac{\pi}{128}\right)=1$.

IIT-JEE – 1999]

34. The maximum value of

$\cos(\alpha_1).\cos(\alpha_2).......\cos(\alpha_n)$

under the restrictions

$0\le\alpha_1, \alpha_2, \alpha_3, ..., \alpha_n \le \frac{\pi}{2}$ and

$\cot(\alpha_1).\cot(\alpha_2).....\cot(\alpha_n)=1$, is

*(a) $\frac{1}{2^{n/2}}$ (b) $\frac{1}{2^n}$

(c) 2^n (d) 1.

[IIT-JEE – 2001]

35. If $\alpha+\beta=\pi/2$ and $\beta+\gamma=\alpha$, then $\tan\alpha$ equals

(a) $2(\tan\beta+\tan\gamma)$

(b) $\tan\beta+\tan\gamma$

(c)* $(\tan\beta+2\tan\gamma)$

(d) $(2\tan\beta+\tan\gamma)$.

[IIT-JEE – 2001]

36. The number of integral values of k for which the equation 7 cos x + 5 sin x = 2 k + 1 has a solution is

(a) 4 (b)* 8

(c) 10 (d) 12

[IIT-JEE – 2002]

37. No questions asked in 2003, 2004.

38. Let $a, b\in[-\pi, \pi]$ be such that

$\cos(a-b)=1$ and $\cos(a+b)=\frac{1}{e}$.

The number of pairs of a, b satisfying the system of equations is

(a) 0 (b) 1

(c) 2 (d)* 4

[IIT-JEE-2005]

39. Find the value of $t\in\left[-\frac{\pi}{2}, \frac{\pi}{2}\right]$ so that

$$2\sin t=\frac{5x^2-2x+1}{5x^2+2x+1},\ x\in R-\left\{1, -\frac{1}{3}\right\}$$

[IIT-JEE-2005]

40. The values of $\theta\in(0,2\pi)$ for which $2\sin^2\theta-5\sin\theta+2>0$ are

(a)* $\left(0,\frac{\pi}{6}\right)\cup\left(\frac{5\pi}{6},2\pi\right)$

(b) $\left(\frac{\pi}{8},\frac{5\pi}{6}\right)$

(c) $\left(0,\frac{\pi}{8}\right)\cup\left(\frac{\pi}{6},\frac{5\pi}{6}\right)$

(d) $\left(\frac{41\pi}{48},\pi\right)$ **[IIT-JEE 2006]**

41. The number of solutions of the pair of equations $2\sin^2\theta-\cos2\theta=0$ and $2\cos^2\theta-3\sin\theta=0$ in the interval $[0, 2\pi]$ is

(a) 0 (b) 1

(c) 2 (d)* 4.

[IIT-JEE-2007]

42. If $\sin\theta=\cos\phi$, then the possible values of $\frac{1}{\pi}\left(\theta\pm\phi-\frac{\pi}{2}\right)$ are............

[IIT-JEE-2008]

43. For $0<\theta<\frac{\pi}{2}$, the solutions of

$$\sum_{m=1}^{6}\operatorname{cosec}\left(\theta+\frac{(m-1)\pi}{4}\right)\operatorname{cosec}\left(\theta+\frac{m\pi}{4}\right)=4\sqrt{2}$$

is (are).

(a) $\frac{\pi}{4}$ (b) $\frac{\pi}{6}$

(c)* $\frac{\pi}{12}$ (d)* $\frac{5\pi}{12}$

[IIT-JEE-2009]

44. The number of all possible values of θ, where $0 < \theta < \pi$ for which the system of equations

$(y+z)\cos 3\theta = (xyz)\sin 3\theta$,

$x \sin 3\theta = \frac{2\cos 3\theta}{y} + \frac{2\sin 3\theta}{z}$ and

$(xyz)\sin 3\theta = (y+2z)\cos 3\theta + y \sin 3\theta$

have a solution (x_0, y_0, z_0) with $y_0, z_0 \neq 0$, is.....................

[IIT-JEE-2010]

45. The number of values of θ in the interval $\left(-\frac{\pi}{2}, \frac{\pi}{2}\right)$ such that $\theta \neq \frac{n\pi}{5}$ for $n = 0, \pm 1, \pm 2$ and $\tan\theta = \cot 5\theta$ as well as $\sin 2\theta = \cos 4\theta$ is...................

[IIT-JEE-2010]

46. No questions asked in 2011.

47. Let $\theta, \phi \in [0, 2\pi]$ be such that

$$2\cos\theta(1-\sin\varphi) = \sin^2\theta\left(\tan\left(\frac{\theta}{2}\right) + \cot\left(\frac{\theta}{2}\right)\right)\cos\varphi - 1$$

and $\tan(2\pi - \theta) > 0$ & $-1 < \sin\theta < -\frac{\sqrt{3}}{2}$. Then ϕ cannot satisfy

(a)* $0 < \phi < \frac{\pi}{2}$

(b) $\frac{\pi}{2} < \phi < \frac{4\pi}{3}$

(c)* $\frac{4\pi}{3} < \phi < \frac{3\pi}{2}$

(d)* $\frac{3\pi}{2} < \phi < 2\pi$

[IIT-JEE-2012]

ANSWERS

1. (b)
2. $-\pi, -\pi/2, -\pi/3, \pi/3, \pi$
4. False
5. (a)
6. (c)
7. $\left(\frac{\pi}{3}, \frac{1}{\sqrt{2}}\right)$
9. $n\pi,\ n\pi + (-1)^n\left(\frac{\pi}{10}\right),\ n\pi + (-1)^n\left(-\frac{3\pi}{10}\right)$
10. $\cos(\log\theta)$
11. $\pi/3, 2\pi/3, -\pi/3, -2\pi/3$
12. No solution
13. (b)
14. (d)
15. (c)
16. (a)
19. $x = 30^o$
20. (d)
21. (d)
22. (d)
23. $p = \frac{\pi}{2\sqrt{2}}$
24. $\theta = \pm\ \pi/3$
25. $n\pi$ or $n\pi \pm \frac{\pi}{3}$

26. $-\pi/2, \pi/2, 0$

27. $\pm\sqrt{2}$

29. (b)

30. (a)

31. (c)

32. (a)

33. (a, b, c, d)

34. (a)

35. (c)

36. (b)

38. (d)

39. $\left[-\frac{\pi}{2}, -\frac{\pi}{10}\right] \cup \left[\frac{3\pi}{2}, \frac{\pi}{2}\right]$

40. (a)

41. (d)

42. 0, 2.

43. (d)

44. 3

45. 3

47. (a) (c)

SELF ASSESMENT TEST - I

Time : 3 Hrs. Ch: Trigonometric Equations M.M. : 100

Give answers of the following questions.

1. If α and β are different values of x satishfying $a \cos x + b \sin x = c$, then $\tan\left(\frac{\alpha+\beta}{2}\right)$ is

   (a) a+b (b) a-b
   (c) b/a (d) a/b.

2. The number of values of x in $(0, \pi)$ satisfying the equation $3\sin^2 x - 7\sin x + 2 = 0$ is

   (a) 0 (b) 5
   (c) 6 (d) 10.

3. The number of integral values of k, for which the equation $7\cos x + 5\sin x = k + 1$ has a solution is

   (a) 4 (b) 8
   (c) 10 (d) 12.

4. The number of distinct real roots of

   $$\begin{vmatrix} \sin x & \cos x & \cos x \\ \cos x & \sin x & \cos x \\ \cos x & \cos x & \sin x \end{vmatrix} = 0 \text{ in } \left[-\frac{\pi}{4}, \frac{\pi}{4}\right] \text{ is}$$

   (a) 0 (b) 2
   (c) 1 (d) 3.

5. Find the smallest positive values of x and y satisfying $x - y = \frac{\pi}{4}$ and $\cot x + \cot y = 2$.

6. The number of all possible triplets (a_1, a_2, a_3) such that $a_1 + a_2 \cos 2x + a_3 \sin^2 x = 0$ for all x is

   (a) 0 (b) 1
   (c) 2 (d) ∞.

7. The number of pairs (x, y) satisfying the equations $\sin x + \sin y = \sin(x + y)$ and $|x| + |y| = 1$ is

   (a) 0 (b) 2
   (c) 4 (d) 6.

8. The number of the point of intersection of the two curves $y = 2\sin x$ and $y = 5x2 + 2x + 3$ is

   (a) ∞ (b) 0
   (c) 1 (d) 12.

9 A solution (x, y) of the system of equations $x - y = \frac{1}{3}$ and $\cos^2(\pi x) - \sin^2(\pi y) = \frac{1}{2}$ is given by

(a) $\left(\frac{7}{6},\frac{5}{6}\right)$ (b) $\left(\frac{2}{3},\frac{1}{3}\right)$

(c) $\left(-\frac{5}{6},-\frac{7}{6}\right)$ (d) $\left(\frac{13}{6},\frac{11}{6}\right)$

10.. If tan (cotx) = cot (tan x), then the value of sin 2 x is

(a) $(2n+1)\frac{\pi}{4}$ (b) $\frac{4}{(2n+1)\pi}$

(c) $(2n+1)\frac{\pi}{2}$ (d) $\frac{4\pi}{(2n+1)}$.

SELF ASSESMENT TEST - II

Time : 3 Hrs. Ch: Trigonometric Equations M.M. : 100

Give answers the following questions :

1. Solve for x : sinx + cosx $= \min_{a\in R}\{1, a^2-4a+6\}$.

2. Solve for x : $|\cos x|$ = cos x - 2 sin x.

3. Find all values of α for which the equation $\sin^4 x+\cos^4 x+\sin 2x+\alpha=0$ is valid. Also, find the general solution of the equation.

4. Find the smallest positive root of the equation $\sqrt{\sin(1-x)}=\sqrt{\cos x}$.

5. Solve for x and y:

$$3^{\sin x+\cos y}=1\ ,\ 25^{\sin^2 x+\cos^2 y}=5.$$

6. Solve for x and y:
$1-2x-x^2=\tan^2(x+y)+\cos^2(x+y)$.

7. Solve for x and y: $x^2+2x\sin(xy)+1=0$.

8. Find the range of y such that the equation in x, $y+\cos x=\sin x$ has a real solution.

For y = 1, find x such that $0<x<2\pi$.

9. For what values of k, the equation sinx + cos (k+ x) + cos (k - x) = 2 has real solutions.

10. If $\sin 5\theta = a\sin^5\theta + b\sin^3\theta + c\sin\theta + d$ for every $\theta \in R$, then prove that

(i) a + b + c + d = 0

(ii) a + b + c = 1

(iii) 5a + 4 b = 0.

(iv) b + 4 c = 0.

ANSWERS

Self Assesment Test-I

1.(a) 2.(d) 3.(a)

4.(b) 5. $x=\frac{5\pi}{2}, y=\frac{\pi}{6}$

6.(d) 7.(d) 8.(b) 9.(a, c, d) 10.(b).

Self Assesment Test-II

1. $x=n\pi+(-1)^n\frac{\pi}{4}-\frac{\pi}{4},\ n\in Z$

2. $x=2m\pi,(2m+1)\pi+\frac{\pi}{4}, m\in Z$.

3. $x=\frac{n\pi}{2}+\frac{(-1)^n}{2}\sin^{-1}\{1-\sqrt{2\alpha+3}\}$, where $\alpha\in\left[-\frac{3}{2},\frac{1}{2}\right]$.

4. $x=\frac{1}{2}+\frac{7\pi}{4}$

5. $\left(\frac{7\pi}{6},\frac{\pi}{3}\right),\left(\frac{7\pi}{6},\frac{5\pi}{3}\right),\left(\frac{11\pi}{6},\frac{\pi}{3}\right),\left(\frac{11\pi}{6},\frac{5\pi}{3}\right)$.

6. $x=-1$, $y=n\pi\pm\frac{\pi}{4}+1$, $n\in Z$

7. $x=\pm1$, $y=2n\pi+\frac{3\pi}{2}$.

8. $x=\pi,\frac{\pi}{2}$.

9. $n\pi-\frac{\pi}{4}\le k\le n\pi+\frac{\pi}{4}, n\in Z$.

CHAPTER 4

TRIGONOMETRIC INEQUALITY

4.1. INTRODUCTION

Before discussing trigonometric inequation, we should know some basic concepts on Real numbers , intervals as well as algebraic inequation. A solid foundation can help you improve your talent and create interest in the subject.

4.2. BASIC CONCEPTS ON THE NUMBER SYSTEM :

(i) **Natural Number** : All counting numbers are called natural numbers. It is denoted as N .

Thus N = { 1, 2, 3, 4, ...}

(ii) **Whole Number :** Natural number, along with zero, is called a whole number . It is denoted as W. Thus W = { 0, 1, 2, 3, 4, ...}

(iii) **Integer** : Natural number along with zero and their negatives is called an integer . It is denoted by I or Z .

Thus I = { ....-3,-2,-1 ,0 , 1, 2, 3, 4,.........}

(iv) **Rational Number** : It is a ratio of two integers , whose denominator is non zero .

It is denoted by Q.

Thus $Q = \left\{\frac{1}{2}, \frac{3}{4}, \frac{6}{4}, \frac{7}{2}, 0, 1, 2, 3, ...\right\}$

(v) **Irrational Number** : It is not expressible as a ratio of two integers . It is denoted by R-

Thus R - Q = $\left\{\sqrt{2}, \sqrt[4]{10}, \sqrt[100]{2}, \sqrt[2013]{10}, ...\right\}$

(vi) **Real Number** : All numbers are said to be real numbers other than even root of a negative number . It is denoted R.

Thus $\{1, 2, 3, \frac{1}{2}, \frac{3}{4}, \sqrt[3]{5}, \sqrt[2013]{10}, \sqrt[3]{-8}, \sqrt[40]{50}, \sqrt[2014]{2013}, ,,,,\}$

(vii) **Imaginary Number** : The square root of a negative number is called an imaginary number.

It is denoted by C - R

Thus C - R = $\left\{\sqrt[2]{-3}, \sqrt[4]{-10}, \sqrt[6]{-16}, \sqrt[2]{-1}......\right\}$

(viii) **Complex Number** : All numbers are said to be complex Numbers. It is denoted by C .

Thus C = { $2 + 3i, 3 + 4i, i, -i, -3 + 2i,$

$1, 2, 3, \sqrt{2}, \sqrt{5}, \sqrt[5]{4}, \sqrt[4]{-3}, ...$ }

(ix) **Prime Number** : A natural number is said to be a prime number if it has only two factors .

Thus , 2, 3, 5, 7, 11, 13 , 17 ,.... are prime numbers .

Note :
1. 1 is not a prime number
2. 2 is only one even prime number

(x) **Composite Number** : A number is said to be a composite number if it has at least 3 distinct factors . Thus , 4 , 6, 8, 9 , 10 , 12 ,..... are composite numbers.

Note :
1. 4 is the smallest composite number
2. Numbers , which are not prime are also known as composite numbers (except 1)

(xi) **Co-prime number** : If the H.C.F of two natural numbers is one , then those numbers are co-prime numbers.

Thus (1, 2) , (2, 3) (1, 3) , (3, 4) , (4, 5) are co-prime numbers .

Note :
1. Consecutive natural numbers are co-prime.
2. Two distinct prime numbers are co-prime.
3. Co-prime numbers need not be a prime no.

(xii) **Twin prime numbers** : If the difference of two prime numbers is two , then those numbers are known as twin prime numbers .

Thus (3, 5), (5, 7), (11, 13), (17, 19), (29, 31) ,..... are twin prime numbers.

4.3. PROPERTIES OF REAL NUMBERS :

Let a, b, c $\in R$. Then

(i) If $a > b$ and $b > c$, then $a > c$

(ii) If $a > b$, then $a + c > b + c$.

(iii) If $a > b$ and $c > d$, then $a + c > b + d$.

(iv) If $a > b$ and $c > d$, then $a - d > b + c$

(v) If $a > b$ and $c > 0$, then $ac > bc$

(vi) If $a > b$ and $c < 0$, then $ac < bc$

(vii) If $a > b > 0$ and $c > d > 0$, then $ac > bd$.

(viii) If $a > b > 0$, then $\frac{1}{b} > \frac{1}{a}$

(ix) If $a > b > 0$ and $c > d > 0$, then $\frac{a}{d} > \frac{b}{c}$.

(x) If $a > b > 0$ and p and q be +ve integers, then $a^{p/q} > b^{p/q}$ and $b^{-p/q} > a^{-p/q}$

4.4. BASIC CONCEPTS ON INTERVALS :

(i) **Interval :** It is the gap between two real numbers. Let a, b $\in$ R such that $a < b$. There are four types of intervals :

(a) **Open interval :** Let $1, 2 \in R$. Then its open interval is denoted by $(1, 2)$.

If $x \in (1, 2)$, then we write it $1 < x < 2$.

(b) **Closed interval :** Let $3, 4 \in R$. Then its closed interval is denoted by $[3, 4]$

If $x \in [3, 4]$, then we write it $3 \le x \le 4$.

(c) **Semi-open interval :** Let $5, 6 \in R$. Then its semi-open interval is denoted by $[5, 6)$.

If $x \in [5, 6)$, then we write it $5 \le x < 6$.

(d) **Semi-closed Interval :** Let $1, 10 \in R$. Then its semi-closed interval is denoted by $(1, 10]$. If $x \in (1, 10]$, then we write it $1 < x \le 10$.

4.5. LINEAR INEQUALITIES :

Let $x + 4 > 3$, $2x + 3 \ge x + 5$, $3x - 4 \le 7$ etc. each is a linear inequality of the variable x.

Ex - 1: Solve : $2x + 3 > 5$

Soln. We have, $2x + 3 > 5$

$\Rightarrow 2x > 5 - 3$

$\Rightarrow 2x > 2$

$\Rightarrow x > 1$

$\Rightarrow x \in (1, \infty)$

Ex-2. Solve : $3x - 4 \ge 2x - 1$

Soln. We have, $3x - 4 \ge 2x - 1$

$\Rightarrow 3x - 2x \ge 4 - 1 = 3$

$\Rightarrow x \ge 3$

$\Rightarrow x \in [3, \infty)$

Ex-3. Solve : $5x - 7 \le 3x + 3$

Soln. We have, $5x - 7 \le 3x + 3$

$\Rightarrow 5x - 3x \le 3 + 7$

$\Rightarrow 2x \le 10$

$\Rightarrow x \le 5$

$\Rightarrow x \in (-\infty, 5]$

4.6. POLYNOMIAL INEQUALITIES :

Type - I : Co-efficient of the highest power of x is +ve

Ex-1: Solve : $x^2 - 3x + 2 > 0$

Soln. We have, $x 2 - 3x + 2 > 0$

$\Rightarrow (x - 1)(x - 2) > 0$

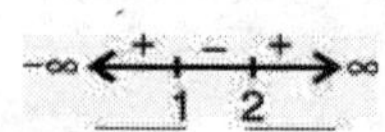

Hence, $x \in (-\infty, 1) \cup (2, \infty)$

Ex-2: Solve : $x^2 - 5x + 6 \ge 0$

Soln. We have, $x^2 - 5x + 6 \ge 0$

$\Rightarrow x^2 - 3x - 2x + 6 \ge 0$

$\Rightarrow \quad (x-3)(x-2) \geq 0$

Hence, $x \in (-\infty, 2] \cup [3, \infty)$

Ex-3. Solve: $^2 - x - 2 < 0$

Soln. We have $x^2 - x - 2 < 0$

$\Rightarrow \quad x^2 - 2x + x - 2 < 0$

$\Rightarrow \quad x(x-2) + 1(x-2) < 0$

$\Rightarrow \quad (x-2)(x+1) < 0$

Hence, $x \in (-1, 2)$

Ex-4. Solve: $x^2 - 6x + 8 \leq 0$

Soln. We have, $x^2 - 6x + 8 \leq 0$

$\Rightarrow \quad x^2 - 4x - 2x + 8 \leq 0$

$\Rightarrow \quad (x-2)(x-4) \leq 0$

Hence, $x \in [2, 4]$

Ex-5. Solve : $x^3 - 6x^2 + 11x - 6 \geq 0$

Soln. We have, $x^3 - 6x^2 + 11x - 6 \geq 0$

$\Rightarrow \quad x^3 - x^2 - 5x^2 + 5x + 6x - 6 \geq 0.$

$\Rightarrow \quad x^2(x-1) - 5x(x-1) + 6(x-1) \geq 0$

$\Rightarrow \quad (x-1)(x^2 - 5x + 6) \geq 0$

$\Rightarrow \quad (x-1)(x-2)(x-3) \geq 0.$

Hence, $x \in [1, 2] \cup [3, \infty)$

Type - II : Reduce the co-efficient of the highest power of the unknown quantity is +ve and then solve as type - I

Ex-1. Solve: $-x^2 + 3x - 2 > 0$

Soln. We have, $-x^2 + 3x - 2 > 0.$

$\Rightarrow \quad x^2 - 3x + 2 < 0$

$\Rightarrow \quad (x-1)(x-2), 0$

Hence, $x \in (1, 2)$

Ex-2. Solve: $-x^3 + 6x^2 - 12x + 8 . \geq 0$

Soln. We have, $-x^3 + 6x^2 - 12x + 8 \geq 0$

$\Rightarrow \quad x^3 - 6x^2 + 12x - 8 \leq 0$

$\Rightarrow \quad x^3 - 3.x^2.2 + 3.x.2^2 - 2^3 \leq 0$

$\Rightarrow \quad (x-2)^3 \leq 0$

Hence, $x \in (-\infty, 2]$

Ex-3. Solve: $-x^3 + 5x^2 - 4 \geq 0$

Soln. We have, $-x^3 + 5x^2 - 4 \leq 0$

$\Rightarrow \quad x^3 - 5x^2 + 4 \geq 0$

$\Rightarrow \quad x(x^2 - 5x + 4) \geq 0$

$\Rightarrow \quad x(x-1)(x-4) \geq 0$

Hence, $x \in [0, 1] \cup [4, \infty)$

Type - III : In an equation , cross multilication is not allowed but transposition is allowed.

Ex-1. Solve: $\frac{1}{x} > 1$

Soln. We have, $\frac{1}{x} > 1$

$\Rightarrow \quad \frac{1}{x} - 1 > 0$

$\Rightarrow \quad \frac{1-x}{x} > 0$

$\Rightarrow \quad \frac{x-1}{x} < 0$

Hence, $x \in (0, 1)$

Ex-2. Solve: $\frac{x-1}{x-2} \geq 1$

Soln. We have, $\frac{x-1}{x-2} \geq 1$

$\Rightarrow \quad \frac{x-2}{x-1} - 1 \geq 0$

$\Rightarrow \quad \frac{x-2-x+1}{x-1} \geq 0$

$\Rightarrow \quad \frac{-1}{x-1} \geq 0$

$$\Rightarrow \quad \frac{1}{x-1} \le 0$$

Hence, $x \in (1, \infty)$

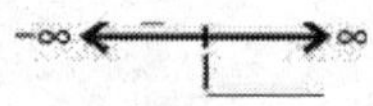

Type - IV : Skip the even power factors on the number line.

Ex-1. Solve: $\dfrac{(x-1)^2(x-2)^3}{(x+1)^5(x-3)^4} \ge 0$

Soln. We have, $\dfrac{(x-1)^2(x-2)^3}{(x+1)^5(x-3)^4} \ge 0$

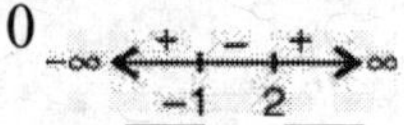

Hence, $x \in (-\infty, -1) \cup [2, \infty) - \{1,3\}$

Ex.-2: Solve: $\dfrac{x^{2013}(x-1)^{2014}}{(x-2)^{2010}(x-3)^{2012}} \ge 0$

Soln. We have, $\dfrac{x^{2013}(x-1)^{2014}}{(x-2)^{2010}(x-3)^{2012}} \ge 0$

Hence, $x \in [0, \infty) - \{1,2,3\}$

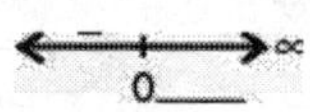

Type - V : In a quadratic polynomial, if a > 0 and D <0 or a < 0 and D < 0, skip the quadratic expression on the number line.

Ex-1. Solve : $x^2 + 2x + 3 > 0$

Soln. Here, $a = 1 > 0$ and $D = 4 - 12 = -8 < 0$

Thus, the given statement is true for every real values of x.

Hence, $x \in R$

Ex.-2. Solve : $(x^2 + x + 1)(x^2 - 4) \le 0$

Soln. We have, $(x^2 + x + 1)(x^2 - 4) \le 0$.

Clearly in $x^2 + x + 1$, $a = 1 > 0$ and $D < 0$.

Thus, the given expression reduces to

$$(x^2 + x + 1)(x+2)(x-2) \le 0$$

$$\Rightarrow \quad (x+2)(x-2) \le 0$$

Hence, $x \in [-2, 2]$

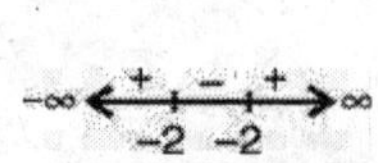

Ex-3. Solve : $x^2 + x + 10 < 0$

Soln. Here, $a = 1 > 0$ and $D < 0$. Thus, $x^2 + x + 10$ is positive for all values of x.

Therefore, the given statement is not true.

Hence, $x \in \varphi$

4.7. TRIGONOMETRIC INEQUALITIES :

Suppose we have to solve $f(x) > k$ or $f(x) < k$

When we solve the inequation, we often use the graphs of the functions $y = f(x)$ and $y = k$.

Then, the solution of the inequality $f(x) > k$ is the values of x for which the point $(x, f(x))$ of the graph of $y = f(x)$ lies above the straight line $y = k$.

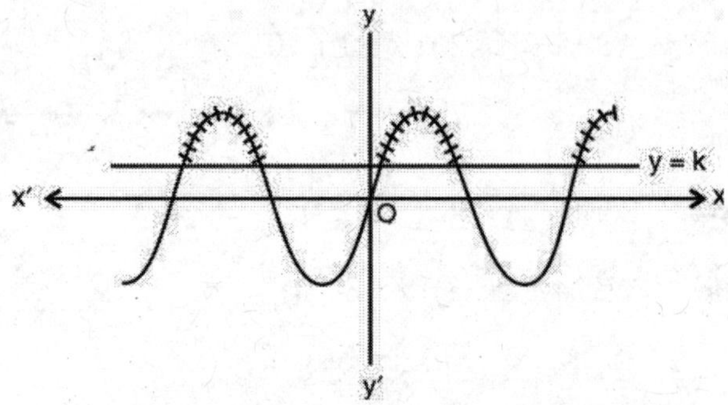

Similarly, when we solve $f(x) < k$, then the solution of the inequality $f(x) < k$ is the values of x for which the point $(x, f(x))$ of the graph of $y = f(x)$ lies below the straight line $y = k$.

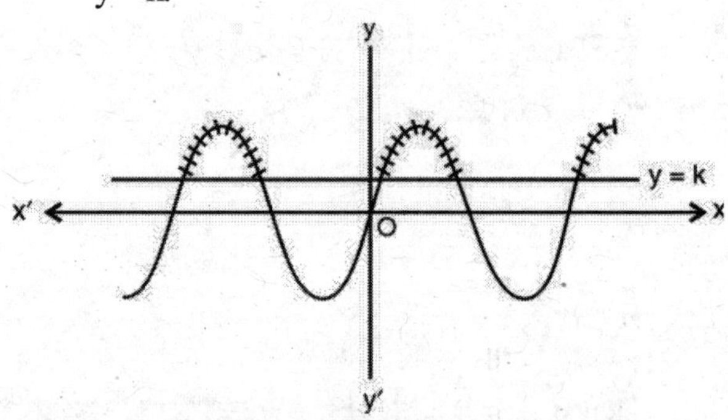

Type - I : An inequation is of the form sin x > k.

Rule : Find the smallest values of x satisfies the given inequation. and then add $2n\pi$ with that values of x

$\sin x > k, -1 < k < 1.$

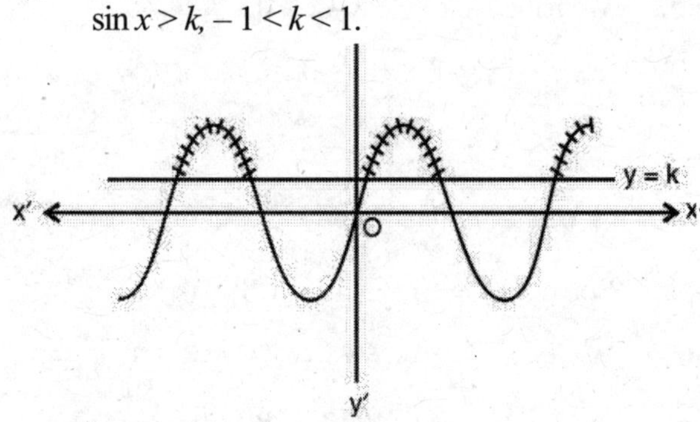

$$\Rightarrow x \in \left(2n\pi + \sin^{-1} k, (2n\pi + 1)\pi - \sin^{-1} k\right), n \in I.$$

Ex-1. Solve $\sin x > 1/2$

Soln. Here , we should construct the graph of $y = \sin x$ and $y = 1/2$

Graph:

Hence, the solution set is

$$x = \bigcup_{n \in I} \left(2n\pi + \frac{\pi}{6}, 2n\pi + \frac{5\pi}{6}\right)$$

Ex-2 : Solve: $\sin x > 1/3$

Soln. Hence, the solution set is

$$x = \bigcup_{n \in I} \left(2n\pi + \sin^{-1}\left(\frac{1}{3}\right), (2n+1)\pi - \sin^{-1}\left(\frac{1}{3}\right)\right)$$

Ex-3. Solve : $\sin x \geq 1$

Soln. We have $\sin x \geq 1$

$$\Rightarrow \quad x = (4n+1)\frac{\pi}{2}, \; n \in I$$

Ex-4. Solve : $\sin x > 0$

Soln. We have , $\sin x > 0$

$$\Rightarrow \quad x = \bigcup_{n \in I} (2n\pi, (2n+1)\pi)$$

Type - II: An in-equation is of the form sin x < k .

Rule : Find the smallest values of x satisfies the given in -equation and then add $2n\pi$ with that values of x.

$\sin x < k, -1\, k < 1.$

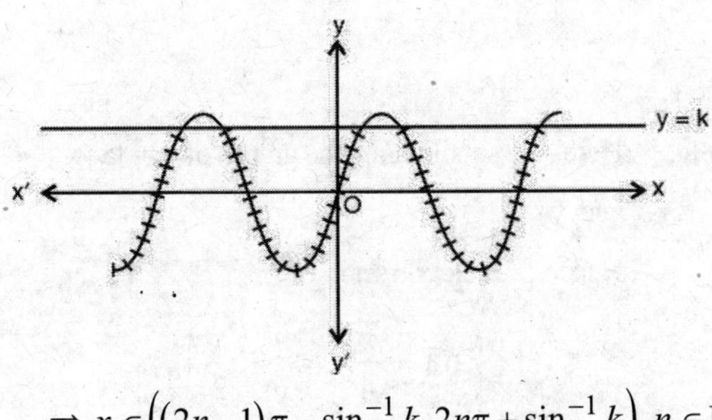

$$\Rightarrow x \in \left((2n-1)\pi - \sin^{-1} k, 2n\pi + \sin^{-1} k\right), n \in I.$$

Ex-1. Solve : $\sin x < 1/2$

Soln. Here, we should draw the graph of $y = \sin x$ and $y = 1/2$.

Hence, the solution set is

$$x = \bigcup_{n \in I} \left(2n\pi - \frac{7\pi}{6}, 2n\pi + \frac{\pi}{6}\right).$$

Ex-2. Solve : $\sin x < 1/5$

Soln. Here , we should draw the graphs of $y = \sin x$ and $y = 1/5$.

Hence, the solution set is

$$x = \bigcup_{n \in I} \left((2n-1)\pi - \sin^{-1}\left(\frac{1}{5}\right), \; 2n\pi + \sin^{-1}\left(\frac{1}{5}\right)\right)$$

Ex-3. Solve : $\sin x \leq \frac{\sqrt{3}}{2}$

Soln. Here, we should draw the graphs of y s= sin x and $y = \frac{\sqrt{3}}{2}$.

Hence, the solution set is

$$x = \bigcup_{n \in I} \left[2n\pi - \frac{4\pi}{3}, 2n\pi + \frac{\pi}{3}\right]$$

Type - III : An in-equation is of the form cos x > k

Rule : First we find the smallest interval for which x satisfies the given in-equation and then add $2n\pi$ with each values of x.

$\cos x > k, -1 < k < 1.$

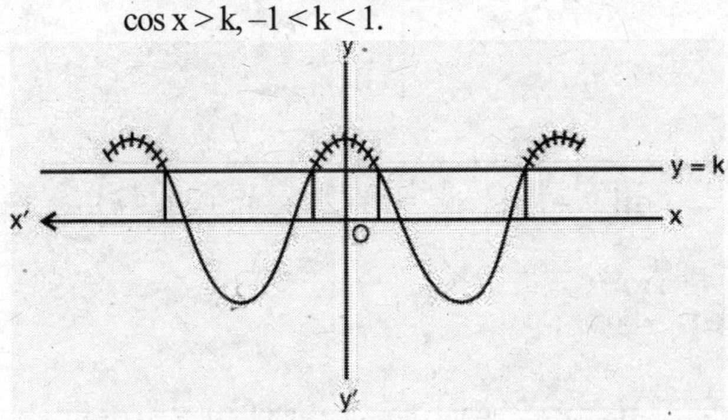

$$\Rightarrow x \in \left(2n\pi + \cos^{-1}(-k), 2n\pi + \cos^{-1}(k)\right), n \in I.$$

Ex-1. Solve : $\cos x > \frac{1}{\sqrt{2}}$

Soln. Here , we should draw the graphs of $y = \cos x$ and $y = \frac{1}{\sqrt{2}}$

Hence, the solution set is

$$x = \bigcup_{n \in I} \left(2n\pi - \frac{\pi}{4}, 2n\pi + \frac{\pi}{4}\right)$$

Ex-2. Solve : $\cos x \geq \frac{1}{2}$

Soln. Here , we should draw the graphs of $y = \cos x$ and $y = 1/2$

Hence, the solution set is

$$x = \bigcup_{n \in I} \left[2n\pi - \frac{\pi}{3}, 2n\pi + \frac{\pi}{3}\right]$$

Type - IV : An in-equation is of the form cos x < k

Rule: First we find the smallest interval for which x satisfies the given inequation and then add $2n\pi$ with each values of x.

$\cos x < k, -1 < k < 1.$

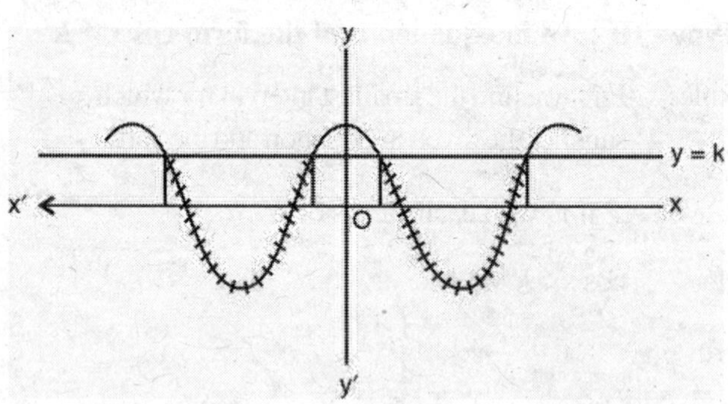

$$\Rightarrow x \in \left((2n+1)\pi + \cos^{-1}(k), (2\pi+2)\pi + \cos^{-1} k\right), n \in I.$$

Ex-1. Solve : $\cos x < \frac{1}{3}$

Soln. Here, we should draw the graphs of $y = \cos x$ and $y = \frac{1}{3}$

Hence , the solution set is

$$x = \bigcup_{n \in I} \left(\begin{array}{c} (2n-1)\pi - \cos^{-1}\left(\frac{1}{3}\right), \\ 2n\pi + \cos^{-1}\left(\frac{1}{3}\right) \end{array} \right)$$

Ex-2. Solve : $\cos x \leq \frac{\sqrt{3}}{2}$

Soln. Here , we should draw the graphs of $y = \cos x$ and $y = \frac{\sqrt{3}}{2}$

Hence, the solution set is

$$x = \bigcup_{n \in I} \left[(2n-1)\pi - \frac{\pi}{6}, 2n\pi + \frac{\pi}{6}\right]$$

Type - V : An in-equation is of the form tan x > k

Rule : First we find the smallest interval for which x satisfies the given in - equation and then add $n\pi$ with each values of x.

$$\tan x > k \Rightarrow x = \bigcup_{n \in I} \left(n\pi + \frac{\pi}{4}, n\pi + \frac{\pi}{2}\right)$$

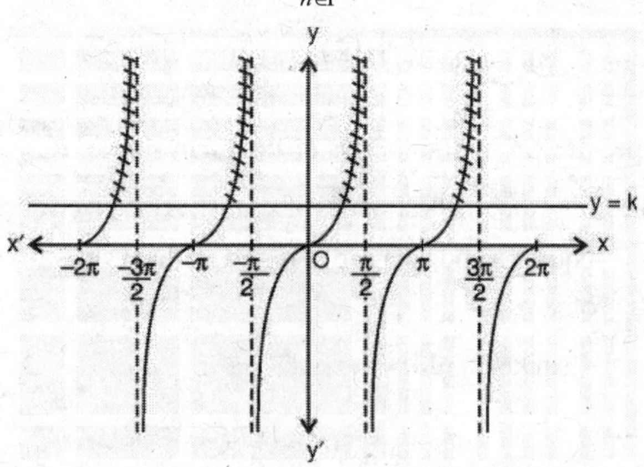

Ex-1. Solve : $\tan x > 1$

Soln. Here , we should draw the graphs of $y = \tan x$ and $y = 1$.

Hence, the solution set is

$$x = \bigcup_{n \in I} \left(n\pi + \frac{\pi}{4}, n\pi + \frac{\pi}{2}\right).$$

Ex-2. Solve : $\tan x \geq \frac{1}{\sqrt{3}}$

Soln. Here , we should draw the graphs of $y = \tan x$

and $y = \frac{1}{\sqrt{3}}$

Hence, the solution set is

$$x = \bigcup_{n \in I} \left[n\pi + \frac{\pi}{6},\ n\pi + \frac{\pi}{2} \right)$$

Ex-3. Solve : $\tan x > 2$

Soln. Here, we should draw the graphs of $y = \tan x$ and $y = 2$.

Hence, the solution set is

$$x = \bigcup_{n \in I} \left(n\pi + \tan^{-1}(2),\ n\pi + \frac{\pi}{2} \right)$$

Type - VI : An in-equation is of the form tan x < k

Rule: First we find the smallest interval for which x satisfies the given in equation and then add $n\pi$ with each values of x.

$$\tan x < k \Rightarrow x = \bigcup_{n \in I} \left(n\pi - \frac{\pi}{4},\ n\pi + \tan^{-1} k \right)$$

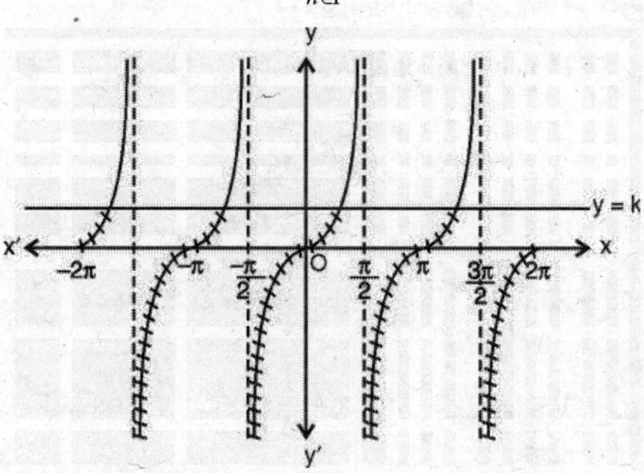

Ex-1. Solve : $\tan x < 1$

Soln. Here , we should draw the graphs of $y = \tan x$ and $y = 1$.

Hence, the solution set is

$$x = \bigcup_{n \in I} \left(n\pi - \frac{\pi}{2},\ n\pi + \frac{\pi}{4} \right)$$

Ex-2. Solve: $\tan x \le \sqrt{3}$

Soln. Here , we should draw the graphs of $y = \tan x$ and $y = \sqrt{3}$

Hence, the solution set is

$$x = \bigcup_{n \in I} \left[n\pi - \frac{\pi}{2},\ n\pi + \frac{\pi}{6} \right]$$

4.8 Some solved examples.

Ex-1. Solve : $\sin x > \cos x$

Soln. We have $\sin x > \cos x$

$$\Rightarrow \sin x - \cos x > 0$$

$$\Rightarrow \frac{1}{\sqrt{2}} \sin x - \frac{1}{\sqrt{2}} \cos x > 0$$

$$\Rightarrow \sin\left(x - \frac{\pi}{4} \right) > 0$$

$$\Rightarrow x \in \left(2n\pi + \frac{\pi}{4},\ (2n+1)\pi + \frac{\pi}{4} \right)$$

Hence, the solution set is

$$x = \bigcup_{n \in I} \left(2n\pi + \frac{\pi}{4},\ (2n+1)\pi + \frac{\pi}{4} \right)$$

Ex-2. Solve : $\cos x > \sin x$

Soln. We have $\cos x > \sin x$

$$\Rightarrow \cos x - \sin x > 0$$

$$\Rightarrow \frac{1}{\sqrt{2}} \cos x - \frac{1}{\sqrt{2}} \sin x > 0$$

$$\Rightarrow \cos\left(x + \frac{\pi}{4} \right) > 0$$

$$\Rightarrow x \in \left(2n\pi - \frac{3\pi}{4},\ 2n\pi + \frac{\pi}{4} \right)$$

Hence, the solution set is

$$x = \bigcup_{n \in I} \left(2n\pi - \frac{3\pi}{4},\ 2n\pi + \frac{\pi}{4} \right)$$

Ex-3. Solve: $-\frac{1}{2} \le \cos x < \frac{1}{\sqrt{2}}$

Soln. We have $\cos x < \frac{1}{\sqrt{2}}$ and $\cos x \ge -\frac{1}{2}$.

$\Rightarrow \quad x \in \left(2n\pi + \frac{\pi}{4}, 2n\pi + \frac{7\pi}{4}\right)$

and

$x \in \left[2n\pi - \frac{\pi}{3}, 2n\pi + \frac{\pi}{3}\right]$

Hence, the solution set is

$$x = \bigcup_{n \in I} \left(2n\pi + \frac{\pi}{4}, 2n\pi + \frac{7\pi}{4}\right) \bigcup \left[2n\pi + \frac{\pi}{3}, 2n\pi + \frac{\pi}{3}\right]$$

Ex-4: Solve: $|\sin x + \cos x| = |\sin x| + |\cos x|$

Soln. We have $|\sin x + \cos x| = |\sin x| + |\cos x|$

As we know that, if

$|f(x) + g(x)| = |f(x)| + |g(x)|$

then $f(x)g(x) \geq 0$

Thus, $\sin x \cos x \geq 0$

$\Rightarrow \quad \sin 2x \geq 0$

$\Rightarrow \quad x \in \left[n\pi, n\pi + \frac{\pi}{2}\right]$

Hence, the solution set is

$$x = \bigcup_{n \in I} \left[n\pi, n\pi + \frac{\pi}{2}\right]$$

Ex-5. Solve: $\sin x \sin 2x < \sin 3x \sin 4x$,

$\forall x \in \left(0, \frac{\pi}{2}\right)$

Soln. We have, $\sin x \sin 2x < \sin 3x \sin 4x$

$\Rightarrow \quad 2 \sin x \sin 2x < 2 \sin 3x \sin 4x$

$\Rightarrow \quad \cos x - \cos 3x < \cos x - \cos 7x$

$\Rightarrow \quad \cos 3x > \cos 7x$

$\Rightarrow \quad \cos 3x - \cos 7x > 0$

$\Rightarrow \quad 2 \sin 5x \sin 2x > 0$

$\Rightarrow \quad \sin 5x > 0$ (since $\sin 2x$ is +ve for $0 < x < \pi/2$)

$\Rightarrow \quad 0 < 5x < \pi$

$\Rightarrow \quad 0 < x < \frac{\pi}{5}$

Hence, the solution set is

$$x = \left(0, \frac{\pi}{5}\right) \cup \left(0, \frac{\pi}{2}\right)$$

Ex-6. Solve: $\cos x - \sin x - \cos 2x > 0$, $\forall x \in (0, 2\pi)$

Soln. We have $\cos x - \sin x - \cos 2x > 0$

$\Rightarrow \quad (\cos x - \sin x) - (\cos^2 x - \sin^2 x) > 0$

$\Rightarrow \quad (\cos x - \sin x)(1 - \cos x - \sin x) > 0.$

$\Rightarrow \quad (\sin x - \cos x)(\sin x + \cos x - 1) > 0$

$\Rightarrow \quad \sin\left(x - \frac{\pi}{4}\right).(\sin x + \cos x - 1) > 0$

Hence, the solution set is

Hence, the solution set is

$$x \in \left(0, \frac{\pi}{4}\right) \cup \left(\frac{5\pi}{4}, 2\pi\right).$$

Ex-7. Solve: $\frac{5}{4}\sin^2 x + \frac{1}{4}\sin^2 2x > \cos 2x$

Soln. We have, $\frac{5}{4}\sin^2 x + \frac{1}{4}\sin^2 2x > \cos 2x$

$\Rightarrow \quad 5(2\sin^2 x) + 2(\sin^2 2x) > 8\cos 2x$

$\Rightarrow \quad 5(1 - \cos 2x) + 2(1 - \cos^2 2x) > 8\cos 2x.$

$\Rightarrow \quad 5 - 5\cos 2x + 2 - 2\cos^2 2x - 8\cos 2x > 0$

$\Rightarrow \quad 2\cos^2 2x + 13\cos 2x - 7 < 0$

$\Rightarrow \quad 2\cos^2 2x + 14\cos 2x - \cos 2x - 7 < 0$

$\Rightarrow \quad 2\cos 2x(\cos 2x + 7) - (\cos 2x + 7) < 0$

$\Rightarrow \quad (\cos 2x + 7)(2\cos 2x - 1) < 0$

$\Rightarrow \quad 2\cos 2x - 1 < 0$

$\Rightarrow \quad \cos 2x < 1/2$

$\Rightarrow \quad x \in \left(n\pi + \frac{\pi}{6}, n\pi + \frac{5\pi}{6}\right)$

Hence, the solution set is

$$x = \bigcup_{n \in I}\left(n\pi + \frac{\pi}{6}, n\pi + \frac{5\pi}{6}\right)$$

Ex.-8 Solve: $6\sin^2 x - \sin x \cos x - \cos^2 x > 2$

Soln. We have, $6\sin^2 x - \sin x \cos x - \cos^2 x > 2$

$\Rightarrow \quad 6\sin^2 x - \sin x \cos x - \cos^2 x > 2(\sin^2 x + \cos^2 x)$

$\Rightarrow \quad 4\sin^2 x - \sin x \cos x - 3\cos^2 x > 0$

$\Rightarrow \quad 4\tan^2 x - \tan x - 3 > 0$

$\Rightarrow \quad 4\tan^2 x - 4\tan x + 3\tan x - 3 > 0$

$\Rightarrow \quad 4\tan x(\tan x - 1) + 3(\tan x - 1) > 0$

$\Rightarrow \quad (\tan x - 1)(4\tan x + 3) > 0$

$\Rightarrow \quad \tan x < -3/4$ and $\tan x > 1$

$\Rightarrow \quad x \in \left(n\pi - \frac{\pi}{2}, n\pi - \tan^{-1}\left(\frac{3}{4}\right)\right)$

and $\quad x \in \left(n\pi + \frac{\pi}{4}, n\pi + \frac{\pi}{2}\right)$

Hence, the solution set is

$$x = \left(n\pi - \frac{\pi}{2}, n\pi - \tan^{-1}\left(\frac{3}{4}\right)\right) \cup \left(n\pi + \frac{\pi}{4}, n\pi + \frac{\pi}{2}\right), n \in I$$

Ex-9. Solve: $\sin^6 x + \cos^6 x > \frac{13}{16}$

Soln. We have, $\sin^6 x + \cos^6 x > \frac{13}{16}$

$\Rightarrow \quad \left(1 - 3\sin^2 x \cos^2 x\right) > \frac{13}{16}$

$\Rightarrow \quad \left(1 - \frac{3}{4}(2\sin^2 x)(2\cos^2 x)\right) > \frac{13}{16}$

$\Rightarrow \quad \left(1 - \frac{3}{4}(1 - \cos 2x)(1 + \cos 2x)\right) > \frac{13}{16}$

$\Rightarrow \quad \left(1 - \frac{3}{4}(1 - \cos^2 2x)\right) > \frac{13}{16}$

$\Rightarrow \quad \left(1 - \frac{3}{4}\sin^2 2x\right) > \frac{13}{16}$

$\Rightarrow \quad \left(1 - \frac{3}{8}(2\sin^2 2x)\right) > \frac{13}{16}$

$\Rightarrow \quad \left(1 - \frac{3}{8}(1 - \cos 4x)\right) > \frac{13}{16}$

$\Rightarrow \quad \left(\frac{5}{8} + \frac{3}{8}\cos 4x\right) > \frac{13}{16}$

$\Rightarrow \quad \frac{3}{8}\cos 4x > \frac{3}{16}$

$\Rightarrow \quad \cos 4x > 1/2$

$\Rightarrow \quad 4x \in \left(2n\pi - \frac{\pi}{3}, 2n\pi + \frac{\pi}{3}\right)$

$\Rightarrow \quad x \in \left(\frac{n\pi}{2} - \frac{\pi}{12}, \frac{n\pi}{2} + \frac{\pi}{12}\right)$

Hence, the solution set is

$$x = \bigcup_{n \in I}\left(\frac{n\pi}{2} - \frac{\pi}{12}, \frac{n\pi}{2} + \frac{\pi}{12}\right)$$

Ex-10. Solve: $\cos^3 x.\cos 3x - \sin^3 x \sin 3x > \frac{5}{8}$

Soln. The given in-equation is

$\cos^3 x.\cos 3x - \sin^3 x \sin 3x > \frac{5}{8}$

$\Rightarrow \quad (\cos 3x + 3\cos x)\cos 3x$

$$(3\sin x-\sin 3x)\sin 3x>\frac{5}{2}$$

$$\Rightarrow \sin^2 3x+\cos^2 3x$$

$$+3\left(\cos 3x\cos x-\sin 3x\sin x\right)>\frac{5}{2}$$

$$\Rightarrow 3\cos 4x+1>\frac{5}{2}$$

$$\Rightarrow \cos 4x>\frac{1}{2}$$

$$\Rightarrow 2n\pi-\frac{\pi}{3}<4x<2n\pi+\frac{\pi}{3},\ n\in I$$

$$\Rightarrow \frac{n\pi}{2}-\frac{\pi}{12}<x<\frac{n\pi}{2}+\frac{\pi}{12},\ n\in I$$

Hence, the solution set is

$$x=\bigcup_{n\in I}\left(\left(\frac{n\pi}{2}-\frac{\pi}{12},\frac{n\pi}{2}+\frac{\pi}{12}\right)\right)$$

Ex-11. Solve the inequality: $\sin^6 x+\cos^6 x>\frac{13}{16}$

Soln. The given inequation is

$$\sin^6 x+\cos^6 x>\frac{13}{16}$$

$$\Rightarrow \left(\sin^2 x+\cos^2 x\right)^3$$

$$-3\sin^2 x\cos^2 x\left(\sin^2 x+\cos^2 x\right)>\frac{13}{16}$$

$$\Rightarrow 1-3\sin^2 x\cos^2 x>\frac{13}{16}$$

$$\Rightarrow 1-\frac{3}{4}\left(\sin^2 2x\right)>\frac{13}{16}$$

$$\Rightarrow 1-\frac{3}{8}\left(2\sin^2 4x\right)>\frac{13}{16}$$

$$\Rightarrow 1-\frac{3}{8}(1-\cos 4x)>\frac{13}{16}$$

$$\Rightarrow \frac{5}{8}+\frac{3}{8}\cos 4x>\frac{13}{16}$$

$$\Rightarrow \frac{3}{8}\cos 4x>\frac{3}{16}$$

$$\Rightarrow \cos 4x>\frac{1}{2}$$

$$\Rightarrow 2n\pi-\frac{\pi}{3}<4x<2n\pi+\frac{\pi}{3},\ n\in I$$

$$\Rightarrow \frac{n\pi}{2}-\frac{\pi}{12}<x<\frac{n\pi}{2}+\frac{\pi}{12},\ n\in I$$

Hence, the solution set is

$$x=\bigcup_{n\in I}\left(\frac{n\pi}{2}-\frac{\pi}{12},\frac{n\pi}{2}+\frac{\pi}{12}\right)$$

Ex-12: Solve; $\frac{5}{4}\sin^2 x+\frac{1}{4}\sin^2 2x>\cos 2x$

Soln. The given inequaton is

$$\frac{5}{4}\sin^2 x+\frac{1}{4}\sin^2 2x>\cos 2x$$

$$\Rightarrow 5\sin^2 x+\sin^2 2x>4\cos 2x$$

$$\Rightarrow 5\left(2\sin^2 x\right)+2\left(\sin^2 2x\right)>4\cos 2x$$

$$\Rightarrow 5(1-\cos 2x)+2\left(1-\cos^2 2x\right)>8\cos 2x$$

$$\Rightarrow 2\cos^2 2x+13\cos 2x-7<0$$

$$\Rightarrow 2\cos^2 2x+14\cos 2x-\cos 2x-7<0$$

$$\Rightarrow 2\cos 2x(\cos 2x+7)-1(\cos 2x+7)<0$$

$$\Rightarrow (2\cos 2x-1)(\cos 2x+7)<0$$

$$\Rightarrow -7<\cos 2x<\frac{1}{2}$$

$$\Rightarrow \cos 2x<\frac{1}{2}$$

$$\Rightarrow 2n\pi-\frac{\pi}{3}<2x<2n\pi+\frac{\pi}{3},\ n\in I$$

$$\Rightarrow n\pi-\frac{\pi}{6}<x<n\pi+\frac{\pi}{6},\ n\in I$$

Hence, the solution set is

$$x = \bigcup_{n \in I} \left(n\pi - \frac{\pi}{6},\ n\pi - \frac{\pi}{6} \right)$$

Ex-13: Solve : $\sin 3x \sin 4x > \sin x \sin 2x$

$$\forall x \in \left(0, \frac{\pi}{2}\right)$$

Soln. The given inequation is

$\sin 3x \sin 4x > \sin x \sin 2x$

$\Rightarrow$ $2 \sin 3x \sin 4x > 2 \sin x \sin 2x$.

$\Rightarrow$ $\cos x - \cos 7x > \cos x - \cos 3x$

$\Rightarrow$ $-\cos 7x > -\cos 3x$

$\Rightarrow$ $\cos 7x < \cos 3x$

$\Rightarrow$ $\cos 7x - \cos 3x < 0$

$\Rightarrow$ $-2 \sin 5x \sin 2x < 0$

$\Rightarrow$ $\sin 5x \sin 2x > 0$

$\Rightarrow$ $\sin 5x > 0$, since $\sin 2x$ is +ve in $\left(0, \frac{\pi}{2}\right)$

$\Rightarrow$ $0 < 5x < \pi$

$\Rightarrow$ $0 < x < \frac{\pi}{5}$

Hence, the solution set is

$$x = \left(0, \frac{\pi}{5}\right)$$

Ex-14: Solve : $|\sin x + \cos x| = |\sin x| + |\cos x|$

Soln. We have, $|\sin x + \cos x| = |\sin x| + |\cos x|$

$\Rightarrow$ $\sin x \cos x \geq 0$

$\Rightarrow$ $2 \sin x \cos x \geq 0$

$\Rightarrow$ $\sin 2x \geq 0$

$\Rightarrow$ $2n\pi \leq 2x \leq 2n\pi + \pi,\ n \in I$

$\Rightarrow$ $n\pi \leq x \leq n\pi + \frac{\pi}{2},\ n \in I$

Hence, the solution set is

$$x = \bigcup_{n \in I} \left(n\pi,\ n\pi + \frac{\pi}{2} \right)$$

Ex-15. Solve; $|\sec x + \tan x| = |\sec x| + |\tan x|$

Soln. We have, $|\sec x + \tan x| = |\sec x| + |\tan x|$

$\Rightarrow$ $\sec x . \tan x \geq 0$

$\Rightarrow$ $\frac{\sin x}{\cos^2 x} \geq 0$

$\Rightarrow$ $\sin x \geq 0,\ \cos^2 x \neq 0$

$\Rightarrow$ $\sin x \geq 0,\ x \neq (2n+1)\frac{\pi}{2}, n \in I$

$\Rightarrow$ $2n\pi \leq x \leq 2n\pi + \pi,\ x \neq (2n+1)\frac{\pi}{2},\ n \in I$

Hence, the solution set is

$$x = \bigcup_{n \in I} (2n\pi, (2n+1)\pi) - (2n+1)\frac{\pi}{2}$$

Ex-16. Solve for x : $\sin^2 x + \sin x - 2 < 0$

and $x^2 - 3x + 2 < 0$

Soln. We have, $\sin^2 x + \sin x - 2 < 0$

$\Rightarrow$ $(\sin x + 2)(\sin x - 1) < 0$

$\Rightarrow$ $-2 < \sin x < 1$

$\Rightarrow$ $\sin x < 1$

$\Rightarrow$ $-\frac{3\pi}{2} < x < \frac{\pi}{2}$

Also, $x^2 - 3x + 2 < 0$

$\Rightarrow$ $(x-1)(x-2) < 0$

$\Rightarrow$ $1 < x < 2$

Hence, the solution set is $1 < x < \frac{\pi}{2}$

Ex-17. Solve for x : $2 \sin^2 x + \sin x - 1 < 0$

and $x^2 + x - 2 < 0$

Soln. We have, $2\sin^2 x + \sin x - 1 < 0$

$\Rightarrow (2\sin x - 1)(\sin x + 1) < 0$

$\Rightarrow -1 < \sin x < 2$

$\Rightarrow -\dfrac{\pi}{2} < x < \dfrac{\pi}{6}$

Also, $x^2 + x - 2 < 0$

$\Rightarrow (x+2)(x-1) < 0$

$\Rightarrow -2 < x < 1$

Hence, the solution set is, $-\dfrac{\pi}{2} < x < \dfrac{\pi}{6}$

Ex-18. Solve for x : $\tan^2 x - 5\tan x + 6 > 0$ and $x^2 - 16 \le 0$

Soln. We have, $\tan^2 x - 5\tan x + 6 > 0$

$\Rightarrow (\tan x - 2)(\tan x - 3) \geq 0$

$\Rightarrow \tan x < 2$ and $\tan x > 3$

$\Rightarrow x < \tan^{-1}(2)$ and $x > \tan^{-1}(3)$

Also, $x^2 - 16 \le 0$

$\Rightarrow (x+4)(x-4) \le 0$

$\Rightarrow -4 \le x \le 4$

Hence, the solution set is

$x \in (-4, \tan^{-1}(2)) \cup (\tan^{-1}(3), 4)$

Ex-19. Solve for x : $\dfrac{x-1}{5-x} < 0$ and

$\tan^2 x + \tan x - 6 < 0$

Soln. We have, $\tan^2 x + \tan x - 6 < 0$

$\Rightarrow (\tan x + 3)(\tan x - 2) < 0$

$\Rightarrow -3 < \tan x < 2$

$\Rightarrow \tan^{-1}(-3) < x < \tan^{-1}(2)$

Also, $\dfrac{x-1}{5-x} < 0$

$\Rightarrow \dfrac{x-1}{x-5} > 0$

$\Rightarrow x < 1$ and $x > 5$

Hence, the solution set is $x \in (1, \tan^{-1}(2))$

Ex-20. Solve for x : $[\sin x] = 0$, where [,] = G.I.F

Soln. We have, $[\sin x] = 0$

$\Rightarrow 0 \le \sin x < 1$

Case - I : When $\sin x \ge 0$

$\Rightarrow 2n\pi \le x \le (2n+1)\pi,\ n \in I$

Case - II : When $\sin x < 1$

$\Rightarrow 2n\pi - \dfrac{3\pi}{2} < x < 2n\pi + \dfrac{\pi}{2}$

$\Rightarrow x \in \left((4n-3)\dfrac{\pi}{2}, \left((4n+1)\dfrac{\pi}{2}\right)\right)$

Hence, the solution set is

$$x = \bigcup_{n \in I} \left([2n\pi, (2n+1)\pi] \cup \left((4n-3)\frac{\pi}{2}, (4n+1)\frac{\pi}{2} \right) \right)$$

Level– I UESTIONS BASED ON FUNDAMENTALS

Q. Solve for x :

1. $\sin x > \dfrac{1}{2}$

2. $\sin x \ge 1$

3. $\sin x \le \dfrac{1}{\sqrt{2}}$

4. $\sin x < \dfrac{\sqrt{3}}{2}$

5. $\cos x > \dfrac{\sqrt{3}}{2}$

6. $\cos x \ge \dfrac{1}{2}$

7. $\cos x < \dfrac{1}{\sqrt{2}}$

8. $\cos x < \dfrac{1}{3}$

9. $\tan x > \dfrac{1}{\sqrt{3}}$

10. $\tan x \geq 1$

11. $\sin 2x < \dfrac{1}{2}$

12. $\sin 3x < \dfrac{\sqrt{3}}{2}$

13. $\cos 5x \geq \dfrac{1}{2}$

14. $\sin x + \cos x > 1$

15. $\sin x - \cos x < 1$

16. $\sqrt{3}\sin x + \cos x > 1$

17. $\sin x - \sqrt{3}\cos x < 1$

18. $\sin^2 x + \sin x - 2 < 0$

19. $\sin^2 x + 3\sin x + 2 < 0$

20. $\cos^2 x - \cos x > 0$

ANSWER

1. $\sin x > \dfrac{1}{2}$

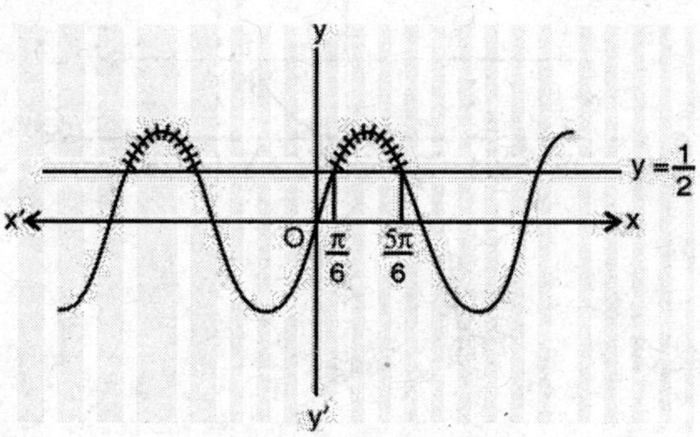

$\Rightarrow 2n\pi + \dfrac{\pi}{6} < x < 2n\pi + \dfrac{5\pi}{6}.\ , n \in \text{I}.$

2. $\sin x \geq 1$

$\Rightarrow \quad \sin x = 1$

$\Rightarrow \quad x = (4n+1)\,\dfrac{\pi}{2}, n \in \text{I}.$

3. $\sin x \leq \dfrac{1}{\sqrt{2}}$

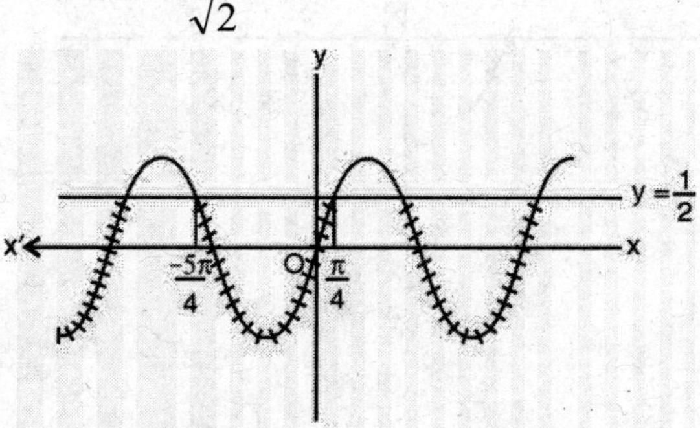

$\Rightarrow 2n\pi - \dfrac{5\pi}{4} \leq x \leq 2n\pi + \dfrac{\pi}{4}.,\ \ n \in \text{I}.$

4. $\sin x < \dfrac{\sqrt{3}}{2}$

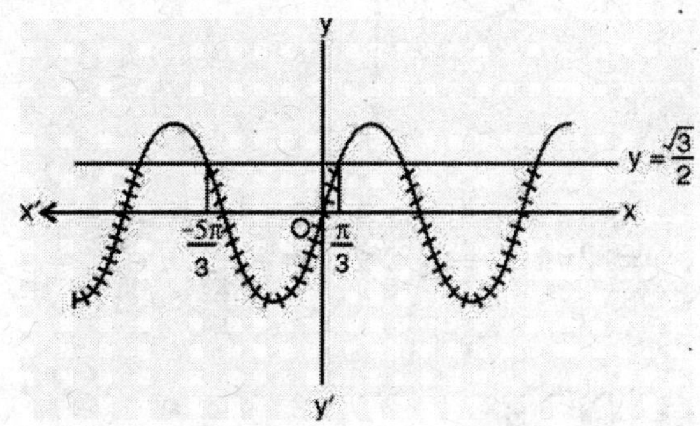

$\Rightarrow 2n\pi - \dfrac{4\pi}{3} \leq x \leq 2n\pi + \dfrac{\pi}{3}.,\ \ n \in \text{I}.$

5. $\cos x > \dfrac{\sqrt{3}}{2}$

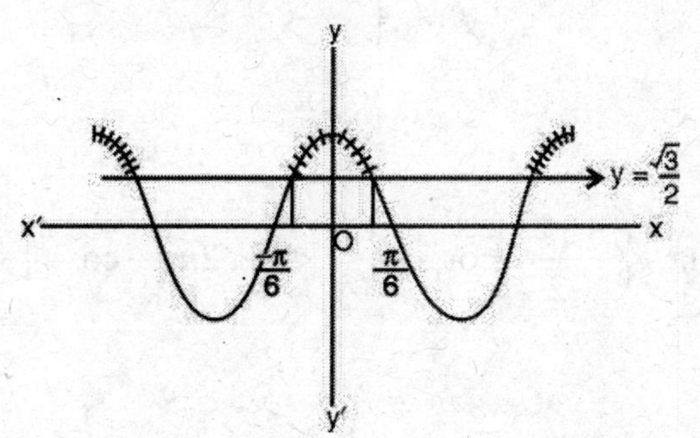

$\Rightarrow 2n\pi - \dfrac{\pi}{6} \leq x \leq 2n\pi + \dfrac{\pi}{6},\ \ n \in \text{I}.$

6. $\cos x \geq \dfrac{1}{2}$

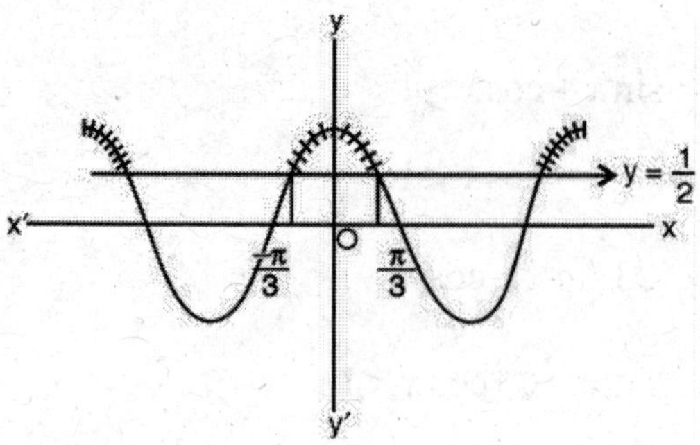

$$\Rightarrow 2n\pi - \frac{\pi}{3} \leq x \leq 2n\pi + \frac{\pi}{3},\ n \in I.$$

7. $\cos x < \dfrac{1}{\sqrt{2}}$

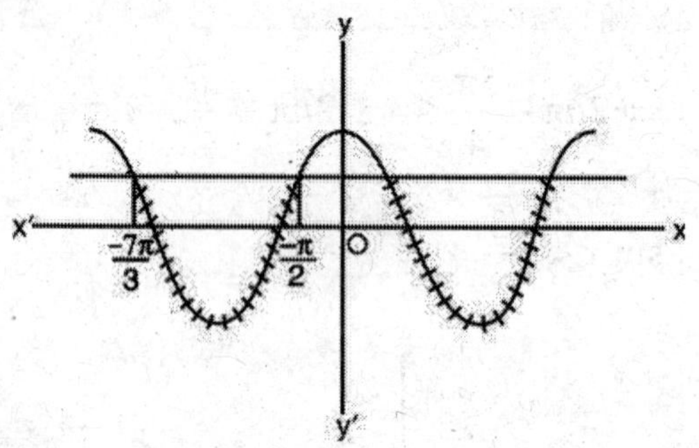

$$\Rightarrow 2n\pi - \frac{7\pi}{3} < x \leq 2n\pi - \frac{\pi}{4},\ n \in I.$$

8. $\cos x < \dfrac{1}{3}$

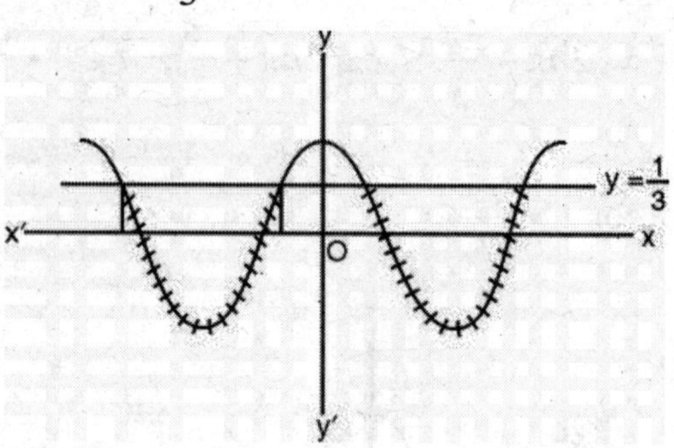

$$\Rightarrow 2n\pi - \left(-\frac{3\pi}{2} - \cos^{-1}\left(\frac{1}{3}\right)\right) < x < 2n\pi - \cos^{-1}\left(\frac{1}{3}\right).$$

9. $\tan x > \dfrac{1}{\sqrt{3}}$

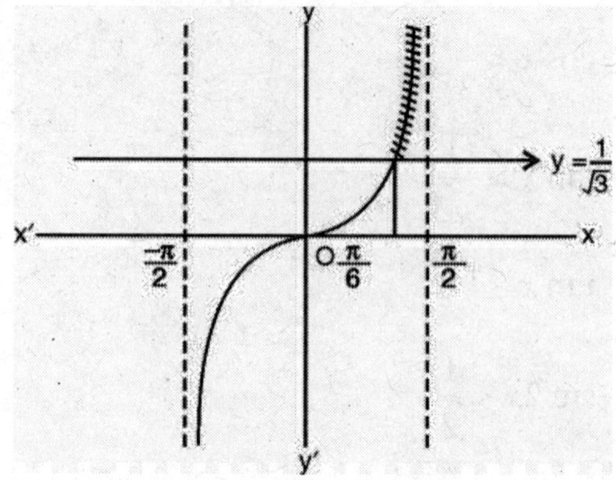

$$\Rightarrow n\pi + \frac{\pi}{6} < x < n\pi + \frac{\pi}{2},\ n \in I.$$

10. $\tan x \geq 1$

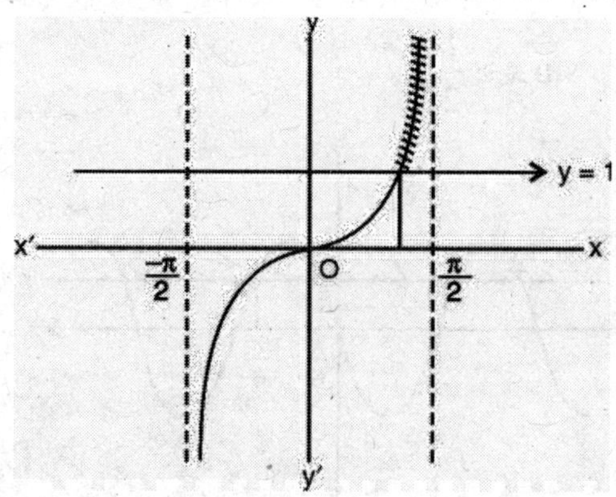

$$\Rightarrow n\pi + \frac{\pi}{4} \leq x < n\pi + \frac{\pi}{2},\ n \in I.$$

11. $\sin 2x < \dfrac{1}{2}$

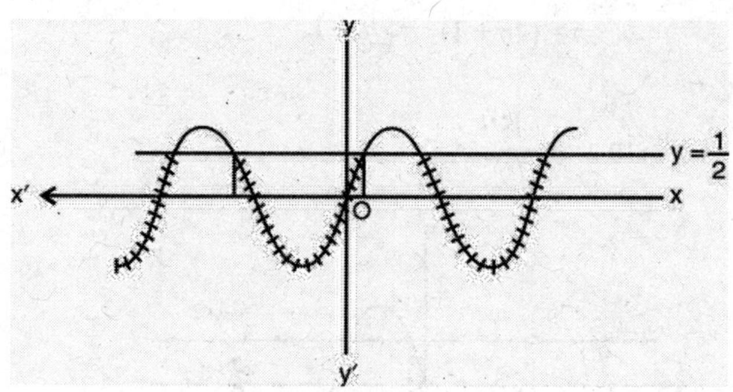

$$\Rightarrow 2n\pi + \frac{\pi}{6} < 2x < 2n\pi + \frac{5\pi}{6}.$$

$$\Rightarrow n\pi + \frac{\pi}{12} < n < n\pi + \frac{5\pi}{12},\ n \in \text{I}.$$

12. $\sin 3x < \frac{\sqrt{3}}{2}$

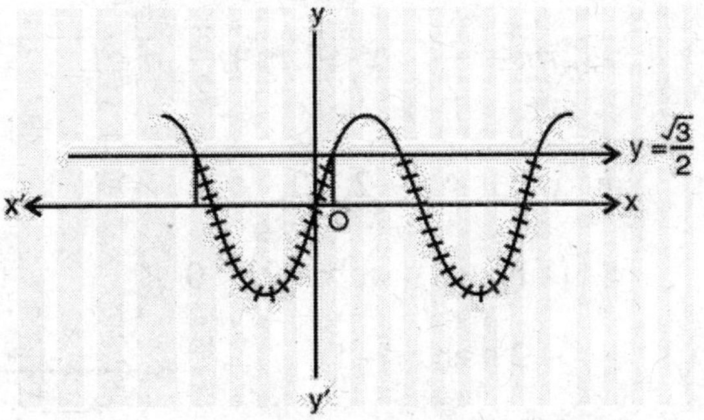

$$\Rightarrow 2n\pi - \frac{4\pi}{3} < 3x < 2n\pi + \frac{\pi}{3}.$$

$$\Rightarrow \frac{2n\pi}{3} - \frac{4\pi}{9} < x < \frac{2n\pi}{3} + \frac{\pi}{9},\ n \in \text{I}.$$

13. $\cos 5x \geq \frac{1}{2}$

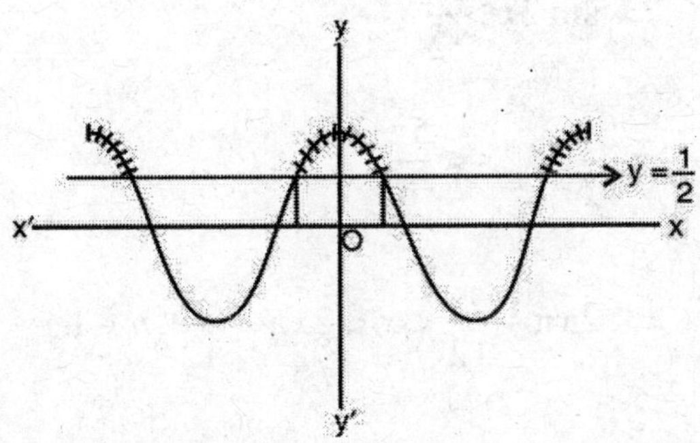

$$\Rightarrow 2n\pi - \frac{\pi}{3} \leq 5x < 2n\pi + \frac{\pi}{3}.$$

$$\Rightarrow \frac{2n\pi}{5} - \frac{\pi}{15} \leq 5x < \frac{2n\pi}{5} + \frac{\pi}{15},\ n \in \text{I}.$$

14. $\sin x + \cos x > 1$

$$\Rightarrow \quad \sin\left(x + \frac{\pi}{4}\right) > \frac{1}{\sqrt{2}}.$$

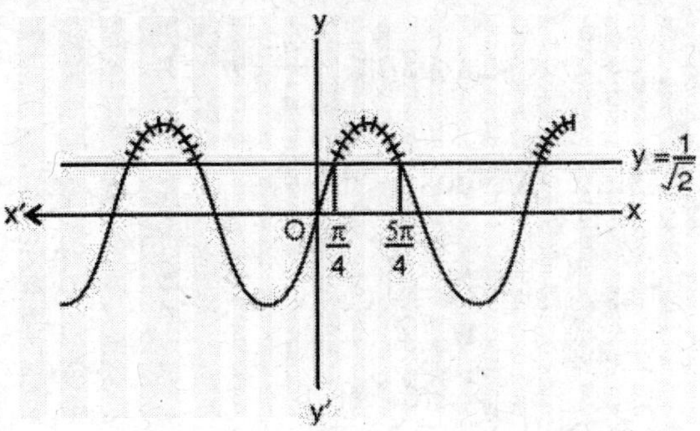

$$\Rightarrow 2n\pi + \frac{\pi}{4} < x + \frac{\pi}{4} < 2n\pi + \frac{5\pi}{4}.$$

$$\Rightarrow 2n\pi < x < 2n\pi + n,\ n \in \text{I}.$$

15. $\sin x - \cos x < 1$

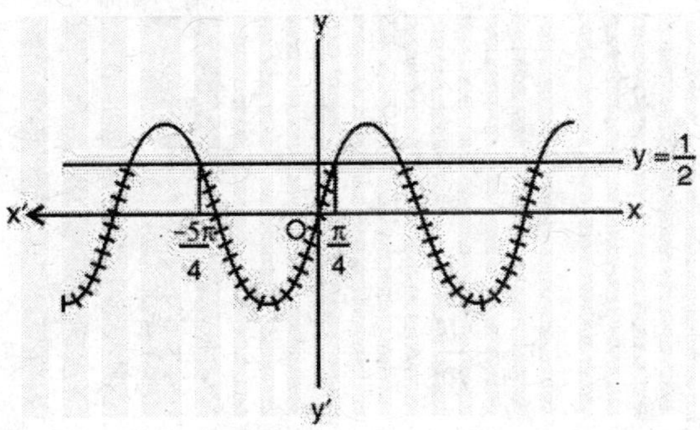

$$\Rightarrow 2n\pi - \frac{5\pi}{4} < x - \frac{\pi}{4} < 2n\pi + \frac{\pi}{4}.$$

$$\Rightarrow (2n-1)\pi < x < 2n\pi + \frac{\pi}{2},\ n \in \text{I}.$$

16. $\sqrt{3}\sin x + \cos x > 1$

$$\Rightarrow \frac{\sqrt{3}}{2}\sin x + \frac{1}{2}\cos x > \frac{1}{2}$$

$$\Rightarrow \sin\left(x + \frac{\pi}{6}\right) > \frac{1}{2}.$$

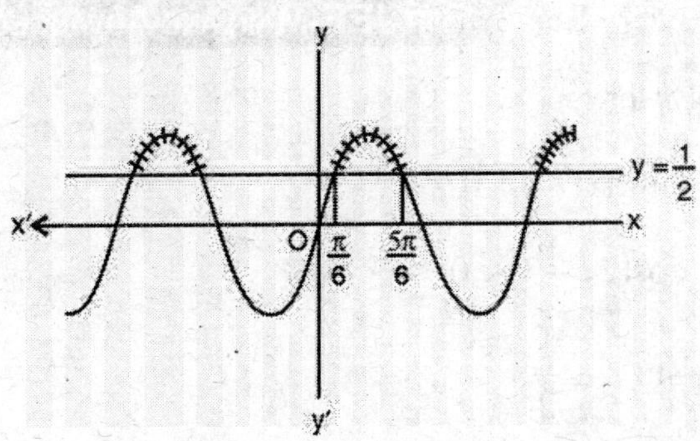

$$\Rightarrow 2n\pi + \frac{\pi}{6} < x + \frac{\pi}{6} < 2n\pi + \frac{5\pi}{6}.$$

$$\Rightarrow 2n\pi < x < 2n\pi + \frac{2\pi}{3},\ n \in \text{I}.$$

17. $\sin x - \sqrt{3}\cos x < 1$

$$\Rightarrow \frac{1}{2}\sin x - \frac{\sqrt{3}}{2}\cos x < \frac{1}{2}.$$

$$\Rightarrow \sin\left(x - \frac{\pi}{3}\right) < \frac{1}{2}.$$

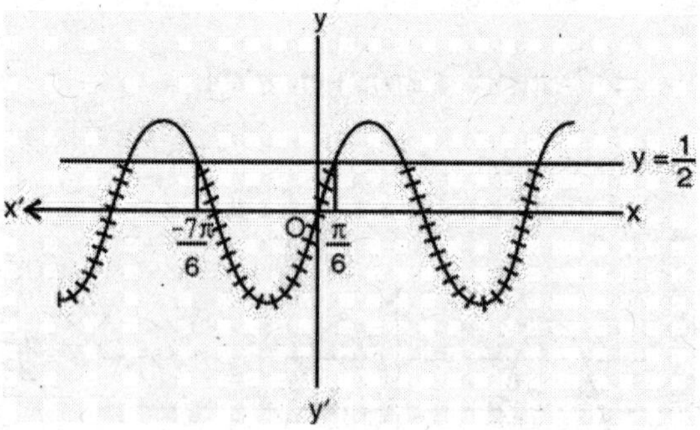

$$\Rightarrow 2n\pi - \frac{7\pi}{6} < x - \frac{\pi}{3} < 2n\pi + \frac{\pi}{6}.$$

$$\Rightarrow 2n\pi - \frac{5\pi}{6} < 2n\pi + \frac{\pi}{2},\ n \in \text{I}.$$

18. $\sin^2 x + \sin x - 2 < 0$

$$\Rightarrow \sin x + 2\sin x - \sin x - 2 < 0$$

$$\Rightarrow \sin x(\sin x + 2) - 1(\sin x + 2) < 0$$

$$\Rightarrow (\sin x + 2)(\sin x - 1) < 0$$

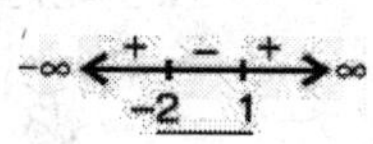

$\therefore \quad -2 < \sin x < 1.$

$\Rightarrow \quad \sin x < 1.$

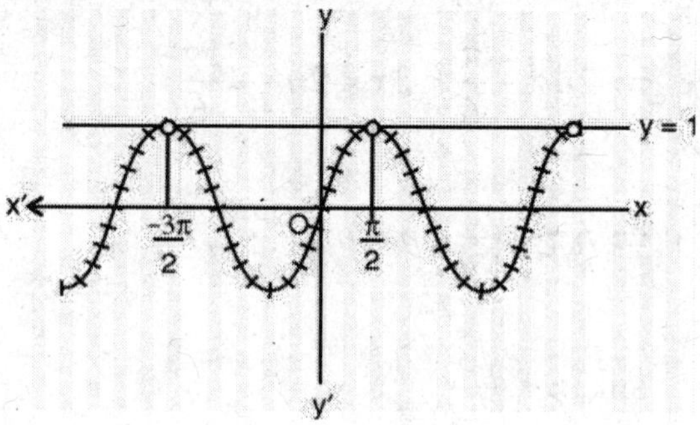

$$\Rightarrow 2n\pi - \frac{3\pi}{2} < x < 2n\pi + \frac{\pi}{2},\ n \in \text{I}.$$

19. $\sin^2 x + 3\sin x + 2 < 0$

$$\Rightarrow (\sin x + 1)(\sin x + 2) < 0$$

$$\Rightarrow \quad -2 < \sin x < -1.$$

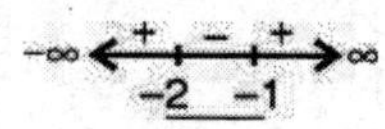

$$\Rightarrow \quad x = \phi.$$

20. $\cos^2 x - \cos x > 0$

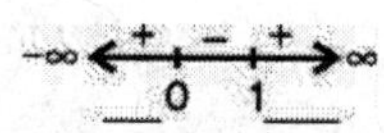

$$\Rightarrow \cos x(\cos x - 1) > 0.$$

$$\Rightarrow \cos x < 0 \text{ and } \cos x > 1.$$

$$\Rightarrow \cos x < 0$$

$$\Rightarrow 2n\pi + \frac{\pi}{2} < x < 2n\pi + \frac{3\pi}{2}, n \in \text{I}.$$

$$\Rightarrow \sin 2x > \frac{1}{2}.$$

$$\Rightarrow \frac{\pi}{6} < 2x < \frac{5\pi}{6}.$$

$$\Rightarrow 2x\pi + \frac{\pi}{12} < x < 2x\pi + \frac{5\pi}{12}, n \in \text{I}.$$

Level– II FOR JEE-MAIN AND ADVANCED EXAM ONLY

Q. Solve for x:

1. $\sin(3x - 1) > 0$
2. $\cos(2x - 3) < 0$
3. $|\sin x| \le \frac{1}{2}$
4. $|\cos x| \le \frac{1}{\sqrt{2}}$
5. $|\sin 2x + \cos 2x| = |\sin 2x| + |\cos 2x|$
6. $2\cos^2 x + \cos x < 1$
7. $4\sin^2 x - 1 \le 0$
8. $4\cos^2 x - 3 \ge 0$
9. $|\sin x| > |\cos x|$

10. $|\cos x| > |\sin x|$

11. $\sin x + \cos x - \cos 2x > 0$

12. $x^2 + x - 2 < 0$ and $\sin x > \frac{1}{2}$ $\quad \forall x \in (0, 2\pi)$

13. $x^2 - 1 \le 0$ and $\cos x < \frac{1}{2}$ $\quad \forall x \in (0, 2\pi)$

14. $4x^2 - 1 \ge 0$ and $\tan x \ge \frac{1}{\sqrt{3}}$ $\quad \forall x \in \left(-\frac{\pi}{2}, \frac{\pi}{2}\right)$

15. $x^2 - 3x + 2 < 0$ and $\sin x^2 - \sin x > 0$ $\quad \forall x \in (0, 2\pi)$

ANSWER

1. $\sin(3x-1) > 0$

$\Rightarrow\ 2n\pi < 3x - 1 < (2n+1)\pi.$

$\Rightarrow\ 2n\pi + 1 < 3x < (2n+1)\pi + 1.$

$\Rightarrow\ \frac{2n\pi+1}{3} < x < \frac{(2n+1)\pi+1}{3}, n \in \text{I}.$

2. $\cos(2x-3) < 0$

$\Rightarrow\ 2n\pi + \frac{\pi}{2} < 2x - 3 < 2n\pi + \frac{3\pi}{2}.$

$\Rightarrow\ (2n+1)\frac{\pi}{2} + 3 < 2n < (4n+3)\frac{\pi}{2} + 3$

$\Rightarrow (4n+1)\frac{\pi}{4} + \frac{3}{2} < x < (4n+3)\frac{\pi}{4} + \frac{3}{2}, n \in \text{I}.$

3. $|\sin x| \le \frac{1}{2}$

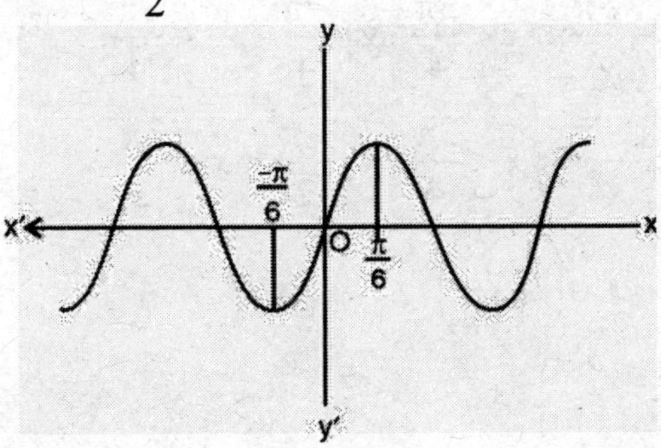

$\Rightarrow n\pi - \frac{\pi}{6} \le x \le n\pi + \frac{\pi}{6}, n \in \text{I}.$

4. $|\cos x| \le \frac{1}{\sqrt{2}}$

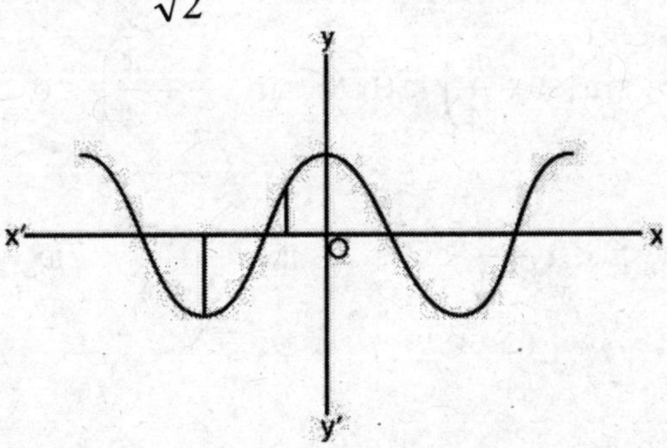

$\Rightarrow\ n\pi - \frac{3\pi}{4} \le x \le n\pi - \frac{\pi}{4}, n \in \text{I}.$

5. $|\sin 2x + \cos 2x| = |\sin 2x| + |\cos 2x|$

6. $2\cos^2 x + \cos x - 1 < 0$

$\Rightarrow 2\cos^2 x + 2\cos n - \cos n - 1 < 0$

$\Rightarrow 2\cos x(\cos x + 1)(2\cos x - 1) < 0$

$\Rightarrow (\cos x + 1)(2\cos x - 1) < 0$

$\therefore\ -1 < \cos x < \frac{1}{2}.$

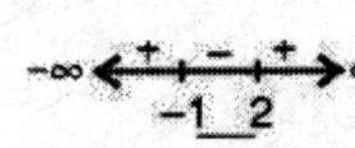

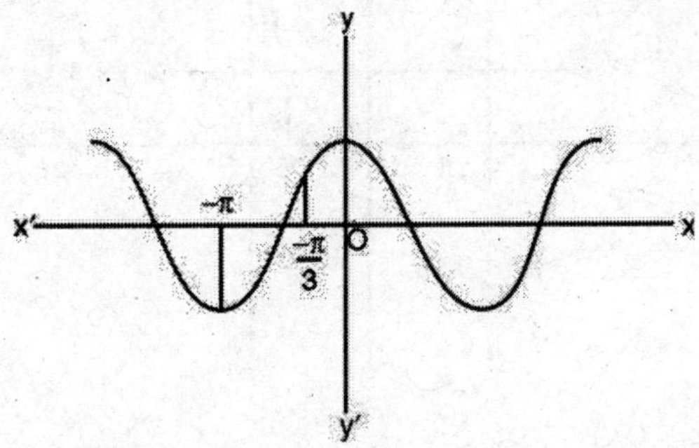

$\Rightarrow\ n\pi - \pi < x < n\pi - \frac{\pi}{3}, n \in \text{I}.$

7. $4\sin^2 x - 1 \le 0$

$\Rightarrow\ \sin^2 x \le \frac{1}{4}.$

$\Rightarrow |\sin x| \le \frac{1}{2}.$

$\Rightarrow\ n\pi - \frac{\pi}{6} \le x \le n\pi + \frac{\pi}{6}, n \in \text{I}.$

8. $4\cos^2 x - 3 \geq 0$

$\Rightarrow \cos^2 x \geq \frac{3}{4}.$

$\Rightarrow |\cos x| \geq \frac{\sqrt{3}}{2}.$

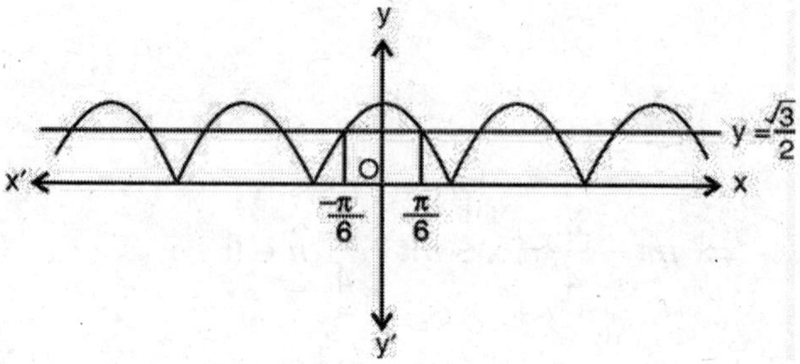

$\Rightarrow n\pi - \frac{\pi}{6} \leq x \leq n\pi + \frac{\pi}{6}, n \in \text{I}.$

9. $|\sin x| > |\cos x|$

$\Rightarrow |\tan x| > 1.$

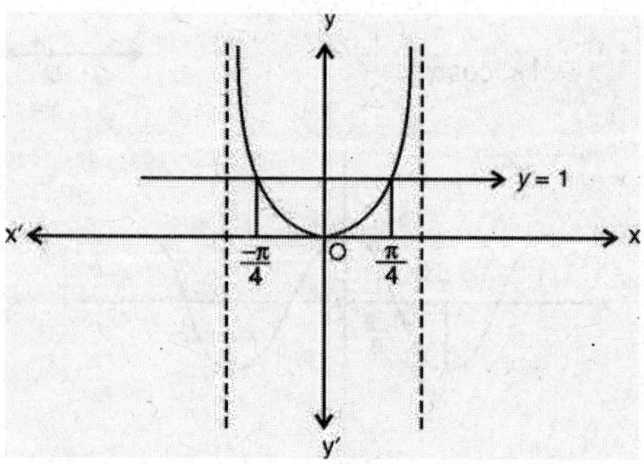

$$\Rightarrow x \in \bigcup_{n \in \text{I}} \left\{ \left(2n - \frac{\pi}{2}\right), n\pi - \frac{\pi}{4} \right.$$

$$\left. \bigcup \left(n\pi + \frac{\pi}{4}, n\pi + \frac{\pi}{2}\right) \right\}.$$

10. $|\cos x| > |\sin x|$

$\Rightarrow |\tan x| < 1.$

$\Rightarrow n\pi - \frac{\pi}{4} < x < n\pi + \frac{\pi}{4}, n \in \text{I}.$

11. $\sin x + \cos x - \cos 2x > 0$

$\Rightarrow (\sin x + \cos x) - (\cos^2 x - \sin^2 x) > 0$

$\Rightarrow (\sin x + \cos x)(1 - \cos x + \sin x) > 0.$

$\Rightarrow (\sin x + \cos x)\left(2\sin^2\frac{x}{2} + \sin\frac{x}{2}\cos\frac{x}{2}\right) > 0.$

$\Rightarrow 2\sin\frac{x}{2}(\sin x + \cos x)\left(\sin\frac{x}{2} + \cos\frac{x}{2}\right) > 0$

$\Rightarrow (\sin x + \cos x)\left(\sin\frac{x}{2} + \cos\frac{x}{2}\right) > 0$

$\left(\because \sin\frac{x}{2} > 0, \forall x \in (0, 2\pi)\right)$

Case-I.

$\sin x + \cos x > 2$

$\sin\frac{x}{2} + \cos\frac{x}{2} > 0.$

$\Rightarrow 0 < x + \frac{\pi}{4} < \pi \ \& \ 0 < \frac{x}{2} + \frac{\pi}{4} < \pi$

$\Rightarrow \frac{-\pi}{4} < x < \frac{3\pi}{4} \ \& \ \frac{-\pi}{4} < \frac{x}{2} < \frac{3\pi}{4}.$

$\Rightarrow \frac{-\pi}{4} < x < \frac{3\pi}{4} \ \& \ \frac{-\pi}{2} < x < \frac{3\pi}{2}.$

$\Rightarrow x \in \left(0, \frac{3\pi}{4}\right).$

Case-II.

$\sin x + \cos x < 0$

$\sin\frac{x}{2} + \cos\frac{x}{2} < 0.$

$\Rightarrow \sin\left(x + \frac{\pi}{4}\right) < 0 \ \& \ \sin\left(\frac{x}{2} + \frac{\pi}{4}\right) < 0$

$\Rightarrow \pi < x + \frac{\pi}{4} < 2\pi \ \& \ \pi < \frac{x}{2} + \frac{\pi}{4} < 2\pi$

$$\Rightarrow \frac{3\pi}{4} < x < \frac{7\pi}{4} \ \& \ \frac{3\pi}{4} < \frac{x}{2} < \frac{7\pi}{4}.$$

$$\Rightarrow \frac{3\pi}{4} < x < \frac{7\pi}{4} \ \& \ \frac{3\pi}{2} < x < \frac{7\pi}{2}.$$

$$\Rightarrow x \in \left(\frac{3\pi}{4}, \frac{7\pi}{4}\right).$$

Hence, the solution set is

$$x \in \left(0, \frac{3\pi}{4}\right) \text{U} \left(\frac{3\pi}{4}, \frac{7\pi}{4}\right).$$

12. $x^2 + x - 2 < 0$

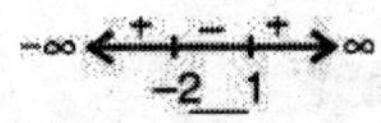

$\Rightarrow (x+2)(x-1) < 0$

$\Rightarrow -2 < x < 1$

Also, $\sin x < \frac{1}{2}.$

$$\Rightarrow x \in \left(\frac{\pi}{6}, \frac{5\pi}{6}\right)$$

Hence, the solution set is $\frac{\pi}{6} < x < 1.$

$$\cos x < \frac{1}{2}.$$

$$\Rightarrow -\frac{11\pi}{6} < x < -\frac{\pi}{3}.$$

Hence, the solution set is $\frac{\pi}{6} < x < 1.$

13. $x^2 - 1 \le 0$ and $\cos x < \frac{1}{2}$

$\Rightarrow (x+1)(x-1) \le 0 \Rightarrow -1 \le n \le 1$

Also, $\cos x < \frac{1}{2}.$

$$\Rightarrow \frac{-11\pi}{6} < x < -\frac{\pi}{3}.$$

Hence, the solution set is $x = \phi.$

14. $4x^2 - 1 \ge 0$ and $\tan x \ge \frac{1}{\sqrt{3}}$

$$\Rightarrow x^2 \ge \frac{1}{4}.$$

$$\Rightarrow |x| \ge \frac{1}{2}.$$

$$\Rightarrow x \in -\infty, -\frac{1}{2} \bigcup \left(\frac{1}{2}, \infty\right).$$

$\tan \ge \frac{1}{\sqrt{3}}$

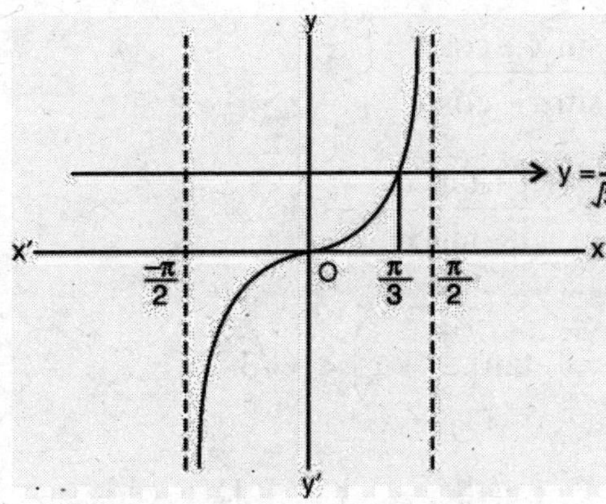

$$\Rightarrow \frac{\pi}{3} \le x < \frac{\pi}{2}.$$

Hence, the solution set is $x \in \left[\frac{\pi}{3}, \frac{\pi}{2}\right].$

15. $x^2 - 3x + 2 < 0$

$\Rightarrow (x-1)(x-2) > 0$

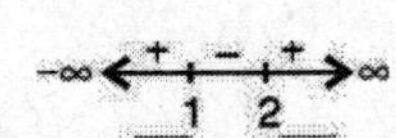

$\Rightarrow n \in (-\infty, 1) \bigcup (2, \infty).$

$\sin x^2 - \sin x > 0$

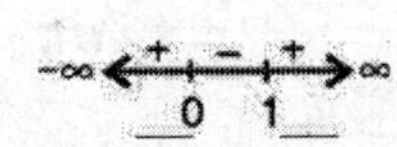

$\Rightarrow \sin x < 0$

$\Rightarrow x \in (\pi, 2\pi)$

Hence, the solution set is $x \in (\pi, 2\pi)$

Level– III OR JEE - ADVANCED EXAM ONLY

Q. Solve for x :

1. $\dfrac{\sin x+\cos x}{\sin x-\cos x}>\sqrt{3}$

2. $|\sin x|>|\cos x|$

3. $\cot x+\dfrac{\sin x}{\cos x-2}\geq 0$

4. $\sin x+\cos x>\sqrt{2}\cos 2x$

5. $4\sin x\sin 2x\sin 3x>\sin 4x$

6. $\dfrac{\cos^2 2x}{\cos^2 x}\geq 3\tan x$

7. $\dfrac{\cos x+2\cos^2 x+\cos 3x}{\cos x+2\cos^2 x-1}>1$

8. $2\left(\sqrt{2}-1\right)\sin x-2\cos 2x+\sqrt{2}\left(\sqrt{2}-1\right)<0$

9. $\sin 2x>\sqrt{2}\sin^2 x+\left(2-\sqrt{2}\right)\cos^2 x$, $x\in(0,\ 2\pi)$

10. $1+\log_4 \sin x+2\log_{16}\cos x>0$

ANSWER

1. $\dfrac{\sin x+\cos x}{\sin x-\cos x}>\sqrt{3}$

$\Rightarrow \dfrac{1+\tan x}{1-\tan x}<-\sqrt{3}$

$\Rightarrow \tan\left(\dfrac{\pi}{4}+x\right)<-\sqrt{3}$

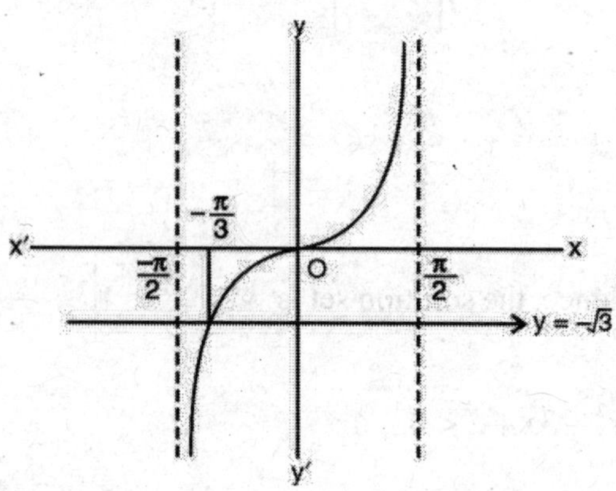

$\Rightarrow n\pi-\dfrac{\pi}{2}<x+\dfrac{\pi}{4}<n\pi-\dfrac{\pi}{3}$

$\Rightarrow n\pi-\dfrac{3\pi}{4}<x<n\pi-\dfrac{7\pi}{12},n\in \text{I}.$

2. $|\sin x|>|\cos x|$

$\Rightarrow |\tan x|>1.$

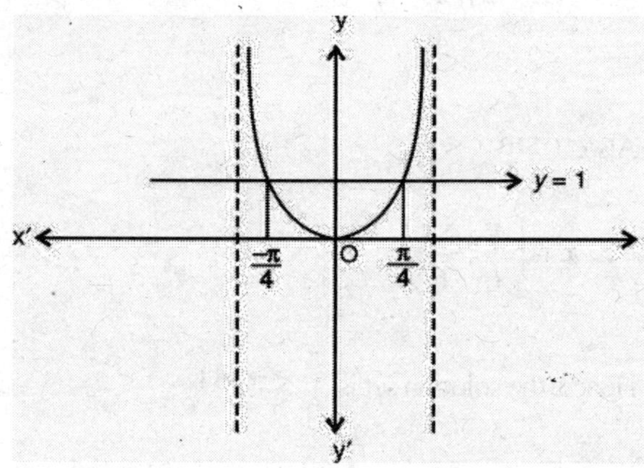

$\Rightarrow x\in\bigcup\left\{n\pi-\dfrac{\pi}{2},n\pi-\dfrac{\pi}{4}\right\}\bigcup\left(n\pi+\dfrac{\pi}{4},n\pi+\dfrac{\pi}{2}\right)$

3. $\cot x+\dfrac{\sin x}{\cos x-2}\geq 0$

$\Rightarrow \dfrac{\cos^2 x-2\cos x+\sin^2 x}{\sin x(\cos x-2)}\geq 0$

$\Rightarrow \dfrac{(1-2\cos x)}{\sin x(\cos x-2)}\geq 0$

$\Rightarrow \dfrac{1-2\cos x}{\sin x}\leq 0.\ (\because \cos x-1<0,\forall n\in R).$

$\Rightarrow \dfrac{2\cos x-1}{\sin x}\geq 0$

Case-II.

$$2\cos x - 1 \geq 0, \quad \sin x > 0$$

$$\Rightarrow \cos x \geq \frac{1}{2}$$

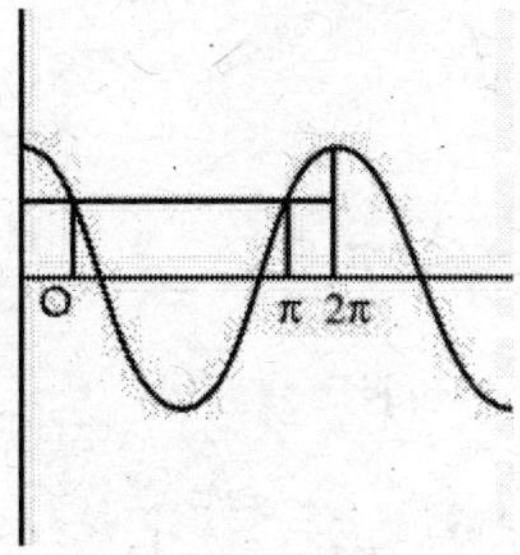

$$\Rightarrow x \in \left[0, \frac{\pi}{3}\right] \bigcup \left[11\frac{\pi}{6}, 2\pi\right]$$

Thus, $x \in \left[\pi, \frac{\pi}{3}\right]$.

Sin $x > 0$

$$\Rightarrow 0 < x < \pi.$$

Case-III.

2cos – 1 0

$$\Rightarrow \cos \leq \frac{1}{2}$$

$$\Rightarrow \cos \leq \frac{1}{2} \quad \Rightarrow \frac{\pi}{3} \leq x \leq x\frac{\pi}{6}.$$

Thus, $x \in \left(\pi, 11\frac{\pi}{6}\right)$.

sin$x$ < 0

$$\Rightarrow \pi < x < 2\pi$$

4. **sin$x$ + cos$x$ > $\sqrt{2}$cos2x**

$$\Rightarrow \sin x + \cos x - \sqrt{2}\cos 2x > 0$$

$$\Rightarrow (\sin x + \cos x) - \sqrt{2}\left(\cos^2 x - \sin^2 x\right) > 0$$

$$\Rightarrow (\sin x + \cos x)\left(1 - \sqrt{2}(\cos x - \sin x)\right) > 0$$

$$\Rightarrow (\sin x + \cos x)\left((\cos x - \sin x) - \frac{1}{\sqrt{2}}\right) < 0.$$

Case-I

sin x + cos x > 0

$$\Rightarrow \sin\left(x + \frac{\pi}{4}\right) > 0,$$

$$\Rightarrow 0 < x + \frac{\pi}{4} < \pi.$$

$$\Rightarrow -\frac{\pi}{4} < x < \frac{3\pi}{4}$$

(cos x – sin x) – $\frac{1}{\sqrt{2}}$ < 0

$$\Rightarrow \cos x - \sin x < \frac{1}{\sqrt{2}}.$$

$$\Rightarrow \cos\left(x + \frac{\pi}{4}\right) < \frac{1}{2}.$$

$$\Rightarrow \frac{\pi}{3} < x + \frac{\pi}{4} < 11\frac{\pi}{6}.$$

$$\Rightarrow \frac{\pi}{12} < x + \frac{19\pi}{12}.$$

Thus, $\frac{\pi}{12} < x + \frac{3\pi}{4}$.

$$\Rightarrow 2x\pi + \frac{\pi}{12} < x < 2x\pi + \frac{3\pi}{4}.$$

Case-II

sin x + cos x < 0

$$\Rightarrow \sin\left(x + \frac{\pi}{4}\right) < 0$$

$$\Rightarrow \pi < x + \frac{\pi}{4} < 2\pi.$$

$$\Rightarrow \frac{3\pi}{4} < x < \frac{7\pi}{4}$$

$$(\cos x - \sin x) - \frac{1}{\sqrt{2}} > 0.$$

$$\Rightarrow \cos x - \sin x > \frac{1}{\sqrt{2}}.$$

$$\Rightarrow \cos\left(x + \frac{\pi}{4}\right) > \frac{1}{\sqrt{2}}.$$

$$\Rightarrow \frac{17\pi}{12} < x < \frac{25 - \pi}{12}$$

Thus, $\frac{17\pi}{12}<x<\frac{7\pi}{4}$.

$\Rightarrow 2n\pi+\frac{17\pi}{12}<x<2n\pi+\frac{7x}{4}$.

Hence, the solution set is

$$x=\bigcup_{n\in I}\left(2n\pi+\frac{\pi}{12},2n\pi+\frac{3\pi}{4}\right)$$

$$\bigcup\left(2n\pi+\frac{17\pi}{12},2n\pi+\frac{7\pi}{4}\right).$$

5. $4\sin x\sin 2x\sin 3x>\sin 4x$

$\Rightarrow 2\sin 2x(2\sin x\sin 3x-\cos 2x)>0.$

$\Rightarrow 2\sin 2x(\cos 2x-\cos 4x-\cos 2x)>0.$

$\Rightarrow -2\sin 2x(-\cos 4x)>0.$

$\Rightarrow \sin 2x\cos 4x>0.$

Case-I.

sin 2x > 0 | **cos 4x < 0**

$\Rightarrow 0<2x<\pi$ | $\Rightarrow \frac{\pi}{2}<4x<\frac{3\pi}{2}$.

$\Rightarrow 0<x<\frac{\pi}{2}$. | $\Rightarrow \frac{\pi}{8}<x<\frac{3\pi}{8}$.

Thus, $\frac{\pi}{8}<x<\frac{3\pi}{8}$.

$\Rightarrow n\pi+\frac{\pi}{8}<x<n\pi+\frac{3\pi}{8}, n\in I.$

Case-II

sin 2x > 0

$\Rightarrow \pi<2x<2\pi$.

$\Rightarrow \frac{\pi}{2}<x<\pi$.

cos 4x > 0

$\Rightarrow \frac{3\pi}{2}<4x<2\pi$.

$\Rightarrow \frac{3\pi}{8}<x<\frac{\pi}{2}$.

$\Rightarrow \frac{\pi}{2}+\frac{3\pi}{8}<x<\pi$.

$\Rightarrow \frac{7\pi}{8}<x<\pi$.

Thus, $\frac{7\pi}{8}<x<\pi$.

$\Rightarrow n\pi+\frac{7\pi}{8}<x<n\pi+\pi$.

Hence, the solution set is

$$x=\bigcup_{n\in I}\left(n\pi+\frac{\pi}{8},\ n\pi+\frac{3\pi}{8}\right)$$

$$\bigcup\left(n\pi+\frac{7\pi}{8},(n+1)\pi\right)$$

7. $\frac{\cos x+2\cos^2 x+\cos 3x}{\cos x+2\cos^2 x-1}>1$

$\Rightarrow \frac{\cos x+2\cos^2 x+4\cos^3 x-3\cos x}{\cos x+2\cos^2 x-1}$

$\Rightarrow \frac{2\cos^2 x+4\cos^3 x-2\cos x}{\cos x+2\cos^2 x-1}>1$

$\Rightarrow 2\cos x\frac{(\cos x+2\cos^2 x-1)}{(\cos x+2\cos^2 x-1)}>1$

$\Rightarrow 2\cos x>1$

$\Rightarrow \cos x>\frac{1}{2}$.

$\Rightarrow 2x\pi-\frac{\pi}{3}<x<2x\pi+\frac{\pi}{3}n\in I.$

8. $2(\sqrt{2}-1)\sin x-2\cos 2x+\sqrt{2}(\sqrt{2}-1)<0$

$\Rightarrow 2(\sqrt{2}-1)\sin x-2(1-2\sin^2 x)+\sqrt{2}(\sqrt{2}-1)<0$

$\Rightarrow (\sqrt{2}-1)\sin x-1+2\sin^2 x+\frac{\sqrt{2}-1}{\sqrt{2}}<0$

$\Rightarrow 2\sin^2 x+(\sqrt{2}-1)\sin x+\left(\frac{\sqrt{2}-1}{\sqrt{2}}-1\right)<0$

$\Rightarrow 2\sqrt{2}\sin^2 x+(2-\sqrt{2})\sin x-1<0.$

$$\Rightarrow 2\sqrt{2}\sin^2 x + 2\sin x - \sqrt{2}\sin x - 1 < 0$$

$$\Rightarrow 2\sin x\left(\sqrt{2}\sin x + 1\right) - 1\left(\sqrt{2}\sin x + 1\right) < 0$$

$$\Rightarrow \left(\sqrt{2}\sin x + 1\right)\left(2\sin x - 1\right) < 0.$$

$$\Rightarrow -\frac{1}{\sqrt{2}} < \sin x < \frac{1}{2}.$$

$$\Rightarrow \frac{5\pi}{6} < x < \frac{5\pi}{4}.$$

$$\Rightarrow 2n\pi + \frac{5\pi}{6} < x < 2n\pi + \frac{5\pi}{4}, n \in \text{I}.$$

Hence, the solution set is

$$x = \bigcup_{n \in \text{I}} \left\{ \left(2n\pi - \frac{\pi}{4},\ 2n\pi + \frac{\pi}{6}\right) \right.$$

$$\left. \bigcup \left(2m\pi + \frac{5\pi}{6},\ 2n\pi + \frac{5\pi}{4}\right) \right\}, n \in \text{I}.$$

9. $\sin 2x > \sqrt{2}\sin^2 x + \left(2 - \sqrt{2}\right)\cos^2 x$

$, x \in (0, 2\pi)$

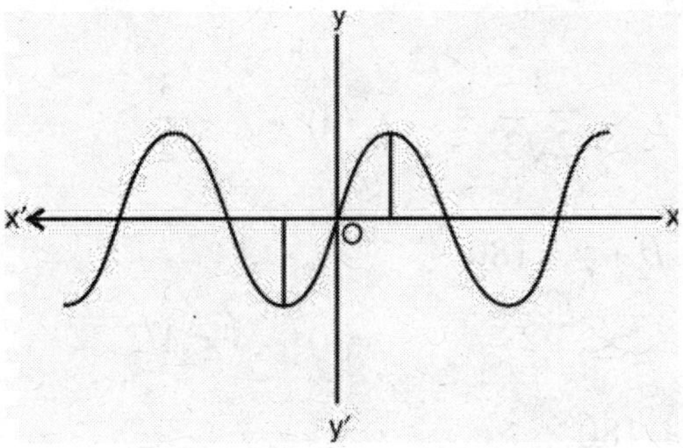

$$\Rightarrow -\frac{\pi}{4} < x < \frac{\pi}{6}$$

$$\Rightarrow 2x\pi - \frac{\pi}{4} < x < 2x\pi + \frac{\pi}{6}$$

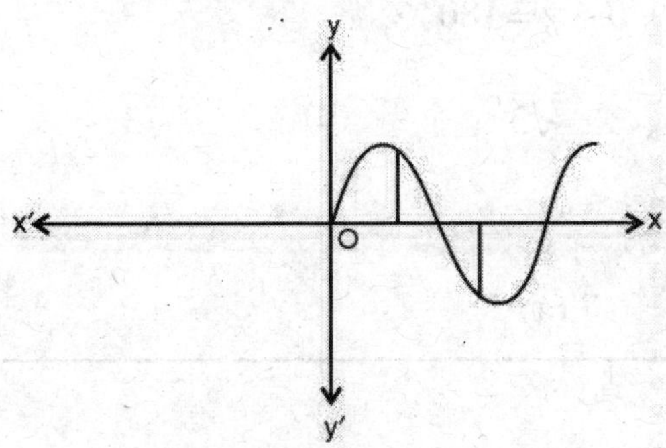

10. $1 + \log_4 \sin x + 2\log_{16} \cos x > 0$

$$\Rightarrow 1 + \frac{1}{2}\log_2 \sin x + \frac{1}{2}\log_2 \cos > 0$$

$$\Rightarrow 2 + \log_2\left(\sin x \cos x\right) > 0$$

$$\Rightarrow 2 + \log_2\left(\frac{\sin 2x}{2}\right) > 0$$

$$\Rightarrow 2 + \log_2\left(\sin 2x\right) - \log_2^2 > 0$$

$$\Rightarrow \log_2\left(\sin 2x\right) - 1.$$

$$\Rightarrow \sin 2x > \frac{1}{2}.$$

$$\Rightarrow 2n\pi + \frac{\pi}{6} < 2x < 2n\pi + \frac{5\pi}{6}.$$

$$\Rightarrow n\pi + \frac{\pi}{12} < x < n + \frac{5\pi}{12}.$$

Hence, the solution set is

$$x = \bigcup_{n \in \text{I}} \left(n\pi + \frac{\pi}{12}, n\pi + \frac{5\pi}{12}\right).$$

Comprehensive Link Passage

PASSAGE - I

If $x_1, x_2, x_3 \in R$, then $\dfrac{f(x_1) + f(x_2) + f(x_3)}{3} \le f\left(\dfrac{x_1 + x_2 + x_3}{3}\right)$

Then

1. The value of $\sin\alpha + \sin\beta + \sin\gamma$ is, where $\alpha + \beta + \gamma = 180^o$

(a) ≤ 1 (b) ≤ 3 (c) $\leq \frac{3\sqrt{3}}{2}$ (d) $\leq \frac{3}{2}$

2. The value of $\cos\alpha + \cos\beta + \cos\gamma$ is, where $\alpha + \beta + \gamma = 180^o$

(a) ≤ 2 (b) $\leq \frac{3}{2}$ (c) ≤ 3 (d) $\leq \frac{\sqrt{3}}{2}$

3. The value of $\cot\alpha + \cot\beta + \cot\gamma$ is, where $\alpha + \beta + \gamma = 180^o$

(a) ≥ 1 (b) $\geq \sqrt{3}$ (c) $\geq 3/2$ (d) $\geq \frac{\sqrt{3}}{2}$

4. The value of $\cot\alpha.\cot\beta.\cot\gamma$ is, where $\alpha + \beta + \gamma = 180^o$

(a) $\leq \frac{1}{3}$ (b) $\leq \frac{1}{3\sqrt{3}}$ (c) $\leq \frac{1}{2\sqrt{3}}$ (d) $\leq \frac{3}{2}$

5. The value of $\sin^2\alpha + \sin^2\beta + \sin^2\gamma$ is, where $\alpha + \beta + \gamma = 180^o$

(a) $\leq 9/4$ (b) $\leq 3/4$ (c) $\leq 3/2$ (d) $\leq 1/2$

6. The value of $\sin\alpha.\sin\beta.\sin\gamma$ is, where $\alpha + \beta + \gamma = 180^o$

(a) $\leq \frac{\sqrt{3}}{4}$ (b) $\leq \frac{3\sqrt{3}}{8}$ (c) $\leq \frac{1}{2\sqrt{3}}$ (d) $\leq \frac{1}{3\sqrt{2}}$

7. The value of $\cot^2\alpha + \cot^2\beta + \cot^2\gamma$ is, where $\alpha + \beta + \gamma = 180^o$

(a) ≥ 1 (b) ≥ 3 (c) $\geq \sqrt{3}$ (d) $\geq \sqrt{3}/2$

ANSWER

1.(c) 2.(b) 3.(b) 4.(b) 5.(a) 6.(b) 7.(a)

MATCH MATRIX

1. Match the following columns :

	Column - I		Column - II
(A)	The number of solutions of $\sin x > \frac{1}{2}$ in $(0, 2\pi)$ is	(P)	6
(B)	The number of solutions of $\lvert\tan x\rvert \le 1$ in $(-\pi, \pi)$ is	(Q)	0
(C)	The number of solutions of $\lvert\cos x\rvert > 1$ in $(0, 2013\pi)$ is	(R)	4
(D)	The number of solutions of $\lvert\sin x + \cos x\rvert = \lvert\sin x\rvert + \lvert\cos x\rvert$ in $(0, 2\pi)$ is	(S)	2

2. Match the following columns :

	Column - I		Column - II
(A)	If $\sin x.\cos^3 x > \cos x.\sin^3 x, x \in [0, 2\pi]$ Then x is	(P)	$\left[-\pi, -\frac{3\pi}{4}\right] \cup \left[-\frac{\pi}{4}, \frac{\pi}{4}\right] \cup \left[\frac{3\pi}{4}, \pi\right]$
(B)	If $4\sin^2 x - 8\sin x + 3 \le 0$, $x \in [0, 2\pi]$ then x is	(Q)	$\left[\frac{3\pi}{2}, 2\pi\right] \cup \{0\}$
(C)	If $\lvert\tan x\rvert \le 1, x \in [-\pi, \pi]$ then x is	(R)	$\left(0, \frac{\pi}{4}\right)$
(D)	If $\cos x - \sin x \ge 1$, $x \in [0, 2\pi]$ then x is	(S)	$\left[\frac{\pi}{6}, \frac{5\pi}{6}\right]$

ANSWERS

1. (A) → (S) ; (B) → (S) ; (C) → (Q) ; (D) → (R).

2. (A) → (R) ; (B) → (S) ; (C) → (P) ; (D) → (Q).

ASSERTION & REASON

Codes :

(A) Both A and R are true and R is the correct explannation of A

(B) Both A and R are true but R is not the correct explanation of A

(C) A is true but R is false.

(D) A is false but R is true.

1. Assertion (A) : The value of $\tan 3\alpha.\cot\alpha$ cannot lie between 3 and 1/3.

Reason (R) : In a triangle ABC, the maximum

value of $\sin\left(\frac{A}{2}\right)\sin\left(\frac{B}{2}\right)\sin\left(\frac{C}{2}\right)$ is $\frac{1}{8}$

2. Assertion (A) : The minimum value of $a^2\tan^2\theta + b^2\cot^2\theta$ is 2ab.

   Reason (R) : For positive real numbers A.M $\geq$ G.M

3. Assertion (A) : The minimum value of $\frac{a^2}{\cos^2 x} + \frac{b^2}{\sin^2 x}$ is $(a+b)^2$

   Reason (R) : The maximum value of $3\sin^2 x + 4\cos^2 x$ is 4

4. Assertion (A) : For all $\theta \in \left[0, \frac{\pi}{2}\right]$, $\cos(\sin\theta) > \sin(\cos\theta)$

   Reason (R) : In a triangle ABC, the maximum value of $\frac{\sin A + \sin B + \sin C}{\cot A + \cot B + \cot C}$ is $\frac{3}{2}$

5. Assertion (A) : $\cot^{-1} x \geq 2 \Rightarrow x \in (-\infty, 2]$

   Reason (R) : $\cot^{-1} x$ is a decreasing function.

ANSWERS

1. (B) 2. (A) 3.(B) 4. (B) 5. (A)

CHAPTER 5

INVERSE TRIGONOMETRIC FUNCTIONS

5.1 INTRODUCTION

Before discussing inverse trigonometric functions, you should know about the concepts of Sets, Real Numbers, Intervals, Functions and also classifications of functions with respect to their images and inverse functions.

Now we shall first discuss about sets theory.

1. **Set** : A well defined collection of distinct objects is called a set. It is denoted by the capital letter of the English alphabet., say, A, B, C ,....etc.

For examples :

(i) A = the set of natural numbers
= { 1, 2, 3,......}

(ii) B = the set of even natural numbers
= { 2, 4, 6, ...}

(iii) C = the set of vowels in english alphabet.
= { a, e, i, o, u }

2. Types of sets :

(i) **Finite set** : A set having finite number of elements is called a finite set .
Such as A = { 1, 2, ..., 10 } ,
B = { 1, 3, 5, ..., 99 } .

(ii) **Infinite set** : A set having infinite number of elelments is called an infinite set .
Such as A = { 2, 4, 6, ...},
B = { 0, 1, 2, 3 , ...} etc.

(iii) **Non empty set** : A set having at least one element is known as non empty set .
Such as A= { 1} , B = { 1, 2, 3, 4}
C = { 2, 4, ..., 100 } etc. are each a non empty set.

(iv) **Empty set** : A set having no element is called an empty set. It is denoted by { , } or ϕ and its read as (phi)

(v) **Subset** :If every element of a set A is also an element of B, then it is called a subset of B. It is denoted as $A \subseteq B$ and read as A is contained in B .If $B \supseteq A$, then we call it B is a super set of A.

Consider A = { 1, 2} and B = { 1, 2, 3 }
Then A is a subset of B .

(a) Let A = { 1, 2 } .
Then the possible subsets of A are
$\varphi, \{1\}, \{2\}, \{1, 2\}$

(b) Let B = { 1, 2, 3 }
Then the possible subsets of B are
$\varphi, \{1\}, \{2\}, \{3\}, \{1,2\}, \{1,3\}, \{2,3\}, \{1,2,3\}$

(c) If a set A contains 'n' elements , then the number of posssible subsets of A is 2^n

(d) **Power set** : The total number of possible subsets of a set is called a power set . If A is a set, then its power set is denoted as P(A) .

Let A = { 1, 2 } .

Then $P(A) = \{\phi, \{1\}, \{2\}, \{1,2\}\}$

(v) **Types of subsets :**

(a) Proper subset (b) Improper subset .

(a) **<u>Proper subset</u>** : A subset A is said to be a proper subset of B , if the number of ele ments of A is less than the number of ele ments of B . Let B = { 1, 2, 3} and A = { 1, 2 }. Then A is a proper subset of B. It is dusted as ACB.

(b) **<u>Improper subset</u>** : A subset is said to be an improper subset of B if either A has no element or A = B .

Let B = { 1, 2} .

Then the subsets of B are $\phi, \{1\}, \{2\}, \{1,2\}$

. Here, ϕ and $\{1,2\}$ are two

improper subsets.

Again, if we consider a set C = {1, 2, 3}

Then ϕ and { 1, 2, 3} are two improper subsets .

Number of subsets :

Thus, if a set contains 'n' elements , then the total number of improper subsets is 2 and the proper subsets = . $2^n - 2$

(vi) **Cardinal Number** :

Cadinal number is the number of distinct elements of a set . If A = { 1, 2, 3, 4, 5 } . Then its cardinal number = n (A) = 5.

(vii) **<u>Union of two sets</u>** : Let A and B be two non empty sets . Then its union is denoted as $A \cup B$ and is defined as $A \cup B =$

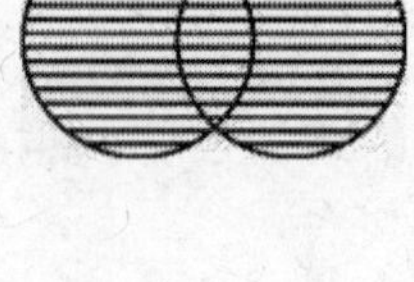

$\{x : x \in A \text{ or } x \in B \text{ or } x \in \text{both A and B}\}$

Let A = { 1, 2 3 } and B = { 2, 3, 4 , 5 }

Then $A \cup B = \{1,2,3,4,5\}$.

(viii) **<u>Intersection of two sets</u>** : Let A and B be two non empty sets .

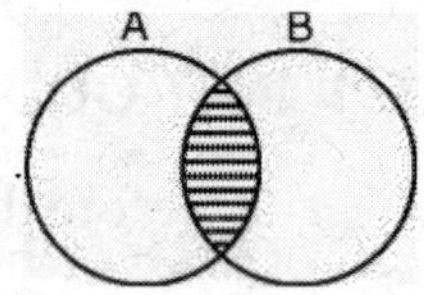

Then its intersetion is denoted as $A \cap B$ and is defined as $A \cap B$

$= \{x : x \in A \text{ and } x \in B\}$

Thus, A intersetion B means we shall count the common number of elements between two sets A and B.

Let A = { 1, 2, 3, 4 } and B = { 3, 4, 5}

Then $A \cap B$ = { 3, 4 }.

(ix) **<u>Universal set</u>** : It is the greatest non-empty sets and all the sets under its consideration are subsets. It is generally denoted as 'U' or 'X'.

Let X = { 1, 2, ..., 20 }

A = { 1, 2, 3} , B = { 2, 3, 4} , C = { 4,5}

D = { 2, 3, ...,10} .

Clearly A, B, C and D are the subsets of X. So X is the universal set .

(x) **<u>Complement of a set</u>** : Let X be a univer sal set and A be a subset of it .Then the complement of A is denoted as A' or A^c and is defined as

$A' = X - A = \{x : x \in X \text{ but } x \notin A\}$.

Let X = { 1,2, 3, ...10} and A = { 3, 4, 5, ...,10 } .

Then A' = X - A = { 1, 2 }

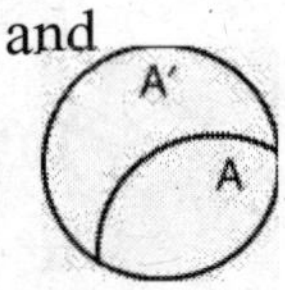

Note : We can find the complement of a set A only when A is a subset of the universal set X.

(xi) **Difference of two sets** : Let A and B be two non empty sets. Then the difference of A and B is denoted as A- B and is defined as A - B = {x : x ∈ A but x ∉ B}

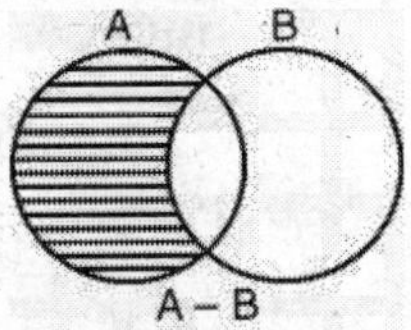

A − B

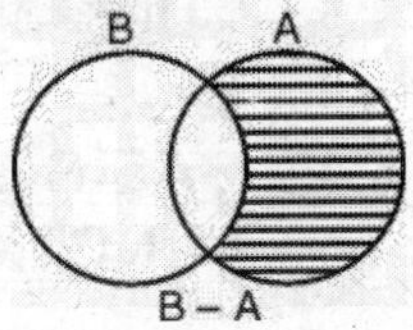

B − A

Let A = { 1, 2, 3, 4 } and B = { 2, 3, 4, 5, 6, 7 , 8}. Then A - B = { 1 } and B - A = { 5, 6, 7 , 8 }.

(xii) **Ordered pair** : Its a pair of objects listed in a specific order. Such as (a, b) , (c, d) , (1, 2) etc. are ordered pair of two objects.

(xiii) **Cartesian product** : Let A and B be two non empty sets . Then the caresian prod uct of A and B is denoted as $A \times B$ and is defined as $A \times B = \{(a, b): a \in A,\ b \in B\}$

Let A = { 1, 2 } and B = { 3, 4 } .

Then $A \times B = \{(1,3),(1,4),(2,3),(2,4)\}$.

and $B \times A = \{(3,1),(3,2),(4,1),(4,2)\}$

Thus, $A \times B \neq B \times A$.

Let A = { 1, 2, 3 } and B = { 2, 3, 4, 5 }

Then the number of elements of $A \times B$ $= n(A \times B) = n(A) \times n(B) = 3 \times 4 = 12.$

Thus, if A contains 'm' elements and B contains 'n' elements , then the number of elements of $A \times B$ is m.n.

It is clear that , the number of subsets of $A \times B$ is 2^{mn}.

(xiv) **Relation** : Let A and B two non-empty sets . Then a relation R from a set A to a set B is a subset of $A \times B$.

Let A = { 1, 2, 3} and B = { 4, 5 } and $R = \{(1,2),(2,3),(3,4),(3,5)\}$

Thus R is a subset of $A \times B$

.i.e. $R \subseteq A \times B$

If a set A contains 'm' elements and B contains 'n' elements , then the number of relations between two sets A and B is 2^{mn}.

5.2 BASIC CONCEPTS OF INTERVALS

(i) **Interval** : It is the gap between two real numbers . Let a, b ∈ R such that a < b . There are four types of intervals :

(a) **Open interval** : Let 1 , 2 ∈ R . Then its open interval is denoted as $(1,2)$.

If $x \in (1,2)$, then we write it $1 < x < 2$.

(b) **Closed interval** : Let 3, 4 ∈ R . Then its closed interval is denoted as $[3, 4]$

If $x \in [3,4]$, then we write it $3 \leq x \leq 4$.

(c) **Semi-open interval** : Let 5, 6 ∈ R . Then its semi-open interval is denoted as $[5, 6)$.

If $x \in [5, 6)$, then we write it $5 \leq x < 6$.

(d) **Semi-closed Interval** : Let 1 , 10 ∈ R . Then its semi-closed interval is denoted as $(1,10]$. If $x \in (1,10]$, then we write it $1 < x \leq 10$.

5.3. BASIC CONCEPTS OF NUMBER SYSTEM :

(i) **Natural Number** : All counting numbers are called natural number . It is denoted as N and is defined as

N = { 1, 2, 3, 4 ,........}.

(ii) **Whole number:** Natural number along with '0' is called a whole number . It is denoted as W and is defined as

W = { 0, 1, 2, 3, ...}

(iii) **Integers** : Natural number along with zero and their negatives is called integers It is denoed as I and I is defined as

I = { ...., -3, -2, -1, 0, 1, 2, 3, ...}

(iv) **Rational number** : It is the ratio of two integers , whose denominator is non zero It is denoted as Q and is defined as

$$Q=\left\{\frac{1}{2},\frac{1}{3},\frac{3}{4}.0,1,2,3,.........\right\}$$

(v) **Irrational numbers** : A number is said to be an irrational number if it is not rational. That is, it is not expressible as a ratio of two intergers. It is denoted as R - Q and is defined as R - Q

$$=\left\{\sqrt{2},\sqrt[4]{5},\sqrt[7]{10},\sqrt[100]{2},..........\right\}$$

(vi) **Real number** : The set of rational and irrational number is called a real number. It is denoted as R and is defined as

$$R=\left\{\frac{1}{2},\frac{2}{5},\frac{3}{4}\sqrt{2},\sqrt[4]{5},0,-2,-3,.....\right\}$$

In another way, we can say, all numbers are said to be a real number, other than even root of a negative number.

(vii) **Complex number** : The union of real and imaginary numbers is called a complex number. It is denoted as C and is defined

as $C=\left\{\frac{1}{2},\frac{2}{5},\sqrt{2},\ 0,-2,2+3i,3+4i,.....\right\}$

Another way, also we can define, all numbers are said to be a complex number Complex number is the universal set. and all the numbers are subsets of it.

Thus, $N\subset W\subset I\subset Q\subset R\subset C$.

5.4. BASIC CONCEPTS OF REAL FUNCTIONS :

(i) **Definition** : Let A and B be two non empty sets. A function 'f' is a rule between two sets A and B in such a way that

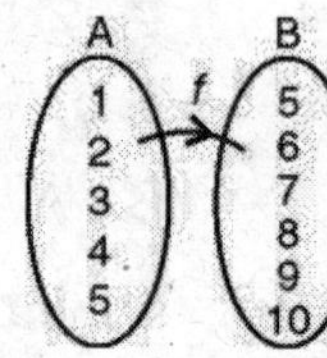

for every element in A there exists a unique element in B. It is denoted as $f:A\to B$ and it is read as f is a function from A to B or f maps from A to B .

Let A = { 1, 2, 3, 4}, B = { 5, 6, 7, 8, 9} and $f(x)=x+4$.

(ii) **Domain** : The first set is called the do main of a function . It is denoted as D_f

Here, $D_f=A$

(iii) **Co-domain** : The second set is called the co-domain of a function . It is denoted as C_f where, $C_f=B$

(iv) **Image** : If f(1) = 5 , then 5 is called in image of 1 under the function 'f' and 1 is called an inverse element of 5 or pre-image of 5.

(v) **Range** : The set of all images is called the range of a function . It is denoted as R_f

Here, $R_f=\{5,6,7,8\}\subseteq B$.

Another way, also we can define , the range of a function is a subset of co-domain.

(vi) **Real function** : If the domain and co-domain are the subsets of a real number , then it is called a real valued function or simply a real function . It is generally denoted as $f : R \rightarrow R$

(vii) **Checking Algebraically :**

1. If $a \in A \Rightarrow f(a) \in B$, then it is a function.

2. If $a \in A \Rightarrow f(a) \notin B$, then it is not a function.

3. If $a \in A \Rightarrow f(a) = bi \notin B$, where i = 1,2,3... then, it is not a function.

(viii) **Checking Geometrically :**

Draw a system of lines parallel to y-axis on the given curve, if those lines intersect the curve in only one point , then it is a function , otherwise a relation.

Ex-1. Let $y = x + 4$

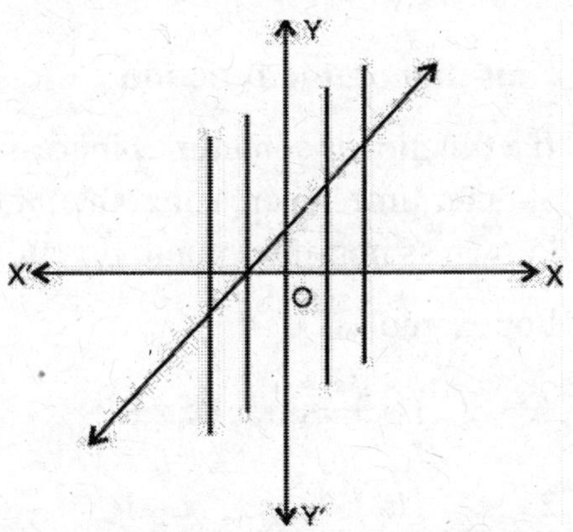

It is a function .

Ex-2. Let $x^2 + y^2 = a^2$

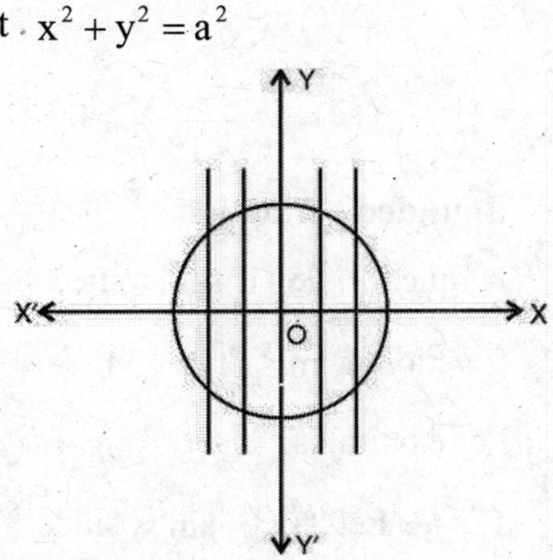

Clearly, it is not a function

(ix) Geometrically, X-axis will represent the domain of a function and Y-axis will represent the co-domain of a function.

(x) Geometrically, again when we find the domain of a function , we shall the con sider the part of f(x) which lies on x-axis and when we find the range of a function, that time, we shall consider the part of f(x) which lies on y-axis.

(xi) **Number of real functions between two sets A and B.**

If the number of elements of a set A contains is m and that of the set B is n , then the number of real functions between two sets A and B is n^m.

Let A = { 1, 2, 3} and B = { 2, 6 }, then the number of functions between two sets A and B is $2^3 = 8$.

(xii) **Some more functions :**

(a) **Even function** : A function f(x) is said to be an even function if $f(-x) = f(x)$ such as $f(x) = x^2, x^4 + x^2, \cos x, \sec x$ etc. are comprising each an even function.

(b) **Odd function** : A function f(x) is said to be an odd function if $f(-x) = -f(x)$ such as $f(x) = x^3$, $x + x^5$, sin x, cosec x , tan x and cot x etc. are each an odd function.

(c) **Periodic function** : A function f(x) is said to be a periodic function if $f(x + T) = f(x)$, where $T \in R$ The least numerical value of T is called the period of a function. All trigonometric functions are periodic.

The periods of

(i) $\sin^n x$, $\cos^n x$, $\operatorname{cosec}^n x$, $\sec^n x$ is 2π when n is odd and π, when n is even.

(ii) $\tan^n x$, $\cot^n x$ is π, whatever n may be.

(iii) $|\sin x|, |\cos x|, |\tan x|, |\operatorname{cosec} x|, |\sec x|, |\cot x|$ is π

(iv) $|\sin x|+|\cos x|$, $|\operatorname{cosec} x|+|\sec x|$ $|\tan x|+|\cot x|$ is $\frac{\pi}{2}$.

(v) $\sin^{2n}x+\cos^{2n}x$, $n\in I(>1)$ is $\frac{\pi}{2}$.

(d) **Strictly increasing function :**

A function f(x) is said to be strictly increasing function if $f'(x)>0 . \forall\ x\in D_f$.

For examples :

1. $f(x)=3x+5$
2. $f(x)=x^3+2$
3. $f(x)=e^x$
4. $f(x)=x^2+1,\ x\in R^+$
5. $f(x)=\frac{x^2}{x^2+1},\ x\in R^+$

Note : If $x>y \Rightarrow f(x)>f(y)$

(e) **Strictly decreasing function :**

A function f(x) is said to be strictly decreasing function if $f'(x)<0, \forall x\in D_f$

For examples :

1. $f(x)=5-3x$
2. $f(x)=\log_e(-x)$
3. $f(x)=e^{-x}$
4. $f(x)=1-x^2,\ x\in R^-$
5. $f(x)=\frac{1}{1+x^2},\ x\in R^+$

Note : If $x>y \Rightarrow f(x)<f(y)$

(f) **Monotonic function :**

If a function f(x) is either strictly inreasing function or strictly decreasing function , then it is known as monotonic function.

For examples :

1. $f(x)=e^x,\ x\in R$
2. $f(x)=\log_e x,\ x\in R^+$
3. $f(x)=\frac{1}{x^2+1},\ x\in R^+$
4. $f(x)=\frac{x}{x^2+1},\ x\in R^-$

(g) **Non-monotonic function :**

If a function is sometimes increasing and at other times decreasing, then it is known as non-monotonic function.

For examples :

1. $f(x)=\sin x,\ x\in R$
2. $f(x)=\cos x,\ x\in R$
3. $f(x)=\frac{1}{x^2+1},\ x\in R$
4. $f(x)=\frac{x^2}{x^2+1},\ x\in R$

(h) **Bounded function**

A function f(x) is said to be a bounded function if $m\le f(x)\le M$

For examples :

1. Let $f(x)=\sin x$, since $-1\le \sin x\le 1$

2. Let $f(x) = 3\cos x + 4$, since

$1 \le 3\cos x + 4 \le 7$

3. Let $f(x) = 3\sin x + 4\cos x$,

since $-5 \le f(x) \le 5$

4. Let $f(x) = \dfrac{x}{x^2+1}$

since $-\dfrac{1}{2} \le f(x) \le \dfrac{1}{2}$

5.5. TRIGONOMETRIC FUNCTION

(i) **Sine function** : A function $f : R \to R$ is defined as $f(x) = \sin x$

Graph of $f(x) = \sin x$

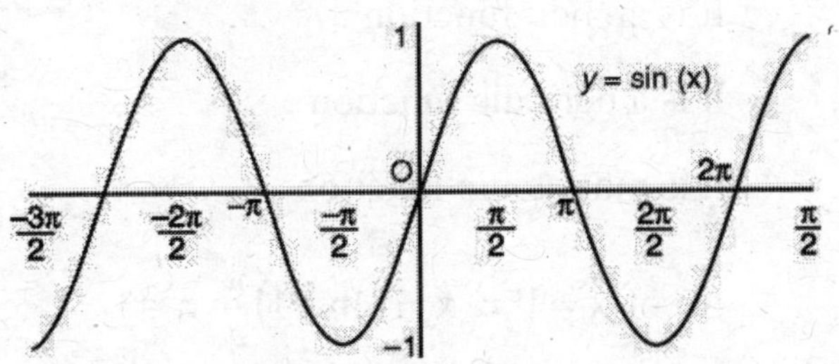

Characteristics of sine function :

1. $D_f = R$
2. $R_f = [-1, 1]$
3. It is an odd function
4. It is a periodic function
5. It is non-monotonic function
6. If $\sin x = 1 \Rightarrow x = (4n+1)\dfrac{\pi}{4}, n \in I$
7. $\sin x = -1 \Rightarrow x = (4n-1)\dfrac{\pi}{4}, n \in I$
8. $\sin x = 0 \Rightarrow x = n\pi, n \in I$
9. If $\sin x > 0 \Rightarrow x \in (2n\pi, (2n+1)\pi), n \in I$
10. If $\sin x < 0 \Rightarrow x \in ((2n-1)\pi, 2n\pi), n \in I$.
11. If $x > y \Rightarrow \sin x > \sin y$, $\forall\, x, y \in \left(-\dfrac{\pi}{2}, \dfrac{\pi}{2}\right)$

(ii) **Cosine function** : A function $f : R \to R$ is defined as $f(x) = \cos x$.

Graph of $f(x) = \cos x$:

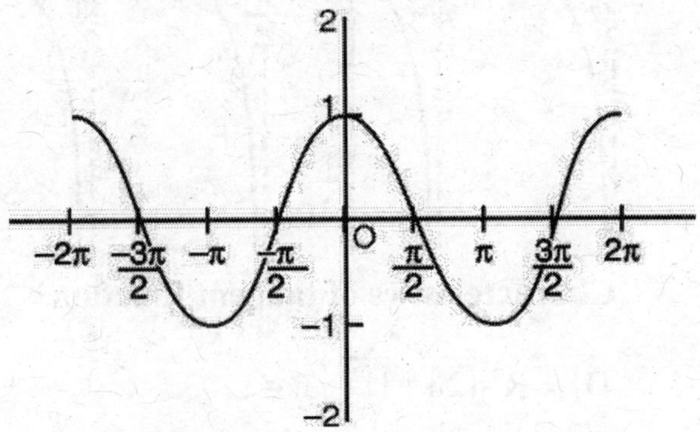

Characteristics of co-sine function :

1. $D_f = R$
2. $R_f = [-1, 1]$
3. It is an even function .
4. It is a periodic function
5. It is non-monotonic function
6. If $\cos x = 1 \Rightarrow x = 2n\pi, n \in I$
7. $\cos x = -1 \Rightarrow x = (2n+1)\pi, n \in I$
8. $\cos x = 0 \Rightarrow x = (2n+1)\dfrac{\pi}{2}, n \in I$
9. If $\cos x > 0$ $\Rightarrow x \in \left((2n-1)\dfrac{\pi}{2}, (2n+1)\dfrac{\pi}{2}\right), n \in I$
10. If $\cos x < 0$ $\Rightarrow x \in \left((2n+1)\dfrac{\pi}{2}, (2n+3)\dfrac{\pi}{2}\right), n \in I$
11. If $x > y \Rightarrow \cos x < \cos y$, $\forall\, x, y \in (0, \pi)$

(*iii*) **Tangent Function** : A function $f : R \to R$ is defined as $f(x) = \tan x$.

Graph of $f(x) = \tan x$

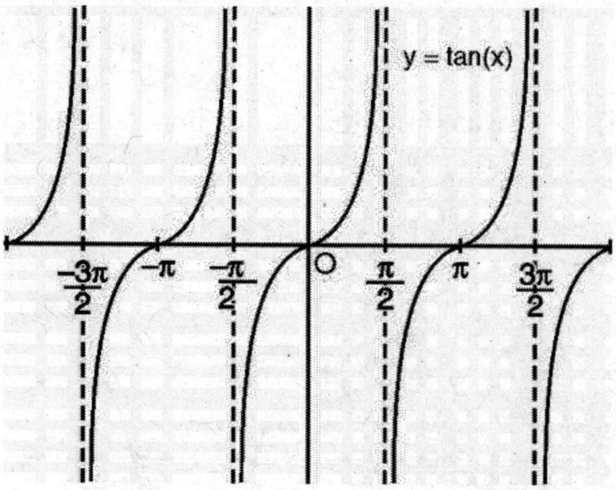

Characteristics of tangent function :

1. $D_f = R - (2n+1)\frac{\pi}{2},\ n \in I$
2. $R_f = R$
3. It is an odd function .
4. It is a periodic function
5. It is monotonic function
6. If $\tan x = 1 \Rightarrow x = (4n+1)\frac{\pi}{4}, n \in I$
7. If $\tan x = -1 \Rightarrow x = (4n-1)\frac{\pi}{4}, n \in I$
8. If $\tan x = 0 \Rightarrow x = n\pi, n \in I$
9. If $\tan x > 0$

   $\Rightarrow x \in \left(n\pi,\ (2n+1)\frac{\pi}{2}\right), n \in I$
10. If $\tan x < 0$

    $\Rightarrow x \in \left((2n-1)\frac{\pi}{2}, n\pi\right), n \in I$
11. If $x > y \Rightarrow \tan x > \tan y$

    $\forall\ x, y \in R - n\pi,\ n \in I$

(iv) **Co-tangent function**: A function $f : R \to R$ is defined as $f(x) = \cot x$

Graph of $f(x) = \cot x$

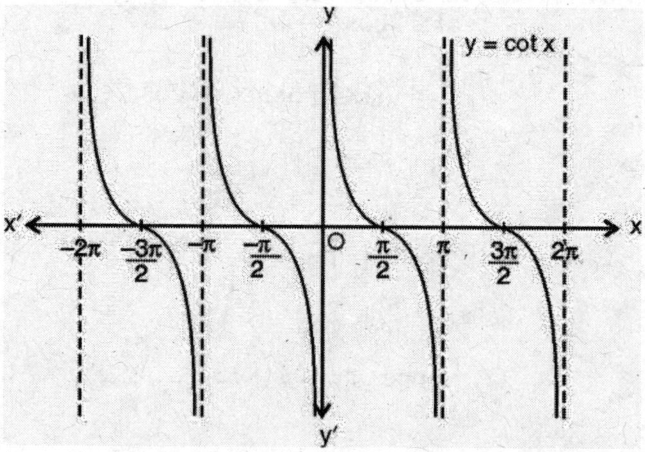

Characteristics of co-tangent function :

1. $D_f = R - n\pi,\ n \in I$
2. $R_f = R$
3. It is an odd function .
4. It is a periodic function
5. It is monotonic function
6. If $\cot x = 1 \Rightarrow x = (4n+1)\frac{\pi}{4}, n \in I$
7. If $\cot x = -1 \Rightarrow x = (4n-1)\frac{\pi}{4}, n \in I$
8. If $\cot x = 0 \Rightarrow x = (2n+1)\frac{\pi}{2}, n \in I$
9. If $\cot x > 0$

   $\Rightarrow x \in \left(n\pi,\ (2n+1)\frac{\pi}{2}\right), n \in I$
10. If $\cot x < 0$

    $\Rightarrow x \in \left((2n-1)\frac{\pi}{2}, n\pi\right), n \in I$

(v) **Co-secant function** : A function $f : R \to R$ is defined as $f(x) = \operatorname{cosec} x$

Graph of $f(x) = \text{cosec } x$

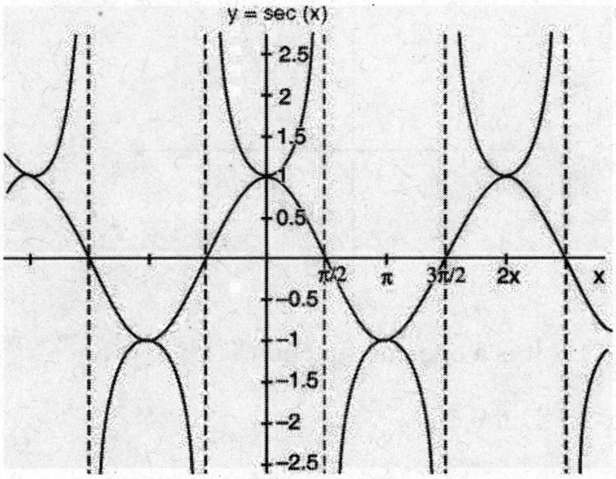

Characteristics of co-secant function :

1. $D_f = R - n\pi,\ n \in I$
2. $R_f = (-\infty, -1] \cup [1, \infty)$
3. It is an odd function
4. It is a periodic function.
5. It is non-monotonic function.
6. If $\text{cosec}x = 1 \Rightarrow x = (4n+1)\dfrac{\pi}{4}, n \in I$
7. If $\text{cosec}x = -1 \Rightarrow x = (4n-1)\dfrac{\pi}{4}, n \in I$
8. cosec x can never be zero.
9. If $\text{cosec}x > 0 \Rightarrow x \in (2n\pi, (2n+1)\pi), n \in I$
10. If $\text{cosec}\,x < 0 \Rightarrow x \in ((2n-1)\pi, 2n\pi), n \in I$.

6. **Secant function** : A function $f : R \to R$ is defined as $f(x) = \sec\ x$

Graph of $f(x) = \sec\ x$

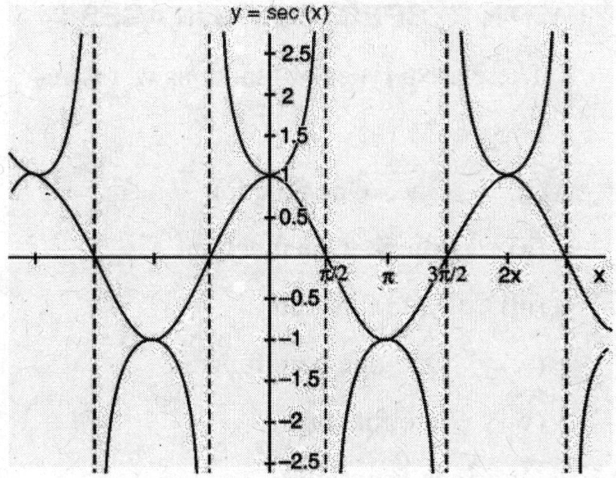

Characteristics of secant function :

1. $D_f = R - (2n+1)\dfrac{\pi}{2},\ n \in I$
2. $R_f = (-\infty, -1] \cup [1, \infty)$
3. It is an even function
4. It is a periodic function.
5. It is non-monotonic function.
6. If $\sec x = 1 \Rightarrow x = 2n\pi, n \in I$
7. If $\sec x = -1 \Rightarrow x = (2n+1)\pi, n \in I$
8. sec x can never be zero.
9. If $\sec x > 0$

   $\Rightarrow x \in \left((4n-1)\dfrac{\pi}{2}, (4n+1)\dfrac{\pi}{2}\right), n \in I$
10. If $\sec\ x < 0$

   $\Rightarrow x \in \left((4n+1)\dfrac{\pi}{2}, (4n+3)\dfrac{\pi}{2}\right), n \in I$

5.6. CLASSIFICATION OF FUNCTIONS WITH RESPECT TO ITS IMAGES :

There are 5 types of functions w. r. t. its images such as

(i) One - One function

(ii) Many - One function

(iii) Onto function

(iv) One-one-onto function

(v) into function .

(i) **One-One function :**

(a) **Definition :** If different elements of the first set provide us different images in the second set, then it is known as one one function.

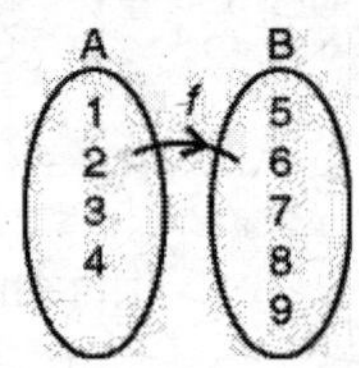

Let A = { 1, 2, 3, 4 }, B = { 5, 6, 7, 8 . 9} and f (x) = x + 4 .

Thus, f(1) = 5, f(2) = 7, ..., f(4) = 8.

Clearly, f is a one-one function .

(b) **Checking algebraically :**

(i) If $x \neq y \Rightarrow f(x) \neq f(y)$, then it is one one function .

(ii) If $x \neq y \Rightarrow f(x) = f(y)$, then it is not a one-one function .

(c) **Checking geometrically :**

Draw a system of lines parallel to x-axis on the given curve. If those lines inter sects the curve at only one point , then it is one one function; otherwise many one function.

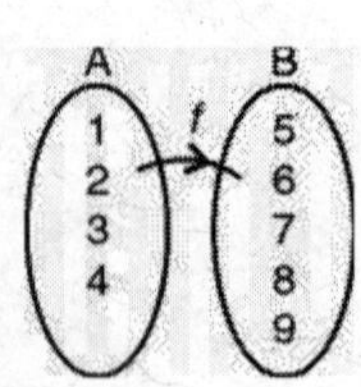

Ex-1. Let $y = x^3$

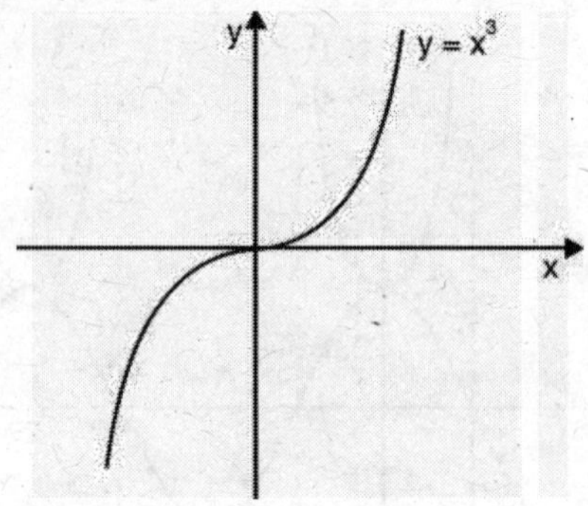

It is a one one function

Ex-2. Let $y = e^x$

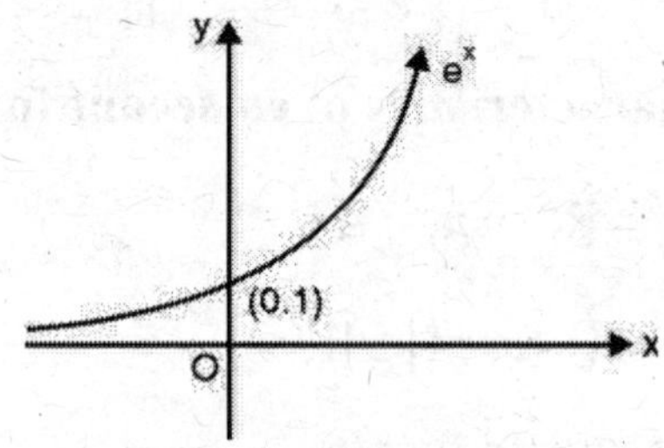

It is one -one function .

Ex-3. Let $y = (1/2)^x$

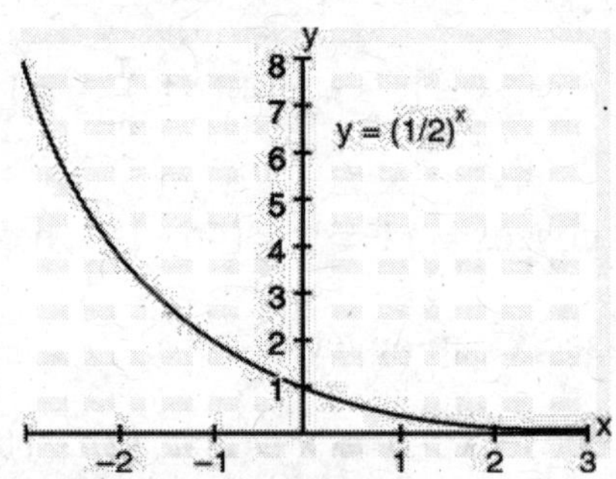

It is one function.

Ex-4. Let $y = \log_e x$

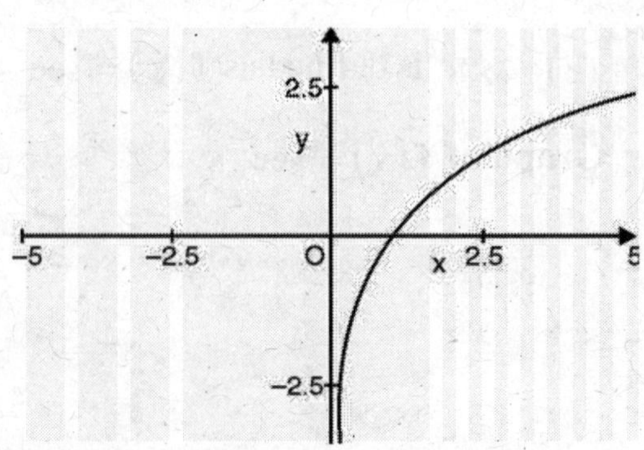

It is one-one function

Ex-5. $y = \sin x, -\frac{\pi}{2} \le x \le \frac{\pi}{2}$

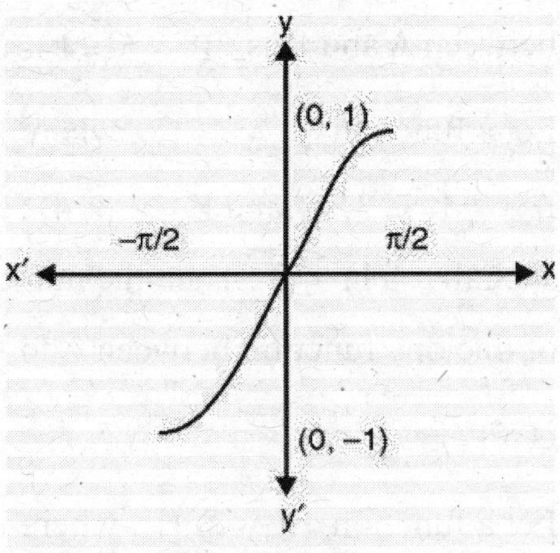

y = sin x

Figure (a)

It is one -one function

Ex-6. Let $y = \cos x, 0 \le x \le \pi$

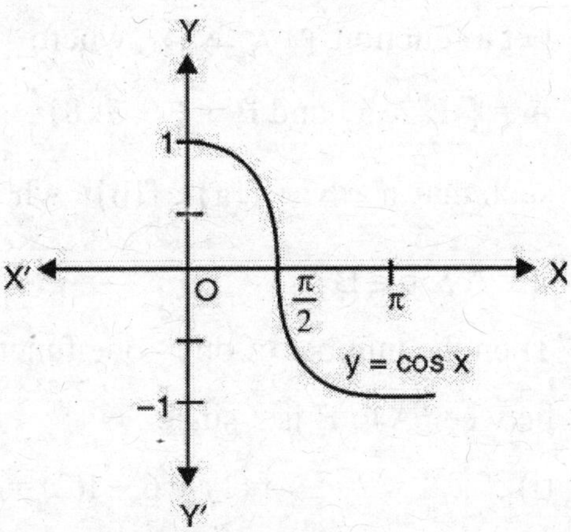

It is one one function.

Ex-7. Let $y = \sqrt{x}$

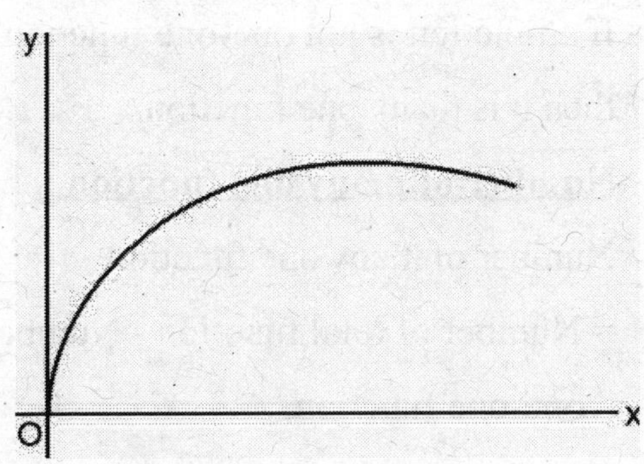

It is one one function.

Ex-8. $y = x^3 + 2x^2 + 3x + 1.$

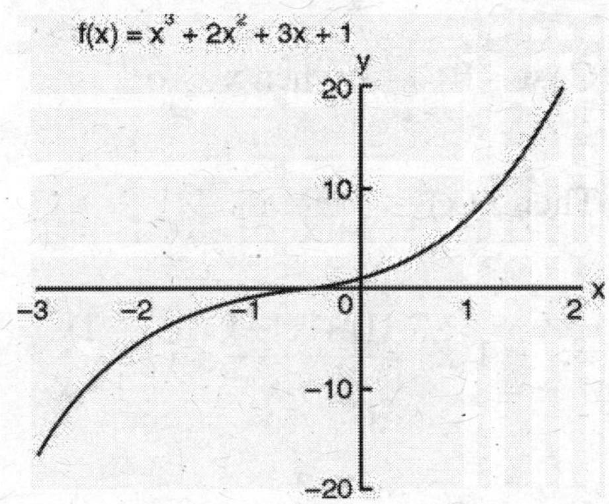

It is one one function.

(d) **Checking by calculus.**

If a function is either strictly increasing or strictly decreasing function , then it is one -one function.

i.e. either $f'(x) > 0$ or $f'(x) < 0$, then it is one one function .

Ex-1. Let $y = f(x) = 3x + 4$

Then $f'(x) = 3 > 0$

$\Rightarrow$ f(x) is strictly oncreasing function.

$\Rightarrow$ f(x) is one one function.

Ex-2. Let $f(x) = \tan^{-1}(3\log(x) + 5)$

Then $f'(x) = \frac{1}{1 + (3\log x + 5)^2} \times \frac{1}{x} > 0$ for all positive x

$\Rightarrow$ f(x) is strictly oncreasing function.

$\Rightarrow$ f(x) is one one function.

Ex-3. Let $f(x) = 2\tan x + 3\sin x + 4\cos x + 10$

Then $f'(x) = 2\sec 2x + 3\cos x - 4\sin x.$

$\Rightarrow$ $f'(x) > 0$ for all x in R.

$\Rightarrow$ f(x) is strictly oncreasing function.

$\Rightarrow$ f(x) is one one function.

Ex-4. Let $f(x) = \dfrac{x}{1+|x|}$

Case - I : when $x \geq o$

Then $f(x) = \dfrac{x}{1+x}$

$\Rightarrow \quad f(x) = \dfrac{(1+x)-1}{1+x} = 1 - \dfrac{1}{1+x}$

$\Rightarrow \quad f'(x) = \dfrac{1}{(1+x)^2} > 0$

Case - II: when $x < 0$

Then $f(x) = \dfrac{x}{1-x}$

$\Rightarrow \quad f(x) = -\dfrac{x-1+1}{x-1} = -1 - \dfrac{1}{x-1}$

$\Rightarrow \quad f'(x) = \dfrac{1}{(x-1)^2} > o$ for all $x < 0$

Thus, f is strictly increasing function.

Hence , f is one-one function.

(e) **Number of one-one function between two sets A and B.**

If A and B having m and n elements respectively , then the total number of functions between two sets A and B is

$$= \begin{cases} 0 : n < m \\ {}^{n}P_{m} : n \geq m \end{cases}$$

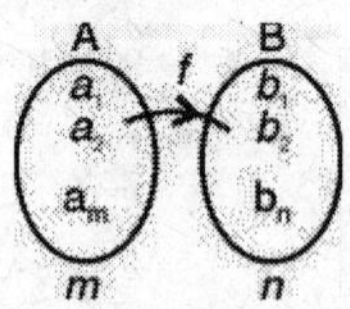

Ex-1. Let A = { 1, 2, 3 } and B = { 4, 5}

Then the total number of one-one function between two sets A and B is 0.

Ex-2. Let A = { 1, 2, 3} and B = { p, q, r }

Then the total number of function between A and B is ${}^{3}P_{3} = 3! = 6$.

Ex-3. Let A = { 1, 2, 3, 4} and B { a, b, c, d, e}

Then the number of one-one function between A and B is ${}^{5}P_{4} = 5! = 120$.

Ex-4. Let a function $f : A \rightarrow B$, where A = { 1, 2, 3, 4} and B = { 3, 4, 5, 6} such that f(1) = 3 . Then the number of one one function between A to B is ${}^{3}P_{3} = 3! = 6$

Ex-5. Let a function $f : A \rightarrow B$, where A ={1, 2, 3} and B = { 4, 5, 6} such that $f(1) \neq 4$. Then the number of one - one function between A to B is $2 + 2 = 4$.

Ex-6. Let a function $f : A \rightarrow B$, where A = { 1, 2, 3} and B = { 6, 7, 8} such that $a < b \Rightarrow f(a) < f(b)$, where $a \in A \,\&\, b \in B$.

Then the number of one - one function between A to B is 3 since

(i) $1 < 2 \Rightarrow f(1) = 6 < f(2) = 7$

(ii) $1 < 3 \Rightarrow f(1) = 6 < f(3) = 8$

(iii) $2 < 3 \Rightarrow f(2) = 7 < f(3) = 8$.

(ii) **Many One function :**

If a function is not one-one function, then it is many one function.

<u>**Number of many one function :**</u>

Number of many one function

= Number of total function - Number of one-one function.

Ex-1. Find the number of many one function between two sets A = { 3, 4, 5} and B = {2, 3, 4 , 5}.

Soln. Number of many one function

= Number of total function - Number of one - one function .

$= 4^3 - {}^4P_3 = 64 - 24 = 40$.

Ex-2. Find the number of many one function between two sets A = { 1, 2, 3, 4 } and B = { 3, 4, 5 }.

Soln. Number of many one function

= Number of total function - Number of one - one function .

$= 3^4 - 0 = 81$

Ex-3. Find the number of many-one function between two sets A = { 1, 2, 3 } and B = { 4, 5, 6} such that $f(1) \neq 4$

Soln. Number of many one function

= Number of total function - Number of one - one function .

$= 3^3 - 2 = 27 - 2 = 25$

(iii) **Onto function :**

(a) **Definition** : If the range of a function is equal to its co-domain , then it is called Onto function. i.e. $R_f = \text{Co-domain}$.

Ex-1. A function $f : R \rightarrow R$ is defined as $f(x) = 3x + 5$. Clearly , the range of f(x) is R . Since $R_f = R$ = Co-domain , so f is onto function.

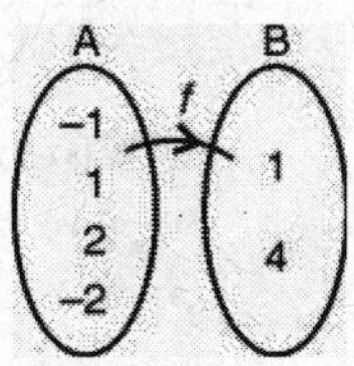

Ex-2. A function $f : R - \{-1\} \rightarrow R - \{1\}$ is defined as $f(x) = \dfrac{x}{x+1}$

Now , let $y = \dfrac{x}{x+1}$

$\Rightarrow \quad yx + y = x$

$\Rightarrow \quad x(y-1) = -y$

$\Rightarrow \quad x = \dfrac{-y}{(y-1)}$

Then, the range of a function is $R - \{1\}$

Thus, $R_f = R - \{1\} = \text{Co-domain}$

Hence, f is onto function.

Ex-3. Let a function $f : R \rightarrow A$ is defined as $f(x) = \dfrac{1}{x^2+1}$. If f is onto function, then find the set of values of A .

Soln. Clearly, the range of a function is = (0, 1]

i.e. $R_f = (0, 1]$. Since f is onto function , so R_f = Co-domain

$\Rightarrow \quad A = R_f = (0, 1]$

Hence, the set A is $(0, 1]$.

Ex-4. Let a function $f : A \rightarrow B$ is defined as $f(x) = \dfrac{x^2}{x^2+1}$. If f is onto function, then find the set of values of $A \cap B$.

Soln. Decideldly, domain of a function is R .

i,e. $D_f = A = R$.

Now , range of the function is [0, 1)

Thus, R_f = [0, 1).

Since f is onto function, so R_f = B.

Thus, B = [0, 1)

Hence, $A \cap B = [0,1)$.

(b) **Number of onto function between two sets A and B.**

If two set A and B having m and n elements respectively, where $m \geq n$, then the number of onto function between two sets A and B

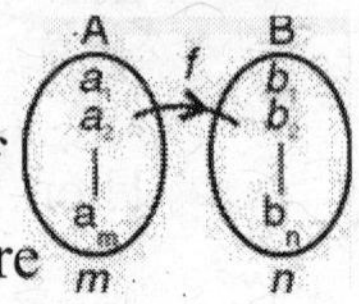

= Number of distribution of m balls into n boxes where no box is remain empty.

Ex-1. Find the number of onto function between two sets A = { 1, 2, 3, 4} and B = { 4, 5 }

Soln. Number of onto function

= Number of distribution of 4 balls into 2 boxes , where no box is to remain empty.

$$= \frac{4!}{3! \times 1!} \times 2! + \frac{4!}{2! \times 2!} \times \frac{2!}{2!}$$

$= 8 + 6 = 14$.

Ex-2. Find the number of onto function between two sets A = {1, 2, 3,4} and B = {4, 5,6 }

Soln. Number of onto function

= Number of distribution of 4 balls into 3 boxes , where no box is to remain empty.

$$= \frac{4!}{3! \times 1!} \times 2!$$

$= 8$

Ex-3. Find the number of onto function between two sets A and B where A = {1,2, 3, 4,5 } and B = { 6, 7, 8}

Soln. Number of onto function

= Number of distribution of 5 balls into 3 boxes , where no box is remain empty.

$$= \frac{5!}{3! \times 1! \times 1!} \times \frac{3!}{2!} + \frac{5!}{2! \times 2! \times 1!} \times \frac{3!}{2!}$$

$= 60 + 90$

$= 150$.

(iv) **Into function** : If a function is not onto function , then it is into function .

Alternatively way also we can say, if the range of a function is a proper subset of a codomain then it is into function.

The number of into functions

= The total number of functions - The number of onto functions.

Ex-1. Find the number of into function between two sets A and B , where A = { 1, 2, 3} and B= { 4, 5}

Soln. The number of into function

= Total number of functions - the number of onto functions

$= 2^3 - 6$

$= 8 - 6$

$= 2$.

Ex-2. Find the number of into function between two sets A and B where A = { 1, 2, 3, 4 } and B = { 5, 6, 7}.

Soln. The number of into functions

= Total number of functions - the number of onto functions

$$= 3^4 - \left(\frac{4!}{2! \times 1! \times 1!} \times \frac{3!}{2!} \right)$$

$= 81 - 36$

$= 45$.

Ex-3. Find the number of into functions be tween two sets A and B where A = { 1, 2, 3, 4, 5} and B = { 6, 7, 8}

Soln. The number of into functions

= The total number of functions - The number of onto functions.

$$= 3^5 - \left(\frac{5!}{3!\times1!\times1!}\times\frac{3!}{2!}+\frac{5!}{2!\times2!\times1!}\times\frac{3!}{2!}\right)$$

= 243 - 150

= 93 .

(v) **One one onto function :**

(a) **Definition** : If a function is one one as well as onto function, then it is one one onto function or bijective function.

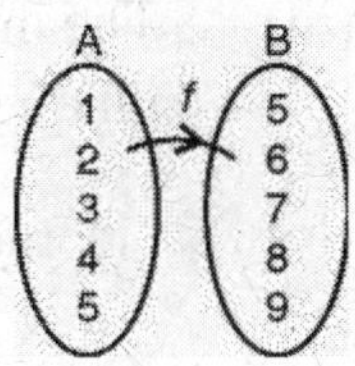

Note : One one onto function exists only when the number of elements of both the sets are same.

Ex-1. A function $f: R \rightarrow R$ is defined as f(x) = 3 x + 5. Prove that f is a bijective function.

Soln. Given f(x) = 3 x + 5.

$\Rightarrow$ $f'(x) = 3 > 0$

$\Rightarrow$ f is strictly increasing function.

$\Rightarrow$ f is one one function.

Also, the range of a function is R

$\Rightarrow$ f is onto function.

Hence, f is a bijective function.

Ex-2. A function $f: R^+ \rightarrow (1, \infty)$ is defined as $f(x) = x^2 + 1$. Prove that the function is bijective.

Soln. Given $f(x) = x^2 + 1$

$\Rightarrow$ $f'(x) = 2x > 0 \ \forall \ x \in R^+$

$\Rightarrow$ f is strictly increasing function .

$\Rightarrow$ f is one one function.

Also , the range of a function is $(1, \infty)$

$\Rightarrow$ $R_f = (1, \infty)$ = Co-domain

$\Rightarrow$ f is onto function .

Hence, the function f is bijective.

Ex-3. A function $f: R^+ \rightarrow (0, 1)$ is defined as $f(x) = \frac{1}{x^2+1}$ .Prove that f is a bijective function.

Soln. Given $f(x) = \frac{1}{x^2+1}$

$\Rightarrow$ $f'(x) = \frac{2x}{(x^2+1)^2} > 0, \ \forall x \in R^+$

$\Rightarrow$ f is strictly increasing function.

$\Rightarrow$ f is one one function.

Also , the range of a function is $(0, 1)$

$\Rightarrow$ $R_f = (0, 1)$ = co-domain

$\Rightarrow$ f is onto function

Thus, f is a bijective function.

Ex-4. A function $f: R \rightarrow \left(-\frac{1}{2}, \frac{1}{2}\right)$ is defined as $f(x) = \frac{x}{x^2+1}, \ \forall x \in (-1, 1)$. Prove that f (x) is a bijective function.

Soln. Given $f(x) = \frac{x}{x^2+1}, \ \forall x \in (-1, 1)$

$\Rightarrow$ $f'(x) = \frac{1-x^2}{(1+x^2)^2} > 0, \ \forall \ x \in (-1, 1)$

$\Rightarrow$ f is strictly increasing function

$\Rightarrow$ f is one one function.

Also, the range of a function is $\left(-\frac{1}{2}, \frac{1}{2}\right)$

$\Rightarrow \quad R_f = \left(-\frac{1}{2}, \frac{1}{2}\right) = \text{Co-domain}$

$\Rightarrow \quad$ f is onto function .

Thus f is a bijective function.

(b) **Number of one-one onto functons between two sets A and B.**

Let the sets A nd B having m and n elements respectively.

Then the number of one one onto functions between two sets A and B

$$= \begin{cases} 0 & : \quad m \neq n \\ {}^{n}P_{m} : & m = n \end{cases}$$

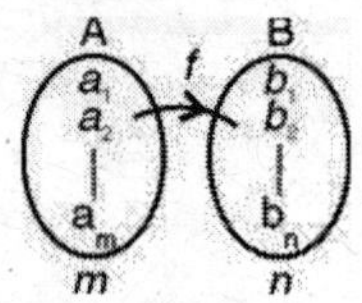

Ex-1: If A = { 1, 2, 3, 4 } and B = { 2, 3, 4 }

Soln. Since the number of elements of A and B are not same , so the number of one one onto function is 0 .

Ex-2. Find the number of bijective functions between two sets A and B, where A = { a, b, c, d, e } and B = { p, q, r, s, t}

Soln. Since the number of elements of both the sets are same, so the number of bijective function is ${}^{5}P_{5} = 5! = 120$.

Ex-3. Find the number of bijective functions between two sets A = { 1, 2, 3, 4}and B = { 5, 6, 7, 8 }such that $f(1) = 5$.

Soln. Hence, the number of bijective functions $= {}^{3}P_{3} = 3! = 6$.

5.7. INVERSE FUNCTION :

Let $f : X \rightarrow Y$ be a bijective function .

If we can make a function g from Y to X , then we shall say that g is the inverse of f

i.e. $g = f^{-1} \neq \frac{1}{f}$.

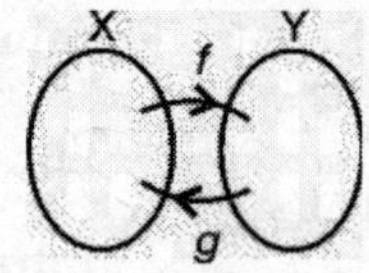

Thus , $f^{-1}(f(x)) = x$

Note :

(i) The inverse of a function exists only when the function f is bijective.

(ii) If the inverse of a function exists , then it is called an invertible function.

(iii) The inverse of a bijective function is unique.

(iv) Geometrically $f^{-1}(x)$ is the image of f(x) with respect to the line $y = x$.

(v) Another way also we can say that $f^{-1}(x)$ is the symmetrical with respect to the line $y = x$.

(vi) A function f(x) is said to be involution if for all x for which f(x) and f(f(x)) are defined such that $f(f(x)) = x$.

(vii) If f is an invertible function , then $(f^{-1})^{-1} = f$.

(viii) If $f : A \rightarrow B$ be a one one function , then $f^{-1} o f = I_A$ and $f o f^{-1} = I_B$, where I_A and I_B are the identity functions of the sets A and B respectively.

(ix) Let $f : A \rightarrow B$, $g : B \rightarrow C$ be two invertible functions , then gof is also invertible with $(gof)^{-1} = (f^{-1} o g^{-1})$.

Rule to find out the inverse of a function:

(i) First, we check the given function is bijective or not.

(ii) If the function is bijective , then inverse exists, otherwise not.

(iii) Find x in terms of y

(iv) And then replace y by x , then we get inverse of f . i.e. $f^{-1}(x)$.

Ex-1. A function $f: R \to R$ is defined as $f(x) = 3x + 5$. Find $f^{-1}(x)$.

Soln. Given $f(x) = 3x + 5$

$\Rightarrow$ $f'(x) = 3 > 0$

$\Rightarrow$ f is strictly increasing function.

$\Rightarrow$ f is one one function

Also , $R_f = R =$ Co-domain

$\Rightarrow$ f is onto function .

Thus, f is a bijective function .

Hence, f^{-1} exists.

Let $y = 3x + 5$

$\Rightarrow$ $x = \dfrac{y-5}{3}$

Thus, $f^{-1}(x) = \dfrac{x-5}{3}$.

Ex-2. A function $f: (0, \infty) \to (2, \infty)$ is defined as $f(x) = x^2 + 2$. Then find $f^{-1}(x)$.

Soln. Given $f(x) = x^2 + 2$

$\Rightarrow$ $f'(x) = 2x > 0$, for every $x > 0$

$\Rightarrow$ f is strictly increasing function.

$\Rightarrow$ f is one one function.

Also, $R_f = (2, \infty) =$ Co-domain

$\Rightarrow$ f is onto function.

Thus, f is a bijective function .

Therefore, the inverse of the given function exists.

Let $y = x^2 + 2$

$\Rightarrow$ $x^2 = y - 2$

$\Rightarrow$ $x = \sqrt{y-2}$

Hence, $f^{-1}(x) = \sqrt{x-2}$

Ex-3. A function $f: R^+ \to [0, 1)$ is defined as $f(x) = \dfrac{x^2}{x^2+1}$. Then find $f^{-1}(x)$.

Soln. Given $f(x) = \dfrac{x^2}{x^2+1}$

$\Rightarrow$ $f(x) = 1 - \dfrac{1}{x^2+1}$

$\Rightarrow$ $f'(x) = \dfrac{2x}{(x^2+1)^2} > 0, \ \forall \ x \in R^+$

$\Rightarrow$ f is strictly increasing function.

$\Rightarrow$ f is one one function.

Also, let $y = \dfrac{x^2}{x^2+1}$

$\Rightarrow$ $y.x^2 + y = x^2$

$\Rightarrow$ $x^2(y-1) = -y$

$\Rightarrow$ $x^2 = -\dfrac{y}{(y-1)} = \dfrac{y}{(1-y)}$

$\Rightarrow$ $x = \sqrt{\dfrac{y}{(1-y)}}$

$\Rightarrow$ $R_f = (0, 1) =$ Co-domain

$\Rightarrow$ f is onto function.

Thus, f is a bijective function .

$\Rightarrow$ $f^{-1}(x)$ exists.

Hence, $f^{-1}(x) = \sqrt{\dfrac{x}{1-x}}$.

Ex-4. A function $f: [1, \infty) \to [1, \infty)$ is defined as $f(x) = 2^{x(x-1)}$. Find $f^{-1}(x)$.

Soln. Given $f(x)=2^{x(x-1)}$.

$\Rightarrow \quad f'(x)=2^{x(x-1)}\times(2x-1)\times\log_e 2>0$

for all x in $[1, \infty)$

$\Rightarrow$ f is strictly increasing function.

$\Rightarrow$ f is one one function.

Also, $R_f=[1, \infty)$

$\Rightarrow \quad R_f=[1, \infty)=$ Co-domain

$\Rightarrow$ f is onto function.

Thus, f is a bijective function.

So its inverse too, is existent.

Let $y=2^{x(x-1)}$

$\Rightarrow \quad y=2^{x^2-x}$

$\Rightarrow \quad x^2-x=\log_2(y)$

$\Rightarrow \quad x^2-x-\log_2(y)=0$

$\Rightarrow \quad x=\dfrac{1\pm\sqrt{1+4\log_2(y)}}{2}$

$\Rightarrow \quad x=\dfrac{1+\sqrt{1+4\log_2(y)}}{2}$

Thus, $f^{-1}(x)=\dfrac{1+\sqrt{1+4\log_2(x)}}{2}$

Ex-5. If a friction f is bijective such that

$f(x)=\dfrac{10^x-10^{-x}}{10^x+10^{-x}}$, then find $f^{-1}(x)$

Soln. Since f is a bijective function, so its inverse is exists.

Let $y=\dfrac{10^x-10^{-x}}{10^x+10^{-x}}=\dfrac{10^{2x}-1}{10^{2x}+1}$.

$\Rightarrow \quad y.10^{2x}+y=10^{2x}-1$

$\Rightarrow \quad 10^{2x}(y-1)=-y-1$

$\Rightarrow \quad 10^{2x}=-\dfrac{y+1}{y-1}=\dfrac{y+1}{1-y}$

$\Rightarrow \quad 2x=\log_{10}\left(\dfrac{y+1}{1-y}\right)$

$\Rightarrow \quad x=\dfrac{1}{2}\log_{10}\left(\dfrac{1+y}{1-y}\right)$

Thus, $f^{-1}(x)=\dfrac{1}{2}\log_{10}\left(\dfrac{1+x}{1-x}\right)$.

Ex-6. A function $f:R\to R$ is defined as $f(x)=x+\sin x$. Find $f^{-1}(x)$

Soln. Given $f(x)=x+\sin x$

$\Rightarrow \quad f'(x)=1+\cos x\geq 0$ for all x in R.

$\Rightarrow$ f is strictly increasing function

$\Rightarrow$ f is one one function.

Also, the range of a function is R

$\Rightarrow$ f is a onto function

Thus, f is a bijective function.

Hence, f^{-1} is also existing.

Therefore, $f^{-1}(x)=x-\sin x$

Ex-7. A function $f:[2, \infty)\to[5, \infty)$ is defined as $f(x)=x^2-4x+9$. Find its inverse.

Soln. Given $f(x)=x^2-4x+9$

$\Rightarrow \quad f'(x)=2x-4\geq 0$ for all x in D_f

$\Rightarrow$ f is strictly increasing function.

$\Rightarrow$ f is one one function.

Also, $R_f=[5, \infty)=$ Co-domain

$\Rightarrow$ f is onto function.

Therefore, f is a bijective function.

Hence, its inverse too, is existent.

Let $y = x^2 - 4x + 5$

$\Rightarrow \quad x^2 - 4x + (5 - y) = 0$

$\Rightarrow \quad x = \dfrac{4 \pm \sqrt{16 - 4(5-y)}}{2}$

$\Rightarrow \quad x = \dfrac{4 \pm \sqrt{4y + 16 - 20}}{2}$

$\Rightarrow \quad x = \dfrac{4 \pm \sqrt{4(y-1)}}{2} = 2 \pm \sqrt{(y-1)}$

$\Rightarrow \quad x = 2 + \sqrt{(y-1)}$, since $x \geq 2$

$\Rightarrow \quad f^{-1}(x) = 2 + \sqrt{(x-1)}$

Ex-8. Find all the real solutions to the equation

$x^2 - \dfrac{1}{4} = \sqrt{x + \dfrac{1}{4}}$.

Soln. Consider the function

$f : [0, \infty) \to \left[-\dfrac{1}{4}, \infty\right)$, where

$f(x) = x^2 - \dfrac{1}{4}$.

Clearly, f is one one and onto function.

So its inverse too, is exists.

Let its inverse is $f^{-1} : \left[-\dfrac{1}{4}, \infty\right) \to [0, \infty)$.

$\Rightarrow \quad f^{-1}(x) = \sqrt{x + \dfrac{1}{4}}$.

Consequently, we can say that, the two sides of the given equation are inverse to each other.

Thus, the intersection point is the solution of the given equation .f (x) = x.

$\Rightarrow \quad x^2 - \dfrac{1}{4} = x$

$\Rightarrow \quad x^2 - x = \dfrac{1}{4}$

$\Rightarrow \quad \left(x - \dfrac{1}{2}\right)^2 = \dfrac{1}{2}$

$\Rightarrow \quad \left(x - \dfrac{1}{2}\right) = \pm \dfrac{1}{\sqrt{2}}$

$\Rightarrow \quad x = \dfrac{1}{2} \pm \dfrac{1}{\sqrt{2}}$

Hence, the solutions are

$\left\{\dfrac{1}{2} + \dfrac{1}{\sqrt{2}}, \dfrac{1}{2} - \dfrac{1}{\sqrt{2}}\right\}$.

Ex-9. For what values of m is $f(x) = (m+2)x^3 - 3mx^2 + 9mx - 1$ is invertible?

Soln. Since the f is invertible , so f is one one function .

Given

$f(x) = (m+2)x^3 - 3mx^2 + 9mx - 1$

$\Rightarrow \quad f'(x) = 3(m+2)x^2 - 6mx + 9m$

$\Rightarrow \quad f'(x) = 3(m+2)x^2 - 6mx + 9m > 0$

Thus, (m+2) > 0 and

$36m^2 - 108m(m+2) < 0$

$\Rightarrow$ m > - 2 & $m^2 - 3m(m+2) < 0$

$\Rightarrow$ m > - 2 & $m^2 + 3m > 0$

$\Rightarrow$ m > - 2 & $m \in (-\infty, -3) \cup (0, \infty)$

$\Rightarrow \quad m \in (-2, -3) \cup (0, \infty)$

Ex-10. Let $f(x) = \sqrt[3]{a - x^3 + 3bx^2 - 3bx + b^3} + b$

Find x, if $f(x)$ is inverse of itself.

Soln. Given $f(x) = f^{-1}(x)$

Thus, the intersection point is the solution of the given equation.

We have f(x) = x

$$x = 3\sqrt{a-(x-b)3}+b$$

$$\Rightarrow \quad (x-b)^3 = a-(a-b)^3$$

$$\Rightarrow \quad x-b = 3\sqrt{\frac{\pi}{2}} \Rightarrow x = b+3\sqrt{\frac{\pi}{2}}.$$

Hence, the solution is $x = b+3\sqrt{\frac{a}{2}}$.

5.7 INVERSE TRIGONOMETRIC FUNCTIONS.

(i) $\sin^{-1}x$: A function $f:[-1, 1] \to \left[-\frac{\pi}{2}, \frac{\pi}{2}\right]$ is defined as $f(x) = \sin^{-1}x = \arcsin x$

Graph of $f(x) = \sin^{-1}x$.

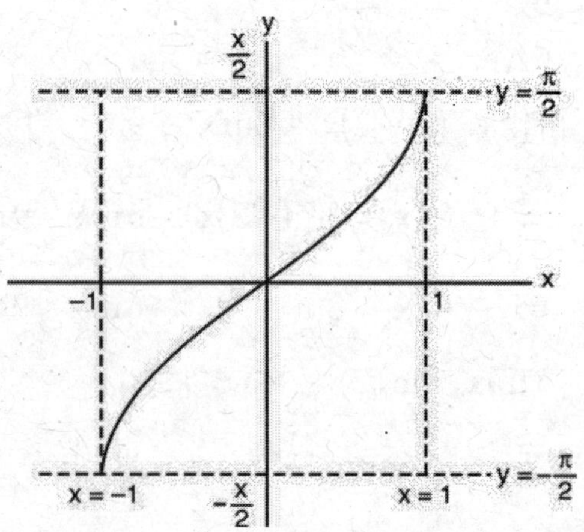

Characteristics of arc sine function :

1. $D_f = [-1, 1]$
2. $R_f = \left[-\frac{\pi}{2}, \frac{\pi}{2}\right]$
3. It is not a periodic function.
4. It is an odd function.

   since, $\sin^{-1}(-x) = -\sin^{-1}x$
5. It is strictly increasing function
6. It is one one function
7. For $0 < x < \frac{\pi}{2}$,

   $\sin x < x < \sin^{-1}x$.

(ii) $\cos^{-1}x$: A function $f:[-1,1] \to [0, \pi]$ is defined as $f(x) = \cos^{-1}x = \text{arc cos } x$

Graph of $f(x) = \cos^{-1}x$

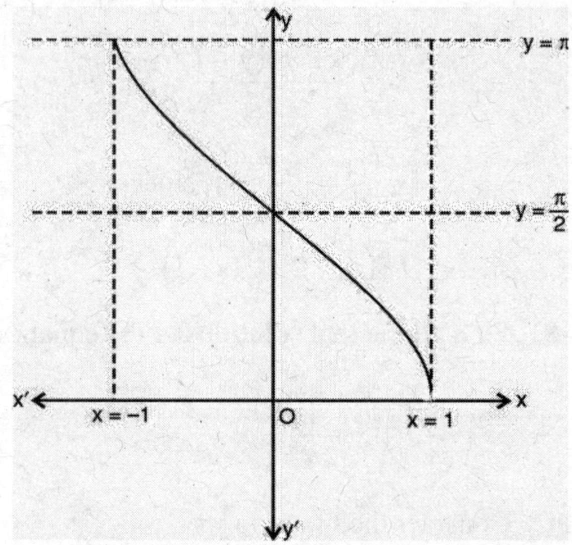

Characteristics of arc cosine function :

1. $D_f = [-1, 1]$
2. $[0, \pi]$
3. It is not a periodic function.
4. It is neither even nor odd function

   since, $\cos^{-1}(-x) = \pi - \cos^{-1}(x)$
5. It is strictly decreasing function.
6. It is one one function.
7. For $0 < x < \frac{\pi}{2}$,

   $\cos^{-1}x < x < \cos x$

(iii) $\tan^{-1}x$: A function $f:R \to \left(-\frac{\pi}{2}, \frac{\pi}{2}\right)$ is defined as $f(x) = \tan^{-1}x$.

Graph of $f(x) = \tan^{-1} x$:

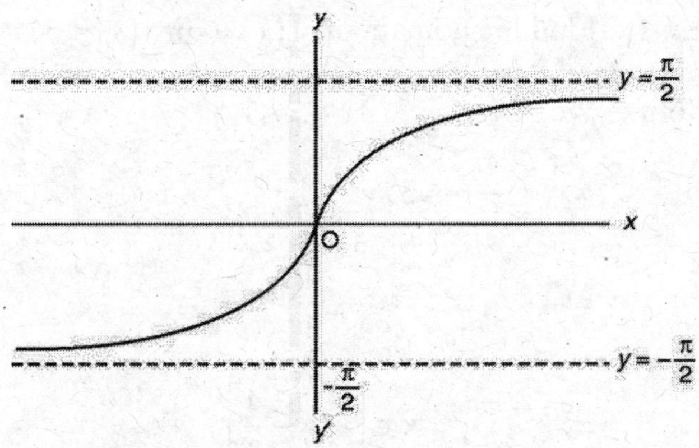

Characteristics of arc tangent function:

1. $D_f = R$
2. $R_f = \left(-\frac{\pi}{2}, \frac{\pi}{2}\right)$
3. It is not a periodic function
4. It is an odd function .

   since, $\tan^{-1}(-x) = -\tan^{-1}x$
5. It is strictly increasing function.
6. It is one one function
7. For $0 < x < \frac{\pi}{2}$,

   $\tan^{-1}x < x < \tan x$.

(iv) $\cot^{-1} x$: A function $f : R \to (0, \pi)$ is defined as $f(x) = \cot^{-1}x$.

Graph of $f(x) = \cot^{-1}x$

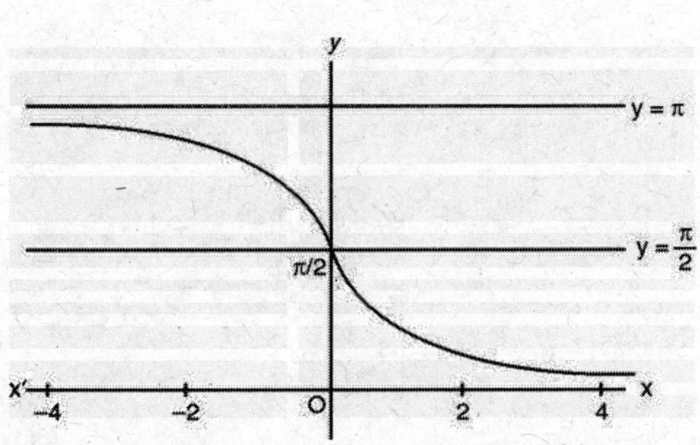

Characteristics of arc co-tangent function :

1. $D_f = R$
2. $R_f = (0, \pi)$
3. It is not a periodic function.
4. It is neither even nor odd function

   since, $\cot^{-1}(-x) = \pi - \cot^{-1}x$
5. It is strictly decreasing function.
6. It is one one function
7. For $0 < x < \frac{\pi}{2}$,

   $\cot x < x < \cot^{-1}x$

(v) $\operatorname{cosec}^{-1}x$: A function $f : (-\infty, -1] \cup [1, \infty) \to \left[-\frac{\pi}{2}, \frac{\pi}{2}\right] - \{0\}$ is defined as $f(x) = \operatorname{cosec}^{-1}x$.

Graph of $f(x) = \operatorname{cosec}^{-1}x$.

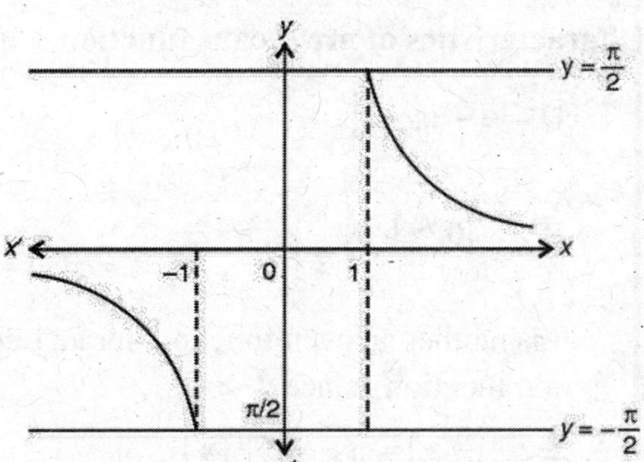

Characteristics of arc co-secant function :

1. $D_f = (-\infty, -1] \cup [1, \infty)$
2. $R_f = \left[-\frac{\pi}{2}, \frac{\pi}{2}\right] - \{0\}$
3. It is an odd function , since

   $\operatorname{cosec}^{-1}(-x) = -\operatorname{cosec}^{-1}(x)$
4. It is non periodic function .
5. It is one one function .

6. It is strictly decreasing function with respect to its domain.

7. For $0 < x < \frac{\pi}{2}$

$\operatorname{cosec}^{-1}x < x < \operatorname{cosec} x$

(v) $\sec^{-1} x$: A function

$f : (-\infty, -1] \cup [1, \infty) \to [0, \pi] - \left\{\frac{\pi}{2}\right\}$

is de fined as $f(x) = \sec^{-1}x$.

Graph of $f(x) = \sec^{-1}x$.

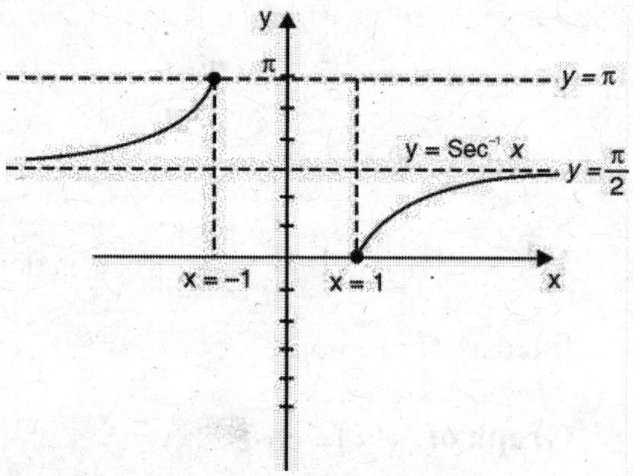

Characteristics of arc secant function :

1. $D_f = (-\infty, -1] \cup [1, \infty)$

2. $R_f = [0, \pi] - \left\{\frac{\pi}{2}\right\}$

3. It is neither an even function nor an odd odd function , since

$\sec^{-1}(-x) = \pi - \sec^{-1}(x)$

4. It is non periodic function .

5. It is one one function .

6. It is strictly decreasing function with respect to its domain.

7. For $0 < x < \frac{\pi}{2}$

$\sec^{-1}x < x < \sec x$

5.8. SOME SOLVED EXAMPLES :

Ex-1. Find the domain of $f(x) = \sin^{-1}(3x+5)$

Soln. We have $-1 \le 3x+5 \le 1$

$\Rightarrow \quad -6 \le 3x \le -4$

$\Rightarrow \quad -2 \le x \le -\frac{4}{3}$

$\Rightarrow \quad D_f = x \in \left[-2, -\frac{4}{3}\right]$

Ex-2. Find the domain of $f(x) = \sin^{-1}\left(\frac{x}{x+1}\right)$

Soln. We have , $-1 \le \frac{x}{x+1} \le 1$

Case - I : $\frac{x}{x+1} \le 1$

$\Rightarrow \quad \frac{x}{x+1} - 1 \le 0$

$\Rightarrow \quad \frac{-1}{x+1} \le 0$

$\Rightarrow \quad \frac{1}{x+1} \ge 0$

$\Rightarrow \quad x > -1$

Case - II : When $\frac{x}{x+1} \ge -1$

$\Rightarrow \quad \frac{x}{x+1} + 1 \ge 0$

$\Rightarrow \quad \frac{2x+1}{x+1} \ge 0$

$\Rightarrow \quad x \in (-\infty, -1) \cup \left[-\frac{1}{2}, \infty\right)$

Hence, $D_f = \left[-\frac{1}{2}, \infty\right)$.

Ex-3. Find the domain of $f(x)=\sin^{-1}\left(\frac{x^2+1}{2x}\right)$

Soln. We have , $-1\le\frac{x^2+1}{2x}\le 1$

$\Rightarrow \quad \left|\frac{x^2+1}{2x}\right|\le 1$

$\Rightarrow \quad \frac{|x^2+1|}{|2x|}\le 1$

$\Rightarrow \quad \frac{|x^2+1|}{2\,|x|}\le 1$

$\Rightarrow \quad x^2+1\le 2\,|x|$

$\Rightarrow \quad |x|^2-2\,|x|+1\le 0$

$\Rightarrow \quad (|x|-1)^2\le 0$

$\Rightarrow \quad (|x|-1)^2=0$

$\Rightarrow \quad (|x|-1)=0$

$\Rightarrow \quad |x|=1$

$\Rightarrow \quad x=\pm 1$

Hence, $D_f=\{-1, 1\}$

Ex-4. Find the domain of $f(x)=\sin^{-1}\left(\frac{|x|-1}{2}\right)$

Soln. We have , $-1\le\frac{|x|-1}{2}\le 1$

$\Rightarrow \quad -2\le |x|-1\le 2$

$\Rightarrow \quad -1\le |x|\le 3$

$\Rightarrow \quad |x|\le 3 \quad (\because\ |x|\ge -1 \text{ is rejected})$

$\Rightarrow \quad -3\le x\le 3$

Hence, $D_f=[-3, 3]$

Ex-5. Find the domain of $f(x)=\sin^{-1}(\log_2 x)$

Soln. We have , $-1\le(\log_2 x)\le 1$

$\Rightarrow \quad 2^{-1}\le x\le 2^1$

$\Rightarrow \quad \frac{1}{2}\le x\le 2$

Hence, $D_f=\left[\frac{1}{2}, 2\right]$

Ex-6. Find the domain of $f(x)=\sin^{-1}(\log_4 x^2)$

Soln. We have , $-1\le\log_4 x^2\le 1$

$\Rightarrow \quad 4^{-1}\le x^2\le 4^1$

$\Rightarrow \quad \frac{1}{4}\le x^2\le 4$

$\Rightarrow \quad \frac{1}{2}\le |x|\le 2$

$\Rightarrow \quad |x|\le 2 \ \&\ |x|\ge\frac{1}{2}$

$\Rightarrow \quad -2\le x\le 2$ and

$x\ge\frac{1}{2}\ \&\ x\le -\frac{1}{2}$

$\Rightarrow \quad x\in\left[-2, -\frac{1}{2}\right]\cup\left[\frac{1}{2}, 2\right]$

Hence, $D_f=\left[-2, -\frac{1}{2}\right]\cup\left[\frac{1}{2}, 2\right]$

Ex-7. Solve for x and y : $\sin^{-1}x+\sin^{-1}y=\pi$

Soln. Given $\sin^{-1}x+\sin^{-1}y=\pi$

It is possible only when each term of the given equation provides the maximum value.

Thus, $\sin^{-1}x = \frac{\pi}{2}$ & $\sin^{-1}y = \frac{\pi}{2}$

$\Rightarrow \quad x = \sin\left(\frac{\pi}{2}\right) = 1$ & $y = \sin\left(\frac{\pi}{2}\right) = 1$

Hence, the solutions are x = 1 and y = 1.

Ex-8. If $\sin^{-1}x + \sin^{-1}y + \sin^{-1}z = \frac{3\pi}{2}$, then find the value of

$$x^{2013} + y^{2013} + z^{2013} - \frac{9}{x^{2014} + y^{2014} + z^{2014}}$$

Soln. Given $\sin^{-1}x + \sin^{-1}y + \sin^{-1}z = \frac{3\pi}{2}$

It is possible only when each term

provide us the maximum value.

Thus, $\sin^{-1}x = \frac{\pi}{2}$, $\sin^{-1}y = \frac{\pi}{2}$

& $\sin^{-1}z = \frac{\pi}{2}$

$\Rightarrow \quad x = 1,\ y = 1$ & $z = 1$

Hence, the value of

$$x^{2013} + y^{2013} + z^{2013} - \frac{9}{x^{2014} + y^{2014} + z^{2014}}$$

$= 1 + 1 + 1 - \frac{9}{1+1+1}$

$= 3 - 3$

$= 0.$

Ex-9. Find the range of

$$f(x) = 2\sin^{-1}(3x+5) + \frac{\pi}{4}.$$

Soln. We have $-\frac{\pi}{2} \le \sin^{-1}(3x+5) \le \frac{\pi}{2}$

$\Rightarrow \quad -\pi \le 2\sin^{-1}(3x+5) \le \pi$

$\Rightarrow -\pi + \frac{\pi}{4} \le 2\sin^{-1}(3x+5) + \frac{\pi}{4} \le \pi + \frac{\pi}{4}$

$\Rightarrow -\frac{3\pi}{4} \le f(x) \le \frac{5\pi}{4}$

Hence, $R_f = \left[-\frac{3\pi}{4}, \frac{5\pi}{4}\right]$

Ex-10. Solve the inequality :

$$\sin^{-1}x > \sin^{-1}(3x-1)$$

Soln. We have, $\sin^{-1}x > \sin^{-1}(3x-1)$

$\Rightarrow \quad x > (3x-1)$

$\Rightarrow \quad 2x - 1 < 0$

$\Rightarrow \quad x < \frac{1}{2}$

$\Rightarrow \quad x \in \left[-1, \frac{1}{2}\right).$

Ex-11. Find the domain of $f(x) = \cos^{-1}(2x+4)$.

Soln. We have, $-1 \le 2x + 4 \le 1$

$\Rightarrow \quad -5 \le 2x \le -3$

$\Rightarrow \quad -\frac{5}{2} \le x \le -\frac{3}{2}$

Hence, $D_f = \left[-\frac{5}{2}, -\frac{3}{2}\right]$

Ex-2. Find the range of

$$f(x) = 2\cos^{-1}(3x+5) + \frac{\pi}{4}$$

Soln. We have, $0 \le \cos^{-1}(3x+5) \le \pi$

$\Rightarrow \ 0 \le 2\cos^{-1}(3x+5) \le 2\pi$

$\Rightarrow \ \frac{\pi}{4} \le 2\cos^{-1}(3x+5) + \frac{\pi}{4} \le 2\pi + \frac{\pi}{4}$

Hence, $R_f = \left[\frac{\pi}{4}, \frac{9\pi}{4}\right]$

Ex-13. Find the range of $f(x) = 3\cos^{-1}(-x^2) - \frac{\pi}{2}$

Soln. We have, $\frac{\pi}{2} \le \cos^{-1}(-x^2) \le \pi$

$$\Rightarrow \frac{3\pi}{2} \le 3\cos^{-1}(-x^2) \le 3\pi$$

$$\Rightarrow \frac{3\pi}{2} - \frac{\pi}{2} \le 3\cos^{-1}(-x^2) - \frac{\pi}{2} \le 3\pi - \frac{\pi}{2}$$

$$\Rightarrow \pi \le f(x) \le \frac{5\pi}{2}$$

Hence, $R_f = \left[\pi, \frac{5\pi}{2}\right]$.

Ex-14. Solve for x : $\cos^{-1}x + \cos^{-1}x^2 = 0$

Soln. Given $\cos^{-1}x + \cos^{-1}x^2 = 0$

It is possible only when each term will provide us the minimum value.

So, $\cos^{-1}x = 0$ & $\cos^{-1}x^2 = 0$

$$\Rightarrow x = 1 \ \& \ x^2 = 1$$

$$\Rightarrow x = 1 \ \& \ x = \pm 1$$

Hence, the solution is $x = 1$.

Ex-15. Solve for x : $[\sin^{-1}x] + [\cos^{-1}x] = 0$, where x is a non negative real number and [,] denotes the greatest integer function.

Soln. Given $[\sin^{-1}x] + [\cos^{-1}x] = 0$ and $x \ge 0$

$$\Rightarrow [\sin^{-1}x] = 0 \ \& \ [\cos^{-1}x] = 0$$

$$\Rightarrow x \in [0, \sin 1) \ \& \ x \in (\cos 1, 1]$$

$$\Rightarrow x \in (\cos 1, \sin 1)$$

Ex-16. Find the domain of $f(x) = \cos^{-1}\left(\frac{x^2}{x^2+1}\right)$

Soln. We have, $-1 \le \frac{x^2}{x^2+1} \le 1$

$$\Rightarrow \left|\frac{x^2}{x^2+1}\right| \le 1$$

$$\Rightarrow \frac{|x^2|}{|x^2+1|} \le 1$$

$$\Rightarrow \frac{x^2}{x^2+1} \le 1$$

$$\Rightarrow x^2 + 1 \ge x^2$$

$$\Rightarrow 1 > 0$$

Hence, $x \in R$

Ex-17. Solve for x : $\cos^{-1}(x) > \cos^{-1}(x^2)$

Soln. We have, $\cos^{-1}(x) > \cos^{-1}(x^2)$

$$\Rightarrow x < x^2$$

$$\Rightarrow x^2 - x > 0$$

$$\Rightarrow x(x-1) > 0$$

$$\Rightarrow x \in [-1, 0)$$

Ex-18. Find the domain of $f(x) = \tan^{-1}\left(\sqrt{9-x^2}\right)$

Soln. Since $\tan^{-1}x$ is defined for all real values of x, so $9 - x^2 \ge 0$

$$\Rightarrow x^2 - 9 \le 0$$

$$\Rightarrow (x+3)(x-3) \le 0$$

$$\Rightarrow -3 \le x \le 3$$

Hence, $D_f = [-3, 3]$

Ex-19. Find the range of the function

$$f(x) = 2\tan^{-1}(1-x^2) + \frac{\pi}{6}$$

Soln. We have, $-\frac{\pi}{2} \le \tan^{-1}(1-x^2) \le \frac{\pi}{4}$

$$\Rightarrow -\pi \le 2\tan^{-1}(1-x^2) \le \frac{\pi}{2}$$

$$\Rightarrow -\pi + \frac{\pi}{6} \le 2\tan^{-1}(1-x^2) + \frac{\pi}{6} \le \frac{\pi}{2} + \frac{\pi}{6}$$

$$\Rightarrow -\frac{5\pi}{6} \le f(x) \le \frac{2\pi}{3}$$

Hence, $R_f = \left[-\frac{5\pi}{6}, \frac{2\pi}{3}\right]$

Ex-20. Find the range of $f(x)=\cot^{-1}(2x-x^2)$

Soln. We have, $f(x)=\cot^{-1}(2x-x^2)$

$\Rightarrow f(x)=\cot^{-1}(1-(x^2-2x+1))$

$\Rightarrow f(x)=\cot^{-1}(1-(x-1)^2)$

since $(1-(x-1)^2)\leq 1$ & $0\leq \cot^{-1}x\leq\pi$

and $\cot^{-1}x$ is strictly decreasing function

so $\cot^{-1}(1)\leq \cot^{-1}(1-(x-1)^2)\leq\cot^{-1}(0)$

$\Rightarrow \frac{\pi}{4}\leq f(x)\leq \pi$

Hence, $R_f=\left[\frac{\pi}{4},\pi\right]$

Ex-21. Solve for x : $[\cot^{-1}x]+[\cos^{-1}x]=0$, where [,] = G. I.F

Soln. We have , $[\cot^{-1}x]+[\cos^{-1}x]=0$

$\Rightarrow [\cot^{-1}x]=0$ & $[\cos^{-1}x]=0$

$\Rightarrow 0\leq\cot^{-1}x<1$ & $0\leq\cos^{-1}x<1$

$\Rightarrow x\in(\cot 1,\infty)$ & $x\in(\cos 1, 1]$

$\Rightarrow x\in(\cot 1, 1]$

Ex-22. Find the number of solutions of

$\sin\{x\}=\cos\{x\},\ \forall\ x\in[0,2\pi]$

Soln. We have $\sin\{x\}=\cos\{x\},\ \forall x\in[0,2\pi]$

$\Rightarrow \tan\{x\}=1$

$\Rightarrow \{x\}=\tan^{-1}(1)=\frac{\pi}{4}$

Hence, the number of solutions = 6

(since $\{x\}$ is a periodic function with period 1, it has one soltion between 0 to 1. So, there is six solutions between 0 to 6.28).

5.8. CONSTANT PROPERTY :

(i) $\sin^{-1}(x)+\cos^{-1}(x)=\frac{\pi}{2},\forall x\in[-1,1]$

(ii) $\tan^{-1}(x)+\cot^{-1}(x)=\frac{\pi}{2},\forall x\in R$

(iii) $\operatorname{cosec}^{-1}(x)+\sec^{-1}(x)=\frac{\pi}{2},\forall x\in R-(-1,1)$

Proof :(i) Let $\sin^{-1}(x)=\theta$, $-\frac{\pi}{2}\leq\theta\leq\frac{\pi}{2}$

$\Rightarrow x=\sin\theta$

$\Rightarrow \cos\left(\frac{\pi}{2}-\theta\right)=x$

$\Rightarrow \cos^{-1}x=\left(\frac{\pi}{2}-\theta\right),\ 0\leq\left(\frac{\pi}{2}-\theta\right)\leq\pi$

Thus, $\sin^{-1}x+\cos^{-1}x=\theta+\frac{\pi}{2}-\theta=\frac{\pi}{2}$

Similarly , we can easily prove (ii) & (iii) .

Ex-1. Find the range of

$f(x)=\sin^{-1}x+\cos^{-1}x+\tan^{-1}x$

Soln. We have $f(x)=\sin^{-1}x+\cos^{-1}x+\tan^{-1}x$

$\Rightarrow f(x)=\frac{\pi}{2}+\tan^{-1}x$

Now, $-\frac{\pi}{2}<\tan^{-1}x<\frac{\pi}{2}$

$\Rightarrow -\frac{\pi}{2}+\frac{\pi}{2}<\frac{\pi}{2}+\tan^{-1}x<\frac{\pi}{2}+\frac{\pi}{2}$

$\Rightarrow 0<f(x)<\pi$

Hence, $R_f=(0,\pi)$.

Ex-2. Solve for x :

$4\sin^{-1}(x-2)+\cos^{-1}(x-2)=\pi$

Soln. We have $4\sin^{-1}(x-2)+\cos^{-1}(x-2)=\pi$

$\Rightarrow 3\sin^{-1}(x-2)+\frac{\pi}{2}=\pi$

$\Rightarrow \quad 3\sin^{-1}(x-2)=\frac{\pi}{2}$

$\Rightarrow \quad \sin^{-1}(x-2)=\frac{\pi}{6}$

$\Rightarrow \quad (x-2)=\sin\left(\frac{\pi}{6}\right)=\frac{1}{2}$

$\Rightarrow \quad x=2+\frac{1}{2}=\frac{5}{2}$

Hence, the solution is x = 1.

Ex-3. Solve for x :

$$\sin^{-1}(x^2-2x+1)+\cos^{-1}(x^2-x)=\frac{\pi}{2}.$$

Soln. As we know that ,

if $\sin^{-1}(f(x))+\cos^{-1}(g(x))=\frac{\pi}{2}$, then

$$f(x)=g(x)$$

$\Rightarrow \quad (x^2-2x+1)=(x^2-x)$

$\Rightarrow \quad 2x-x=1$

$\Rightarrow \quad x=1$

Hence, the solution is x = 1.

Ex-4. Find the number of real solutions of

$$\tan^{-1}\sqrt{x(x+1)}+\sin^{-1}\sqrt{x^2+x+1}=\frac{\pi}{2}$$

Soln. We have,

$$\tan^{-1}\sqrt{x(x+1)}+\sin^{-1}\sqrt{x^2+x+1}=\frac{\pi}{2}$$

$\Rightarrow \quad \cos^{-1}\left(\frac{1}{\sqrt{x^2+x+1}}\right)+\sin^{-1}\sqrt{x^2+x+1}=\frac{\pi}{2}$

$\Rightarrow \quad \left(\frac{1}{\sqrt{x^2+x+1}}\right)=\sqrt{x^2+x+1}$

$\Rightarrow \quad x^2+x+1=1$

$\Rightarrow \quad x^2+x=0$

$\Rightarrow \quad x(x+1)=0$

$\Rightarrow \quad x=0\ \&\ -1$

Hence, the number of solutions is 2.

Ex-5. If $\sin^{-1}\left(x-\frac{x^2}{2}+\frac{x^3}{4}-........\right)$

$+\cos^{-1}\left(x^2-\frac{x^4}{2}+\frac{x^6}{4}-.......\right)=\frac{\pi}{2}$, for

$0<|x|<\sqrt{2}$, then find x .

Soln. As we know, that

if $\sin^{-1}(f(x))+\cos^{-1}(g(x))=\frac{\pi}{2}$, then

$$f(x)=g(x)$$

$$\left(x-\frac{x^2}{2}+\frac{x^3}{4}-...\right)=\left(x^2-\frac{x^4}{2}+\frac{x^6}{4}-...\right)$$

$\Rightarrow \quad x\left(1-\frac{x}{2}+\frac{x^2}{4}-...\right)=x^2\left(1-\frac{x^2}{2}+\frac{x^4}{4}-...\right)$

$\Rightarrow \quad x\left(\frac{1}{1+\frac{x}{2}}\right)=x^2\left(\frac{1}{1+\frac{x^2}{2}}\right)$

$\Rightarrow \quad \left(\frac{2x}{x+2}\right)=\left(\frac{2x^2}{x^2+2}\right)$

$\Rightarrow \quad x\left\{\left(\frac{1}{x+2}\right)-\left(\frac{x}{x^2+2}\right)\right\}=0$

$\Rightarrow \quad x=0\ \&\left(\frac{1}{x+2}\right)=\left(\frac{x}{x^2+2}\right)$

$\Rightarrow \quad x=0\ \&\ x=1$

Ex-6. Solve for x : $\sin^{-1}x>\cos^{-1}x$

Soln. We have $\sin^{-1}x>\cos^{-1}x$

$\Rightarrow \quad 2\sin^{-1}x>\sin^{-1}x+\cos^{-1}x$

$\Rightarrow \quad 2\sin^{-1}x > \frac{\pi}{2}$

$\Rightarrow \quad \sin^{-1}x > \frac{\pi}{4}$

$\Rightarrow \quad x > \sin\left(\frac{\pi}{4}\right)$

$\Rightarrow \quad x > \frac{1}{\sqrt{2}}$

$\Rightarrow \quad x \in \left(\frac{1}{\sqrt{2}}, 1\right]$

5.9. CONVERSION OF INVERSE TRIGONOMETRIC FUNCTIONS

Case - I: when $x > 0$

	Functions	**Principal values**
1.	$\sin^{-1}x$	$\left[0, \frac{\pi}{2}\right]$
2.	$\cos^{-1}x$	$\left[0, \frac{\pi}{2}\right]$
3.	$\tan^{-1}x$	$\left(0, \frac{\pi}{2}\right)$
4.	$\cot^{-1}x$	$\left(0, \frac{\pi}{2}\right]$
5.	$\operatorname{cosec}^{-1}x$	$\left(0, \frac{\pi}{2}\right]$
6.	$\sec^{-1}x$	$\left[0, \frac{\pi}{2}\right)$

Case - II : when $x < 0$

	Functions	Principal values
1.	$\sin^{-1}x$	$\left[-\frac{\pi}{2}, 0\right]$
2.	$\operatorname{cosec}^{-1}x$	$\left[-\frac{\pi}{2}, 0\right)$
3.	$\tan^{-1}x$	$\left(-\frac{\pi}{2}, 0\right)$
4.	$\cos^{-1}x$	$\left[\frac{\pi}{2}, \pi\right]$
5.	$\sec^{-1}x$	$\left(\frac{\pi}{2}, \pi\right]$
6.	$\cot^{-1}x$	$\left[\frac{\pi}{2}, \pi\right)$.

Here, we shall discuss, any inverse trigonometric function can be expressed in terms of any other inverse trigonometric functions.

Step - I :

(i) $\sin^{-1}(x) = \operatorname{cosec}^{-1}\left(\frac{1}{x}\right), \ x \in [-1,1]-\{0\}$

(ii) $\operatorname{cosec}^{-1}(x) = \sin^{-1}\left(\frac{1}{x}\right), \ |x| \geq 1$

(iii) $\cos^{-1}(x) = \sec^{-1}\left(\frac{1}{x}\right), \ x \in [-1,1]-\{0\}$

(iv) $\sec^{-1}x = \cos^{-1}\left(\frac{1}{x}\right), \ |x| \geq 1$

(v) $\tan^{-1}(x) = \begin{cases} \cot^{-1}\left(\frac{1}{x}\right) & : x > 0 \\ -\pi + \cot^{-1}\left(\frac{1}{x}\right) & : x < 0 \end{cases}$

Proof: Let $\tan^{-1}x = \theta$

Case - I : when $x > 0$

Then $\tan(\theta) > 0$

$\Rightarrow \quad 0 < \theta < \frac{\pi}{2}$

Now, $\cot\theta = \frac{1}{\tan\theta} = \frac{1}{x}$

$\Rightarrow \quad \theta = \cot^{-1}\left(\frac{1}{x}\right)$

Thus, $\tan^{-1}x = \cot^{-1}\left(\frac{1}{x}\right)$

Case - II : when $x < 0$

Then $\tan\theta < 0$

$\Rightarrow \quad -\frac{\pi}{2} < \theta < 0$

$\Rightarrow \quad 0 < -\theta < \frac{\pi}{2}$

$\Rightarrow \quad -\pi < -\pi - \theta < -\pi + \frac{\pi}{2}$

$\Rightarrow \quad -\pi < -\pi - \theta < -\frac{\pi}{2}$

Now, $\cot(-\pi - \theta) = \cot\theta$

and $\cot\theta = \frac{1}{\tan\theta} = \frac{1}{x}$

Thus, $(-\pi - \theta) = \cot^{-1}\left(\frac{1}{x}\right)$

$\Rightarrow \quad (-\pi - \tan^{-1}x) = \cot^{-1}\left(\frac{1}{x}\right)$

$\Rightarrow \quad \tan^{-1}x = -\pi + \cot^{-1}\left(\frac{1}{x}\right)$.

(vi) $$\cot^{-1}(x) = \begin{cases} \tan^{-1}\left(\frac{1}{x}\right) & : x > 0 \\ \pi + \tan^{-1}\left(\frac{1}{x}\right) & : x < 0 \end{cases}$$

Proof: <u>Case - I</u>: when $x > 0$

Let $\cot^{-1}(x) = \theta$

$\Rightarrow \quad \cot\theta = x$

Here, $\cot\theta > 0 \Rightarrow \quad 0 < \theta < \frac{\pi}{2}$

Now, $\tan\theta = \frac{1}{\cot\theta} = \frac{1}{x}$

$\Rightarrow \quad \theta = \tan^{-1}\left(\frac{1}{x}\right)$

$\Rightarrow \quad \cot^{-1}x = \tan^{-1}\left(\frac{1}{x}\right)$

Case - II : when $x < 0$

Then $\cot\theta < 0$

$\Rightarrow \quad \frac{\pi}{2} < \theta < \pi$

Now, $\tan(-\pi + \theta) = \tan\theta = \frac{1}{x}$

$\Rightarrow \quad (-\pi + \theta) = \tan^{-1}\left(\frac{1}{x}\right)$

$\Rightarrow \quad -\pi + \cot^{-1}x = \tan^{-1}\left(\frac{1}{x}\right)$

$\Rightarrow \quad \cot^{-1}x = \pi + \tan^{-1}\left(\frac{1}{x}\right)$

Step-II:

(i) $$\sin^{-1}x = \begin{cases} \cos^{-1}\left(\sqrt{1-x^2}\right) & : 0 \le x \le 1 \\ -\cos^{-1}\left(\sqrt{1-x^2}\right) & : -1 \le x < 0 \end{cases}$$

(ii) $$\sin^{-1}x = \begin{cases} \sec^{-1}\left(\frac{1}{\sqrt{1-x^2}}\right) & : 0 \le x \le 1 \\ -\sec^{-1}\left(\frac{1}{\sqrt{1-x^2}}\right) & : -1 \le x < 0 \end{cases}$$

(iii) $$\cos^{-1}x = \begin{cases} \sin^{-1}\left(\sqrt{1-x^2}\right) & : 0 \le x \le 1 \\ \pi - \sin^{-1}\left(\sqrt{1-x^2}\right) & : -1 \le x < 0 \end{cases}$$

(iv) $$\cos^{-1}x = \begin{cases} \operatorname{cosec}^{-1}\left(\frac{1}{\sqrt{1-x^2}}\right) & : 0 < x \le 1 \\ \pi - \operatorname{cosec}^{-1}\left(\frac{1}{\sqrt{1-x^2}}\right) & : -1 \le x < 0 \end{cases}$$

(v) $$\sin^{-1}x = \tan^{-1}\left(\frac{x}{\sqrt{1-x^2}}\right) : -1 < x < 1$$

(vi) $\sin^{-1}x = \begin{cases} \cot^{-1}\left(\dfrac{\sqrt{1-x^2}}{x}\right) & : 0 < x \le 1 \\ -\pi + \cot^{-1}\left(\dfrac{\sqrt{1-x^2}}{x}\right) & : -1 \le x < 0 \end{cases}$

Ex-1. Find the value of $\cos\left(\frac{1}{2}\cos^{-1}\left(\frac{3}{5}\right)\right)$

Soln. Let $\frac{1}{2}\cos^{-1}\left(\frac{3}{5}\right) = \theta$

$\Rightarrow \quad \cos^{-1}\left(\frac{3}{5}\right) = 2\theta$

$\Rightarrow \quad \cos(2\theta) = \frac{3}{5}$

$\Rightarrow \quad 2\cos^2\theta - 1 = \frac{3}{5}$

$\Rightarrow \quad 2\cos^2\theta = 1 + \frac{3}{5} = \frac{8}{5}$

$\Rightarrow \quad \cos^2\theta = \frac{4}{5}$

$\Rightarrow \quad \cos\theta = \frac{2}{\sqrt{5}}$

Ex-2. Find the value of $\sin\left(\frac{\pi}{4} + \sin^{-1}\left(\frac{1}{2}\right)\right)$

Soln. We have, $\sin\left(\frac{\pi}{4} + \sin^{-1}\left(\frac{1}{2}\right)\right)$

$= \sin\left(\frac{\pi}{4} + \theta\right), \; \theta = \sin^{-1}\left(\frac{1}{2}\right)$

$= \sin\left(\frac{\pi}{4} + \theta\right), \; \sin\theta = \frac{1}{2}$

$= \sin\left(\frac{\pi}{4}\right)\cos(\theta) + \cos\left(\frac{\pi}{4}\right)\sin(\theta)$

$= \frac{1}{\sqrt{2}}\cos(\theta) + \frac{1}{\sqrt{2}}\sin(\theta)$

$= \frac{1}{\sqrt{2}} \times \frac{\sqrt{3}}{2} + \frac{1}{\sqrt{2}} \times \frac{1}{2}$

$= \frac{\sqrt{3}+1}{2\sqrt{2}}$

Ex-4. If m is a root of $x^2 + 3x + 1 = 0$, then find the value of $\tan^{-1}(m) + \tan^{-1}\left(\frac{1}{m}\right)$.

Soln. Let m_1 and m_2 are two roots of

$x^2 + 3x + 1 = 0$.

Thus, $m_1 + m_2 = -3 < 0$

and $m_1.m_2 = 1$

It is possible only when both are negative.

Thus, $\tan^{-1}(m) + \tan^{-1}\left(\frac{1}{m}\right)$

$= \tan^{-1}(m) - \pi + \cot^{-1}(m)$

$= \tan^{-1}(m) + \cot^{-1}(m) - \pi$

$= \frac{\pi}{2} - \pi$

$= -\frac{\pi}{2}$

Ex-5. Prove that $\cos(\tan^{-1}(\sin(\cot^{-1}x)))$

$= \sqrt{\frac{x^2+1}{x^2+2}}$

Soln. We have, $\cos(\tan^{-1}(\sin(\cot^{-1}x)))$

$= \cos(\tan^{-1}(\sin\theta)), \; \cot\theta = x$

$= \cos\left(\tan^{-1}\left(\frac{1}{\sqrt{1+x^2}}\right)\right)$

$= \cos\varphi, \; \tan\varphi = \left(\frac{1}{\sqrt{1+x^2}}\right)$.

$$= \sqrt{\frac{x^2+1}{x^2+2}}$$

5.10. COMPOSITION OF TRIGONOMETRIC FUNCTIONS AND ITS INVERSE .

Let $y = \sin^{-1}x$

$\Rightarrow \quad x = \sin y$

$\Rightarrow \quad x = \sin\left(\sin^{-1}x\right)$

$\Rightarrow \quad \sin\left(\sin^{-1}x\right) = x$

Therefore, $\sin\left(\sin^{-1}x\right)$ provides us a real value which lies in [-1, 1]

Hence,

(i) $\sin\left(\sin^{-1}x\right) = x, \ |x| \le 1$

(ii) $\cos\left(\cos^{-1}x\right) = x, \ |x| \le 1$

(iii) $\tan\left(\tan^{-1}x\right) = x, \ x \in R$

(iv) $\cot\left(\cot^{-1}x\right) = x, \ x \in R$

(v) $\text{cosec}\left(\text{cosec}^{-1}x\right) = x, \ |x| \ge 1$

(vi) $\sec\left(\sec^{-1}x\right) = x, \ |x| \ge 1$

(i) **Graph of $y = \sin\left(\sin^{-1}x\right)$**

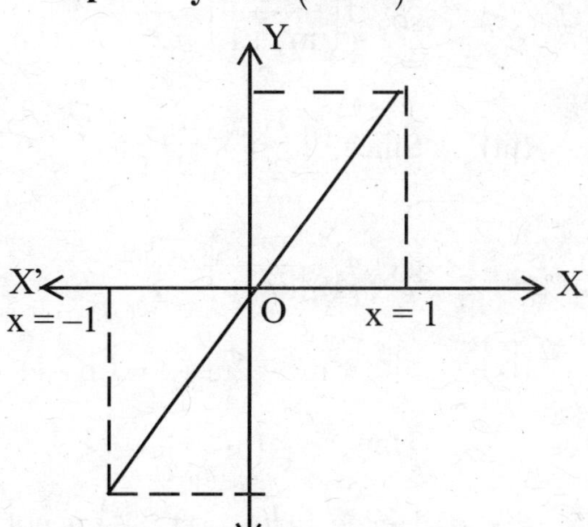

1. $D_f = [-1, 1]$
2. $R_f = [-1, 1]$
3. It is non periodic function

(ii) **Graph of $y = \cos\left(\cos^{-1}x\right)$**

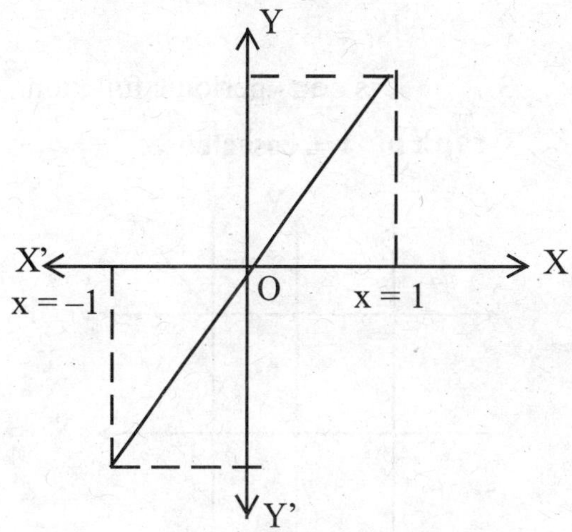

1. $D_f = [-1, 1]$
2. $R_f = [-1, 1]$
3. It is non periodic function

(iii) **Graph of $y = \tan\left(\tan^{-1}x\right)$**

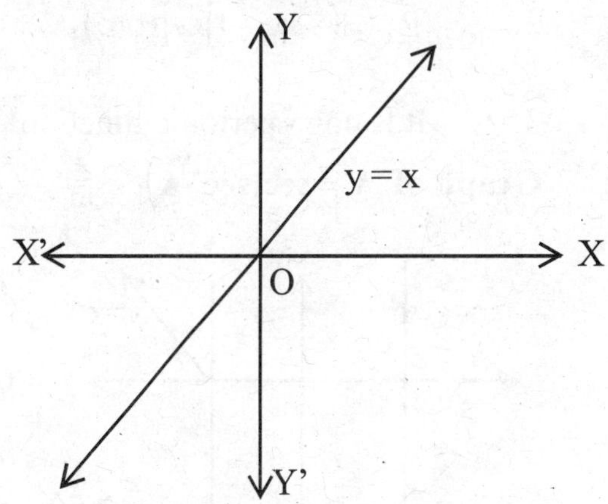

1. $D_f = R$
2. $R_f = R$
3. It is non - periodic function.

(iv) **Graph of $y = \cot\left(\cot^{-1}x\right)$**

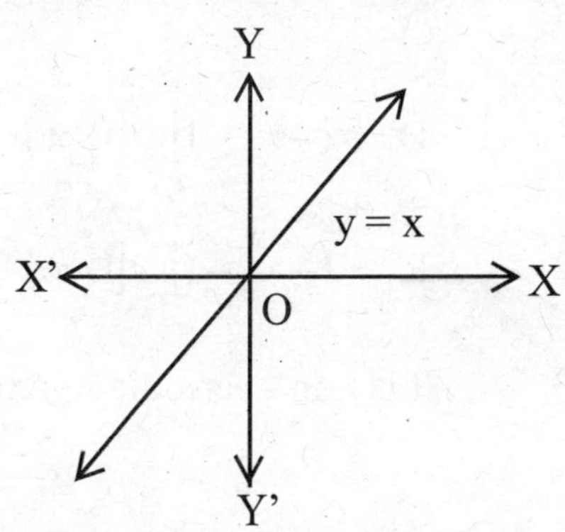

1. $D_f = R$
2. $R_f = R$
3. It is non - periodic function.

(v) **Graph of $y = \text{cosec}(\text{cosec}^{-1}x)$**

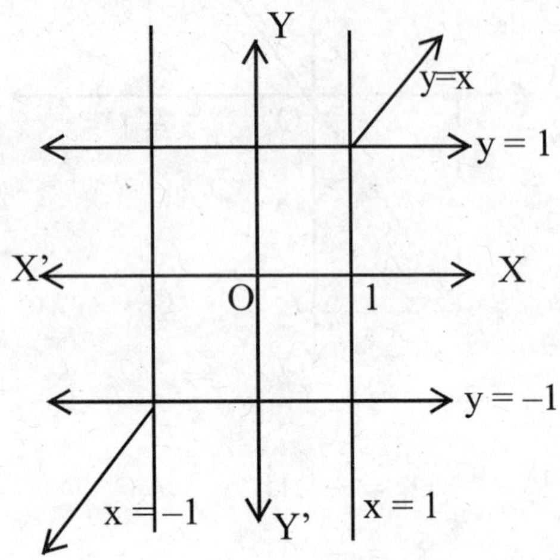

1. $D_f = (-\infty, -1] \cup [1, \infty)$
2. $R_f = (-\infty, -1] \cup [1, \infty)$
3. It is non - periodic function

(vi) **Graph of $y = \sec(\sec^{-1}x)$**

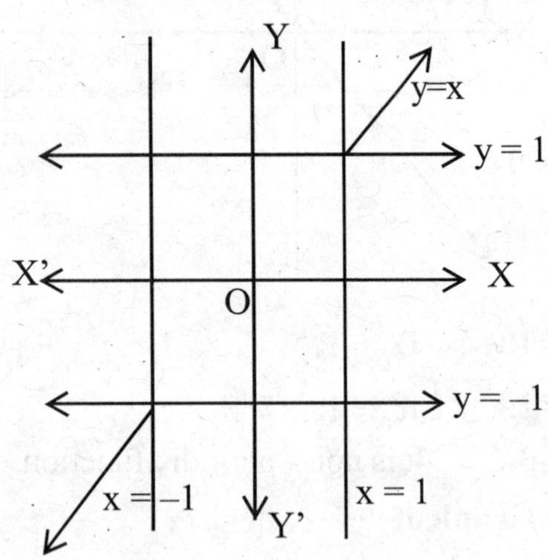

1. $D_f = (-\infty, -1] \cup [1, \infty)$
2. $R_f = (-\infty, -1] \cup [1, \infty)$
3. It is non - periodic function

Ex-1. Let $f(x) = \sin^{-1}x + \cos^{-1}x$

Then find the value of

(i) $f\left(\dfrac{1}{m^2+1}\right)$, $m \in R$

(ii) $f\left(\dfrac{m^2}{m^2+1}\right)$, $m \in R$

(iii) $f\left(\dfrac{m}{m^2+1}\right)$, $m \in R$

(iv) $f(m^2 - 2m + 6)$, $m \in R$

(v) $f(m^2+1)$, $m \in R$

Soln. As we know, $\sin^{-1}x + \cos^{-1}x = \dfrac{\pi}{2}$, for every x in [-1, 1]

(i) since $0 < \dfrac{1}{m^2+1} \le 1$

so, $f\left(\dfrac{1}{m^2+1}\right) = \dfrac{\pi}{2}$

(ii) Since, $0 \le \dfrac{m^2}{m^2+1} < 1$,

so $f\left(\dfrac{m^2}{m^2+1}\right) = \dfrac{\pi}{2}$.

(iii) Since, $-\dfrac{1}{2} \le \dfrac{m}{m^2+1} \le \dfrac{1}{2}$

so, $f\left(\dfrac{m}{m^2+1}\right) = \dfrac{\pi}{2}$

(iv) Since $m^2 - 2m + 6 = (m-1)^2 + 5$

Thus, $5 \le (m-1)^2 + 5 < \infty$

Hence, $f((m-1)^2 + 5)$ is not defined.

(v) Also, $1 \le m^2 + 1 < \infty$

So, $f(m^2+1)$ is not defined.

Ex-2. If $\cos^{-1}x+\cos^{-1}y=\frac{2\pi}{3}$, then find the value of $\sin^{-1}x+\sin^{-1}y$.

Soln. Given, $\cos^{-1}x+\cos^{-1}y=\frac{2\pi}{3}$

Now, $\sin^{-1}x+\sin^{-1}y$

$$=\frac{\pi}{2}-\cos^{-1}x+\frac{\pi}{2}-\cos^{-1}y$$

$$=\pi-(\cos^{-1}x+\cos^{-1}y)$$

$$=\pi-\frac{2\pi}{3}$$

$$=\frac{\pi}{3}$$

Ex-3. If m is the root of $x^2+3x+1=0$, then find the value of $\tan^{-1}(m)+\tan^{-1}\left(\frac{1}{m}\right)$.

Soln. Let m_1 and m_2 are the two roots of the given equation.

Now, $m_1+m_2=-3$ and $m_1.m_2=1$

$\Rightarrow$ m_1 and m_2 are two -ve roots.

Now, $\tan^{-1}(m)+\tan^{-1}\left(\frac{1}{m}\right)$

$$=\tan^{-1}(m)-\pi+\cot^{-1}(m)$$

$$=-\pi+\tan^{-1}(m)+\cot^{-1}(m)$$

$$=-\pi+\frac{\pi}{2}$$

$$=-\frac{\pi}{2}$$

Ex-4. solve for x :

$$\sin^{-1}\left(\sin\left(\frac{2x^2+5}{x^2+2}\right)\right)>\sin^{-1}(\sin 3)$$

Soln. Let $m=\frac{2x^2+5}{x^2+2}=2+\frac{1}{x^2+2}$

Thus, $m\in\left[2,\frac{5}{2}\right]$

now, $\sin^{-1}\left(\sin\left(\frac{2x^2+5}{x^2+2}\right)\right)>\sin^{-1}(\sin 3)$

$$\Rightarrow \sin^{-1}\left(\sin\left(\pi-\frac{2x^2+5}{x^2+2}\right)\right)$$

$$<\sin^{-1}(\sin(\pi-3))$$

$$\Rightarrow \pi-\left(\frac{2x^2+5}{x^2+2}\right)>\pi-3$$

$$\Rightarrow \left(\frac{2x^2+5}{x^2+2}\right)<3$$

$$\Rightarrow \left(\frac{2x^2+5}{x^2+2}-3\right)<0$$

$$\Rightarrow \left(\frac{-x^2-5}{x^2+2}\right)<0$$

$$\Rightarrow \left(\frac{x^2+5}{x^2+2}\right)>0$$

$$\Rightarrow x\in R$$

Ex-5. Solve for x : $x^2-3x<\sin^{-1}(\sin 2)$

Soln. We have, $x^2-3x<\sin^{-1}(\sin 2)$

$$\Rightarrow x^2-3x<\pi-2$$

$$\Rightarrow x\in\left(\frac{3-\sqrt{1+4\pi}}{2},\frac{3+\sqrt{1+4\pi}}{2}\right)$$

5.11. COMPOSITION OF INVERSE TRIGONOMETRIC FUNCTIONS AND TRIGONOMETRIC FUNCTIONS.

(i) A function $f:R\to\left[-\frac{\pi}{2},\frac{\pi}{2}\right]$ is defined as $f(x)=\sin^{-1}(\sin x)$

Graph of $f(x)=\sin^{-1}(\sin x)$

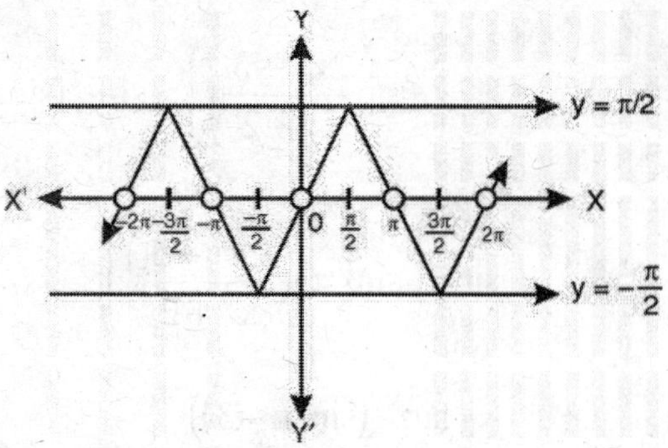

1. $D_f = R$

2. $R_f = \left[-\frac{\pi}{2}, \frac{\pi}{2}\right]$

3. It is an odd function .

4. It is a periodic function with period 2π

5. $\sin^{-1}(\sin x) = \begin{cases} x & : -\frac{\pi}{2} \le x \le \frac{\pi}{2} \\ \pi - x & : \frac{\pi}{2} \le x \le \frac{3\pi}{2} \\ x - 2\pi & : \frac{3\pi}{2} \le x \le \frac{5\pi}{2} \\ -\pi - x & : -\frac{3\pi}{2} \le x \le -\frac{\pi}{2} \end{cases}$

(ii) $\cos^{-1}(\cos x)$: A function $f : R \to [0, \pi]$ is defined as $f(x) = \cos^{-1}(\cos x)$

Graph of $f(x) = \cos^{-1}(\cos x)$:

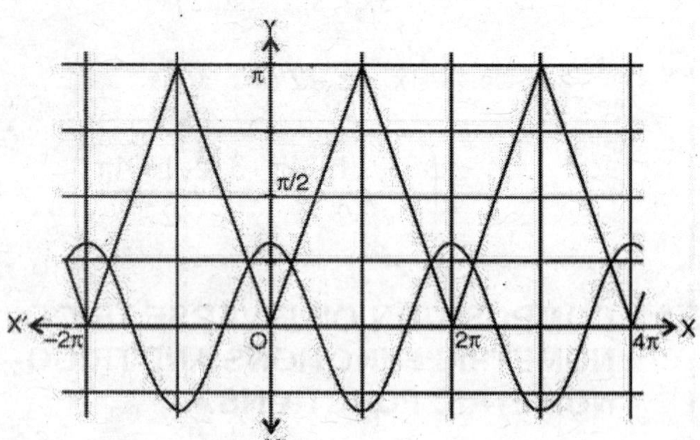

1. $D_f = R$

2. $R_f = [0, \pi]$

3. It is neither odd nor even function

4. It is periodic function with period 2π

5. $\cos^{-1}(\cos x) = \begin{cases} x & : 0 \le x \le \pi \\ 2\pi - x & : \pi \le x \le 2\pi \\ x - 2\pi & : 2\pi \le x \le 3\pi \\ -x & : -\pi \le x \le 0 \end{cases}$

(iii) $\tan^{-1}(\tan x)$: A function

$f : R - (2n+1)\frac{\pi}{2} \to \left(-\frac{\pi}{2}, \frac{\pi}{2}\right)$ is defined

as $f(x) = \tan^{-1}(\tan x)$

Graph of $f(x) = \tan^{-1}(\tan x)$:

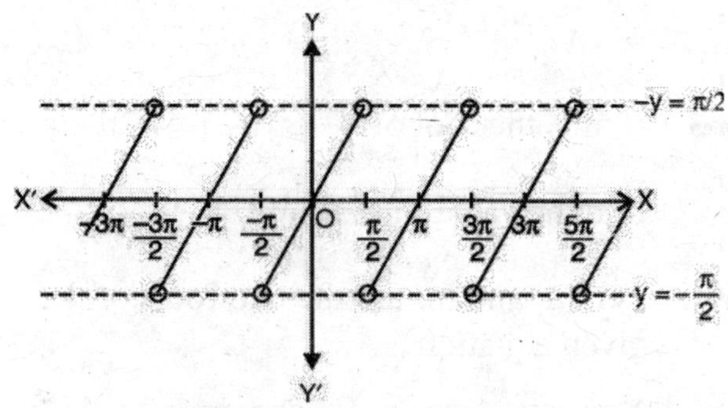

1. $D_f = R - (2n+1)\frac{\pi}{2}, \; n \in I$

2. $R_f = \left(-\frac{\pi}{2}, \frac{\pi}{2}\right)$

3. It is an odd function

4. It is a periodic function with period π

5. $\tan^{-1}(\tan x) = \begin{cases} x & : \frac{-\pi}{2} < x < \frac{\pi}{2} \\ x - \pi & : \frac{\pi}{2} < x < \frac{3\pi}{2} \\ x - 2\pi & : \frac{\pi}{2} < x < \frac{3\pi}{2} \\ x + \pi & : \frac{-3\pi}{2} < x < \frac{-\pi}{2} \end{cases}$

(iv) $\cot^{-1}(\cot x)$: A function

$f : R - (n\pi) \to (0, \pi)$ is defined

as $f(x) = \cot^{-1}(\cot x)$

Graph of $f(x) = \cot^{-1}(\cot x)$:

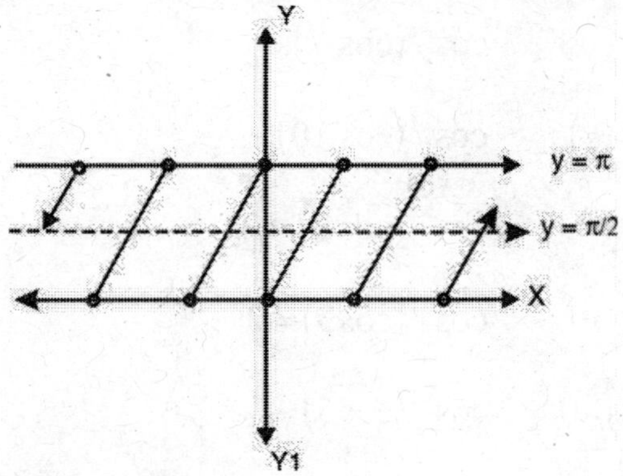

1. $D_f = R - n\pi$, $n \in I$

2. $R_f = (0, \pi)$

3. It is neither even nor odd function.

4. It is a periodic function with period π

5. $$\cot^{-1}(\cot x) = \begin{cases} x & : 0 < x < \pi \\ x - \pi & : \pi < x < 2\pi \\ x - 2\pi & : 2\pi < x < 3\pi \\ \pi + x & : -\pi < x < 0 \end{cases}$$

(v) $\text{cosec}^{-1}(\text{cosec } x)$: A function

$f : R - (n\pi) \to \left[-\frac{\pi}{2}, \frac{\pi}{2}\right] - \{0\}$ is defined

as $f(x) = \text{cosec}^{-1}(\text{cosec } x)$

Graph of $f(x) = \text{cosec}^{-1}(\text{cosec } x)$

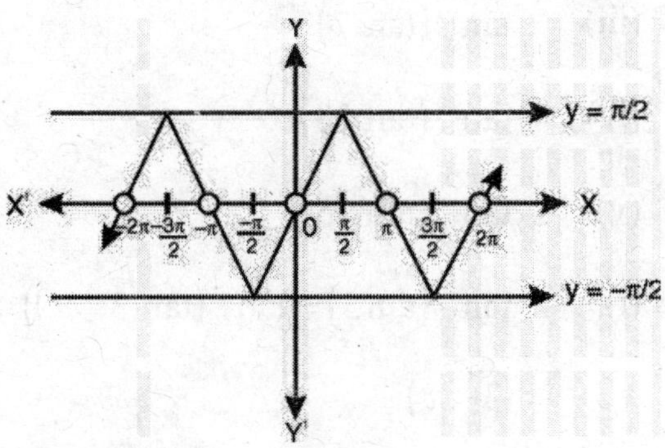

1. $D_f = R - n\pi$, $n \in I$

2. $R_f = \left[-\frac{\pi}{2}, \frac{\pi}{2}\right] - \{0\}$

3. It is an odd function

4. It is a periodic function with period 2π

5. $\text{cosec}^{-1}(\text{cosec } x)$

$$= \begin{cases} x & : \frac{-\pi}{2} \le x \le \frac{\pi}{2} \\ \pi - x & : \frac{\pi}{2} \le x \le \frac{3\pi}{2} \\ x - 2\pi & : \frac{3\pi}{2} \le x \le \frac{5\pi}{2} \\ -x - \pi & : \frac{-3\pi}{2} \le x \le \frac{-\pi}{2} \end{cases}$$

(vi) $\sec^{-1}(\sec x)$: A function

$f : R - (2n+1)\frac{\pi}{2} \to [0, \pi] - \left\{\frac{\pi}{2}\right\}$ is

defined as $f(x) = \sec^{-1}(\sec x)$

Graph of $f(x) = \sec^{-1}(\sec x)$

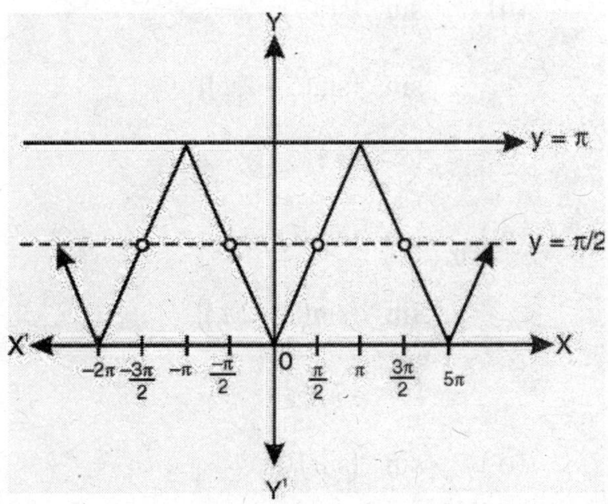

1. $D_f = R - (2n+1)\frac{\pi}{2}$, $n \in I$

2. $R_f = [0, \pi] - \left\{\frac{\pi}{2}\right\}$

3. It is neither even nor odd function.

4. It is a periodic function with period 2π

5. $$\sec^{-1}(\sec x) = \begin{cases} x & : 0 \le x \le \pi \\ 2\pi - x & : \pi \le x \le 2\pi \\ x - 2\pi & : 2\pi \le x \le 3\pi \\ -x & : -\pi \le x \le 0 \end{cases}$$

5.12. SOME SOLVED EXAMPLES :

Ex-1. Find the value of

(i) $\sin^{-1}(\sin 3)$

(ii) $\sin^{-1}(\sin 5)$

(iii) $\sin^{-1}(\sin 7)$

(iv) $\sin^{-1}(\sin 10)$

(v) $\sin^{-1}(\sin 20)$

Soln. (i) $\sin^{-1}(\sin 3)$

$= \sin^{-1}(\sin(\pi - 3))$

$= (\pi - 3)$

(ii) $\sin^{-1}(\sin 5)$

$= \sin^{-1}(\sin(5 - 2\pi))$

$= (5 - 2\pi)$

(iii) $\sin^{-1}(\sin 7)$

$= \sin^{-1}(\sin(7 - 2\pi))$

$= (7 - 2\pi)$

(iv) $\sin^{-1}(\sin 10)$

$= \sin^{-1}(\sin(3\pi - 10))$

$= (3\pi - 10)$

(v) $\sin^{-1}(\sin 20)$

$= \sin^{-1}(\sin(20 - 6\pi))$

$= (20 - 6\pi)$

Ex.2. Find the value of

(i) $\cos^{-1}(\cos 2)$

(ii) $\cos^{-1}(\cos 3)$

(iii) $\cos^{-1}(\cos 5)$

(iv) $\cos^{-1}(\cos 7)$

(v) $\cos^{-1}(\cos 10)$

Soln. (i) $\cos^{-1}(\cos 2) = 2$

(ii) $\cos^{-1}(\cos 3) = 2$

(iii) $\cos^{-1}(\cos 5)$

$= \cos^{-1}(\cos(2\pi - 5))$

$= (2\pi - 5)$

(iv) $\cos^{-1}(\cos 7)$

$= \cos^{-1}(\cos(7 - 2\pi))$

$= (7 - 2\pi)$

(v) $\cos^{-1}(\cos 10)$

$= \cos^{-1}(\cos(4\pi - 10))$

$= (4\pi - 10)$

Ex-3. Find the value of

(i) $\tan^{-1}(\tan 3)$

(ii) $\tan^{-1}(\tan 5)$

(iii) $\tan^{-1}(\tan 7)$

(iv) $\tan^{-1}(\tan 10)$

(v) $\tan^{-1}(\tan 15)$

Soln. (i) $\tan^{-1}(\tan 3) = \tan^{-1}(\tan(3 - \pi))$

$= (3 - \pi)$

(ii) $\tan^{-1}(\tan 5)$

$= \tan^{-1}(\tan(5 - 2\pi))$

$= (5 - 2\pi)$

(iii) $\tan^{-1}(\tan 7)$

$= \tan^{-1}(\tan(7-2\pi))$

$= (7-2\pi)$

(iv) $\tan^{-1}(\tan 10)$

$= \tan^{-1}(\tan(10-3\pi))$

$= (10-3\pi)$

(v) $\tan^{-1}(\tan 15)$

$= \tan^{-1}(\tan(15-5\pi))$

$= (15-5\pi)$.

Ex-4. Find the value of $\cos^{-1}(\sin(-5))$.

Soln. We have , $\cos^{-1}(\sin(-5))$

$= \cos^{-1}(-\sin 5)$

$= \cos^{-1}\left(\cos\left(5-\frac{3\pi}{2}\right)\right)$

$= \left(5-\frac{3\pi}{2}\right)$

Ex-5. Find the value of $\tan\left(\frac{1}{2}\cos^{-1}\left(\frac{\sqrt{5}}{3}\right)\right)$

Soln. We have , $\tan\left(\frac{1}{2}\cos^{-1}\left(\frac{\sqrt{5}}{3}\right)\right)$

$= \tan(\theta)$, where $\theta = \frac{1}{2}\cos^{-1}\left(\frac{5}{\sqrt{3}}\right)$

Now, $\cos(2\theta) = \frac{5}{\sqrt{3}}$

$\Rightarrow \frac{1-\tan^2\theta}{1+\tan^2\theta} = \frac{5}{\sqrt{3}}$

$\Rightarrow \tan^2(\theta) = \frac{5-\sqrt{3}}{5+\sqrt{3}}$

$\Rightarrow \tan(\theta) = \sqrt{\frac{5-\sqrt{3}}{5+\sqrt{3}}}$

$\Rightarrow \tan(\theta) = \frac{(5-\sqrt{3})}{\sqrt{22}}$

Ex-6. Find f '(x) , where $f(x) = \sin^{-1}(\sin x)$ and $-2\pi \le x \le \pi$

Soln. We have, $f(x)$

$= \sin^{-1}(\sin x)$

$= x + 2\pi - \pi - x + x + \pi - x$

$= 2\pi$

$\Rightarrow f'(x) = 0$

Ex-7. Find $f'(x)$, where $f(x) = \cos^{-1}(\cos x)$ and $-\pi \le x \le 2\pi$

Soln. We have f (x)

$= \cos^{-1}(\cos x)$

$= -x + x + 2\pi - x$

$= 2\pi - x$

$\Rightarrow f'(x) = -1$

Ex-8. Solve for x : $\sin^{-1}\left(\sin\left(\frac{2x^2+5}{x^2+1}\right)\right) < \pi - 3$

Soln. We have , $\sin^{-1}\left(\sin\left(\frac{2x^2+5}{x^2+1}\right)\right) < \pi - 3$

$\Rightarrow \sin^{-1}\left(\sin\left(\pi - \left(\frac{2x^2+5}{x^2+1}\right)\right)\right) > \pi - 3$

$\Rightarrow \left(\pi - \left(\frac{2x^2+5}{x^2+1}\right)\right) < \pi - 3$

$\Rightarrow -\left(\frac{2x^2+5}{x^2+1}\right) < -3$

$$\Rightarrow \left(\frac{2x^2+5}{x^2+1}\right) > 3$$

$$\Rightarrow \left(\frac{2x^2+5}{x^2+1} - 3\right) > 0$$

$$\Rightarrow \left(\frac{2x^2+5-3x^2-3}{x^2+1}\right) > 0$$

$$\Rightarrow x^2 < 2$$

$$\Rightarrow -\sqrt{2} < x < \sqrt{2}$$

Ex-9. Find the integral values of x satisfying the inequality, $x^2 - 3x < \sin^{-1}(\sin 2)$

Soln. We have, $x^2 - 3x < \sin^{-1}(\sin 2)$

$$\Rightarrow x^2 - 3x < \sin^{-1}(\sin(\pi - 2))$$

$$\Rightarrow x^2 - 3x < (\pi - 2)$$

$$\Rightarrow x^2 - 3x + (2 - \pi) < 0$$

$$\Rightarrow \left(x - \frac{3+\sqrt{1+4\pi}}{2}\right)\left(x - \frac{3-\sqrt{1+4\pi}}{2}\right) < 0$$

$$\Rightarrow \frac{3-\sqrt{1+4\pi}}{2} < x < \frac{3+\sqrt{1+4\pi}}{2}$$

Ex-10. Find the value of

$$\sin^{-1}(\sin 50) + \cos^{-1}(\cos 50) + \tan^{-1}(\tan 50)$$

Soln. We have

$$= \sin^{-1}(\sin 50) + \cos^{-1}(\cos 50) + \tan^{-1}(\tan 50)$$

$$= \sin^{-1}(\sin(50 - 16\pi)) + \cos^{-1}(\cos(16\pi - 50)) + \tan^{-1}(\tan(50 - 16\pi))$$

$$= (50 - 16\pi) + (16\pi - 50) + (50 - 16\pi)$$

$$= (50 - 16\pi)$$

5.13. SUM OF ANGLES

(i) $\sin^{-1} x + \sin^{-1} y$

$$= \begin{cases} \alpha : & x^2 + y^2 \le 1 \\ \pi - \alpha : & x > 0, y > 0, x^2 + y^2 > 1 \\ \alpha : & xy < 0,\ x^2 + y^2 > 1 \\ -\pi - \alpha : & x < 0,\ y > 0,\ x^2 + y^2 > 1 \end{cases}$$

where $\alpha = \sin^{-1}\left(x\sqrt{1-y^2} + y\sqrt{1-x^2}\right)$

Proof : Let $\sin^{-1}x = A$ & $\sin^{-1}y = B$

$$\Rightarrow x = \sin A \ \& \ y = \sin B$$

and $A, B \in \left[-\frac{\pi}{2}, \frac{\pi}{2}\right]$

$$\Rightarrow \cos A = \sqrt{1-x^2}, \ \cos B = \sqrt{1-y^2}$$

Now, $\sin(A+B)$

$$= \sin A \cos B + \cos A . \sin B$$

$$= x\sqrt{1-y^2} + y\sqrt{1-x^2}$$

Case - I : when $-1 \le x, y \le 1$ & $x^2 + y^2 \le 1$

In this case, $x^2 + y^2 \le 1$

$$\Rightarrow 1 - x^2 \ge y^2 \ \& \ 1 - y^2 \ge x^2$$

$$\Rightarrow (1-x^2)(1-y^2) \ge x^2y^2$$

$$\Rightarrow \sqrt{(1-x^2)(1-y^2)} - xy \ge 0$$

$$\Rightarrow \cos(A+B) \ge 0$$

$\Rightarrow$ A+B lies either in the the first or the fourth quadrant.

$$\Rightarrow A + B \in \left(-\frac{\pi}{2}, \frac{\pi}{2}\right)$$

$$\sin(A+B) = x\sqrt{1-y^2} + y\sqrt{1-x^2}$$

$$\Rightarrow A + B = \sin^{-1}\left(x\sqrt{1-y^2} + y\sqrt{1-x^2}\right)$$

$$\Rightarrow \sin^{-1}x + \sin^{-1}y = \sin^{-1}\left\{x\sqrt{1-y^2} + y\sqrt{1-x^2}\right\}$$

Case - II : when $xy<0$ and $x^2+y^2>1$

In this case, we have , $xy<0$

$\Rightarrow$ $(x>0 \text{ and } y<0)$ or $(x<0 \text{ and } y>0)$

$\Rightarrow$ $A\in\left(0,\frac{\pi}{2}\right]$ and $B\in\left[-\frac{\pi}{2}.0\right)$

$\Rightarrow$ $-\frac{\pi}{2}\le A+B\le\frac{\pi}{2}$

Also, $x^2+y^2>1$

$\Rightarrow$ $1-x^2>y^2$ and $1-y^2>x^2$

$\Rightarrow$ $(1-x^2)(1-y^2)>x^2y^2$

$\Rightarrow$ $\sqrt{(1-x^2)(1-y^2}<|xy|$

$\Rightarrow$ $-xy<\sqrt{(1-x^2)(1-y^2)}<xy$

$\Rightarrow$ $\sqrt{(1-x^2)(1-y^2)}-xy>0, (\because xy<0)$

$\Rightarrow$ $\cos(A+B)>0$

$\Rightarrow$ A+B lies either in the first quadrant or in the fourth quadrant..

$\Rightarrow$ $A+B\in\left[-\frac{\pi}{2},\frac{\pi}{2}\right]$

$\Rightarrow$ $\sin(A+B)=x\sqrt{1-y^2}+y\sqrt{1-x^2}$

$\Rightarrow$ $A+B=\sin^{-1}\left(x\sqrt{1-y^2}+y\sqrt{1-x^2}\right)$

$\Rightarrow$ $\sin^{-1}x+\sin^{-1}y=\sin^{-1}\left(x\sqrt{1-y^2}+y\sqrt{1-x^2}\right)$

Case - III : when $0<x,y\le1$ and $x^2+y^2<1$

In this case we have $0<x,y\le1$

$\Rightarrow$ $A\in\left[0,\frac{\pi}{2}\right]$ and $B\in\left[0,\frac{\pi}{2}\right]$

$\Rightarrow$ $A+B\in[0,\pi]$

Also, $x^2+y^2<1$

$\Rightarrow$ $1-x^2<y^2$ and $1-y^2<x^2$

$\Rightarrow$ $(1-x^2)(1-y^2)<x^2y^2$

$\Rightarrow$ $\sqrt{(1-x^2)(1-y^2)}<xy$

$\Rightarrow$ $\sqrt{(1-x^2)(1-y^2)}-xy<0$

$\Rightarrow$ $\cos(A+B)<0$

$\Rightarrow$ A+B lies either in II quadrant or in III quadrant .

$\Rightarrow$ $\frac{\pi}{2}\le A+B\le\pi$, since $A+B\in(0,\pi)$

$\Rightarrow$ $-\pi\le-(A+B)\le-\frac{\pi}{2}$

$\Rightarrow$ $0\le\pi-(A+B)\le\frac{\pi}{2}$

Now, $\sin(A+B)=x\sqrt{1-y^2}+y\sqrt{1-x^2}$

$\Rightarrow$ $\sin(\pi-(A+B))=x\sqrt{1-y^2}+y\sqrt{1-x^2}$

$\Rightarrow$ $(\pi-(A+B))=\sin^{-1}\left(x\sqrt{1-y^2}+y\sqrt{1-x^2}\right)$

$\Rightarrow$ $A+B=\pi-\sin^{-1}\left(x\sqrt{1-y^2}+y\sqrt{1-x^2}\right)$

$\Rightarrow \sin^{-1}x+\sin^{-1}y=\pi-\sin^{-1}\left(x\sqrt{1-y^2}+y\sqrt{1-x^2}\right)$

Case - IV : when $-1\le x,y<0$ and $x^2+y^2>1$

In this case , we have , $-1\le x,\ y<0$

$\Rightarrow$ $A\in\left[-\frac{\pi}{2},0\right)$ and $B\in\left[-\frac{\pi}{2},0\right)$

$\Rightarrow$ $A+B\in[-\pi,0)$

Also, $x^2+y^2>1$

$\Rightarrow$ $1-x^2<y^2$ and $1-y^2<x^2$

$\Rightarrow$ $(1-x^2)(1-y^2)<x^2y^2$

$\Rightarrow$ $\sqrt{(1-x^2)(1-y^2)}<xy$ (since $xy>0$)

$\Rightarrow \quad \sqrt{(1-x^2)(1-y^2)} - xy < 0$

$\Rightarrow \quad \cos(A+B) < 0$

$\Rightarrow$ A + B lies either in the first quadrant or in the third quadrant .

$\Rightarrow \quad -\pi \le A+B \le \frac{\pi}{2}$

$\Rightarrow \quad -\frac{\pi}{2} \le -(A+B) \le \pi$

$\Rightarrow \quad -\frac{\pi}{2} \le -\pi-(A+B) \le 0$

$\Rightarrow \quad \sin(A+B) = x\sqrt{1-y^2} + y\sqrt{1-x^2}$

$\Rightarrow \quad -\sin\{\pi+(A+B)\} = x\sqrt{1-y^2} + y\sqrt{1-x^2}$

$\Rightarrow \quad \sin\{-\pi-(A+B)\} = \left(x\sqrt{1-y^2} + y\sqrt{1-x^2}\right)$

$\Rightarrow \quad -\pi-(A+B) = \sin^{-1}\left(x\sqrt{1-y^2} + y\sqrt{1-x^2}\right)$

$\Rightarrow \quad A+B = -\pi - \sin^{-1}\left(x\sqrt{1-y^2} + y\sqrt{1-x^2}\right)$

$\Rightarrow \quad \sin^{-1}x + \sin^{-1}y$

$= -\pi - \sin^{-1}\left(x\sqrt{1-y^2} + y\sqrt{1-x^2}\right)$

(ii) $\sin^{-1}x - \sin^{-1}y$

$$= \begin{cases} \alpha : & x^2+y^2 \le 1 \\ \pi-\alpha : & x>0, y<0, x^2+y^2>1 \\ \alpha : & xy>0,\ x^2+y^2>1 \\ -\pi-\alpha : & x<0, y>0, x^2+y^2>1 \end{cases}$$

Where $\alpha = \sin^{-1}\left(x\sqrt{1-y^2} - y\sqrt{1-x^2}\right)$

Proof : Do yourself.

(iii) $\cos^{-1}x + \cos^{-1}y$

$$= \begin{cases} \alpha & : x+y \ge 0 \\ 2\pi-\alpha & : x+y<0 \end{cases}$$

where $\alpha = \cos^{-1}\left(xy - \sqrt{1-x^2}\sqrt{1-y^2}\right)$

Proof: Let $\cos^{-1}x = A$ and $\cos^{-1}y = B$

Then x = cos A and y = cos B

$\Rightarrow \quad A \in [0, \pi]$ and $B \in [0, \pi]$

Now, $\sin A = \sqrt{1-x^2}$ and $\sin B = \sqrt{1-y^2}$

$\cos(A+B) = xy - \sqrt{1-x^2}\sqrt{1-y^2}$

$\cos(A-B) = xy + \sqrt{1-x^2}\sqrt{1-y^2}$

Case - I : $-1 \le x, y \le 1$ and $x+y \ge 0$

In this case , $-1 \le x, y \le 1$

$\Rightarrow \quad 0 \le A+B \le 2\pi$ and $x+y \ge 0$

$\Rightarrow \quad 0 \le A+B \le 2\pi$ and $\cos A + \cos B \ge 0$

$\Rightarrow \quad \cos A \ge -\cos B$

$\Rightarrow \quad \cos A \ge \cos(\pi - B)$

$\Rightarrow \quad A \le \pi - B$

$\Rightarrow \quad 0 \le A+B \le \pi$

$\Rightarrow \quad \cos(A+B) = xy - \sqrt{1-x^2}\sqrt{1-y^2}$

$\Rightarrow \quad A+B = \cos^{-1}\left(xy - \sqrt{1-x^2}\sqrt{1-y^2}\right)$

$\Rightarrow \quad \cos^{-1}x + \cos^{-1}y = \cos^{-1}\left(xy - \sqrt{1-x^2}\sqrt{1-y^2}\right)$

Case - II: when $-1 \le x, y < 0$ and $x+y<0$

In this case $-1 \le x, y < 0$

$\Rightarrow \quad A, B \in [0, \pi]$

$\Rightarrow \quad 0 \le A+B \le 2\pi$ and $x+y \le 0$

$\Rightarrow \quad \cos A + \cos B \le 0$

$\Rightarrow \quad \cos A \le \cos(\pi - B)$

$\Rightarrow \quad A \ge \pi - B$

$\Rightarrow \quad A+B \ge \pi$

Thus, $\pi \le A+B \le 2\pi$

$\Rightarrow \quad -2\pi \le -(A+B) \le -\pi$

$\Rightarrow \quad 0 \le 2\pi-(A+B) \le \pi$

Now, cos (A+ B) = $xy-\sqrt{1-x^2}\sqrt{1-y^2}$

$\Rightarrow \quad \cos(2\pi-(A+B)) = xy-\sqrt{1-x^2}\sqrt{1-y^2}$

$\Rightarrow \quad (2\pi-(A+B)) = \cos^{-1}\left(xy-\sqrt{1-x^2}\sqrt{1-y^2}\right)$

$\Rightarrow \quad A+B = 2\pi-\cos^{-1}\left(xy-\sqrt{1-x^2}\sqrt{1-y^2}\right)$

$\Rightarrow \quad \cos^{-1}x+\cos^{-1}y$

$= 2\pi-\cos^{-1}\left(xy-\sqrt{1-x^2}\sqrt{1-y^2}\right)$

(iv) $\cos^{-1}x-\cos^{-1}y$

$$= \begin{cases} \alpha & : \ x \le y \\ -\alpha & : \ x > y \end{cases}$$

where $\alpha = \cos^{-1}\left(xy+\sqrt{1-x^2}\sqrt{1-y^2}\right)$.

Proof : Do yourself.

(v) $\tan^{-1}x+\tan^{-1}y$

$$= \begin{cases} \alpha : & xy<1 \\ \pi+\alpha & : x>0,\ y>0,\ xy>1 \\ -\pi+\alpha & : x<0, y<0,\ xy>1 \\ \dfrac{\pi}{2} & : x>0,\ y>0,\ xy=1 \\ \dfrac{\pi}{2} & : x<0,\ y<0,\ xy=1 \end{cases}$$

where $\alpha = \tan^{-1}\left(\dfrac{x+y}{1-xy}\right)$

Proof: Let $\tan^{-1}x = A$ and $\tan^{-1}y = B$

Then x = tan A and y = tan B .

$\Rightarrow \quad A \in \left(-\dfrac{\pi}{2}, \dfrac{\pi}{2}\right)$ and $B \in \left(-\dfrac{\pi}{2}, \dfrac{\pi}{2}\right)$

$$\tan(A+B) = \frac{\tan A+\tan B}{1-\tan A.\tan B} = \frac{x+y}{1-xy}$$

Case - I : when x > 0, y > 0 and xy < 1

In this case , x >0 , y>0 and xy < 1

$\Rightarrow \quad \dfrac{x+y}{1-xy} < 0$

$\Rightarrow \quad$ tan (A+ B) > 0

$\Rightarrow \quad$ A+B lies either in the first quadrant or in third quadrant .

$\Rightarrow \quad 0 < A+B < \pi$

$\Rightarrow \quad \tan(A+B) = \dfrac{x+y}{1-xy}$

$\Rightarrow \quad (A+B) = \tan^{-1}\left(\dfrac{x+y}{1-xy}\right)$

$\Rightarrow \quad \tan^{-1}x+\tan^{-1}y = \tan^{-1}\left(\dfrac{x+y}{1-xy}\right)$

Case - II : when x < 0 , y < 0 and xy < 1

In this case , x < 0 y < 0 and xy < 1

$\Rightarrow \quad \dfrac{x+y}{1-xy} < 0$

$\Rightarrow \quad$ tan (A+ B) < 0

$\Rightarrow \quad$ A+ B lies either in II quadrant or in IV quadrant

$\Rightarrow \quad$ A + B lies in the IV quadrant

$\Rightarrow \quad -\pi < A+B < 0$

Now, $\tan(A+B) = \dfrac{x+y}{1-xy}$

$\Rightarrow \quad A+B = \tan^{-1}\left(\dfrac{x+y}{1-xy}\right)$

$\Rightarrow \quad \tan^{-1}x+\tan^{-1}y = \tan^{-1}\left(\dfrac{x+y}{1-xy}\right)$

Case - III : when (x > 0 , y < 0) or (x < 0 and y > 0)

In this case , x >0 and y < 0

$\Rightarrow \quad A \in \left(0, \frac{\pi}{2}\right)$ and $B \in \left(-\frac{\pi}{2}, 0\right)$

$\Rightarrow \quad A+B \in \left(-\frac{\pi}{2}, \frac{\pi}{2}\right)$

Now, $\tan(A+B) = \frac{x+y}{1-xy}$

$\Rightarrow \quad \tan^{-1}x + \tan^{-1}y = \tan^{-1}\left(\frac{x+y}{1-xy}\right)$

similarly, if $x < 0$ and $y > 0$, we have

$\tan^{-1}x + \tan^{-1}y = \tan^{-1}\left(\frac{x+y}{1-xy}\right)$

It follows from above three cases that

$\tan^{-1}x + \tan^{-1}y = \tan^{-1}\left(\frac{x+y}{1-xy}\right)$

Case - IV: when $x > 0$, $y > 0$ and $xy > 1$

In this case, we have $x, y > 0$ and $xy > 1$

$\Rightarrow \quad \frac{x+y}{1-xy} < 0$

$\Rightarrow \quad \tan(A+B) < 0$

$\Rightarrow$ A+ B either lies in the II quadrant or in the IV quadrant.

$\Rightarrow$ A + B lies in the II quadrant.

$\Rightarrow \quad \frac{\pi}{2} < A+B < \pi$

$\Rightarrow \quad \frac{\pi}{2} - \pi < (A+B) - \pi < 0$

$\Rightarrow \quad -\frac{\pi}{2} < (A+B) - \pi < 0$

Now, $\tan(A+B) = \frac{x+y}{1-xy}$

$\Rightarrow \quad -\tan(\pi - (A+B)) = \frac{x+y}{1-xy}$

$\Rightarrow \quad A+B-\pi = \tan^{-1}\left(\frac{x+y}{1-xy}\right)$

$\Rightarrow \quad A+B = \pi + \tan^{-1}\left(\frac{x+y}{1-xy}\right)$

$\Rightarrow \quad \tan^{-1}x + \tan^{-1}y = \pi + \tan^{-1}\left(\frac{x+y}{1-xy}\right)$

Case- V: when $x < 0$, $y < 0$ and $xy > 1$

In this case $x, y < 0$ and $xy > 1$

$\Rightarrow \quad \frac{x+y}{1-xy} > 0$

$\Rightarrow \quad \tan(A+B) > 0$

$\Rightarrow$ A + B lies either in the I quadrant or in the III quadrant

$\Rightarrow$ A + B lies in the third quadrant

$\Rightarrow \quad -\pi < A+B < -\frac{\pi}{2}$

$\Rightarrow \quad \pi - \pi < \pi + (A+B) < \pi - \frac{\pi}{2}$

$\Rightarrow \quad 0 < \pi + (A+B) < \frac{\pi}{2}$

Now, $\tan(A+B) = \frac{x+y}{1-xy}$

$\Rightarrow \quad \tan(\pi + (A+B)) = \frac{x+y}{1-xy}$

$\Rightarrow \quad (\pi + (A+B)) = \tan^{-1}\left(\frac{x+y}{1-xy}\right)$

$\Rightarrow \quad A+B = -\pi + \tan^{-1}\left(\frac{x+y}{1-xy}\right)$

$\Rightarrow \quad \tan^{-1}x + \tan^{-1}y = -\pi + \tan^{-1}\left(\frac{x+y}{1-xy}\right)$

(vi) $\tan^{-1} - \tan^{-1} y$

$$= \begin{cases} \alpha & : \quad xy > -1 \\ \pi + \alpha & : xy < -1, x > 0, y < 0 \\ -\pi + \alpha & : xy < -1, x < 0, y > 0 \\ \dfrac{\pi}{2} & : xy = -1, x > 0, y < 0 \\ -\dfrac{\pi}{2} & : xy = -1, x < 0, y > 0 \end{cases}$$

where $\alpha = \tan^{-1}\left(\dfrac{x-y}{1+xy}\right)$

Proof : Do yourself.

5.14. SOME SOLVED EXAMPLES

Ex-1. Find the value of $\tan^{-1}\left(\dfrac{1}{2}\right) + \tan^{-1}\left(\dfrac{1}{3}\right)$

Soln. We have $\tan^{-1}\left(\dfrac{1}{2}\right) + \tan^{-1}\left(\dfrac{1}{3}\right)$

$$= \tan^{-1}\left(\dfrac{\frac{1}{2}+\frac{1}{3}}{1-\frac{1}{2}\times\frac{1}{3}}\right)$$

$$= \tan^{-1}\left(\dfrac{5/6}{1-1/6}\right)$$

$$= \tan^{-1}\left(\dfrac{5/6}{5/6}\right) = \tan^{-1}(1) = \pi/4$$

Ex-2. Find the value of $\tan^{-1}(1) + \tan^{-1}(2) + \tan^{-1}(3)$.

Soln. We have, $\tan^{-1}(1) + \tan^{-1}(2) + \tan^{-1}(3)$

$$= \dfrac{\pi}{4} + \pi + \tan^{-1}\left(\dfrac{2+3}{1-2.3}\right)$$

$$= \dfrac{\pi}{4} + \pi + \tan^{-1}(-1)$$

$$= \dfrac{\pi}{4} + \pi - \dfrac{\pi}{4}$$

$$= \pi$$

Ex-3. Find the value of $\tan^{-1}(9) + \tan^{-1}\left(\dfrac{5}{4}\right)$

Soln. We have, $\tan^{-1}(9) + \tan^{-1}\left(\dfrac{5}{4}\right)$

$$= \pi + \tan^{-1}\left(\dfrac{9+\frac{5}{4}}{1-9.\frac{5}{4}}\right)$$

$$= \pi + \tan^{-1}\left(\dfrac{\frac{41}{4}}{-\frac{41}{4}}\right)$$

$$= \pi + \tan^{-1}(-1)$$

$$= \pi - \dfrac{\pi}{4} = \dfrac{3\pi}{4}.$$

Ex-4. Find the value of

$$\sin^{-1}\left(\dfrac{4}{5}\right) + \sin^{-1}\left(\dfrac{5}{13}\right) - \sin^{-1}\left(\dfrac{63}{65}\right)$$

Soln. We have ,

$$\sin^{-1}\left(\dfrac{4}{5}\right) + \sin^{-1}\left(\dfrac{5}{13}\right) - \sin^{-1}\left(\dfrac{63}{65}\right)$$

$$= \sin^{-1}\left(\dfrac{4}{5}.\sqrt{1-\left(\dfrac{5}{13}\right)^2} + \dfrac{5}{13}.\sqrt{1-\left(\dfrac{4}{5}\right)^2}\right) - \sin^{-1}\left(\dfrac{63}{65}\right)$$

$$= \sin^{-1}\left(\dfrac{4}{5}.\dfrac{12}{13} + \dfrac{5}{13}.\dfrac{3}{13}\right) - \sin^{-1}\left(\dfrac{63}{65}\right)$$

$$= \sin^{-1}\left(\dfrac{48}{65} + \dfrac{15}{65}\right) - \sin^{-1}\left(\dfrac{63}{65}\right)$$

$$= \sin^{-1}\left(\dfrac{63}{65}\right) - \sin^{-1}\left(\dfrac{63}{65}\right)$$

$$= 0$$

Ex-5. Prove that $2\tan^{-1}\left(\dfrac{1}{3}\right) + \tan^{-1}\left(\dfrac{1}{7}\right) = \dfrac{\pi}{4}$

Soln. We have , $2\tan^{-1}\left(\dfrac{1}{3}\right) + \tan^{-1}\left(\dfrac{1}{7}\right)$

$$= \tan^{-1}\left(\dfrac{2.\frac{1}{3}}{1-\left(\frac{1}{3}\right)^2}\right) + \tan^{-1}\left(\dfrac{1}{7}\right)$$

$= \tan^{-1}\left(\dfrac{\frac{2}{3}}{\frac{8}{9}}\right)+\tan^{-1}\left(\dfrac{1}{7}\right)$

$= \tan^{-1}\left(\dfrac{3}{4}\right)+\tan^{-1}\left(\dfrac{1}{7}\right)$

$= \tan^{-1}\left(\dfrac{\frac{3}{4}+\frac{1}{7}}{1-\frac{3}{4}.\frac{1}{7}}\right)$

$= \tan^{-1}\left(\dfrac{25}{25}\right)$

$= \dfrac{\pi}{4}$

Ex-6. If $\sin^{-1}x+\sin^{-1}y+\sin^{-1}z=\pi$, then prove that

$x\sqrt{1-x^2}+y\sqrt{1-y^2}+z\sqrt{1-z^2}=2xyz$

Soln. Let $\sin^{-1}x = A,\ \sin^{-1}y = B,\ \sin^{-1}z = C$

Then $x = \sin A,\ y = \sin B,\ z = \sin C$

we have, $x\sqrt{1-x^2}+y\sqrt{1-y^2}+z\sqrt{1-z^2}$

$= \sin A.\cos A+\sin B.\cos B$

$+ \ +\sin C.\cos C$

$= \dfrac{1}{2}(\sin 2A+\sin 2B+\sin 2C)$

$= \dfrac{1}{2}(4\sin A.\sin B.\sin C)$

$= 2\sin A.\sin B.\sin C$

$= 2\,x\,y\,z.$

Ex-7. If $\cos^{-1}x+\cos^{-1}y+\cos^{-1}z=\pi$

then prove that $x^2+y^2+z^2+2xyz=1$

Soln. We have, $\cos^{-1}x+\cos^{-1}y+\cos^{-1}z=\pi$

$\Rightarrow \cos^{-1}x+\cos^{-1}y=\pi-\cos^{-1}z$

$\Rightarrow \cos^{-1}x+\cos^{-1}y=\cos^{-1}(-z)$

$\Rightarrow \cos^{-1}\left(xy-\sqrt{1-x^2}\sqrt{1-y^2}\right)$

$= \cos^{-1}(-z)$

$\Rightarrow (xy+z)^2=(1-x^2)(1-y^2)$

$\Rightarrow x^2y^2+2xyz+z^2$

$= 1-x^2-y^2+x^2y^2$

$\Rightarrow x^2+y^2+z^2+2xyz=1$

Ex-8. If $\cos^{-1}\left(\dfrac{x}{2}\right)+\cos^{-1}\left(\dfrac{y}{3}\right)=\theta$, prove that,

$9x^2+12xy\cos\theta+4y^2=36\sin^2\theta$

Soln. Given $\cos^{-1}\left(\dfrac{x}{2}\right)+\cos^{-1}\left(\dfrac{y}{3}\right)=\theta$

$\Rightarrow \cos^{-1}\left(\dfrac{x}{2}.\dfrac{y}{3}-\sqrt{1-\dfrac{x^2}{4}}\sqrt{1-\dfrac{y^2}{9}}\right)=\theta$

$\Rightarrow \left(\dfrac{xy}{6}-\sqrt{1-\dfrac{x^2}{4}}\sqrt{1-\dfrac{y^2}{9}}\right)=\cos\theta$

$\Rightarrow \left(\dfrac{xy}{6}-\cos\theta\right)^2=\left(1-\dfrac{x^2}{4}\right)\left(1-\dfrac{y^2}{9}\right)$

$\Rightarrow \dfrac{x^2y^2}{36}-\dfrac{xy}{3}\cos\theta+\cos^2\theta$

$= 1-\dfrac{x^2}{4}-\dfrac{y^2}{9}+\dfrac{x^2y^2}{36}$

$\Rightarrow \dfrac{x^2}{4}+\dfrac{y^2}{9}-\dfrac{xy}{3}\cos\theta=1-\cos^2\theta$

$\Rightarrow \dfrac{x^2}{4}+\dfrac{y^2}{9}-\dfrac{xy}{3}\cos\theta=\sin^2\theta$

$\Rightarrow 9x^2+4y^2-12\,xy\,\cos\theta=36\sin^2\theta$

Ex-9. Let $m=\dfrac{\tan^{-1}(1)+\tan^{-1}(2)+\tan^{-1}(3)}{\cot^{-1}(1)+\cot^{-1}(2)+\cot^{-1}(3)}$

Then find the value of $(m-1)^{2013}$.

Soln. We have ,

$\tan^{-1}(1)+\tan^{-1}(2)+\tan^{-1}(3)=\pi$

and $\cot^{-1}(1)+\cot^{-1}(2)+\cot^{-1}(3)$

$= \tan^{-1}(1)+\tan^{-1}\left(\frac{1}{2}\right)+\tan^{-1}\left(\frac{1}{3}\right)$

$= \frac{\pi}{4}+\frac{\pi}{4}$

$= \frac{\pi}{2}$

Hence, $m=\frac{\pi}{\pi/2}=2$

Ex-10. Solve for x: $\tan^{-1}(2x)+\tan^{-1}(3x)=\frac{3\pi}{4}$

Soln. **<u>Case - I</u>**: when $x\le 0$

Then , $\tan^{-1}(2x)\le 0,\ \tan^{-1}(3x)\le 0,$

$\Rightarrow \quad x\le 0$

So, it has no solution.

<u>Case-II</u> : when x >0 , 2 x. 3 x = $6x^2<1$

$\Rightarrow \quad x<\frac{1}{\sqrt{6}}$

Then $\frac{3\pi}{4}=\tan^{-1}(2x)+\tan^{-1}(3x)<\frac{\pi}{2}$

So, it is not possible .

<u>Case - III</u> : when x > 0 , 2x .3x >1

$\Rightarrow \quad x>\frac{1}{\sqrt{6}}$

Then $\frac{3\pi}{4}=\tan^{-1}(2x)+\tan^{-1}(3x)$

$\Rightarrow \quad \pi+\tan^{-1}\left(\frac{5x}{1-6x^2}\right)=\frac{3\pi}{4}$

$\Rightarrow \quad \tan^{-1}\left(\frac{5x}{1-6x^2}\right)=\frac{3\pi}{4}-\pi=-\frac{\pi}{4}$

$\Rightarrow \quad \left(\frac{5x}{1-6x^2}\right)=-1$

$\Rightarrow \quad 6x^2-5x-1=0$

$\Rightarrow \quad$ x = 1 , - 1/ 6

Thus , x = 1 is a solution.

Ex-11. Solve for x : $\sin^{-1}(x)+\sin^{-1}(2x)=\frac{\pi}{3}$

Soln. We have , $\sin^{-1}(x)+\sin^{-1}(2x)=\frac{\pi}{3}$

$\Rightarrow \quad \sin^{-1}x+\sin^{-1}(2x)=\sin^{-1}\left(\frac{\sqrt{3}}{2}\right)$

$\Rightarrow \quad \sin^{-1}x-\sin^{-1}\left(\frac{\sqrt{3}}{2}\right)=-\sin^{-1}(2x)$

$\Rightarrow \quad \sin^{-1}\left(\frac{x}{2}-\frac{\sqrt{3}}{2}\sqrt{1-x^2}\right)=\sin^{-1}(-2x)$

$\Rightarrow \quad \left(\frac{x}{2}-\frac{\sqrt{3}}{2}\sqrt{1-x^2}\right)=-2x$

$\Rightarrow \quad 5x=\sqrt{3}\sqrt{1-x^2}$

$\Rightarrow \quad 25x^2=3(1-x^2)$

$\Rightarrow \quad 28x^2=3$

$\Rightarrow \quad x=\pm\frac{\sqrt{3}}{2\sqrt{7}}$

$\Rightarrow \quad x=\frac{\sqrt{3}}{2\sqrt{7}}$, negative value of x does not satisfy the given equation.

Ex-12. Let $f(x)=\cos^{-1}(x)$

$+\cos^{-1}\left(\frac{x}{2}+\frac{\sqrt{3-3x^2}}{2}\right)$, for $\frac{1}{2}\le x\le 1$

Then find $f(2013)$

Soln. We have $f(x)$

$= \cos^{-1}(x)+\cos^{-1}\left(\frac{x}{2}+\frac{\sqrt{3-3x^2}}{2}\right)$

$= \cos^{-1}(x)+\cos^{-1}\left(\frac{1}{2}\right)-\cos^{-1}(x)$

$= \frac{\pi}{3}$

Now, $f(2013)=\frac{\pi}{3}$

5.15. MULTIPLE ANGLES :

(i) $\sin^{-1}\left(2x\sqrt{1-x^2}\right)$

$$=\begin{cases} 2\sin^{-1}x & : -\frac{1}{\sqrt{2}} \le x \le \frac{1}{\sqrt{2}} \\ \pi - 2\sin^{-1}x & : \frac{1}{\sqrt{2}} < x \le 1 \\ -\pi - 2\sin^{-1}x & : -1 \le x < -\frac{1}{\sqrt{2}} \end{cases}$$

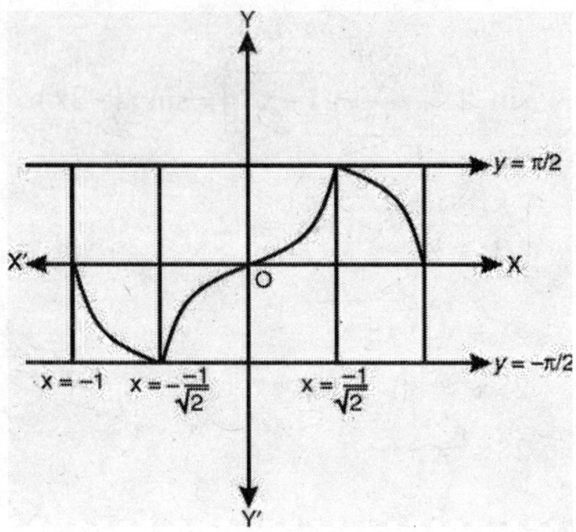

(ii) $$\cos^{-1}\left(2x^2-1\right)=\begin{cases} 2\cos^{-1}x & :0 \le x \le 1 \\ 2\pi - 2\cos^{-1}x & :-1 \le x < 0 \end{cases}$$

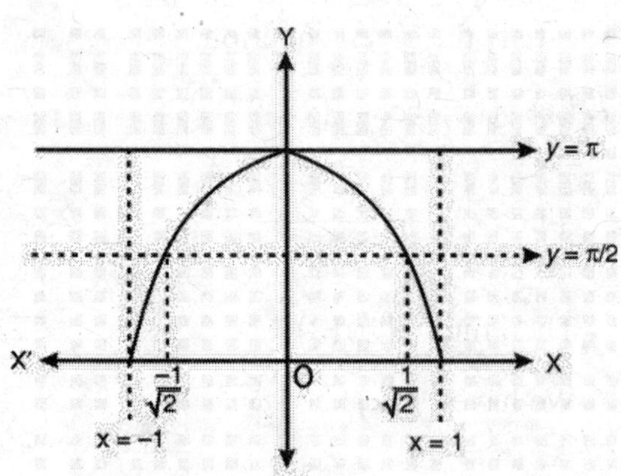

(iii) $$\tan^{-1}\left(\frac{2x}{1-x^2}\right)=\begin{cases} \alpha & : -1 < x < \\ -\pi + \alpha & : x > 1 \\ \pi + \alpha & : x < -1 \end{cases}$$

where $\alpha = 2\tan^{-1}(x)$.

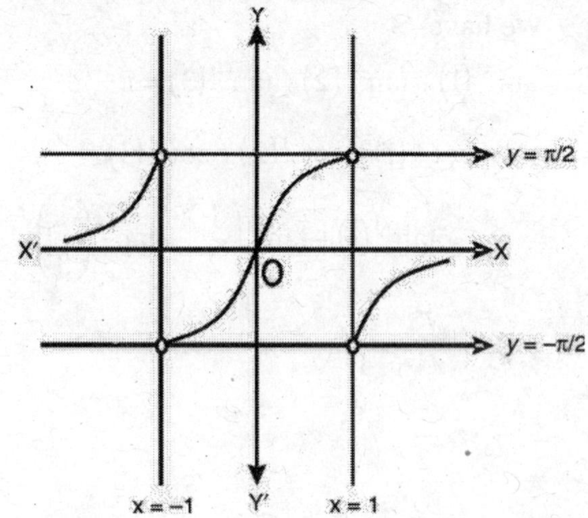

(iv) $$\sin^{-1}\left(\frac{2x}{1+x^2}\right)=\begin{cases} \alpha & : -1 \le x \le 1 \\ \pi - \alpha & : x > 1 \\ -\pi - \alpha & : x < -1 \end{cases}$$

where $\alpha = 2\tan^{-1}(x)$

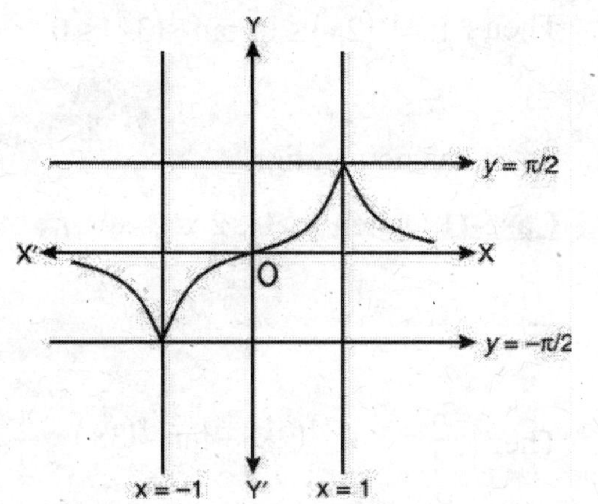

(v) $$\cos^{-1}\left(\frac{1-x^2}{1+x^2}\right)=\begin{cases} 2\tan^{-1}(x) & : x \ge 0 \\ -2\tan^{-1}(x) & : x \le 0 \end{cases}$$

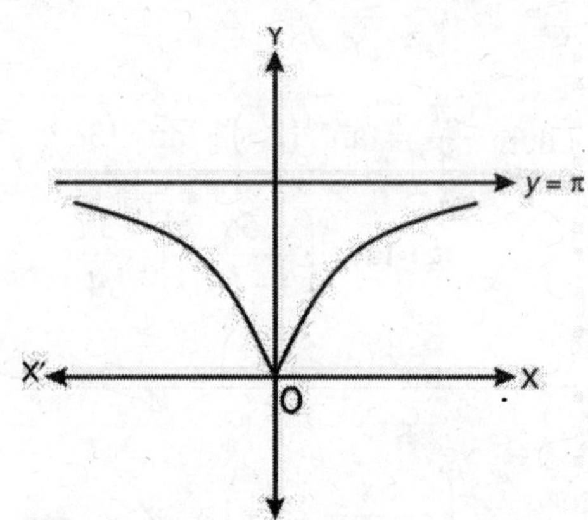

Ex-1. Find the value of $\sin\left(2\sin^{-1}\left(\frac{1}{4}\right)\right)$

Soln. Let $\sin^{-1}\left(\frac{1}{4}\right)=\theta$

$\Rightarrow \quad \sin\theta=\frac{1}{4}$

Now, $\sin(2\theta)$

$= \quad 2\sin\theta.\cos\theta$

$= \quad 2\times\frac{1}{4}\times\sqrt{1-\frac{1}{16}}$

$= \quad \frac{\sqrt{15}}{8}$

Ex-2. Find the value of $\cos\left(2\cos^{-1}\left(\frac{1}{3}\right)\right)$

Soln. Let $\cos^{-1}\left(\frac{1}{3}\right)=\theta$

$\Rightarrow \quad \cos\theta=\frac{1}{3}$

Now, $\cos(2\theta)$

$= \quad 2\cos^2\theta-1$

$= \quad \frac{2}{9}-1$

$= \quad -\frac{7}{9}$

Ex-3. Find the value of $\cos\left(2\tan^{-1}\left(\frac{1}{3}\right)\right)$

Soln. Let $\tan^{-1}\left(\frac{1}{3}\right)=\theta$

$\Rightarrow \quad \tan\theta=\frac{1}{3}$

Now, $\cos(2\theta)$

$= \quad \frac{1-\tan^2\theta}{1+\tan^2\theta}$

$= \quad \frac{1-\left(\frac{1}{3}\right)^2}{1+\left(\frac{1}{3}\right)^2}$

$= \quad \frac{9-1}{9+1}$

$= \quad \frac{8}{10}=\frac{4}{5}$

Ex-4. Find the value of $\sin\left(\frac{1}{2}\cot^{-2}\left(\frac{3}{4}\right)\right)$

Soln. Let $\cot^{-1}\left(\frac{3}{4}\right)=\theta$

$\Rightarrow \quad \cot(\theta)=\left(\frac{3}{4}\right)$

$\Rightarrow \quad \tan(\theta)=\left(\frac{4}{3}\right)$

Now, $\sin\left(\frac{\theta}{2}\right)$

$= \quad \sqrt{\frac{1-\cos\theta}{2}}$

$= \quad \sqrt{\frac{1-\frac{\cos\theta}{\sin\theta.\text{cosec}\,\theta}}{2}}$

$= \quad \sqrt{\frac{1-\frac{\cot\theta}{\text{cosec}\theta}}{2}}$

$= \quad \sqrt{\frac{1-\frac{\cot\theta}{\sqrt{1+\cot^2\theta}}}{2}}$

$= \quad \frac{1}{\sqrt{2}}\sqrt{1-\frac{3/4}{\sqrt{1+9/16}}}$

$= \quad \frac{1}{\sqrt{2}}\sqrt{1-\frac{3}{5}}$

$= \quad \frac{2}{\sqrt{5}}$

Ex-5. Find the value of

$$\tan^{-1}\left(\frac{3\pi}{4}-2\tan^{-1}\left(\frac{3}{4}\right)\right)$$

Soln. Let $\tan^{-1}\left(\frac{3}{4}\right)=\theta$

$$\Rightarrow \quad \tan\theta = \frac{3}{4}$$

We have, $\tan\left(\frac{3\pi}{4} - 2\theta\right)$

$$= \frac{\tan\left(\frac{3\pi}{4}\right) - \tan(2\theta)}{1 + \tan\left(\frac{3\pi}{4}\right).\tan(2\theta)}$$

$$= \frac{-1 - \tan(2\theta)}{1 - .\tan(2\theta)}$$

$$= \frac{-1 - \frac{2\tan\theta}{1 - \tan^2\theta}}{1 - \frac{2\tan\theta}{1 - \tan^2\theta}}$$

$$= \frac{\tan^2\theta - 2\tan\theta - 1}{1 - \tan^2\theta - 2\tan\theta}$$

$$= \frac{\frac{9}{16} - \frac{6}{4} - 1}{1 - \frac{9}{16} - \frac{6}{4}}$$

$$= \frac{41}{17}$$

5.16. MORE MULTIPLE ANGLES :

(i) $\sin^{-1}(3x - 4x^3)$

$$= \begin{cases} 3\sin^{-1}x & : -\frac{1}{2} \le x \le \frac{1}{2} \\ \pi - 3\sin^{-1}x & : \frac{1}{2} < x \le 1 \\ -\pi - 3\sin^{-1}x & : -1 \le x < -\frac{1}{2} \end{cases}$$

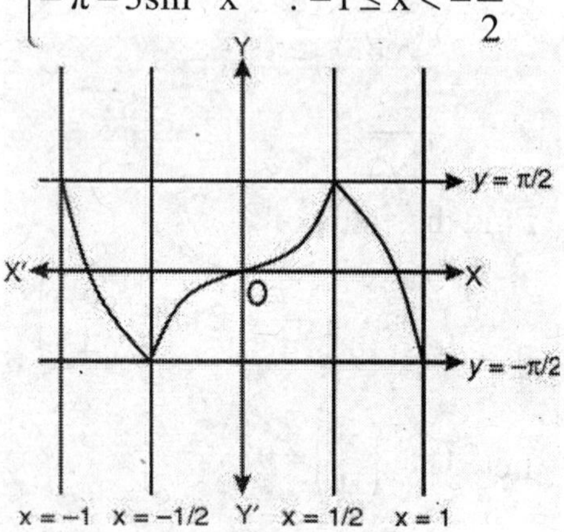

(ii) $\cos^{-1}(4x^3 - 3x)$

$$= \begin{cases} 3\cos^{-1}x & : \frac{1}{2} \le x \le 1 \\ 2\pi - 3\cos^{-1}x & : -\frac{1}{2} \le x < \frac{1}{2} \\ -2\pi + 3\cos^{-1}x & : -1 \le x < -\frac{1}{2} \end{cases}$$

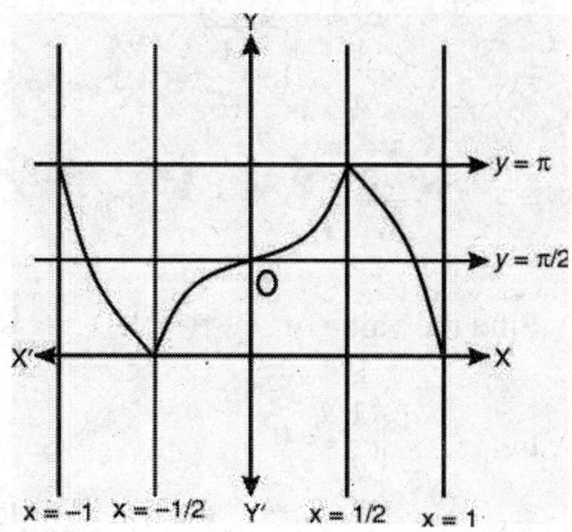

(iii) $\tan^{-1}\left(\frac{3x - x^3}{1 - 3x^2}\right)$

$$= \begin{cases} 3\tan^{-1}x & : -\frac{1}{\sqrt{3}} < x < \frac{1}{\sqrt{3}} \\ \pi + 3\tan^{-1}x & : -\infty < x < -\frac{1}{\sqrt{3}} \\ -\pi + 3\tan^{-1}x & : \frac{1}{\sqrt{3}} < x < \infty \end{cases}$$

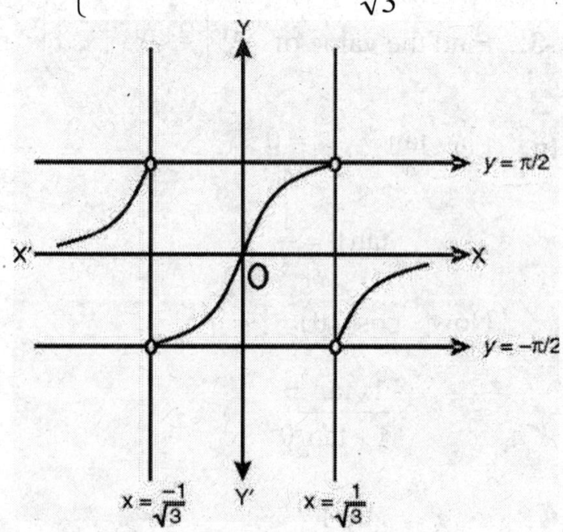

5.17. SOME SOLVED EXAMPLES :

Ex-1. Let $f(x) = \sin^{-1}\left(\frac{2x}{1+x^2}\right) + 2\tan^{-1}(x)$, x>1 then find the value of $f(2013)$

Soln. We have, $f(x)$

$$= \sin^{-1}\left(\frac{2x}{1+x^2}\right)+2\tan^{-1}(x)$$

$$= \pi-2\tan^{-1}(x)+2\tan^{-1}(x)$$

$$= \pi$$

Hence, the value of $f(2013) = \pi$

Ex-2. Let $f(x)=2\tan^{-1}\left(\frac{1+x}{1-x}\right)+\sin^{-1}\left(\frac{1-x^2}{1+x^2}\right)$ for $0\le x<1$, then find the value of $f\left(\frac{1}{2014}\right)$.

Soln. we have, $f(x)$

$$= 2\tan^{-1}\left(\frac{1+x}{1-x}\right)+\sin^{-1}\left(\frac{1-x^2}{1+x^2}\right)$$

$$= 2\left(\tan^{-1}(1)+\tan^{-1}(x)\right)+\frac{\pi}{2}-\cos^{-1}\left(\frac{1-x^2}{1+x^2}\right)$$

$$= 2\left(\frac{\pi}{4}+\tan^{-1}(x)\right)+\frac{\pi}{2}-\cos^{-1}\left(\frac{1-x^2}{1+x^2}\right)$$

$$= \left(\frac{\pi}{2}+2\tan^{-1}(x)\right)+\left(\frac{\pi}{2}-2\tan^{-1}(x)\right)$$

$$= \pi$$

Hence, the value of $f\left(\frac{1}{2014}\right)=\pi$

Ex-3. Let $f(x)=\sin^{-1}\left(\frac{6x}{x^2+9}\right)+2\tan^{-1}\left(-\frac{x}{3}\right)$ is independent of x, then find the value of x.

Soln. We have, $f(x)$

$$= \sin^{-1}\left(\frac{6x}{x^2+9}\right)+2\tan^{-1}\left(-\frac{x}{3}\right)$$

$$= \sin^{-1}\left(\frac{2.\left(\frac{x}{3}\right)}{1+\left(\frac{x}{3}\right)^2}\right)-2\tan^{-1}\left(\frac{x}{3}\right)$$

$$= 2\tan^{-1}\left(\frac{x}{3}\right)-2\tan^{-1}\left(\frac{x}{3}\right)$$

$$= 0$$

It will happen when $\left|\frac{x}{3}\right|\le 1$

$$\Rightarrow \quad |x|\le 3$$

$$\Rightarrow \quad -3\le x\le 3$$

Ex-4. Find the interval of x for which the function $f(x)=\cos^{-1}\left(\frac{1-x^2}{1+x^2}\right)+2\tan^{-1}(x)$ is a constant function.

Soln. We have, $f(x)$

$$= \cos^{-1}\left(\frac{1-x^2}{1+x^2}\right)+2\tan^{-1}(x)$$

$$= -2\tan^{-1}(x)+2\tan^{-1}(x),\ x\le 0$$

$$= 0$$

It is possible only when $x\le 0$

$$\Rightarrow \quad x\in(-\infty,\ 0]$$

Ex-5. Find the interval of x for which the function $f(x)$ $= 3\cos^{-1}(2x^2-1)+2\cos^{-1}(4x^3-3x)$ is independent of x.

Soln. We have f (x)

$$= 3\cos^{-1}(2x^2-1)+2\cos^{-1}(4x^3-3x)$$

$$= 3\left(2\cos^{-1}x\right) + 2\left(2\pi-3\cos^{-1}x\right)$$

$$(\text{for } 0\le x<1) \quad \left(\text{for } -\frac{1}{2}\le x\le\frac{1}{2}\right)$$

It is possible only when $0 \le x \le \frac{1}{2}$

$\Rightarrow \quad x \in \left[0, \frac{1}{2}\right]$

Als0, f(x)

$= 3\cos^{-1}(2x^2-1) + 2\cos^{-1}(4x^3-3x)$

$= 3(2\pi - 2\cos^{-1}x) + 2(-2\pi + 3\cos^{-1}x)$

(for $-1 \le x \le 0$) (for $-1 \le x < -\frac{1}{2}$)

$= \quad 2\pi$

It is possible only when $x \in \left[-1, -\frac{1}{2}\right)$

Hence, the value of x is

$\left[-1, -\frac{1}{2}\right) \cup \left[0, \frac{1}{2}\right]$

Ex-6. If $\tan^{-1} y : \tan^{-1} x = 4:1$, then express y as algebraic function of x . Also, prove that $\tan\left(22\frac{1}{2}^{\circ}\right)$ is a root of $x^4 - 6x^2 + 1 = 0$.

Soln. We have , $\tan^{-1}y = 4\tan^{-1}x$

$\Rightarrow \quad \tan^{-1}y = \tan^{-1}\left(\frac{4x-4x^3}{1-6x^2+x^4}\right)$

$\Rightarrow \quad y = \frac{4x(1-x^2)}{1-6x^2+x^4}$

Which is a function of x .

Let $\tan^{-1}x = \frac{\pi}{8}$

$\Rightarrow \quad x = \tan\left(\frac{\pi}{8}\right)$

$\Rightarrow \quad \tan^{-1}y = 4\tan^{-1}x = \frac{\pi}{2}$

$\Rightarrow \quad \frac{4x(1-x^2)}{1-6x^2+x^4} \to \infty$

$\Rightarrow \quad 1-6x^2+x^4 = 0$

$\Rightarrow \quad x = \tan\left(\frac{\pi}{8}\right)$ is a root of $1+x^4 = 6x^2$

5.18. GRAPHS OF SOME ADDITIONAL FUNCTIONS :

(i) Graph of $y = \tan(\cos^{-1} x) = \frac{\sqrt{1-x^2}}{x}$

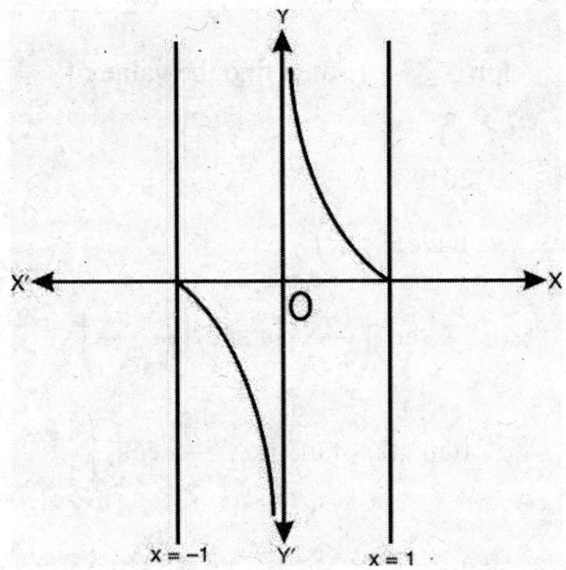

(ii) Graph of $y = \tan(\cot^{-1} x) = 1/x$

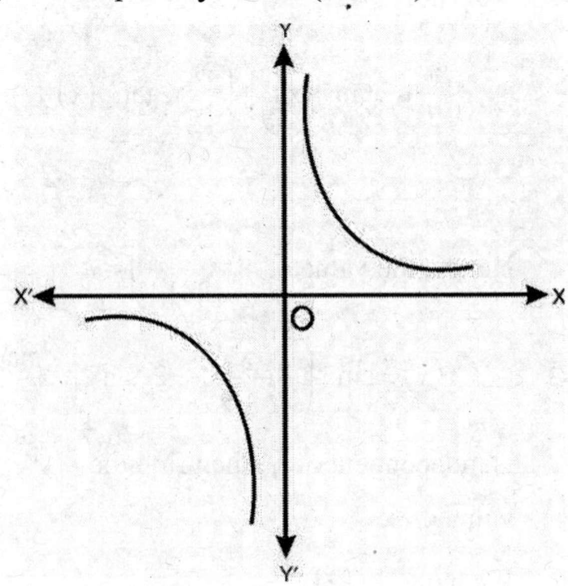

(iii) Graph of $y = \sin(\tan^{-1} x) = \frac{x}{\sqrt{1+x^2}}$

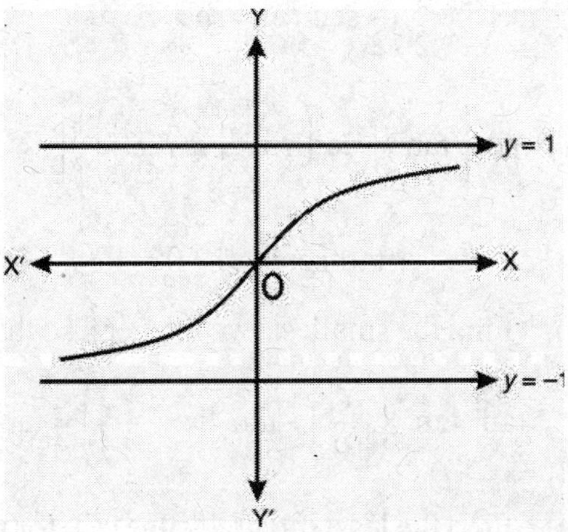

(iv) Graph of $y = \cos(\tan^{-1} x) = \frac{1}{\sqrt{1+x^2}}$

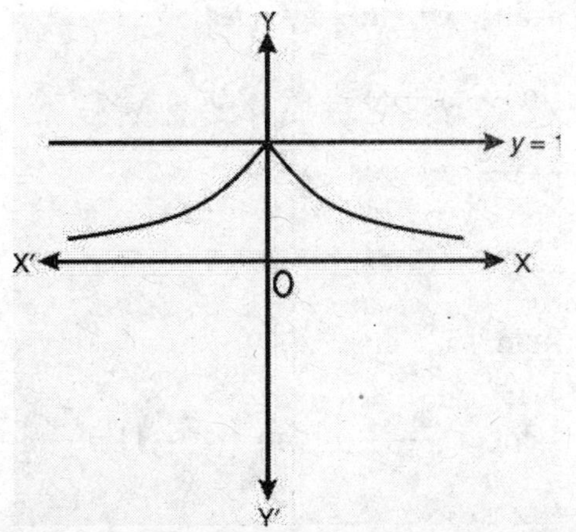

Level– I QUESTIONS BASED ON FUNDAMENTALS

Q.1 Find the domains of

(i). $f(x) = \sin^{-1}(2x-3)$

(ii). $f(x) = \sin^{-1}\left(\frac{x}{x-1}\right)$

(iii). $f(x) = \cos^{-1}(3x+4)$

(iv). $f(x) = \sin^{-1}\left(\frac{|x|-1}{2}\right)$

(v) $f(x) = \cos^{-1}\left(\frac{1-|x|}{5}\right)$

(vi) $f(x) = \sin^{-1}\left(\frac{|x|-2}{3}\right) + \cos^{-1}\left(\frac{1-|x|}{4}\right)$

(vii). $f(x) = \sin^{-1}(2x^2-1)$

(viii). $f(x) = \sqrt{5\pi\sin^{-1}x - 6(\sin^{-1}x)^2}$

(ix). $f(x) = \log_2\left(\frac{3\tan^{-1}x + \pi}{\pi - 4\tan^{-1}x}\right)$

(x). $f(x) = \cos^{-1}\left(\frac{3}{2+\sin x}\right)$

(xi) $f(x) = \sin^{-1}\left(\frac{x^2+1}{2x}\right)$

(xii) $f(x) = \cos^{-1}\left(\frac{x^2+1}{x^2}\right)$

(xiii) $f(x) = \sin^{-1}(\log_2(x^2+3x+4))$

(xiv). $f(x) = \sin^{-1}\left(\log_2\left(\frac{x^2}{2}\right)\right)$

(xv) $f(x) = \sin^{-1}[2-3x^2]$

(xvi) $f(x) = \frac{1}{x} + 3^{\sin^{-1}x} + \frac{1}{\sqrt{x-2}}$

(xvii). $f(x) = \sin^{-1}(\log_2 x^2)$

(xviii). $f(x) = e^x + \sin^{-1}\left(\frac{x}{2}-1\right) + \frac{1}{x}$

(xix). $f(x) = \sqrt{\sin^{-1}(\log_x 2)}$

(xx). $f(x) = \sqrt{\sin^{-1}(\log_2 x)}$

Q.2 Solve for x :

(i). $(\sin^{-1}x)^2 - 3\sin^{-1}x + 2 = 0$

(ii). $\sin^{-1}x + \sin^{-1}2y = \pi$

(iii). $\cos^{-1}x + \cos^{-1}x^2 = 2\pi$

(iv). $\cos^{-1}x + \cos^{-1}x^2 = 0$

(v). $4\sin^{-1}(x-1) + \cos^{-1}(x-1) = \pi$

(vi). $\sin^{-1}x > \cos^{-1}x$

(vii) $\cot^{-1}\left(\dfrac{1}{x^2-1}\right) + \tan^{-1}(x^2-1) = \dfrac{\pi}{2}$

(viii) $\cot^{-1}\left(\dfrac{x^2-1}{2x}\right) + \tan^{-1}\left(\dfrac{2x}{x^2-1}\right) = \dfrac{2\pi}{3}$

(ix). $4\sin^{-1}x + \cos^{-1}x = \dfrac{3\pi}{4}$

(x). $5\tan^{-1}x + 3\cot^{-1}x = \dfrac{7\pi}{4}$

(xi). $5\tan^{-1}x + 4\cot^{-1}x = 2\pi$

(xii) $\cot^{-1}x - \cot^{-1}(x+1) = \dfrac{\pi}{2}$

(xiii). $3x^2 + 8x < 2\sin^{-1}(\sin 4) - \cos^{-1}(\cos 4)$

(xiv). $\sin^{-1}\left(\sin\left(\dfrac{2x^2+4}{x^2+1}\right)\right) < \pi - 3$

(xv). $[\sin^{-1}x] + [\cos^{-1}x] = 0$

(xvi). $[\tan^{-1}x] + [\cot^{-1}x] = 0$

(xvii) $[\sin^{-1}\cos^{-1}\sin^{-1}\tan^{-1}x] = 0$

(xviii). $[\sin^{-1}\cos^{-1}\sin^{-1}\tan^{-1}x] = 1$

(xix). $(\tan^{-1}x)^2 + (\cot^{-1}x)^2 = \dfrac{5\pi^2}{8}$

(xx). $\sin^{-1}x + \sin^{-1}(1-x) = \cos^{-1}x$

(xxi). $x^2 - 4x > \sin^{-1}(\sin[\pi^{3/2}]) + \cos^{-1}(\cos[\pi^{3/2}])$

(xxii). $\cos(\tan^{-1}x) = x$

(xxiii). $\sin(\tan^{-1}x) = \cos(\cot^{-1}(x+1))$

(xxiv) $\sec^{-1}\left(\dfrac{x}{2}\right) - \sec^{-1}x = \sec^{-1}2$

(xxv) $\cos\left(\tan^{-1}\left(\cot\left(\sin^{-1}\left(x+\dfrac{3}{2}\right)\right)\right)\right) + \tan(\sec^{-1}x) = 0$

(xxvi). Find the smallest +ve integer x so that

$$\tan\left(\tan^{-1}\left(\frac{x}{10}\right) + \tan^{-1}\left(\frac{1}{x+1}\right)\right) = \tan\left(\frac{\pi}{4}\right).$$

(xxvii) Find the least integral value of k for which

$$(k-2)x^2 + 8x + k + 4 > \sin^{-1}(\sin 12) + \cos^{-1}(\cos 12)$$

holds for all x in R.

(xxvii) If $\alpha = 2\tan^{-1}\left(\dfrac{1+x}{1-x}\right)$ and

$\beta = \sin^{-1}\left(\dfrac{1-x^2}{1+x^2}\right)$ for $0 < x < 1$,

then prove that $\alpha + \beta = \pi$.

Q.3. Solve for x :

(i). $6(\sin^{-1}x)^2 - \pi\sin^{-1}x \le 0$

(ii). $\dfrac{2\tan^{-1}x + \pi}{4\tan^{-1}x - \pi} \le 0$

(iii). $\sin^{-1}x < \sin^{-1}x^2$

(iv). $\cos^{-1}x > \cos^{-1}x^2$

(v). $\log_2(\tan^{-1}x) > 1$

(vi). $(\cot^{-1}x)^2 - 5\cot^{-1}x + 6 > 0$

(vii) $\sin^{-1}x < \cos^{-1}x$

(viii). $\sin^{-1}x > \sin^{-1}(1-x)$

(ix). $\sin^{-1}2x > \operatorname{cosec}^{-1}x$

(x). $\tan^{-1}3x < \cot^{-1}x$

(xi). $\cos^{-1} 2x \geq \sin^{-1} x$

(xii). $x^2 - 2x < \sin^{-1}(\sin 2)$

(xiii). $\sin^{-1}\left(\frac{x}{2}\right) < \cos^{-1}(x+1)$

(xiv). $\tan^{-1} 2x > 2\tan^{-1} x$

(xv). $\tan(\cos^{-1} x) \leq \sin\left(\cot^{-1}\left(\frac{1}{2}\right)\right)$

Q4. Find the ranges of

(i) $f(x) = \sin^{-1}(2x-3)$

(ii). $f(x) = 2\sin^{-1}(2x-1) - \frac{\pi}{4}$

(iii). $f(x) = 2\cos^{-1}(-x^2) - \pi$

(iv) $f(x) = \frac{1}{2}\tan^{-1}(1-x^2) - \frac{\pi}{4}$

(v). $f(x) = \cot^{-1}(2x - x^2)$

(vi). $f(x) = \sin^{-1} x + \cos^{-1} x + \tan^{-1} x$

(vii). $f(x) = \sin^{-1} x + \sec^{-1} x + \tan^{-1} x$

(viii). $f(x) = 3\cot^{-1} x + 2\tan^{-1} x + \frac{\pi}{4}$.

(ix) $f(x) = \text{cosec}^{-1}[1 + \sin^2 x]$.

(x) $f(x) = \sin^{-1}(\log_2(x^2 + 3x + 4))$

Q.5 (i) Let $f(x) = \sin^{-1}(\sin x)$, $\forall\, x \in [-\pi, 2\pi]$.

Then find $f'(x)$.

(ii) Let $f(x) = \cos^{-1}(\cos x)$, $\forall\, x \in [-2\pi, \pi]$.
Then find $f'(x)$.

(iii) Let $f(x) = \tan^{-1}(\tan x)$, $\forall\, x \in \left[-\frac{3\pi}{2}, \frac{5\pi}{2}\right]$ .. Then find $f'(x)$.

Q.6 Evaluate to each of the following in verse trigonometric functions :

(i) $\sin^{-1}\left(\frac{1}{2}\right)$

(ii) $\sin^{-1}\left(-\frac{1}{2}\right)$

(iii) $\cos^{-1}\left(-\frac{1}{\sqrt{2}}\right)$

(iv) $\tan^{-1}\left(-\frac{1}{\sqrt{3}}\right)$

(v) $\sec^{-1}\left(\frac{2}{\sqrt{3}}\right)$

(vi) $\cot^{-1}(-\sqrt{3})$

(vii) $\sin\left(\frac{\pi}{3} - \sin^{-1}\left(-\frac{1}{2}\right)\right)$

(viii) $\sin\left(\frac{\pi}{3} - \sin^{-1}\left(-\frac{\sqrt{3}}{2}\right)\right)$

(ix) $\cos\left(\frac{\pi}{6} + \cos^{-1}\left(-\frac{\sqrt{3}}{2}\right)\right)$

(x) $\tan\left(\frac{\pi}{2} - \sec^{-1} 3\right)$

(xi) $\operatorname{cosec}\left(\sin^{-1}\left(-\frac{1}{\sqrt{3}}\right)\right)$

(xii) $\sin\left(\frac{\pi}{6} + \cos^{-1}\left(\frac{1}{4}\right)\right)$

(xiii) $\sin\left(\frac{\pi}{4} - \sin^{-1}\left(-\frac{1}{2}\right)\right)$

(xiv) $\tan\left(\cos^{-1}\left(\frac{1}{2}\right) + \tan^{-1}\left(-\frac{1}{\sqrt{3}}\right)\right)$

(xv) $\sin^{-1}\left(\cos\left(\sin^{-1}\left(\frac{\sqrt{3}}{2}\right)\right)\right)$

(xvi) $\sin\left(\cos^{-1}\left(\frac{3}{5}\right)\right)$

(xvii) $\tan\left(\cos^{-1}\left(\frac{1}{3}\right)\right)$

(xviii) $\cos\left(\sin^{-1}\left(\frac{4}{5}\right)+\cos^{-1}\left(\frac{2}{3}\right)\right)$

Q.7 Evaluate the following :

(i) $\sin^{-1}\left(\sin\left(\frac{7\pi}{3}\right)\right)$

(ii) $\sin^{-1}\left(\sin\left(\frac{11\pi}{3}\right)\right)$

(iii) $\sin^{-1}\left(\sin\left(\frac{49\pi}{8}\right)\right)$

(iv) $\sin^{-1}(\sin 3)$

(v) $\sin^{-1}(\sin 5)$

(vi) $\sin^{-1}(\sin 7)$

(vii) $\sin^{-1}(\sin 10)$

(viii) $\sin^{-1}(\sin 12)$

(ix) $\sin^{-1}(\sin 20)$

(x) $\sin^{-1}(\sin 50)$

(xi) $\sin^{-1}(\sin 80)$

(xii) $\sin^{-1}(\sin 100)$

Q.8 Evaluate the following :

(i) $\cos^{-1}\left(\cos\left(\frac{7\pi}{6}\right)\right)$

(ii) $\cos^{-1}\left(\cos\left(\frac{31\pi}{7}\right)\right)$

(iii) $\cos^{-1}\left(\cos\left(\frac{103\pi}{7}\right)\right)$

(iv) $\cos^{-1}(\cos(3))$

(v) $\cos^{-1}(\cos(5))$

(vi) $\cos^{-1}(\cos(7))$

(vii) $\cos^{-1}(\cos(10))$

(viii) $\cos^{-1}(\cos(12))$

(ix) $\cos^{-1}(\cos(15))$

(x) $\cos^{-1}(\cos(40))$

(xi) $\cos^{-1}(\cos(60))$

(xii) $\cos^{-1}(\cos(100))$

Q.9 Find the value of

(i) $\sin^{-1}(\sin 1)+\sin^{-1}(\sin 2)+\sin^{-1}(\sin 3)$

(ii) $\sin^{-1}(\sin 10)+\sin^{-1}(\sin 20)$
$+\sin^{-1}(\sin 30)+\sin^{-1}(\sin 40)$

(iii) $\cos^{-1}(\cos 1)+\cos^{-1}(\cos 2)$
$+\cos^{-1}(\cos 3)+\cos^{-1}(\cos 4)$

(iv) $\cos^{-1}(\cos 10)+\cos^{-1}(\cos 20)$
$+\cos^{-1}(\cos 30)+\cos^{-1}(\cos 40)$

(v) $\sin^{-1}(\sin 10)+\cos^{-1}(\cos 10)$

(vi) $\sin^{-1}(\sin 50)+\cos^{-1}(\cos 50)$

(vii) $\sin^{-1}(\sin 100)+\cos^{-1}(\cos 100)$

(viii) $\cos^{-1}(\sin(-5))+\sin^{-1}(\cos(-5))$

Q.10 Find the number of ordered pairs of (x, y) satisfying the equations $y=|\sin x|$ and $y=\cos^{-1}(\cos x)$, where $x\in[-2\pi,2\pi]$

Q.11 Let $f(x)=\cos^{-1}(\cos x)-\sin^{-1}(\sin x)$ in $[0,\pi]$. Find the area bounded by f(x) and x - axis.

Q12. Evaluate the following :

(i) $\tan^{-1}\left(\tan\left(\frac{2\pi}{3}\right)\right)$

(ii) $\tan^{-1}\left(\tan\left(\frac{29\pi}{5}\right)\right)$

(iii) $\tan^{-1}\left(\tan\left(\frac{121\pi}{12}\right)\right)$

(iv) $\tan^{-1}(\tan 3)$

(v) $\tan^{-1}(\tan 5)$

(vi) $\tan^{-1}(\tan 7)$

(vii) $\tan^{-1}(\tan 10)$

(viii) $\tan^{-1}(\tan 20)$

(ix) $\tan^{-1}(\tan 50)$

(x) $\tan^{-1}(\tan 1)+\tan^{-1}(\tan 2)$
$+\tan^{-1}(\tan 3)+\tan^{-1}(\tan 4)$

(xi) $\tan^{-1}(\tan 20)+\tan^{-1}(\tan 40)$
$+\tan^{-1}(\tan 60)+\tan^{-1}(\tan 80)$

(xii) $\sin^{-1}(\sin 15)+\cos^{-1}(\cos 15)+\tan^{-1}(\tan 15)$

(xiii) $\sin^{-1}(\sin 50)+\cos^{-1}(\cos 50)$
$-\tan^{-1}(\tan 50)$

Q13. Prove that each of the following :

(i) $\tan^{-1}\left(\frac{1}{7}\right)+\tan^{-1}\left(\frac{1}{13}\right)=\tan^{-1}\left(\frac{2}{9}\right)$

(ii) $\tan^{-1}(9)+\tan^{-1}\left(\frac{5}{4}\right)=\frac{3\pi}{4}$

(iii) $\tan^{-1}\left(\frac{3}{4}\right)+\tan^{-1}\left(\frac{3}{5}\right)-\tan^{-1}\left(\frac{8}{19}\right)=\frac{\pi}{4}$

(iv) $\tan^{-1}\left(\frac{3}{4}\right)+\tan^{-1}\left(\frac{1}{5}\right)$
$+\tan^{-1}\left(\frac{1}{7}\right)+\tan^{-1}\left(\frac{1}{8}\right)=\frac{\pi}{4}$

(v) $2\tan^{-1}\left(\frac{1}{3}\right)+\tan^{-1}\left(\frac{1}{7}\right)=\frac{\pi}{4}$

(vi) $2\tan^{-1}\left(\frac{1}{5}\right)+\tan^{-1}\left(\frac{1}{7}\right)+2\tan^{-1}\left(\frac{1}{8}\right)=\frac{\pi}{4}$

(vii) $\sin^{-1}\left(\frac{3}{5}\right)+\sin^{-1}\left(\frac{8}{17}\right)=\sin^{-1}\left(\frac{77}{85}\right)$

(ix) $\cos^{-1}\left(\frac{4}{5}\right)+\cos^{-1}\left(\frac{12}{13}\right)=\cos^{-1}\left(\frac{33}{65}\right)$

(x) $\cos^{-1}\left(\frac{1}{2}\right)+\cos^{-1}\left(-\frac{1}{7}\right)+\cos^{-1}\left(\frac{13}{14}\right)=\pi$

(xi) $\sin^{-1}\left(\frac{1}{5}\right)+\cot^{-1}(3)=\frac{\pi}{4}$.

(xiv) $\left(\frac{\tan^{-1}(1)+\tan^{-1}(2)+\tan^{-1}(3)}{\cot^{-1}(1)+\cot^{-1}(2)+\cot^{-1}(3)}\right)$

(xii) $2\tan^{-1}\left(\frac{3}{2}\right)+\tan^{-1}\left(\frac{12}{5}\right)=\pi$.

Q14. Prove that each of the following inverse trigonometric functions :

(i) $\sin\left(2\sin^{-1}\left(\frac{1}{2}\right)\right)=\frac{\sqrt{3}}{2}$

(ii) $\sin\left(3\sin^{-1}\left(\frac{1}{3}\right)\right)=\frac{23}{27}$

(iii) $\cos\left(\frac{1}{2}\cos^{-1}\left(\frac{1}{8}\right)\right)=\frac{3}{4}$

(iv) $\cos\left(\frac{1}{2}\cos^{-1}\left(-\frac{1}{10}\right)\right)=\frac{3\sqrt{5}}{10}$

(v) $\sin\left(\frac{1}{2}\cos^{-1}\left(\frac{1}{9}\right)\right)=\frac{2}{3}$

(vi) $\sin\left(\frac{1}{4}\tan^{-1}\left(\sqrt{63}\right)\right)=\frac{1}{2\sqrt{2}}$

(vii) $\cos\left(\frac{1}{4}\left(\tan^{-1}\left(\frac{24}{7}\right)\right)\right)=\frac{3}{\sqrt{10}}$

(viii) $\tan\left(\frac{1}{2}\cos^{-1}\left(\frac{2}{3}\right)\right)=\frac{1}{\sqrt{5}}$

(ix) $\tan\left(2\tan^{-1}\left(\frac{1}{5}\right)-\frac{\pi}{4}\right)=-\frac{7}{17}$

(x) $\tan\left(\frac{3\pi}{4}-\frac{1}{4}\sin^{-1}\left(-\frac{4}{5}\right)\right)=\frac{1-\sqrt{5}}{2}$

Q15. Find the integral values of x satisfying the inequation $x^2-3x<\sin^{-1}(\sin 2)$.

Q16. Find the value of x satisfying the inequation $3x^2+8x<2\sin^{-1}(\sin 4)-\cos^{-1}(\cos 4)$.

Q17. For what value of x,

$$f(x)=\cos^{-1}x+\cos^{-1}\left\{\frac{x}{2}+\frac{\sqrt{1-3x^2}}{2}\right\}$$

is a constant function.

Q18. Write the simplest form of each of the following inverse trigonometric functions :

(i) $\cos^{-1}\sqrt{1-x^2}$

(ii) $\cos^{-1}(1-2x^2)$

(iii) $\cos^{-1}(2x^2-1)$

(iv) $\tan^{-1}\left(\frac{\cos x}{1+\sin x}\right),\ -\frac{\pi}{2}<x<\frac{\pi}{2}$

(v) $\tan^{-1}\left(\frac{\cos x}{1-\sin x}\right),\ -\frac{\pi}{2}<x<\frac{\pi}{2}$

(vi) $\tan^{-1}\left(\frac{\cos x-\sin x}{\cos x+\sin x}\right), x<\pi$

(vii) $\tan^{-1}\left(\frac{x}{\sqrt{a^2-x^2}}\right)$, -a < x < a

(viii) $\sin^{-1}\left(\sqrt{\frac{x}{1+x}}\right)$

(ix) $\tan^{-1}\left(\frac{a\cos x-b\sin x}{b\cos x+a\sin x}\right)$

(x) $\tan^{-1}\left(\frac{\sqrt{1+x^2}-1}{x}\right)$

(xi) $\tan^{-1}\left(\sqrt{\frac{1-\cos x}{1+\cos x}}\right), x<\pi$

(xii) $\tan^{-1}\left(\frac{3a^2x-x^3}{a^3-3ax^2}\right)$

(xiii) $\cot^{-1}\left(\frac{\sqrt{1+\sin x}+\sqrt{1-\sin x}}{\sqrt{1+\sin x}-\sqrt{1-\sin x}}\right)$

(xiv) $\sin^{-1}\left(x\sqrt{1-x}-\sqrt{x}\sqrt{1-x}\right)$

(xv) $\sin^{-1}\left(\frac{\sin x+\cos x}{\sqrt{2}}\right), -\frac{\pi}{4}<x<\frac{\pi}{4}$

(xvi) $\cos^{-1}\left(\frac{\sin x+\cos x}{\sqrt{2}}\right), \frac{\pi}{4}<x<\frac{5\pi}{4}$

(xvii) $\tan^{-1}\left(\frac{\sqrt{1-x^2}+\sqrt{1-x^2}}{\sqrt{1-x^2}-\sqrt{1-x^2}}\right)$

(xviii) $\sin^{-1}\left(\frac{3}{5}\cos x+\frac{4}{5}\sin x\right)$.

19. Prove that

(i) $\tan^{-1}\left(\frac{\sqrt{1+x^2}+\sqrt{1-x^2}}{\sqrt{1+x^2}-\sqrt{1-x^2}}\right)=\frac{\pi}{4}+\frac{1}{2}\cos^{-1}x^2$

(ii) $\tan\left(\frac{\pi}{4}+\frac{1}{2}\cos^{-1}\frac{a}{b}\right)+\tan\left(\frac{\pi}{4}-\frac{1}{2}\cos^{-1}\frac{a}{b}\right)=\frac{2b}{a}$

(iii) $\tan^{-1}\frac{p-q}{1+pq}+\tan^{-1}\frac{q-r}{1+qr}+\tan^{-1}\frac{r-p}{1+rp}$ $= \pi$, where $p>q>0$ and $pr<-1<qr$,

(iv) $\cot^{-1}\left(\frac{ab+1}{a-b}\right)+\cot^{-1}\left(\frac{bc+1}{b-c}\right)+\cot^{-1}\left(\frac{ac+1}{c-a}\right)=0.$

(v) $\tan\left(\frac{1}{2}\sin^{-1}\left(\frac{2x}{1+x^2}\right)+\frac{1}{2}\cos^{-1}\left(\frac{1-y^2}{1+y^2}\right)\right)=\frac{x+y}{1-xy}$, $xy<1$.

(vi). $\tan^{-1}\left(\frac{1-x}{1+x}\right)-\tan^{-1}\left(\frac{1-y}{1+y}\right)=\sin^{-1}\left(\frac{y-x}{\sqrt{1+x^2}\sqrt{1+y^2}}\right)$

(vii) $\tan^{-1}\left(\frac{1}{2}\tan 2A\right)+\tan^{-1}(\cot A)+\tan^{-1}(\cot^3 A)=0$

(viii) $2\tan^{-1}\left[\sqrt{\frac{a-b}{a+b}}\tan\left(\frac{\theta}{2}\right)\right]=\cos^{-1}\left(\frac{b+a\cos\theta}{a+b\cos\theta}\right)$

(ix) $\tan(2\tan^{-1}a)=2\tan(\tan^{-1}a+\tan^{-1}a^3)$

(x) $\cos^{-1}x+\cos^{-1}\left(\frac{x}{2}+\frac{\sqrt{3-3x^2}}{2}\right)=\frac{\pi}{3}$, $\frac{1}{2}<x<1$

20. If $\sin^{-1}x+\sin^{-1}y+\sin^{-1}z=\pi$, then prove that

$$x\sqrt{1-x^2}+y\sqrt{1-y^2}+z\sqrt{1-z^2}=2xyz$$

21 If $\cos^{-1}x+\cos^{-1}y+\cos^{-1}z=\pi$, then prove that, $x^2+y^2+z^2+2xyz=1$

22 If $\cos^{-1}\left(\frac{x}{2}\right)+\cos^{-1}\left(\frac{y}{3}\right)=\theta$, then prove that, $9x^2-12xy\cos\theta+4y^2=36\sin^2\theta$.

23 If $\sin^{-1}x+\sin^{-1}y+\sin^{-1}z=\frac{3\pi}{2}$, then prove that, $x^2+y^2+z^2-2xyz=1$.

24 If $\sin^{-1}x+\sin^{-1}y+\sin^{-1}z=\frac{3\pi}{2}$, then prove that $xy+yz+zx=3$.

25 ...... $^{-1}x+\sin^{-1}y+\sin^{-1}z=\frac{3\pi}{2}$, then find the value of

$$x^{2012}+y^{2012}+z^{2012}-\frac{9}{x^{2013}+y^{2013}+z^{2013}}$$

26 If $\cos^{-1}x+\cos^{-1}y+\cos^{-1}z=3\pi$, then prove that, $xy+yz+zx=3$.

27. If $\cos^{-1}x+\cos^{-1}y+\cos^{-1}z=3\pi$, then find the value of

$$\left(\frac{x^{2013}+y^{2013}+z^{2013}+6}{x^{2014}+y^{2014}+z^{2014}}\right).$$

28 If $\tan^{-1}x+\tan^{-1}y+\tan^{-1}z=\frac{\pi}{2}$, then prove that $xy+yz+zx=1$.

29. If $\tan^{-1}x+\tan^{-1}y=\frac{\pi}{4}$, then prove that $x+y+xy=1$.

30. If $\tan^{-1}x+\tan^{-1}y+\tan^{-1}z=\pi$, then prove that $x+y+z=xyz$

31 If $\tan^{-1}\left(\frac{\sqrt{1+x^2}-\sqrt{1-x^2}}{\sqrt{1+x^2}+\sqrt{1-x^2}}\right)=\alpha$, then prove that $x^2=\sin 2\alpha$

32. Let $m=\tan^2(\sec^{-1}2)+\cot^2(\operatorname{cosec}^{-1}3)$ Then find the value of (m^2+m+10).

33 $\frac{1}{2}\sin^{-1}\left(\frac{3\sin 2\theta}{5+4\cos 2\theta}\right)=\frac{\pi}{4}$, then find the value of $\tan\theta$.

34. Let $m=\frac{(\tan^{-1}1+\tan^{-1}2+\tan^{-1}3)}{(\cot^{-1}1+\cot^{-1}2+\cot^{-1}3)}$, then prove that $(m+2)^{m+1}=64$.

Q. Solve for x :

35. $\tan^{-1}(2x)+\tan^{-1}(3x)=\frac{3\pi}{4}$

36. $\tan^{-1}\left(\frac{x+1}{x-1}\right)+\tan^{-1}\left(\frac{x-1}{x}\right)=\tan^{-1}(-7)$

37. $\sin^{-1}(2x)+\sin^{-1}(x)=\frac{\pi}{3}$

38. $\sin^{-1}\left(\frac{1}{\sqrt{5}}\right)+\cos^{-1}x=\frac{\pi}{4}$

39. $\sin^{-1}(x)+\sin^{-1}(3x)=\frac{\pi}{3}$

40. $\tan^{-1}\left(\frac{1}{1+2x}\right)+\tan^{-1}\left(\frac{1}{1+4x}\right)=\tan^{-1}\left(\frac{2}{x^2}\right)$

41 $2\tan^{-1}(2x+1)=\cos^{-1}x$

42. $\cos^{-1}x-\sin^{-1}x=\cos^{-1}(x\sqrt{3})$

43. If $\tan^{-1}y:\tan^{-1}x=4:1$, express y as an algebraic function of x.

Hence, prove that $\tan\left(\frac{\pi}{8}\right)$ is a root of $x^4+1=6x^2$

ANSWER

1. (i). $[1, 2]$ (ii). $\left(-\infty, \frac{1}{2}\right]$ (iii). $\left[-\frac{5}{3}, -1\right]$

(iv). $[-3, 3]$ (v). $[-6, 6]$ (vi). $[-5, 5]$

(vii) $[-1, 1]$ (viii) $\left[0, \frac{1}{2}\right]$ (ix) $[-\sqrt{3}, 1]$

(x) $x=(4n+1)\frac{\pi}{2}, n\in I$ (xi) $x=\{-1, 1\}$ (xii) $x=\phi$

(xiii) $[-2, -1]$ (xiv) $[-2, -1]\cup[1, 2]$ (xv) $\left[-1, -\frac{1}{\sqrt{3}}\right]\cup\left[\frac{1}{\sqrt{3}}, 1\right]$

(xvi) $x=\phi$ (xvii) $\left[-\frac{1}{\sqrt{2}}, -\sqrt{2}\right]\cup\left[\frac{1}{\sqrt{2}}, \sqrt{2}\right]$ (xviii) $(0, 4]$

(xix) $(0,1)\cup(1, \infty)$ (xx) $[1, 2]$

2 (i) $x=\frac{\pi}{2}$ (ii) $x=1$, $y=1/2$ (iii) $x=1$

(iv) $x=1$ (v) $x=3/2$ (vi) $x=\left(\frac{1}{\sqrt{2}}, 1\right]$

(vii) $x=\pm\sqrt{2}$ (viii) $x=\sqrt{3}, -\frac{1}{\sqrt{3}}$ (ix) $x=\frac{\sqrt{3}-1}{2\sqrt{2}}$

(x) $x=\sqrt{2}-1$ (xi) $x=0$ (xii) $x=\dfrac{-1\pm\sqrt{5}}{2}$

(xiii) $\left(-2,-\dfrac{3}{2}\right)$ (xiv) $(-1, 1)$ (xv) $(\cos 1, \sin 1)$

(xvi) $(\cot 1, \tan 1)$ (xvi) $[\tan(\sin(\sin 1)), \tan(\sin 1))$

(xvii) $[\tan(\sin(\cos(\sin 1))), \tan(\sin(\cos(\sin 1)))]$ (xviii) $x=-1$

(xix) $x=1/2$ (xx) $x=1/2$ (xxi) $(-\infty, 0)\cup(4, \infty)$

(xxii) $x=\pm\sqrt{\dfrac{\sqrt{5}-1}{2}}$ (xxiii) No solution (xxiv)

(xxiv) $x=\pm 2$ (xxv) $x=-13/12$ (xxvi) $x=8$.

(xxvii) $k=5$

3. (i) $\left[1, \dfrac{1}{2}\right]$ (ii) $(-\infty, 1]$ (iii) $[-1, 0)$

(iv) $[-1, 0)$ (v) $(\tan 2, \infty)$ (vi) $(\tan 3, \cot 2)$

(vii) $\left[-1, -\dfrac{1}{\sqrt{2}}\right)$ (viii) $\left(\dfrac{1}{2}, 1\right]$ (ix) $\left(-\dfrac{1}{\sqrt{2}}, 0\right)\cup\left(\dfrac{1}{\sqrt{2}}, 1\right)$

(x) $\left(-\infty, -\dfrac{1}{\sqrt{3}}\right)\cup\left(0, \dfrac{1}{\sqrt{3}}\right)$ (xi) $\left[-1, \dfrac{1}{2}\right]$ (xii) $\left(1-\sqrt{\pi-1}, 1+\sqrt{\pi-1}\right)$

(xiii) $\left(-\dfrac{8}{5}, 0\right)$ (xiv) $(-\infty, -1)\cup(1, \infty)$ (xv) $-\sqrt{\dfrac{5}{6}}\le x\le\sqrt{\dfrac{5}{6}}$

4. (i) $\left[-\dfrac{\pi}{2}, \dfrac{\pi}{2}\right]$ (ii) $\left[-\dfrac{5\pi}{4}, \dfrac{3\pi}{4}\right]$ (iii) $[-\pi, \pi]$

(iv) $\left(-\dfrac{\pi}{2}, 0\right)$ (v) $(0, \pi)$ (vi) $(0, \pi)$

(vii) $(0, \pi)$ (viii) $\left(\dfrac{5\pi}{4}, \dfrac{9\pi}{4}\right)$ (ix) $[1. 2]$

(x) $\{\text{cosec}^{-1}(1)\}$

5. (i) 0 (ii) . 1 (iii) 4.

6. (i) $\dfrac{\pi}{6}$ (ii) $-\dfrac{\pi}{4}$ (iii) $\dfrac{3\pi}{4}$

(iv) $-\frac{\pi}{6}$ (v) $\frac{\pi}{6}$ (vi) $\frac{5\pi}{6}$

(vii) 1 (viii) $\frac{\sqrt{3}}{2}$ (ix) -1

(x) $\frac{1}{2\sqrt{2}}$ (xi) $-\sqrt{3}$ (xii) $\left(\frac{1+3\sqrt{5}}{8}\right)$

(xiii) $\left(\frac{\sqrt{3}+1}{2\sqrt{2}}\right)$ (xiv) $\frac{1}{\sqrt{3}}$ (xv) $\frac{\pi}{6}$

(xvi) $4/5$ (xvii) $2\sqrt{2}$ (xviii) $\left(\frac{2}{\sqrt{5}}-\frac{4}{3\sqrt{5}}\right)$

7. (i) $-\frac{\pi}{3}$ (ii) $-\frac{\pi}{3}$ (iii) $\frac{\pi}{8}$

(iv) $(\pi-3)$ (v) $(5-2\pi)$ (vi) $(7-2\pi)$

(vii) $(3\pi-10)$ (viii) $(12-4\pi)$ (ix) $(20-6\pi)$

(x) $(50-16\pi)$ (xi) $(25\pi-80)$ (xii) $(32\pi-100)$

8. (i) $\frac{5\pi}{6}$ (ii) $\frac{3\pi}{7}$ (iii) $\frac{5\pi}{7}$

(iv) 3 (v) $(2\pi-5)$ (vi) $(7-2\pi)$

(vii) $(4\pi-10)$ (viii) $(4\pi-12)$ (ix) $(15-4\pi)$

(x) $(40-12\pi)$ (xi) $(32\pi-100)$ (xii) $(32\pi-100)$

9. (i) $(2\pi-4)$ (ii) 0 (iii) $(10-\pi)$

(iv) $20-4\pi$ (v) $7\pi-20$ (vi) 0

(vii) 0 (viii) $-\pi$

10. 3 11. π^2.

12. (i) $-\frac{\pi}{3}$ (ii) $-\frac{\pi}{5}$ (iii) $\frac{\pi}{12}$ (iv) $(3-\pi)$

(v) $(5-2\pi)$ (vi) $(7-2\pi)$ (vii) $(10-3\pi)$ (viii) $(20-6\pi)$

(ix) $(50-16\pi)$ (x) $(10-3\pi)$ (xi) $(200-63\pi)$ (xii) $(15-4\pi)$

(xiii) $(16\pi-50)$ (xiv) 2.

15. $x \in \left(\frac{3-\sqrt{1+4\pi}}{2}, \frac{3+\sqrt{1+4\pi}}{2}\right)$ 16. $x \in \left(-2, -\frac{2}{3}\right)$

17. $x \in \left[\frac{1}{2}, 1\right]$ 25. 0 32. 142

33. 3 35. $x = 1$ 36. $x = 2$

37. $x = 2\sqrt{\frac{3}{7}}$ 38. $\frac{3}{\sqrt{10}}$ 39. $\frac{1}{2}\sqrt{\frac{3}{13}}$

40. $x = 3$ 41. $x = 0$ 42. $x = 0, \pm\left(\frac{1}{2}\right)$

Level– II FOR JEE-MAIN AND ADVANCED EXAM ONLY

1. The set of values of k for which $x^2 - kx + \sin^{-1}(\sin 4) > 0$ for all real x is
   (a) {0} (b) (-2,2)
   (c) R (d)* None of these.

2. If $x < 0$ then value of
   $\tan^{-1}(x) + \tan^{-1}\left(\frac{1}{x}\right) =$
   (a) $\frac{\pi}{2}$ (b)* $-\frac{\pi}{2}$
   (c) 0 (d) None of these.

3. If $\sin^{-1} x + \sin^{-1} y = \frac{2\pi}{3}$, then $\cos^{-1}x + \cos^{-1}y$ is
   (a) $\frac{2\pi}{3}$ (b)* $\frac{\pi}{3}$
   (c) $\frac{\pi}{6}$ (d) π

4. Let $f(x) = \sin^{-1}x + \cos^{-1}x$. Then $\frac{\pi}{2}$ is equal to
   (a)* $f\left(\frac{1}{2}\right)$ (b) $f(k^2-2k+3), k \varepsilon R$
   (c)* $f\left(\frac{1}{1+k^2}\right), k \varepsilon R$ (d) $f(-2)$

5. Which one of the following is correct ?
   (a)* $\tan 1 > \tan^{-1} 1$ (b) $\tan 1 < \tan^{-1} 1$
   (c) $\tan 1 = \tan^{-1} 1$ (d) None of these.

6. If $a \sin^{-1}x - b \cos^{-1}x = c$, then the value of $a \sin^{-1}x + b \cos^{-1}x$ is
   (a) 0 (b) $\frac{\pi ab + c(b-a)}{a+b}$
   (c) $\frac{\pi ab - c(b-a)}{a+b}$ (d) $\frac{\pi}{2}$

7. The number of solutions of the equation
   $\sin^{-1}(1-x) - 2\sin^{-1}x = \frac{\pi}{2}$
   (a) 0 (b) 1
   (c) 2 (d) More than two.

8. The smallest and the largest values of
   $\tan^{-1}\left(\frac{1-x}{1+x}\right), 0 \le x \le 1$ are
   (a) $0, \pi$ (b) $0, \frac{\pi}{4}$
   (c) $-\frac{\pi}{4}, \frac{\pi}{4}$ (d) $\frac{\pi}{4}, \frac{\pi}{2}$.

9. The equation $\sin^{-1}x - \cos^{-1}x = \cos^{-1}\left(\frac{\sqrt{3}}{2}\right)$ has
   (a) No solution

(b) Unique Solution

(c) Infinite No of soln (d)None

10. If $-\pi \le x \le 2\pi$, then $\cos^{-1}(\cos x)$ is

(a) x (b) $\pi - x$

(c) $2\pi + x$ (d) $2\pi - x$.

11. If $\sin^{-1} x + \cot^{-1}(\frac{1}{2}) = \frac{\pi}{2}$, then x is equal to

(a) 0 (b) $\frac{1}{\sqrt{5}}$

(c) $\frac{2}{\sqrt{5}}$ (d) $\frac{\sqrt{3}}{2}$

12. If $\cos[\tan^{-1}\{\sin(\cot^{-1}\sqrt{3})\}] = y$,

then the value of y is

(a) $y = \frac{4}{5}$ (b)* $y = \frac{2}{\sqrt{5}}$

(c) $y = -\frac{2}{\sqrt{5}}$ (d) $\frac{\sqrt{3}}{2}$

13. If $x = \frac{1}{5}$, then the value of

$\cos(\cos^{-1}x + 2\sin^{-1}x)$ is

(a) $\sqrt{\frac{24}{25}}$ (b) $-\sqrt{\frac{24}{25}}$

(c) $\frac{1}{5}$ (d)* $-\frac{1}{5}$

14. $\tan^{-1}(\frac{1}{2}) + \tan(\frac{1}{3})$ is equal to

(a)* $\frac{\pi}{4}$ (b) $\frac{\pi}{2}$

(c) $\frac{\pi}{3}$ (d) None of these.

15. $\tan^{-1} a + \tan^{-1} b$, where $a > 0, b > 0$

$ab > 1$ is equal to

(a) $\tan^{-1}\left(\frac{a+b}{1-ab}\right)$ (b) $\tan^{-1}\left(\frac{a+b}{1-ab}\right) - \pi$

(c)* $\pi + \tan^{-1}\left(\frac{a+b}{1-ab}\right)$ (d) $\pi - \tan^{-1}\left(\frac{a+b}{1-ab}\right)$

16. A solution to the equation

$\tan^{-1}(1+x) + \tan^{-1}(1-x) = \frac{\pi}{2}$ is

(a) $x = 1$ (b) $x = -1$

(c) $x = 0$ (d) $x = \pi$

17. All possible values of p and q for which

$\cos^{-1}\sqrt{p} + \cos^{-1}\sqrt{1-p} + \cos^{-1}\sqrt{1-q}$

$= \frac{3\pi}{4}$ holds , is

(a) $p = 1, q = 1/2$

(b) $q>1, p = 1/2$

(c)* $0<p<1, q = 1/2$

(d) None.

18. $\tan\left(\frac{\pi}{4} + \frac{1}{2}\cos^{-1}x\right) +$

$\tan\left(\frac{\pi}{4} - \frac{1}{2}\cos^{-1}x\right), x \neq 0,$ is equal to

(a) x (b) 2x

(c)* $\frac{2}{x}$ (d) $\frac{x}{2}$.

19. the value of $\cot^{-1} 3 + \text{cosec}^{-1}\sqrt{5}$ is

(a) $\frac{\pi}{2}$ (b) $\frac{\pi}{3}$

(c) $\frac{\pi}{4}$ (d) $\frac{\pi}{6}$

20. If $\sum_{i=1}^{2n} \sin^{-1}x_i = n\pi$, then $\sum_{i=1}^{2n} x_i$ is

(a)* n (b) 2n

(c) $\frac{n(n+1)}{2}$ (d) $\frac{n(n-1)}{2}$

21. If $u = \cot^{-1}\sqrt{\tan\alpha} - \tan^{-1}\sqrt{\tan\alpha}$, then

$\tan\left(\frac{\pi}{4} - \frac{u}{2}\right)$ is equal to

(a)* $\sqrt{\tan\alpha}$ (b) $\sqrt{\cot\alpha}$

(c) $\tan\alpha$ (d) $\cot\alpha$.

22. The value of $\tan^{-1}\left(\frac{a}{b+c}\right)+\tan^{-1}\left(\frac{b}{a+c}\right)$, if $\angle C = 90$, in triangle ABC is

(a)* $\frac{\pi}{4}$ (b) $\frac{\pi}{3}$

(c) $\frac{\pi}{2}$ (d) π

23. If $\cot^{-1}\frac{n}{\pi} > \frac{\pi}{6}$, $n \varepsilon N$, then the maximum value of 'n' is

(a) 1 (b) 5

(c) 9 (d) None of these.

24. $\sin^{-1}x > \cos^{-1}x$ holds for

(a) all values of x (b) $x \varepsilon \left(0, \frac{1}{\sqrt{2}}\right)$

(c)* $\left(\frac{1}{\sqrt{2}}, 1\right)$ (d) $x = 0.75$.

25. The value of $\cos\left(\frac{1}{2}\cos^{-1}\frac{1}{8}\right)$ is equal to

(a) $\frac{3}{4}$ (b) $-\frac{3}{4}$

(c) $\frac{1}{16}$ (d) 4.

26. The values of x satisfying

$$\tan\left(\sec^{-1}x\right)=\sin\left(\cos^{-1}\left(\frac{1}{\sqrt{5}}\right)\right) \text{ is}$$

(a) $\pm\frac{\sqrt{5}}{3}$ (b)* $\pm\frac{3}{\sqrt{5}}$

(c) $\pm\frac{\sqrt{3}}{5}$ (d) $\pm\frac{3}{5}$

27. If $\left(\tan^{-1}x\right)^2+\left(\cot^{-1}x\right)^2=\frac{5\pi^2}{8}$, then the value of x is

(a) 0 (b)* -1

(c) -2 (d) -3.

28. The number of real solutions of $\cos^{-1}x+\cos^{-1}2x=-\pi$ is

(a)* 0 (b) 1

(c) 2 (d) infinitely many.

If $\theta = \tan^{-1}\left(\sqrt{\frac{a(a+b+c)}{bc}}\right) + \tan^{-1}\left(\sqrt{\frac{b(a+b+c)}{ca}}\right) + \tan^{-1}\sqrt{\frac{c(a+b+c)}{ab}}$, then the value of $\tan\theta$ is

(a)* 0 (b) 1

(c) -1 (d) None.

30. The set of values of x satisfying the inequation $\tan^2\left(\sin^{-1}x\right)>1$ is

(a) [-1, 1] (b) $\left[-\frac{1}{\sqrt{2}}, \frac{1}{\sqrt{2}}\right]$

(c)* $(-1,1)-\left[-\frac{1}{\sqrt{2}}, \frac{1}{\sqrt{2}}\right]$

(d) $[-1,1]-\left(-\frac{1}{\sqrt{2}}, \frac{1}{\sqrt{2}}\right)$

31. The value of a for which

$$ax^2+\sin^{-1}\left(x^2-2x+2\right)+\cos^{-1}\left(x^2-2x+2\right)=0$$

has a real solution, is

(a) $\pi/2$ (b)* $-\pi/2$

(c) $2/\pi$ (d) $-2/\pi$

32. The value of

$$\sin^{-1}\left[\cot\left(\sin^{-1}\sqrt{\left(\frac{2-\sqrt{3}}{4}\right)}+\cos^{-1}\left(\frac{\sqrt{12}}{4}\right)+\sec^{-1}\sqrt{2}\right)\right]$$

is

(a)* 0 (b) $\pi/4$

(c) $\pi/6$ (d) $\pi/2$

33. The number of positive integral solutions of

$\tan^{-1}x+\cot^{-1}\left(\frac{1}{y}\right)=\sin^{-1}\left(\frac{3}{\sqrt{10}}\right)$ is

(a)* 0 (b) 1

(c) 2 (d) 3

34. The value of

$\cos\left[\frac{1}{2}\cos\left\{\cos\left(\sin^{-1}\left(\frac{\sqrt{63}}{8}\right)\right)\right\}\right]$ is

(a) $\frac{3}{16}$ (b) $\frac{3}{8}$

(c)* $\frac{3}{4}$ (d) $\frac{3}{2}$

35. If $\tan^{-1}x+\tan^{-1}y+\tan^{-1}z=\pi$, then the value of $\frac{1}{yz}+\frac{1}{zx}+\frac{1}{xy}$ is

(a) 0 (b)* 1

(c) $\frac{1}{xyz}$ (d) xyz.

36. If $x<0$, then $\tan^{-1}\left(\frac{1}{x}\right)$ is

(a) $\cot^{-1}(x)$ (b) $-\cot^{-1}(x)$

(c)* $-\pi+\cot^{-1}(x)$ (d) None

37. The number of triplets satisfying $\sin^{-1}x+\cos^{-1}y+\sin^{-1}z=2\pi$, is

(a) 0 (b) 2

(c)* 1 (d) infinite.

38. If $x^2+y^2+z^2=r^2$, then

$\tan^{-1}\left(\frac{xy}{zr}\right)+\tan^{-1}\left(\frac{zy}{xr}\right)+\tan^{-1}\left(\frac{zx}{yr}\right)$ is equal to ....

(a) π (b)* $\pi/2$

(c) 0 (d) None.

39. If $\tan^{-1}x+\tan^{-1}2x+\tan^{-1}3x=\pi$, then the value of x is

(a) 0 (b) -1

(c)* 1 (d) ϕ

40. The number of solutions of the equation $1+x^2+2x\sin\left(\cos^{-1}y\right)=0$ is

(a)* 1 (b) 2

(c) 3 (d) 4

41. If α is the only real root of the equation $x^2+bx^2+cx+1=0$ $(b<c)$, then the value of $\tan^{-1}\alpha+\tan^{-1}\left(\frac{1}{\alpha}\right)$ is equal to

(a) $\pi/2$ (b)* $-\pi/2$

(c) 0 (d) None.

42. If α, β, γ are the roots of $x^3+px^2+2x+p=0$, the the general value of $\tan^{-1}\alpha+\tan^{-1}\beta+\tan^{-1}\gamma$ is

(a)* $n\pi$ (b) $n\pi/2$

(c) $(2n+1)\pi/2$ (d) depend on p.

43.. If $\left[\sin^{-1}\left(\cos^{-1}\left(\sin^{-1}\left(\tan^{-1}x\right)\right)\right)\right]=1$, where [,] = G. I. F, the value of x lies in

(a)* $[\tan\sin\cos 1, \tan\sin\cos\sin 1]$

(b) $(\tan\sin\cos 1, \tan\sin\cos\sin 1)$

(c) $[-1, 1]$

(d) $[\sin\cos\tan 1, \sin\cos\sin\tan 1]$

29. Let a, b, c be positive real numbers and

$\theta=\tan^{-1}\sqrt{\frac{a(a+b+c)}{bc}}+\tan^{-1}\sqrt{\frac{b(a+b+c)}{ac}}+\tan^{-1}\sqrt{\frac{c(a+b+c)}{ab}}$, then the value of $\tan\theta$ is

(a)* 0 (b) 1

(c) -1 (d) None.

30. The set of values of x satisfying the inequation $\tan^2\left(\sin^{-1}x\right)>1$ is

(a) [-1, 1] (b) $\left[-\frac{1}{\sqrt{2}}, \frac{1}{\sqrt{2}}\right]$

(c)* $(-1,1)-\left[-\frac{1}{\sqrt{2}}, \frac{1}{\sqrt{2}}\right]$

(d) $[-1,1]-\left(-\frac{1}{\sqrt{2}}, \frac{1}{\sqrt{2}}\right)$

31. The value of a for which $ax^2+\sin^{-1}(x^2-2x+2)+\cos^{-1}(x^2-2x+2)=0$ has a real solution, is

(a) $\pi/2$ (b)* $-\pi/2$

(c) $2/\pi$ (d) $-2/\pi$

32. The value of

$$\sin^{-1}\left[\cot\left(\sin^{-1}\sqrt{\left(\frac{2-\sqrt{3}}{4}\right)}+\cos^{-1}\left(\frac{\sqrt{12}}{4}\right)+\sec^{-1}\sqrt{2}\right)\right]$$

is

(a)* 0 (b) $\pi/4$

(c) $\pi/6$ (d) $\pi/2$

33. The number of positive integral solutions of $\tan^{-1}x+\cot^{-1}\left(\frac{1}{y}\right)=\sin^{-1}\left(\frac{3}{\sqrt{10}}\right)$ is

(a)* 0 (b) 1

(c) 2 (d) 3

34. The value of

$$\cos\left[\frac{1}{2}\cos\left\{\cos\left(\sin^{-1}\left(\frac{\sqrt{63}}{8}\right)\right)\right\}\right]$$ is

(a) $\frac{3}{16}$ (b) $\frac{3}{8}$

(c)* $\frac{3}{4}$ (d) $\frac{3}{2}$

35. If $\tan^{-1}x+\tan^{-1}y+\tan^{-1}z=\pi$, then the value of $\frac{1}{yz}+\frac{1}{zx}+\frac{1}{xy}$ is

(a) 0 (b)* 1

(c) $\frac{1}{xyz}$ (d) xyz.

ANSWER

1. (d)	**2.** (b)	**3.** (b)	**4.** (b)	**5.** (d)	**6.** (b)	**7.** (b)	**8.** (d)	**9.** (a)	**10.** (c)
11. (c)	**12.** (c)	**13.** (a, c)	**14.** (a)	**15.** (d)	**16.** (b)	**17.** (b)	**18.** (c)	**19.** (c)	**20.** (b)
21. (a)	**22.** (a)	**23.** (b)	**24.** (c, d)	**25.** (a)	**26 (b)**	**27 (b)**	**28 (a)**	**29 (a)**	**30 (c)**
31 (b)	**32 (a)**	**33 (a)**	**34 (c)**	**35 (b)**	**36 (c)**	**37 (c)**	**38 (b)**	**39 (c)**	**40 (a)**
41 (b)	**42 (a)**	**43 (a)**							

Level– III FOR JEE - ADVANCED EXAM ONLY

1. Prove that $\sin^{-1}(\cos(\sin^{-1}x))+\cos^{-1}(\sin(\cos^{-1}x))=\frac{\pi}{2}$

2. Prove that $\tan^{-1}\{\operatorname{cosec}(\tan^{-1}x)-\tan(\cot^{-1}x)\}=\tan^{-1}x$, where $x\neq 0$

3. Prove that $\tan(\tan^{-1}x+\tan^{-1}y+\tan^{-1}z)=\cot(\cot^{-1}x+\cot^{-1}y+\cot^{-1}z)$.

4. Prove that $\sin(\cot^{-1}(\tan(\cos^{-1}x)))=x$, $\forall x\in(0,1]$

5. Prove that $\sin(\operatorname{cosec}^{-1}(\cot(\tan^{-1}x)))=x$, $\forall x\in(0,1]$

6. Find the value of $\sin^{-1}(\sin 5)+\cos^{-1}(\cos 10)+\tan^{-1}(\tan(-6))$

$+\cot^{-1}(\cot(-10))$

7. Find the simplest value of

$$\cos^{-1} x+\cos^{-1}\left(\frac{x}{2}+\frac{\sqrt{3-3x^2}}{2}\right),$$

$$\forall x \in\left(\frac{1}{2},1\right)$$

8. If U= $\cot^{-1}\left(\sqrt{\cos 2\theta}\right)-\tan^{-1}\left(\sqrt{\cos 2\theta}\right)$, then prove that Sin U= $\tan^2\theta$

9. Prove that $\tan\left(\frac{\pi}{4}+\frac{1}{2}\cos^{-1}\frac{a}{b}\right)$

$$+\tan\left(\frac{\pi}{4}-\frac{1}{2}\cos^{-1}\frac{a}{b}\right)=\frac{2b}{a}$$

10. Prove that $\cos^{-1}\left(\frac{\cos x+\cos y}{1+\cos x\cos y}\right)$

$$=2\tan^{-1}\left(\tan\left(\frac{x}{2}\right)\tan\left(\frac{y}{2}\right)\right)$$

11 Prove that $2\tan^{-1}\left(\sqrt{\frac{a-b}{a+b}}\tan\left(\frac{x}{2}\right)\right)$

$$=\cos^{-1}\left(\frac{b+a\cos x}{a+b\cos x}\right).$$

12 If $\tan^{-1}x, \tan^{-1}y, \tan^{-1}z$ are in A.P. then prove that $y^2(x+z)+2y(1-xz)=x+z$, where $y\in(0,1)$, $xz<1$, $x>0$ and $z>0$.

13. Prove that

$$\sin^{-1}\left(\sin\frac{33\pi}{7}\right)+\cos^{-1}\left(\cos\frac{46\pi}{7}\right)$$

$$+\tan^{-1}\left(-\tan\frac{13\pi}{8}\right)+\cot^{-1}\left(\cot\left(-\frac{19\pi}{8}\right)\right)$$

$$=\frac{13\pi}{7}$$

14 If $\cos^{-1}\left(\frac{x}{a}\right)+\cos^{-1}\left(\frac{y}{b}\right)=\alpha$, then prove that, $\frac{x^2}{a^2}-\frac{2xy}{ab}\cos\alpha+\frac{y^2}{b^2}=\sin^2\alpha$.

15 If $\sin^{-1}x+\sin^{-1}y+\sin^{-1}z=\pi$, then prove that,

$$x\sqrt{1-x^2}+y\sqrt{1-y^2}+z\sqrt{1-z^2}=2xyz.$$

16. Find the greatest and least value of the function $f(x)=\left(\sin^{-1}x\right)^3+\left(\cos^{-1}x\right)^3$.

17. Solve for x: $\sin^{-1}x+\sin^{-1}2x=\frac{\pi}{2}$

18. Solve for x:

$$\tan^{-1}\left(\frac{1}{1+2x}\right)+\tan^{-1}\left(\frac{1}{1+4x}\right)$$

$$=\tan^{-1}\left(\frac{2}{x^2}\right)$$

19 Solve for x :.

$$\tan^{-1}(x-1)+\tan^{-1}(x)+\tan^{-1}(x+1)$$

$$=\tan^{-1}(3x)$$

20 Solve for x :. $\sin^{-1}\left(\frac{1}{\sqrt{5}}\right)+\cos^{-1}x=\frac{\pi}{4}$.

21 Solve for x :.

$$\cos^{-1}\left(\frac{x^2-1}{x^2+1}\right)+\tan^{-1}\left(\frac{2x}{x^2-1}\right)=\frac{2\pi}{3}.$$

22. Solve for x :. $2\tan^{-1}x=$

$$\cos^{-1}\left(\frac{1-a^2}{1+a^2}\right)-\cos^{-1}\left(\frac{1-b^2}{1+b^2}\right), a>0, b\ \ 0$$

23. Solve for x :.

$$\cot^{-1}x+\cot^{-1}(n^2-x+1)=\cot^{-1}(n-1)$$

24. Solve for x :

$$\tan^{-1}\left(\frac{x-1}{x+1}\right)+\tan^{-1}\left(\frac{2x-1}{2x+1}\right)=\tan^{-1}\left(\frac{23}{26}\right).$$

25. Solve for x :.

$$\sec^{-1}\left(\frac{x}{a}\right)-\sec^{-1}\left(\frac{x}{b}\right)=\sec^{-1}b-\sec^{-1}a.$$

26. $\sin^{-1}x+\sin^{-1}y=\frac{2\pi}{3}$,

$\cos^{-1}x-\cos^{-1}y=\frac{\pi}{3}$.

27. Find all the positive integral solutions of

$\tan^{-1}x+\cos^{-1}\left(\frac{y}{\sqrt{1+y^2}}\right)=\sin^{-1}\left(\frac{3}{\sqrt{10}}\right)$

28. Let $\alpha=\sin^{-1}\left(\frac{36}{85}\right)$, $\beta=\cos^{-1}\left(\frac{4}{5}\right)$,

$\gamma=\tan^{-1}\left(\frac{8}{15}\right)$, then find $\alpha+\beta+\gamma$ and hence prove that

(i) $\sum\cot\alpha=\prod\cot\alpha$.

(ii) $\sum\tan\alpha\tan\beta=1$.

29. If $\alpha=2\tan^{-1}\left(\frac{1-x}{1+x}\right)$ and

$\beta=\sin^{-1}\left(\frac{1-x^2}{1+x^2}\right)$ for $0<x<1$, then

prove that $\alpha+\beta=\pi$. What will be the value of $\alpha+\beta$ if $x>1$.

30. If $y=\tan^{-1}\left(\frac{\sqrt{1+x^2}-\sqrt{1-x^2}}{\sqrt{1+x^2}+\sqrt{1-x^2}}\right)$, then

prove that $x^2=\sin 2y$

31. Prove that

$\frac{\beta^3}{2}\operatorname{cosec}^2\left(\frac{1}{2}\tan^{-1}\left(\frac{\beta}{\alpha}\right)\right)$

$+\frac{\alpha^3}{2}\sec^2\left(\frac{1}{2}\tan^{-1}\left(\frac{\alpha}{\beta}\right)\right)$

$=(\alpha+\beta)\left(\alpha^2+\beta^2\right)$.

Solve the following inequalities :

32. $\sin^{-1}x>\cos^{-1}x$

33. $\cos^{-1}x>\sin^{-1}x$

34. $\left(\cot^{-1}x\right)^2-5\left(\cot^{-1}x\right)+6>0$

35. $\tan^2\left(\sin^{-1}x\right)>1$

36. $4\left(\tan^{-1}x\right)^2-8\left(\tan^{-1}x\right)+3<0$

37. $4\cot^{-1}x-\left(\cot^{-1}x\right)^2-3\geq 0$

38. $\sin^{-1}\left(\sin\left(\frac{2x^2+4}{1+x^2}\right)\right)<\pi-2$

39. Find the maximum value of n, if

$\cot^{-1}\left(\frac{n}{\pi}\right)>\frac{\pi}{6}, n\in N$

40. Find the minimum value of n, if

$\cot^{-1}\left(\frac{n^2-10n+21.6}{\pi}\right)>\frac{\pi}{6}, n\in N$

41.. Find the set of values of k for which

$x^2-kx+\sin^{-1}(\sin 4)>0$, for all real x.

42. If A= $2\tan^{-1}\left(2\sqrt{2}-1\right)$ and

B= $3\sin^{-1}\left(\frac{1}{3}\right)+\sin^{-1}\left(\frac{3}{5}\right)$,

then prove that A>B.

43. Prove that

$\sin^{-1}\left\{\cot\left\{\sin^{-1}\left(\sqrt{\frac{2-\sqrt{3}}{4}}\right)+\cos^{-1}\left(\frac{\sqrt{12}}{4}\right)+\sec^{-1}\left(\sqrt{2}\right)\right\}\right\}$

$=0$.

44. Solve for x ;

$\left[\sin^{-1}\left(\cos^{-1}\left(\sin^{-1}\left(\tan^{-1}x\right)\right)\right)\right]=1$,

where [,] = G.I.F

45 Find the interval for which

$2\tan^{-1}x+\sin^{-1}\left(\frac{2x}{1+x^2}\right)$ is independent

of x.

46 Prove that $\cos^{-1}\left(\frac{\cos x+\cos y}{1+\cos x\cos y}\right)$

$= 2\tan^{-1}\left(\tan\left(\frac{x}{2}\right)\tan\left(\frac{y}{2}\right)\right)$.

47. Prove that $2\tan^{-1}\left\{\tan\left(\frac{\alpha}{2}\right)\tan\left(\frac{\pi}{4}-\frac{\beta}{2}\right)\right\}$

$= \tan^{-1}\left(\frac{\sin\alpha\cos\beta}{\cos\alpha+\cos\beta}\right)$, .

where $\alpha, \beta \in \left(0, \frac{\pi}{2}\right)$

48. If $p > q > 0$, $pr < -1 < qr$, then prove that

$\tan^{-1}\left(\frac{p-q}{1+pq}\right)+\tan^{-1}\left(\frac{q-r}{1+qr}\right)+\tan^{-1}\left(\frac{r-p}{1+rp}\right)=\pi$

49. Consider the equation

$\left(\sin^{-1}x\right)^3+\left(\cos^{-1}x\right)^3=\pi a^3$,

find the values of 'a' so that the given equation has a solution.

50. If $x = \operatorname{cosec}\left(\tan^{-1}\left(\cos\left(\cot^{-1}\left(\sec\left(\sin^{-1}a\right)\right)\right)\right)\right)$

and

$y = \sec\left(\cot^{-1}\left(\sin\left(\tan^{-1}\left(\operatorname{cosec}\left(\cos^{-1}a\right)\right)\right)\right)\right)$,

where $a \in [0,1]$, then find the relation between x and y.

51. If $\tan^{-1}y = 4\tan^{-1}x$, $\left(|x| < \tan\left(\frac{\pi}{8}\right)\right)$, find y as an algebraix function of x and hence prove that $\tan\left(\frac{\pi}{8}\right)$ is a root of the equation $x^4-6x^2+1=0$.

52. Prove that

$\tan^{-1}\left(\sqrt{\frac{a(a+b+c)}{bc}}\right)+\tan^{-1}\left(\sqrt{\frac{b(a+b+c)}{ac}}\right)+$

$\tan^{-1}\left(\sqrt{\frac{c(a+b+c)}{ab}}\right)=\pi$

where a, b, c > 0

53. Solve :

$\theta = \tan^{-1}\left(2\tan^2\theta\right)-\frac{1}{2}\sin^{-1}\left(\frac{3\sin 2\theta}{5+4\cos 2\theta}\right)$

54. Simplify:

$\tan^{-1}\left(\frac{x\cos\theta}{1-x\sin\theta}\right)-\cot^{-1}\left(\frac{\cos\theta}{x-\sin\theta}\right)$

55. Solve: $\cos^{-1}\left(\frac{x^2-1}{x^2+1}\right)+\sin^{-1}\left(\frac{2x}{x^2+1}\right)$

$+\tan^{-1}\left(\frac{2x}{x^2-1}\right)=\frac{2\pi}{3}$

56. Prove that

$\tan^{-1}\left(\frac{yz}{xr}\right)+\tan^{-1}\left(\frac{xz}{yr}\right)+\tan^{-1}\left(\frac{xy}{zr}\right)=\frac{\pi}{2}$.

Find the sum of the following series :

57. $\tan^{-1}\left(\frac{1}{3}\right)+\tan^{-1}\left(\frac{1}{7}\right)+\tan^{-1}\left(\frac{1}{13}\right)+\ldots\ldots$

58. $\tan^{-1}\left(\frac{1}{3}\right)+\tan^{-1}\left(\frac{2}{9}\right)+---$

$---+\tan^{-1}\left(\frac{2^{n-1}}{1+2^{n-1}}\right)+--------\text{to}\infty$

59. $\tan^{-1}\left(\frac{1}{n^2+n+1}\right)+\tan^{-1}\left(\frac{1}{n^2+3n+3}\right)$

$+\tan^{-1}\left(\frac{1}{n^2+5n+7}\right)--\text{to - n - terms}$

60 $\tan^{-1}\left(\frac{1}{2}\right)+\tan^{-1}\left(\frac{1}{8}\right)$

$+\tan^{-1}\left(\frac{1}{18}\right)+---\text{to}\infty$

61 $\cot^{-1}(7)+\cot^{-1}(13)$

$+\cot^{-1}(21)+\cot^{-1}(31)+$

........to – n – temrs

62. $\cot^{-1}(2.1^2)+\cot^{-1}(2.2^2)+\cot^{-1}(2.3^2)+\ldots\ldots\ldots\text{to}\infty$

$+\ldots\ldots+\sin^{-1}\left(\frac{\sqrt{n}-\sqrt{n-1}}{\sqrt{n}\times\sqrt{n-1}}\right)+\ldots\ldots\text{to }\infty$

63. $\sin^{-1}\left(\frac{1}{\sqrt{2}}\right)+\sin^{-1}\left(\frac{\sqrt{2}-1}{\sqrt{6}}\right)+\sin^{-1}\left(\frac{\sqrt{3}-\sqrt{2}}{\sqrt{12}}\right)$

64. $\tan^{-1}\left(\frac{c_1x-y}{x+c_1y}\right)+\tan^{-1}\left(\frac{c_2-c_1}{1+c_1c_2}\right)+\tan^{-1}\left(\frac{c_3-c_2}{1+c_3c_2}\right)+\ldots\ldots\ldots+\tan^{-1}\left(\frac{1}{c_n}\right)$

ANSWERS

6. $8\pi-21$

7. $\frac{\pi}{3}$

16. Max Value $\frac{7\pi^3}{8}$, when x = -1 and Min Value = $\frac{\pi^3}{32}$, when $x=\frac{1}{\sqrt{2}}$.

17 $x=\frac{1}{2}\sqrt{\frac{3}{7}}$

18. $x=3$

19. x = 0, 1/2, -1/2

20. $x=\frac{3}{\sqrt{10}}$

21. $x=2-\sqrt{3},\sqrt{3}$

22. $x=\frac{a-b}{a+b}$

23. $x=n^2-n+1$, n

24.. x = 4/3

25. x = ab,

26. x = 1/2 , y = 1.

27. x = 1, y = 2 ; x = 2, y = 7

29. $-\pi$

32. $x\in\left(\frac{1}{\sqrt{2}},1\right]$

33. $x\in\left(-1,\frac{1}{\sqrt{2}}\right]$

34. $x\in(-\infty,\cot 3)\cup(\cot 2,\infty)$

36 $x\in\left(\frac{1}{\sqrt{2}},1\right]$

37. $x\in\left(\frac{1}{\sqrt{2}},1\right)\cup\left(-1,-\frac{1}{\sqrt{2}}\right)$

38. $x\in(-1,1)$

39. n=5

40. n = 3

41. $k=\phi$.

44. $x\in[\tan(\sin(\cos 1)),\tan(\sin(\cos(\sin 1)))]$.

45. $[1,\infty)$

49. $a\in\left[\frac{1}{32},\frac{7}{8}\right]$

50. $x = y = \sqrt{3-a^2}$

53. $\theta \in n\pi + \tan^{-1}(-2), n \in Z$

54. $\{\theta\}$

55. $x = \frac{1}{\sqrt{3}}$, $\cot 20^o$

57. $\frac{\pi}{4}$

58. $\frac{\pi}{4}$

59. $\tan^{-1}(x+n) - \tan^{-1} x$

60. $\frac{\pi}{4}$

61. $\cot^{-1}\left(\frac{2n+5}{n}\right)$

62 $\frac{\pi}{4}$

63.. $\frac{\pi}{2}$

64.. $\tan^{-1}\left(\frac{x}{y}\right)$

COMPREHENSIVE LINK PASSAGE

In these questions, a passage (paragraph) has been given followed by questions based on each of the passage. You have to answer the questions based on the passage given.

PASSAGE–1

Function	Domain	Co-domain
$\sin^{-1}x$	$[-1, 1]$	$\left[\frac{\pi}{2}, \frac{3\pi}{2}\right]$
$\tan^{-1}x$	R	$\left(\frac{\pi}{2}, \frac{3\pi}{2}\right)$
$\cos^{-1}x$	$[-1, 1]$	$[\pi, 2\pi]$
$\cot^{-1}x$	R	$[\pi, 2\pi]$

1. $\sin^{-1}(-x)$ is

(a) $-\sin^{-1} x$

(b) $\pi + \sin^{-1} x$

(c) $2\pi - \sin^{-1} x$

(d) $2\pi - \cos^{-1}\sqrt{1-x^2}$, $x > 0$

2. If $f(x) = 3\sin^{-1}x - 2\cos^{-1}x$, then f(x) is

(a) even function (b) odd function

(c) Neither even nor odd (d) even as well as odd function.

3. The minimum value of $(\sin^{-1} x)^3 - (\cos^{-1} x)^3$ is

(a) $-\frac{63\pi^3}{8}$ (b) $\frac{63\pi^3}{8}$ (c) $\frac{125\pi^3}{32}$ (d) $-\frac{125\pi^3}{32}$

4. The value of $\sin^{-1}x+\cos^{-1}x$ is

(a) $\frac{\pi}{2}$ (b) $\frac{3\pi}{2}$ (c) $\frac{5\pi}{2}$ (d) $\frac{7\pi}{2}$

5. If the co-domain of $\sin^{-1}x$ is $\left[-\frac{5\pi}{2},-\frac{3\pi}{2}\right]$ such that

$\sin^{-1}x+\cos^{-1}x=\frac{5\pi}{2}$, then the co-domain of $\cos^{-1}x$ is

(a) $[4\pi, 5\pi]$ (b) $[3\pi, 4\pi]$ (c) $[6\pi, 7\pi]$ (d) $[5\pi, 6\pi]$

PASSAGE–II

We know that corresponding to every bijection function $f: A\rightarrow B$, there exist a bijection $g: B\rightarrow A$ defined by $g(y)=x$ if and only if $f(x)=y$ The function $g: B\rightarrow A$ is called the inverse of function $f: A\rightarrow B$ and is denoted by f^{-1}.

Thus, we have $f(x)=y\Rightarrow f^{-1}(y)=x$ We know that trigonometric functions are periodic functions and hence, in general all trigonometric functions are not bijectives.

Consequently, their inverse do not exist.

However, if we restrict their domains and co-domains, they we can make then bijectives and also we can find their inverse..

Now, answer the following questions.

1. $\sin^{-1}(\sin\theta)=\theta$, for all θ belonging to

(a) $[0,\pi]$ (b) $[-\frac{\pi}{2},\frac{\pi}{2}]$ (c) $[-\frac{\pi}{2},0]$ (d) None of theses

2. $\cos^{-1}(\cos\theta)=\theta$, for all θ belonging to

(a) $[0,\pi]$ (b) $(-\frac{\pi}{2},\frac{\pi}{2})$ (c) $[-\frac{\pi}{2},\frac{\pi}{2}]$ (d) None of theses.

3. $\tan^{-1}(\tan\theta)=\theta$, for all θ belonging to

(a) $[0,\pi]$ (b) $[-\frac{\pi}{2},\frac{\pi}{2}]-\{0\}$ (c) $(-\frac{\pi}{2},\frac{\pi}{2})$ (d) None of theses

4. $\text{cosec}^{-1}(\text{cosec}\,\theta)=\theta$, for all θ belonging to

(a) $[-\frac{\pi}{2},\frac{\pi}{2}]$ (b) $(-\frac{\pi}{2},\frac{\pi}{2})-\{0\}$ (c) $[0,\pi]$ (d) $(0,\pi)$.

5. $\sec^{-1}(\sec\theta)=\theta$, for all θ belonging to

(a) $[0,\pi]-\{\frac{\pi}{2}\}$ (b) $(0,\pi)-\{\frac{\pi}{2}\}$ (c) $(0,\pi)$ (d) None of These.

6. $\sin^{-1}(\sin x) = x$, for all x belonging to

(a) R (B) φ (c) [-1,1] (d) $[\frac{\pi}{2}, \frac{3\pi}{2}]$.

7. The value of $\sin^{-1}(\sin 2) + \cos^{-1}(\cos 2)$ is

(a) 0 (b) $\frac{\pi}{2}$ (c) $-\frac{\pi}{2}$ (d) None of these

PASSAGE - III

Let $f(x) = \sin\{\cot^{-1}(x+1)\} - \cos(\tan^{-1} x)$, and $a = \cos(\tan^{-1}(\sin(\cot^{-1} x)))$ &

$b = \cos(2\cos^{-1}x + \sin^{-1}x)$

1. The value of x for which $f(x) = 0$ is

(a) -1/2 (b) 0 (c) 1/2 (d) 1

2. If $f(x) = 0$, then a^2 is equal to

(a) 1/2 (b) 2/3 (c) 5/9 (d) 9/5

3. If $a^2 = 26/51$, then b^2 is equal to

(a) 1/25 (b) 24/25 (c) 25/26 (d) 50/51

PASSAGE - IV

Every bijective (one - one onto function) $f : A \to B$ there exists a bijection $g : B \to A$ is defined by $g(y) = x$ if and only if $f(x) = y$.

The function $g : B \to A$ is called the inverse of function $f : A \to B$ and is denoted by f^{-1}. If no branch of an inverse trigonometric function is mentioned, then it means the principal value branch of that function.

1. The value of $\cos\{\tan^{-1}(\tan 2)\}$ is

(a) $1/\sqrt{5}$ (b) $-1/\sqrt{5}$ (c) cos 2 (d) -cos2

2. If x takes negative permissible value then $\sin^{-1} x$ is

(a) $\cos^{-1}(\sqrt{1-x^2})$ (b) $-\cos^{-1}(\sqrt{1-x^2})$ (c) $\cos^{-1}(\sqrt{x^2-1})$ (d) $\pi - \cos^{-1}(\sqrt{1-x^2})$.

3. If $x + \frac{1}{x} = 2$, then the value of $\sin^{-1} x$ is

(a) $\frac{\pi}{4}$ (b) $\frac{\pi}{2}$ (c) π (d) $\frac{3\pi}{2}$.

PASSAGE - V

Let $\cos^{-1}x + (\sin^{-1}y)^2 = \frac{a\pi^2}{4}$ ....(i)

and $\cos^{-1}x.(\sin^{-1}y)^2 = \frac{\pi^2}{16}$ ....(ii) where $-1 \le x, y \le 1$. Then

1. The set of values of 'a' for which the equation (i) holds good is

(a) $\left(0,\ 2+\frac{4}{\pi}\right)$ (b) $\left[0,\ 1+\frac{4}{\pi}\right)$ (c) R (d) $\left[0,\ -1+\frac{4}{\pi}\right)$

2. The set of values of 'a' for which equations (i) and (ii) posses solutions

(a) $(-\infty, 2]\cup[2, \infty)$ (b) $(-2, 2)$ (c) $\left[2, 1+\frac{4}{\pi}\right]$ (d) R

3. The values of x and y, the system of equations (i) and (ii) posses solutions for integral values of 'a'

(a) $\left\{\cos\left(\frac{\pi^2}{4}\right), 1\right\}$ (b) $\left\{\cos\left(\frac{\pi^2}{4}\right), -1\right\}$ (c) $\left\{\cos\left(\frac{\pi^2}{4}\right), \pm 1\right\}$ (d) $\{(x, y): x \in R, y \in R\}$

ANSWER

Passage - I

1. (c) 2.(b) 3. (a) 4.(c) 5. (a)

Passage - II

1.(b) 2.(a) 3. (c) 4. (a) 5. (a) 6. (c) 7. (a)

Passage - III

1 (a) 2. (c) 3. (b)

Passage - IV

1(d) 2. (b) 3.(b)

Passage - V

1.(b) 2.(c) 3.(c)

MATCH MATRIX

Given below are Matching type questions, with two columns (each having some items) each. Each item of column I has to be matched with the items of Column II, by encircling the correct match(es).

Note : An item of Column I can be matched with more than one items of Column II. All the items of Column II have to be matched.

1. **Match the following columns :**

	Column - I		Column - II
(A)	The principal value of $\sin^{-1}(\sin 20)$ is	(P)	$(20-6\pi)$
(B)	The principal value of $\sin^{-1}(\sin 10)$ is	(Q)	$(3\pi-10)$
(C)	The principal value of $\cos^{-1}(\cos 10)$ is	(R)	$(4\pi-10)$
(D)	The principal value of $\cos^{-1}(\cos 20)$ is	(S)	$(5\pi-20)$

2. **Match the following columns :**

	Column - I		Column - II
(A)	The range of $f(x)= 3\sin^{-1}x+2\cos^{-1}x$ is	(P)	$(0, \pi)$
(B)	The range of $f(x)=\sin^{-1}x+\cos^{-1}x+\tan^{-1}x$ is	(Q)	$\left[\frac{\pi}{2}, \frac{3\pi}{2}\right]$
(C)	The range of $f(x)=\sqrt{\sin^{-1}x+\pi}$ is	(R)	$\left(-\frac{\pi}{2}, \frac{3\pi}{2}\right)$
(D)	The range of $f(x)=2\tan^{-1}x+\sin^{-1}x+\sec^{-1}\left(\frac{1}{x}\right)$ is	(R)	$[0, \pi]$

3. **Match the following columns :**

	Column - I		Column - II
(A)	$\sin(\sin^{-1}x)=\sin^{-1}(\sin x)$, if	(P)	$-1\le x\le 1$
(B)	$\cos(\cos^{-1}x)=\cos^{-1}(\cos x)$, if	(Q)	$0\le x\le 1$
(C)	$\tan(\tan^{-1}x)=\tan^{-1}(\tan x)$,if	(R)	$0<x<\pi$
(D)	$\cot(\cot^{-1}x)=\cot^{-1}(\cot x)$, if	(R)	$-\frac{\pi}{2}\le x\le\frac{\pi}{2}$

4. **Match the following columns:**

	Column - I		Column - II
(A)	$\sin^{-1}\left(\frac{1}{2}\right)+\cos^{-1}\left(-\frac{1}{2}\right)$	(P)	$7\pi/6$
(B)	$\sin^{-1}\left(-\frac{1}{2}\right)+\cos^{-1}\left(\frac{1}{2}\right)$	(Q)	$5\pi/6$
(C)	$\tan^{-1}(\sqrt{3})+\cot^{-1}(-\sqrt{3})$	(R)	$\pi/6$
(D)	$\sin^{-1}\left(\frac{1}{2013}\right)+\cos^{-1}\left(\frac{1}{2013}\right)$	(S)	$\pi/2.$

5 **Match the following columns :**

	Column - I		Column - II
(A)	The value of $\tan^{-1}1+\tan^{-1}2+\tan^{-1}3$ is	(P)	$\frac{3\pi}{4}$
(B)	The value of $\tan^{-1}1+\tan^{-1}\left(\frac{1}{2}\right)+\tan^{-1}\left(\frac{1}{3}\right)$ is	(Q)	$\frac{\pi}{2}$
(C)	The value of $\tan^{-1}(9)+\tan^{-1}\left(\frac{5}{4}\right)$ is	(R)	π
(D)	The value of $2\tan^{-1}x-\tan^{-1}\left(\frac{2x}{1-x^2}\right)$, $x>1$ is	(S)	$-\frac{\pi}{2}$

6. **Match the following columns :**

	Column - I		Column - II
(A)	$(\sin^{-1}x)^3+(\cos^{-1}x)^3$ is maximum at	(P)	$x=\frac{1}{\sqrt{2}}$
(B)	$(\sin^{-1}x)^2+(\cos^{-1}x)^2$ is minimum at	(Q)	$x=1$
(C)	$(\sin^{-1}x)-(\cos^{-1}x)$ is minimum at	(R)	$x=-1$
(D)	$(\tan^{-1}x)^2+(\cot^{-1}x)^2$ is minimum at	(S)	$x=0$.

ANSWERS

1. (A)→(P) ; (B)→(Q) ; (C) → (R) ; (D) →(P)

2. (A)→(S) ; (B)→(P) ; (C) → (Q) ; (D) →(R)

3. (A) → (P) ;(B) →(Q); (C) →(S); (D) →(R).

4. (A) →(Q) ; (B) →(R) ;(C) →(P) ;(D) →(S).

5. (A) → (R) ;(B) →(Q); (C) →(P); (D) →(R).

6. (A) → (R) ;(B) →(P); (C) →(Q); (D) →(Q).

ASSERTION & REASON

Codes :

(A) Both A and R are individually true and R is the correct explanation of A

(B) Both A and R are individually true and R is not the correct explananation of A.

(C) A is true but R is false.

(D) A is false but R is true .

1. Assertion (A) : If

$$\sin^{-1}x+\sin^{-1}y+\sin^{-1}z=\frac{3\pi}{2},$$

then the value of $\left(x^{2013}+y^{2013}+z^{2013}\right)-\frac{9}{\left(x^{2014}+y^{2014}+z^{2014}\right)}$ is zero.

Reason (R) : Maximum value of $\sin^{-1}x$

is $\frac{\pi}{2}$

(a) A (b) B
(c) C (d) D

2. Assertion (A): The value of

$$2\tan^{-1}x-\tan^{-1}\left(\frac{2x}{1-x^2}\right)$$

is π

Reason (R): $x>1$

(a) A (b) B
(c) C (d) D

3. Assertion (A): The value of

$$\tan^{-1}(p)+\tan^{-1}\left(\frac{1}{p}\right) \text{ is } -\frac{\pi}{2}$$

Reason (R): P is the root of $x^2+2013x+2014=0$.

(a) A (b) B
(c) C (d) D

4. Assertion (A): If $\sin^{-1}x+\sin^{-1}y=\frac{2\pi}{3}$; then the value of

$$\cos^{-1}x+\cos^{-1}y \text{ is } \frac{\pi}{3}$$

Reason (R): $\sin^{-1}x+\cos^{-1}x=\frac{\pi}{2}$, when $x\in[-1,1]$

(a) A (b) B
(c) C (d) D

5. Assertion (A): The value of $\cos^{-1}(\cos 10)$ is $(2\pi-5)$

Reason(R): The range of $\cos^{-1}x$ is $[0,\pi]$

(a) A (b) B
(c) C (d) D

6. Assertion (A): If

$$\cos^{-1}x+\cos^{-1}y+\cos^{-1}z=\pi,$$

then $x^2+y^2+z^2+2xyz=1$

Reason(R): For $-1\le x, y, z\le 1$.

(a) A (b) B
(c) C (d) D

7. Assertion (A): If

$$\tan^{-1}(2x)+\tan^{-1}(3x)=\frac{\pi}{4}, \text{ then}$$

x is $\frac{1}{6}$

Reason(R): For $0<2x, 3x<1$

(a) A (b) B
(c) C (d) D

8. Assertion (A): $\cos\left(\frac{1-x^2}{1+x^2}\right)=2\tan^{-1}x$

Reason(R): for $x\ge 0$

(a) A (b) B
(c) C (d) D

9. Assertion (A):

$$\sin^{-1}(3x-4x^3)=\pi-3\sin^{-1}x$$

Reason(R): for $\frac{1}{2}<x\le 1$

(a) A (b) B
(c) C (d) D

10. Assertion (A):

$$\cos^{-1}(4x^3-3x)=2\pi-3\cos^{-1}(x)$$

Reason(R): For $-\frac{1}{2}\le x<\frac{1}{2}$

(a) A (b) B
(c) C (d) D

11. Assertion (A):

$$\cot^{-1}(x) = \tan^{-1}\left(\frac{1}{x}\right), x > 0$$

Reason (R):

$$\cot^{-1}(x) = \pi + \tan^{-1}\left(\frac{1}{x}\right), x < 0$$

(a) A (b) B
(c) C (d) D

12. Assertion (A): If α, β are the roots of $x^2 - 3x + 2 = 0$, then $\sin^{-1}\alpha$ is exist but not $\sin^{-1}\beta$, where $\alpha > \beta$

Reason (R): Domain of $\sin^{-1}x$ is $[-1, 1]$

(a) A (b) B
(c) C (d) D

ANSWERS

1.(a) 2.(a) 3.(a) 4.(a) 5.(a) 6.(a) 7.(a) 8.(a) 9.(a) 10.(a)
11.(b) 12.(a)

QUESTIONS ASKED IN ROORKEE EXAM

1. If $\cos^{-1}\left(\frac{x}{2}\right) + \cos^{-1}\left(\frac{y}{3}\right) = \theta$, then prove that $9x^2 - 12xy\cos\theta + 4y^2 = 36\sin^2\theta$

[Roorkee - 1984]

2. No questions asked in 1985.

3. Evaluate : $\tan\left(\frac{1}{2} - \cos^{-1}\left(\frac{\sqrt{5}}{3}\right)\right)$

[Roorkee - 1986]

4. No questions asked in 1987, 1988, 1989, 1990 and 1991.

5. Solve for x : $\sin[2\cos^{-1}\{\cot(2\tan^{-1}x)\}] = 0$
[Roorkee - 1992]

6. Find all positive integral solutions of

$$\tan^{-1}x + \cos^{-1}\left(\frac{y}{\sqrt{1+y^2}}\right) = \sin^{-1}\left(\frac{3}{\sqrt{10}}\right)$$

[Roorkee Main -1993]

7. If $\cos^{-1}x + \cos^{-1}y + \cos^{-1}z = \pi$, then find the value of $x^2 + y^2 + z^2 + 2xyz$

[Roorkee Main - 1994]

8. Convert the trigonometric function $\sin[2\cos^{-1}\{\cot(2\tan^{-1}x)\}]$ into an algebraic function f(x)

Then from the algebraic function f(x), find all values of x for which f(x) is zero. Also, express the values of x in the form of $a \pm \sqrt{b}$, where a and b are rational numbers. **[Roorkee Main - 1995]**

9. No questions asked in 1996.

10. If

$$\theta = \tan^{-1}(2\tan^2\theta) - \frac{1}{2}\sin^{-1}\left(\frac{3\sin 2\theta}{5 + 4\cos 2\theta}\right)$$

, then find the general value of θ

[Roorkee Main - 1997]

11. No questions asked in 1998.

12. Using the principal values, express the following expression as a single angle.

$$3\tan^{-1}\left(\frac{1}{2}\right) + 2\tan^{-1}\left(\frac{1}{5}\right) + \sin^{-1}\left(\frac{142}{65\sqrt{5}}\right)$$

[Roorkee Main - 1999]

13. Solve for x :

$$\sin^{-1}\left(\frac{ax}{c}\right) + \sin^{-1}\left(\frac{bx}{c}\right) = \sin^{-1}x \text{ where}$$

$a^2 + b^2 = c^2,\ c \neq 0$

[Roorkee Main - 2000]

14. Solve for x : $\cos^{-1}(x\sqrt{6})+\cos^{-1}(3\sqrt{3}\,x^2)=\frac{\pi}{2}$

[Roorkee Main - 2001]

ANSWERS

3. $\pm\left(\frac{3-\sqrt{5}}{2}\right)$

5. $x=\pm\sqrt{3\pm 2\sqrt{2}}$ and $x=\pm 1$

6. 7. 1

8. $x=0,\ \pm\sqrt{1},\ 1\pm\sqrt{2},\ -1\pm\sqrt{2}$

10. $\theta=n\pi\pm(\pi/4)$,

$m\pi+p\pi+\tan^{-1}(-2)$

, m, n, p ∈ I

12. π 13 $x=0$

14. x = 1/3 , - 1/3 . $i/\sqrt{3}$

QUESTIONS ASKED IN IIT-JEE

1. Let a, b, c positive real numbers such that

$\theta=\tan^{-1}\sqrt{\frac{a(a+b+c)}{bc}}$

$+\tan^{-1}\sqrt{\frac{b(a+b+c)}{ca}}+\tan^{-1}\sqrt{\frac{c(a+b+c)}{ab}}$

. Then $\tan\theta$ is equal to ..

[IIT-JEE-1981]

2. The numerical value of

$\tan^{-1}\left\{2\tan^{-1}\left(\frac{1}{5}\right)-\frac{\pi}{4}\right\}$ [IIT-JEE -1981]

3. Find the value of $\cos(2\cos^{-1}x+\sin^{-1}x)$ at $x=1/5$, where $0\le\cos^{-1}x\le\pi$ and

$-\frac{\pi}{2}\le\sin^{-1}x\le\frac{\pi}{2}$ [IIT-JEE-1981]

4. The value of $\tan\left[\cos^{-1}\left(\frac{4}{5}\right)+\tan^{-1}\left(\frac{2}{3}\right)\right]$ is

(a) 6/17 (b) 17/6

(c) - 17 / 6 (d) - 6 / 17

[IIT-JEE-1983]

5. The principal value of $\sin^{-1}\left(\sin\left(\frac{2\pi}{3}\right)\right)$ is

(a) $-\frac{2\pi}{3}$ (b) $\frac{2\pi}{3}$

(c) $\frac{\pi}{3}$ (d) $\frac{5\pi}{3}$

[IIT-JEE - 1986]

6. The greater of the two angles

$A=2\tan^{-1}(2\sqrt{2}-1)$ and

$B=3\sin^{-1}\left(\frac{1}{3}\right)+\sin^{-1}\left(\frac{3}{5}\right)$ is ....

[IIT-JEE-1989]

7. The number of real solutions of

$\tan^{-1}\sqrt{x(x+1)}+\sin^{-1}\sqrt{x^2+x+1}=\frac{\pi}{2}$ is

(a) 0 (b) 1

(c) 2 (d) ∞

[IIT-JEE-1999]

8. If $\sin^{-1}\left(x-\frac{x^2}{2}+\frac{x^3}{4}-.......\right)$

$+\cos^{-1}\left(x^2-\frac{x^4}{2}+\frac{x^6}{4}-....\right)=\frac{\pi}{2}$

, for $0<|x|<\sqrt{2}$, then x is

(a) 1/2 (b) 1
(c) -1/2 (d) -1

[IIT-JEE-2002]

9. Prove that

$$\cos\left(\tan^{-1}\left(\sin\left(\cot^{-1}x\right)\right)\right)=\sqrt{\frac{x^2+1}{x^2+2}}$$

[IIT-JEE-2002]

10. The domain of $f(x)=\sqrt{\sin^{-1}(2x)+\frac{\pi}{6}}$ is

(a) $\left(-\frac{1}{2},\frac{1}{2}\right]$ (b) $\left[-\frac{1}{4},\frac{3}{4}\right)$

(c) $\left[-\frac{1}{4},\frac{1}{4}\right]$ (d) $\left[-\frac{1}{4},\frac{1}{2}\right]$

[IIT-JEE-2004]

11. If $\sin\left(\cot^{-1}(x+1)\right)=\cos\left(\tan^{-1}x\right)$, then the value of x is

(a) -1/2 (b) 1/2
(c) 0 (d) 9/4

[IIT-JEE-2004]

12. Match the following columns :

Let (x, y) be such that $\sin^{-1}(a\,x)+\cos^{-1}(y)+\cos^{-1}(b\,x\,y)=\frac{\pi}{2}$

Column - I	**Column - II**
(A) If a = 1 and b = 0, then (x, y)	(P) lies on the circle $x^2+y^2=1$
(B) If a = 1 and b = 1 , then (x, y)	(Q) lies on $(x^2-1)(y^2-1)=0$
(C) If a = 1 and b = 2 , then (x, y)	(R) lies on the line y = x
(D) If a = 2 and b = 2 , then (x, y)	(S) lies on $(4x^2-1)(y^2-1)=0$

[IIT-JEE-2007]

13. If $0<x<1$, then
$\sqrt{1+x^2}\times\left[\left\{x\cos\left(\cos^{-1}x\right)+\sin\left(\cot^{-1}x\right)\right\}^2-1\right]^{1/2}$
equals

(a) $\frac{x}{\sqrt{1+x^2}}$ (b) x

(c) $x\sqrt{1+x^2}$ (d) $\sqrt{1+x^2}$

[IIT-JEE-2008]

ANSWERS

1. '0'
2. - 7 / 17
3. $-\frac{2\sqrt{6}}{5}$
4. (b)
5. (c)
6. A > B
7. (c)
8. (b)
10. (d)
11 (a)
12. (A) → (P) ; (B) → (Q) ; (C) → (P) ; (D) → (S)
13. (c)

SELF ASSESMENT TEST

Time : 3 Hrs. **Ch: Inverse Trigonometric Function** **M.M. : 100**

Give answers of the following questions.

1. Solve for x : $\sin^{-1}\left(\sin\left(\frac{2x^2+4}{x^2+1}\right)\right) < \pi - 3$

2. If $a\sin^{-1}x - b\cos^{-1}x = c$,

   then find the value of $a\sin^{-1}x + b\cos^{-1}x$

3. If $\sin^{-1}x + \sin^{-1}y = \frac{2\pi}{3}$, then find

   the value of $\cos^{-1}x + \cos^{-1}y$

4. If m is a root of $x^2 + 3x + 1 = 0$,

   then find the value of $\tan^{-1}(m) + \tan^{-1}\left(\frac{1}{m}\right)$

5. Find the value of

   $\sin^{-1}(\sin 3) + \cos^{-1}(\cos 7) - \tan^{-1}(\tan 5) + 10$

6. Solve for x :

   $\sin^{-1}(x^2 - 2x + 1) + \cos^{-1}(x^2 - x) = \frac{\pi}{2}$

7. Let $\cos^{-1}(x) + \cos^{-1}(2x) + \cos^{-1}(3x) = \pi$.

   If x satisfies the cubic equation

   $ax^3 + bx^2 + cx + d = 0$, then find the value of $a + b + c + d$.

8. Consider α, β, γ are the roots of $x^3 - x^2 + 3x + 4 = 0$ such that $\tan^{-1}\alpha + \tan^{-1}\beta + \tan^{-1}\gamma = \theta$.

   If the positive value of $\sec(\theta)$ is $\frac{\sqrt{p}}{q}$, where p and q are natural numbers, then find the value of $p + q + 69$.

9. If M is the number of real solution of $\cos^{-1}x + \cos^{-1}(2x) + \pi = 0$ and N is the number of values of x satisfying the equation $\sin^{-1}\left(\frac{5}{x}\right) + \sin^{-1}\left(\frac{12}{x}\right) = \frac{\pi}{2}$,

   then find the value of M + N + 4

10. Solve for x :

    $\sin(\sin^{-1}(\log_{1/2}x)) + 2\left|\cos\left(\sin^{-1}\left(\frac{x}{2} - 1\right)\right)\right| = 0$

ANSWERS

1. $x \in (-1, 1)$
2. $\frac{\pi ab + c(a-b)}{a+b}$
3. $\frac{\pi}{3}$
4. $-\frac{\pi}{2}$
5. $\pi + 9$
6. x = 1
7. 26
8. 100
9. M = 0, N = 1 , 5
10. $x = \phi$.

CHAPTER 6

PROPERTIES OF TRIANGLES

6.1 INTRODUCTION :

In any Δ, the three sides and the three angles are generally called the elements of the triangle.

A triangle which does not contain a right angle is called an oblique triangle.

In any Δ ABC, the measures of the angles $\angle$BAC, $\angle$CBA and $\angle$ACB are denoted by the letters A, B and C respectively and the sides BC, CA and AB opposite to the angles A, B and C are respectively denoted by a, b and c. These six elements of a triangle are not independent and are connected by the relations.

i) $A+B+C=\pi$

ii) $a+b>c;\ b+c>a;\ c+a>b$

6.2 SINE RULE :

Statement : The sides of a triangle are in proportion to the sines of the angles opposite to them

i.e. In a Δ ABC, $\dfrac{a}{\sin A}=\dfrac{b}{\sin B}=\dfrac{c}{\sin C}$

The above rule may also be expressed as

$$\frac{\sin A}{a}=\frac{\sin B}{b}=\frac{\sin C}{c}.$$

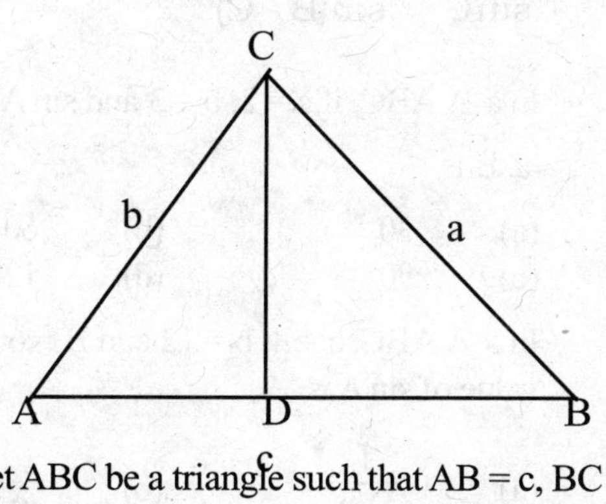

Proof: Let ABC be a triangle such that AB = c, BC = a, CA = b

Draw CD which is perpendicular on AB, jointing the vertex (C).

From ΔACD, $\sin A=\dfrac{CD}{AC}=\dfrac{CD}{b}$

$\Rightarrow$ $CD=b\ \sin A$ ....(i)

From ΔBCD, $\sin B=\dfrac{CD}{BC}=\dfrac{CD}{a}$

$\Rightarrow$ $CD=a\sin B$ ....(ii)

From (i) and (ii), we get,

$b\sin A=a\sin B$

$\Rightarrow$ $\dfrac{\sin A}{a}=\dfrac{\sin B}{b}$ ....(iii)

Similarly, we can prove that,

$\dfrac{\sin A}{a}=\dfrac{\sin C}{c}$ ....(iv)

From (iii) and (iv), we get,

$\dfrac{\sin A}{a}=\dfrac{\sin B}{b}=\dfrac{\sin C}{c}$

$\Rightarrow$ $\dfrac{a}{\sin A}=\dfrac{b}{\sin B}=\dfrac{c}{\sin C}$

Ex-1: In a triangle ABC, prove that,

$$a\sin\left(\frac{A}{2}+B\right)=(b+c)\sin\left(\frac{A}{2}\right)$$

Soln. Let $\dfrac{a}{\sin A}=\dfrac{b}{\sin B}=\dfrac{c}{\sin C}=k$

We have, $\dfrac{b+c}{a}=\dfrac{k(\sin B+\sin C)}{k\ \sin A}$

$$=\frac{(\sin B+\sin C)}{\sin A}$$

$$=\frac{2\sin\left(\frac{B+C}{2}\right)\cos\left(\frac{B-C}{2}\right)}{2\sin\left(\frac{A}{2}\right)\cos\left(\frac{A}{2}\right)}$$

$$= \frac{2\sin\left(\frac{\pi}{2}-\frac{A}{2}\right)\cos\left(\frac{B-C}{2}\right)}{2\sin\left(\frac{A}{2}\right)\cos\left(\frac{A}{2}\right)}$$

$$= \frac{2\cos\left(\frac{A}{2}\right)\cos\left(\frac{B-C}{2}\right)}{2\sin\left(\frac{A}{2}\right)\cos\left(\frac{A}{2}\right)}$$

$$= \frac{\cos\left(\frac{B-C}{2}\right)}{\sin\left(\frac{A}{2}\right)}$$

$$= \frac{\sin\left(\frac{A}{2}+B\right)}{\sin\left(\frac{A}{2}\right)}$$

Ex-2. In a triangle ABC , prove that , $\frac{a\sin(B-C)}{b^2-c^2}=\frac{b\sin(C-A)}{c^2-a^2}=\frac{c\sin(A-B)}{a^2-b^2}$

Soln. Let $\frac{a}{\sin A}=\frac{b}{\sin B}=\frac{c}{\sin C}=k$

We have, $\frac{a\sin(B-C)}{b^2-c^2}$

$$= \frac{k\sin A.\sin(C-A)}{k^2(\sin^2C-\sin^2A)}$$

$$= \frac{\sin(B+C).\sin(B-C)}{k(\sin^2B-\sin^2C)}$$

$$= \frac{(\sin^2B-\sin^2C)}{k(\sin^2B-\sin^2C)} = \frac{1}{k}$$

$\frac{b\sin(C-A)}{c^2-a^2}$

$$= \frac{k\sin(C+A).\sin(C-A)}{k^2(\sin^2C-\sin^2A)}$$

$$= \frac{\sin^2C-\sin^2A}{k(\sin^2C-\sin^2A)} = \frac{1}{k}$$

Also, $\frac{c\sin(A-B)}{a^2-b^2}$

$$= \frac{k\sin C\sin(A-B)}{k^2(\sin^2A-\sin^2B)}$$

$$= \frac{\sin(A+B)\sin(A-B)}{k(\sin^2A-\sin^2B)}$$

$$= \frac{\sin^2 A-\sin^2 B}{k(\sin^2A-\sin^2B)} = \frac{1}{k}$$

Hence, the result.

Exercise - 1

(Questions based on Sine Rule :)

1. In a Δ ABC, prove that

i) $\frac{\sin(B-C)}{\sin(B+C)} = \frac{b^2-c^2}{a^2}$

ii) $a\sin(B-C)+b\sin(C-A)+c\sin(A-B)=0$

2. In Δ ABC, prove that

i) $\frac{a^2\sin(B-C)}{\sin B+\sin C}+\frac{b^2\sin(C-A)}{\sin C+\sin A}+\frac{c^2\sin(A-B)}{\sin A+\sin B}=0$

ii) $\frac{b^2-c^2}{\cos B+\cos C}+\frac{c^2-a^2}{\cos C+\cos A}+\frac{a^2-b^2}{\cos A+\cos B}=0$

3. In a Δ ABC, prove that

i) $\frac{1+\cos(A-B)\cos C}{1+\cos(A-C)\cos B}=\frac{a^2+b^2}{a^2+c^2}$

ii) ^{2}, b^2, c^2 are in A.P., if

$\frac{\sin A}{\sin C}=\frac{\sin(A-B)}{\sin(B-C)}$

4. In a Δ ABC, if a = 2, b = 3 and sin A = $\frac{2}{3}$, then $\angle$B is

(a) 30 (b) 60

(c) 90° (d) 120°

5. In a Δ ABC, a = 4, b = 12 and B = 60°, then the value of sin A is

(a) $\frac{1}{2\sqrt{3}}$ (b) $\frac{1}{3\sqrt{2}}$

(c) $\frac{2}{\sqrt{3}}$ (d) $\frac{\sqrt{3}}{2}$

6. In a Δ ABC such that $\angle A = 45°$, $\angle B = 75°$, then $a + c\sqrt{2}$ is

(a) 0 (b) b
(c) 2b (d) –b

7. If $\cot\frac{A}{2} = \frac{b+c}{a}$, then Δ ABC is

(a) Isosceles (b) Equilateral
(c) Rt. angled (d) None

8. If in a Δ ABC, $\frac{a^2-b^2}{a^2+b^2} = \frac{\sin(A-B)}{\sin(A+B)}$, then the triangle is

(a) Rt. angled or isosceles
(b) Rt. angled and isosceles
(c) Equilateral
(c) None

Answers

4.(b) 5.(a) 6.(c) 7.(c) 8.(a).

6.3 COSINE RULE :

Statement : In any triangle ABC

i) $\cos A = \frac{b^2+c^2-a^2}{2bc}$

ii) $\cos B = \frac{a^2+c^2-b^2}{2ac}$

iii) $\cos C = \frac{a^2+b^2-c^2}{2ab}$

Proof :

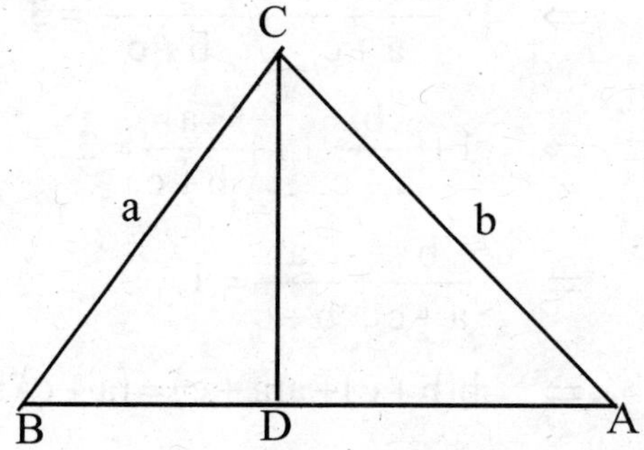

(i) From ΔBCD, $\sin A = \frac{CD}{AC} = \frac{CD}{b}$

$\Rightarrow$ $CD = b\sin A$

From ΔACD, $\cos A = \frac{AD}{AC} = \frac{AD}{b}$

$\Rightarrow$ $AD = b\cos A$

$\Rightarrow$ $AB - BD = b\cos A$

$\Rightarrow$ $c - BD = b\cos A$

$\Rightarrow$ $BD = c - b\cos A$

From ΔBCD, $BC^2 = BD^2 + CD^2$

$\Rightarrow$ $a^2 = (c - b\cos A)^2 + (b\sin A)^2$

$\Rightarrow$ $a^2 = c^2 - 2bc\cos A + b^2\cos^2 A + b^2\sin^2 A$

$= c^2 - 2bc\cos A + b^2$

$\Rightarrow$ $2bc\cos A = b^2 + c^2 - a^2$

$\Rightarrow$ $\cos A = \frac{b^2+c^2-a^2}{2bc}$

Similarly, we easily proved that, (ii) and (iii)

Ex-1. In a triangle ABC, the angles are in A.P, then prove that, $2\cos\left(\frac{A-C}{2}\right) = \frac{a+c}{\sqrt{a^2-ac+c^2}}$

Soln. Since the angles are in A.P, so $A + C = 2B$

$\Rightarrow$ $A + B + C = 3B$

$\Rightarrow$ $3B = 180^\circ$

$\Rightarrow$ $B = 60^\circ$

$\Rightarrow$ $\cos B = \cos(60^\circ) = \frac{1}{2}$

$\Rightarrow$ $\frac{a^2+c^2-b^2}{2ac} = \frac{1}{2}$

$\Rightarrow$ $a^2 + c^2 - b^2 = ac$

$\Rightarrow$ $a^2 + c^2 - ac = b^2$

Now, R.H.S $= \frac{a+c}{\sqrt{a^2-ac+c^2}}$

$= \frac{a+c}{b} = \frac{k(\sin A + \sin C)}{k\sin B}$

$= \frac{2\sin\left(\frac{A+C}{2}\right)\cos\left(\frac{A-C}{2}\right)}{2\sin\left(\frac{B}{2}\right)\cos\left(\frac{B}{2}\right)}$

$$= \frac{\cos\left(\frac{B}{2}\right)\cos\left(\frac{A-C}{2}\right)}{\sin\left(\frac{B}{2}\right)\cos\left(\frac{B}{2}\right)}$$

$$= \frac{\cos\left(\frac{A-C}{2}\right)}{\sin\left(\frac{B}{2}\right)}$$

$$= \frac{\cos\left(\frac{A-C}{2}\right)}{\sin\left(\frac{60^o}{2}\right)}$$

$$= 2\cos\left(\frac{A-C}{2}\right).$$

Ex-2. In a triangle ABC, prove that

$$\left(\frac{b^2-c^2}{a^2}\right)\sin 2A + \left(\frac{c^2-a^2}{b^2}\right)\sin 2B + \left(\frac{a^2-b^2}{c^2}\right)\sin 2C = 0$$

Soln. We have, $\left(\frac{b^2-c^2}{a^2}\right)\sin 2A$

$$= \left(\frac{b^2-c^2}{a^2}\right)(2\sin A.\cos A)$$

$$= \left(\frac{b^2-c^2}{a^2}\right)\left(2ka\left(\frac{b^2+c^2-a^2}{2bc}\right)\right)$$

$$= \left(\frac{(b^2-c^2)(b^2+c^2-a^2)}{kabc}\right)$$

$$= \frac{1}{kabc}\times\{(b^4-c^4)-a^2(b^2-c^2)\}$$

Similarly, $\left(\frac{c^2-a^2}{b^2}\right)\times\sin 2B$

$$= \frac{1}{kabc}\times\{(c^4-a^4)-b^2(c^2-a^2)\}$$

$\left(\frac{a^2-b^2}{c^2}\right)\sin 2C$ and

$$= \frac{1}{kabc}\times\{(a^4-b^4)-c^2(a^2-b^2)\}$$

Thus, $\left(\frac{b^2-c^2}{a^2}\right)\sin 2A + \left(\frac{c^2-a^2}{b^2}\right)\sin 2B + \left(\frac{a^2-b^2}{c^2}\right)\sin 2C$

$$= \frac{1}{kabc}\times\{(b^4-c^4)-a^2(b^2-c^2)\} + \frac{1}{kabc}\times\{(c^4-a^4)-b^2(c^2-a^2)\} + \frac{1}{kabc}\times\{(a^4-b^4)-c^2(a^2-b^2)\}$$

$$= \frac{1}{kabc}\times\left\{\begin{matrix}(b^4-c^4+c^4-a^4+a^4-b^4)\\ +\left\{\begin{matrix}a^2(c^2-b^2)+b^2(a^2-c^2)\\ +c^2(b^2-a^2)\end{matrix}\right\}\end{matrix}\right\}$$

$$= 0$$

Ex-3. If $\frac{1}{a+c}+\frac{1}{b+c}=\frac{3}{a+b+c}$, then find $\angle C$

Soln. Given $\frac{1}{a+c}+\frac{1}{b+c}=\frac{3}{a+b+c}$

$$\Rightarrow \frac{a+b+c}{a+c}+\frac{a+b+c}{b+c}=3$$

$$\Rightarrow 1+\frac{b}{a+c}+1+\frac{a}{b+c}=3$$

$$\Rightarrow \frac{b}{a+c}+\frac{a}{b+c}=1$$

$$\Rightarrow b(b+c)+a(a+c)=(a+c)(b+c)$$

$$\Rightarrow b^2+bc+a^2+ac=ab+ac+bc+c^2$$

$$\Rightarrow a^2+b^2-c^2=ab$$

Now, $\cos(C)=\left(\frac{a^2+b^2-c^2}{2ab}\right)=\frac{ab}{2ab}=\frac{1}{2}$

$$\Rightarrow C=\frac{\pi}{3}$$

Exercise - 2

(Questions based on cosine Rule :)

1. In any Δ ABC, prove that

i) $a(b\cos C - c\cos B) = b^2 - c^2$

ii) $\dfrac{\cos A}{a} + \dfrac{\cos B}{b} + \dfrac{\cos C}{c} = \dfrac{a^2+b^2+c^2}{2abc}$

2. In any Δ ABC, prove that

i) $(a-b)^2\cos^2\dfrac{C}{2} + (a+b)^2\sin^2\dfrac{C}{2} = c^2.$

ii) $(b^2-c^2)\cot A + (c^2-a^2)\cot B$

$+ (a^2-b^2)\cot C = 0$

3. In any Δ ABC, if $2\cos B = \dfrac{a}{c}$, then the triangle is

(a) right angled (b) equilateral
(c) isosceles (d) None

4. In a Δ ABC, $a = 4$, $b = 3$, $\angle A = 60°$. Then c is the root of the equation.

(a) $c^2 - 3c - 7\ 0$
(b) $c^2 + 3c + 7 = 0$
(c) $c^2 - 3c + 7 = 0$
(d) $c^2 + 3c - 7 = 0$

5. The expression

$$\frac{(a+b+c)(b+c-a)(c+a-b)(a+b-c)}{4b^2c^2}$$

is

(a) $\cos^2\theta$ (b) $\sin^2\theta$
(c) cosA cosB cosC (d) None

6. In a Δ ABC, $(a+b+c)(b+c-a) = \lambda bc$ if

(a) $\lambda < 0$ (b) $\lambda > 0$
(c) $0 < \lambda < 4$ (d) $\lambda > 4$

7. In a Δ ABC, if $a\cos A = b\cos B$, then the triangle is

(a) Isoceles
(b) Right angle
(c) Isoscele or right angle
(d) Right angle isosceles

8. If the sides of a Δ are in the ratio $2 : \sqrt{6} : (\sqrt{3}+1)$, then the largest angles of the Δ will be

(a) 60° (b) 75°
(c) 90° (d) 120°

Answers

3.(c) 4.(a) 5.(b)
6.(a) 7.(d) 8.(b).

6.4 PROJECTION FORMULAE :

Statement : In any triangle ABC

i) $a = b\cos C + c\cos B$

ii) $b = c\cos A + a\cos C$

iii) $c = a\cos B + b\cos A$

Proof:

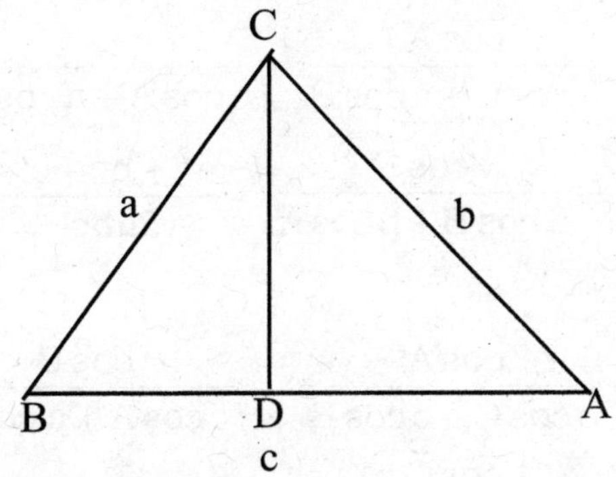

From ΔABD, we have,

$a = BC = BD + DC = c\cos B + b\cos C$

$\Rightarrow \quad a = b\cos C + c\cos B$

Similarly, we can easily prove that in (ii) and (iii)

Ex-1. In a triangle ABC, prove that,

$$\frac{\sin B}{\sin C} = \frac{c - a\cos B}{b - a\cos C}$$

Soln. We have, $\dfrac{c - a\cos B}{b - a\cos C}$

$$= \frac{a\cos B + b\cos A - a\cos B}{c\cos A + a\cos C - a\cos C}$$

$$= \frac{b\cos A}{c\cos A}$$

$$= \frac{b}{c}$$

$$= \frac{k\sin B}{k\sin C}$$

$$= \frac{\sin B}{\sin C}$$

Ex-2. In any triangle ABC, prove that,

$$2\left(a\sin^2\frac{C}{2} + c\sin^2\frac{A}{2}\right) = a + c - b$$

Soln. We have, $2\left(a\sin^2\frac{C}{2}+c\sin^2\frac{A}{2}\right)$

$= a\left(2\sin^2\left(\frac{C}{2}\right)\right)+c\left(2\sin^2\left(\frac{A}{2}\right)\right)$

$= a(1-\cos C)+c(1-\cos A)$

$= a+c-(a\cos C+c\cos A)$

$= a+c-b$

Ex-3. In any triangle ABC, prove that,

$$\frac{\cos A}{b\cos C+c\cos B}+\frac{\cos B}{c\cos A+a\cos C}+\frac{\cos C}{a\cos B+b\cos A}=\frac{a^2+b^2+c^2}{2abc}$$

Soln. We have,

$$\frac{\cos A}{b\cos C+c\cos B}+\frac{\cos B}{c\cos A+a\cos C}+\frac{\cos C}{a\cos B+b\cos A}$$

$$=\frac{\cos A}{a}+\frac{\cos B}{b}+\frac{\cos C}{c}$$

$$=\left(\frac{b^2+c^2-a^2}{2abc}\right)+\left(\frac{c^2+a^2-b^2}{2abc}\right)+\left(\frac{a^2+b^2-c^2}{2abc}\right)$$

$$=\frac{1}{2abc}\left\{\begin{array}{l}(b^2+c^2-a^2)+\\(c^2+a^2-b^2)\\+(a^2+b^2-c^2)\end{array}\right\}$$

$$=\frac{(a^2+b^2+c^2)}{2abc}$$

Exercise - 3

Questions based on Projection Formulae:

1. In any ΔABC, prove that

i) $\frac{\sin B}{\sin C}=\frac{c-a\cos B}{b-a\cos C}$

ii) $(b+c)\cos A+(c+a)\cos B+(a+b)\cos C = a+b+c$

2. In any ΔABC, prove that

i) $2\left(a\sin^2\frac{C}{2}+c\sin^2\frac{A}{2}\right)=a+c-b$

ii) $2\left(b\cos^2\frac{C}{2}+c\cos^2\frac{B}{2}\right)=a+b+c$

3. In any ΔABC, prove that

i) $\frac{\cos A}{b\cos C+c\cos B}+\frac{\cos B}{c\cos A+a\cos C}+\frac{\cos C}{a\cos B+b\cos A}=\frac{a^2+b^2+c^2}{2abc}$

ii) $2[bc\cos A+ca\cos B+ab\cos C] = a^2+b^2+c^2$

4. If $\frac{1}{a+c}+\frac{1}{b+c}=\frac{3}{a+b+c}$, then $\angle C$ is

(a) 90 (b) 60
(c) 45 (d) 30.

5. If the angles A, B and C of the triangle ABC be in A. P., then $\frac{a+c}{\sqrt{a^2-ac+c^2}}$ is

(a) $2\cos\left(\frac{A+C}{2}\right)$

(b) $2\sin\left(\frac{A+C}{2}\right)$

(c) $2\cos\left(\frac{A-C}{2}\right)$

(d) $2\sin\left(\frac{A-C}{2}\right)$.

Answers

4.(b) 5.(c).

6.5 NAPIER'S ANALOGY (LAW OF TANGENTS) :

Statement : In any ΔABC, prove that

i) $\tan\left(\frac{B-C}{2}\right)=\frac{b-c}{b+c}\cot\left(\frac{A}{2}\right)$

ii) $\tan\left(\frac{C-A}{2}\right)=\left(\frac{c-a}{c+a}\right)\cot\frac{B}{2}$

iii) $\tan\left(\frac{A-B}{2}\right)=\left(\frac{a-b}{a+b}\right)\cot\frac{C}{2}$

Proof: (i) Let $\frac{a}{\sin A}=\frac{b}{\sin B}=\frac{c}{\sin C}=k$

We have, $\left(\frac{b-c}{b+c}\right)\cot\left(\frac{A}{2}\right)$

$$=\left(\frac{\sin B-\sin C}{\sin B+\sin C}\right)\times\cot\left(\frac{A}{2}\right)$$

$$=\left(\frac{2\cos\left(\frac{B+C}{2}\right)\sin\left(\frac{B-C}{2}\right)}{2\sin\left(\frac{B+C}{2}\right)\cos\left(\frac{B-C}{2}\right)}\right)\times\cot\left(\frac{A}{2}\right)$$

$$=\left(\frac{2\cos\left(\frac{B+C}{2}\right)\sin\left(\frac{B-C}{2}\right)}{2\sin\left(\frac{B+C}{2}\right)\cos\left(\frac{B-C}{2}\right)}\right)\times\cot\left(\frac{A}{2}\right)$$

$$=\left(\frac{\sin\left(\frac{A}{2}\right)\sin\left(\frac{B-C}{2}\right)}{\cos\left(\frac{A}{2}\right)\cos\left(\frac{B-C}{2}\right)}\right)\times\cot\left(\frac{A}{2}\right)$$

$$=\left(\frac{\sin\left(\frac{A}{2}\right)\sin\left(\frac{B-C}{2}\right)}{\cos\left(\frac{A}{2}\right)\cos\left(\frac{B-C}{2}\right)}\right)\times\frac{\cos\left(\frac{A}{2}\right)}{\sin\left(\frac{A}{2}\right)}$$

$$=\tan\left(\frac{B-C}{2}\right)$$

Similarly, we can easily prove that in (ii) and (iii)

Ex-1. In a ΔABC, $b=\sqrt{3}+1$, $c=\sqrt{3}-1$ and $\angle A=60°$, then find the value of $\tan\left(\frac{B-C}{2}\right)$.

Soln. As we know that,

$$\tan\left(\frac{B-C}{2}\right)=\left(\frac{b-c}{b+c}\right)\cot\left(\frac{A}{2}\right)$$

$$=\left(\frac{(\sqrt{3}+1)-(\sqrt{3}-1)}{(\sqrt{3}+1)+(\sqrt{3}-1)}\right)\cot\left(\frac{60^o}{2}\right)$$

$$=\frac{1}{\sqrt{3}}\cot(30^o)$$

$$=\frac{1}{\sqrt{3}}\times\sqrt{3}$$

$$= 1$$

Ex-2. If in a triangle ABC, $a=6$, $b=3$ and $\cos(A-B)=4/5$, then find the angle C.

Soln. Given $\cos(A-B)=\frac{4}{5}$

$$\Rightarrow \frac{1-\tan^2\left(\frac{A-B}{2}\right)}{1+\tan^2\left(\frac{A-B}{2}\right)}=\frac{4}{5}$$

$$\Rightarrow 4+4\tan^2\left(\frac{A-B}{2}\right)=5-5\tan^2\left(\frac{A-B}{2}\right)$$

$$\Rightarrow 4+4\tan^2\left(\frac{A-B}{2}\right)=5-5\tan^2\left(\frac{A-B}{2}\right)$$

$$\Rightarrow 9\tan^2\left(\frac{A-B}{2}\right)=5-4=1$$

$$\Rightarrow \tan\left(\frac{A-B}{2}\right)=\frac{1}{3}$$

$$\Rightarrow \left(\frac{a-b}{a+b}\right)\cot\left(\frac{C}{2}\right)=\frac{1}{3}$$

$$\Rightarrow \left(\frac{6-3}{6+3}\right)\cot\left(\frac{C}{2}\right)=\frac{1}{3}$$

$$\Rightarrow \frac{1}{3}\cot\left(\frac{C}{2}\right)=\frac{1}{3}$$

$$\Rightarrow \cot\left(\frac{C}{2}\right)=1$$

$$\Rightarrow \left(\frac{C}{2}\right)=\frac{\pi}{4}$$

$$\Rightarrow C=\frac{\pi}{2}$$

Hence, the value of C is $\frac{\pi}{2}$.

Exercise - 4

(Questions based on Napiers Analogy :)

1. In a ΔABC, $b=\sqrt{3}$, $c=1$ and $B-C=90°$ then $\angle A$ is

(a) 30 (b) 45°
(c) 75° (d) 15°

2. If in a triangle ABC, $a = 6$, $b = 3$ and $\cos(A - B) = 4/5$, then angle C is

(a) 30 (b) 45°
(c) 75° (d) 90°

3. In a triangle ABC, if

$$x = \tan\left(\frac{B-C}{2}\right)\tan\left(\frac{A}{2}\right),$$

$$y = \tan\left(\frac{C-A}{2}\right)\tan\left(\frac{B}{2}\right) \text{ and}$$

$$z = \tan\left(\frac{A-B}{2}\right)\tan\left(\frac{C}{2}\right), \text{ then the value of}$$

$x + y + z$ is

(a) xyz (b) $2xyz$
(c) $3xyz$ (d) $-xyz$

4. In a triangle ABC, if $a = 5$, $b = 4$ and $\cos(A - B) = \frac{31}{32}$, then the side 'c' is

(a) 4 (b) 6
(c) 8 (d) 10.

Answers :

1.(d). 2.(d) 3(d) 4.(b)

6.6 HALF-ANGLED FORMULAE :

In this section, we shall desire to the formulas for the sine, cosine and tangent to the half of the angles of any triangle in terms of its sides. The perimeter of ΔABC will be denoted by 2s i.e.

$$a + b + c = 2s$$

and the areas is denoted by Δ.

i.e. $\Delta = \sqrt{s(s-a)(s-b)(s-c)}$

1) Formulae for $\sin\left(\frac{A}{2}\right)$, $\sin\left(\frac{B}{2}\right)$, $\sin\left(\frac{C}{2}\right)$:

In any ΔABC,

i) $\sin\frac{A}{2} = \sqrt{\frac{(s-b)(s-c)}{bc}}$

ii) $\sin\frac{B}{2} = \sqrt{\frac{(s-c)(s-a)}{ca}}$

iii) $\sin\frac{C}{2} = \sqrt{\frac{(s-a)(s-b)}{ab}}$

Proof: (i) We have, $2\sin^2\left(\frac{A}{2}\right) = 1 - \cos A$

$$\Rightarrow 2\sin^2\left(\frac{A}{2}\right) = 1 - \left(\frac{b^2+c^2-a^2}{2bc}\right)$$

$$= \left(\frac{2bc - b^2 - c^2 + a^2}{2bc}\right)$$

$$= \left(\frac{a^2 - (b^2 + c^2 - 2bc)}{2bc}\right)$$

$$= \left(\frac{a^2 - (b-c)^2}{2bc}\right)$$

$$= \left(\frac{\{a + (b-c)\} \times \{a - (b-c)\}}{2bc}\right)$$

$$= \left(\frac{(a+b-c) \times (a+c-b)}{2bc}\right)$$

$$= \left(\frac{(a+b+c-2c) \times (a+b+c-2b)}{2bc}\right)$$

$$= \left(\frac{(2s-2c) \times (2s-2b)}{2bc}\right)$$

$$= \left(\frac{2(s-c) \times (s-b)}{bc}\right)$$

$$\Rightarrow 2\sin^2\left(\frac{A}{2}\right) = \left(\frac{2(s-c) \times (s-b)}{bc}\right)$$

$$\Rightarrow \sin^2\left(\frac{A}{2}\right) = \left(\frac{(s-c) \times (s-b)}{bc}\right)$$

$$\Rightarrow \sin\left(\frac{A}{2}\right) = \sqrt{\frac{(s-c) \times (s-b)}{bc}}$$

Similarly, we can easily prove that in (ii) and (iii)

2) Formulae for $\cos\left(\frac{A}{2}\right)$, $\cos\left(\frac{B}{2}\right)$, $\cos\left(\frac{C}{2}\right)$:

In any ΔABC,

i) $\cos\frac{A}{2} = \sqrt{\frac{s(s-a)}{bc}}$

ii) $\cos\frac{B}{2} = \sqrt{\frac{s(s-b)}{ca}}$

iii) $\cos\dfrac{C}{2} = \sqrt{\dfrac{s(s-c)}{ab}}$

Proof: (i) We have, $2\cos^2\left(\dfrac{A}{2}\right) = 1 + \cos A$

$$\Rightarrow \quad 2\cos^2\left(\frac{A}{2}\right) = 1 + \left(\frac{b^2+c^2-a^2}{2bc}\right)$$

$$= \left(\frac{2bc+b^2+c^2-a^2}{2bc}\right)$$

$$= \left(\frac{(b+c)^2-a^2}{2bc}\right)$$

$$= \left(\frac{(a+b+c)(b+c-a)}{2bc}\right)$$

$$= \left(\frac{(a+b+c)(b+c+a-2a)}{2bc}\right)$$

$$= \left(\frac{2s(2s-2a)}{2bc}\right)$$

$$= \left(\frac{2s(s-a)}{bc}\right)$$

$$\Rightarrow \quad 2\cos^2\left(\frac{A}{2}\right) = \left(\frac{2s(s-a)}{bc}\right)$$

$$\Rightarrow \quad \cos^2\left(\frac{A}{2}\right) = \left(\frac{s(s-a)}{bc}\right)$$

$$\Rightarrow \quad \cos\left(\frac{A}{2}\right) = \sqrt{\left(\frac{s(s-a)}{bc}\right)}$$

Similarly, we can easily prove that in (ii) and (iii).

3) Formulae for $\tan\left(\frac{A}{2}\right), \tan\left(\frac{B}{2}\right), \tan\left(\frac{C}{2}\right)$:

In any ΔABC,

i) $\tan\dfrac{A}{2} = \sqrt{\dfrac{(s-b)(s-c)}{s(s-a)}}$

ii) $\tan\dfrac{B}{2} = \sqrt{\dfrac{(s-c)(s-a)}{s(s-b)}}$

iii) $\tan\dfrac{C}{2} = \sqrt{\dfrac{(s-a)(s-b)}{s(s-c)}}$

Proof: (i) We have, $\tan\left(\dfrac{A}{2}\right) = \dfrac{\sin\left(\frac{A}{2}\right)}{\cos\left(\frac{A}{2}\right)}$.

$$\Rightarrow \quad \tan\left(\frac{A}{2}\right) = \frac{\sqrt{\dfrac{(s-b)(s-c)}{bc}}}{\sqrt{\dfrac{s(s-a)}{bc}}}$$

$$\Rightarrow \quad \tan\left(\frac{A}{2}\right) = \sqrt{\frac{(s-b)(s-c)}{s(s-a)}}$$

Similarly, we can easily prove that in (ii) and (iii)

Ex-1. In any Δ ABC, prove that ,

$$b\cos^2\left(\frac{C}{2}\right) + c\cos^2\left(\frac{B}{2}\right) = s.$$

Soln. We have, $b\cos^2\left(\dfrac{C}{2}\right) + c\cos^2\left(\dfrac{B}{2}\right)$

$$= b\left(\frac{s(s-c)}{ab}\right) + c\left(\frac{s(s-b)}{ac}\right)$$

$$= \frac{s}{a}(s-c+s-b)$$

$$= \frac{s}{a}(2s-c-b)$$

$$= \frac{s}{a}(a+b+c-c-b)$$

$$= s$$

Ex-2. In a Δ ABC, $3a = b + c$, then find the value of $\cot\dfrac{B}{2}\cot\dfrac{C}{2}$.

Soln. We have, $\cot\left(\dfrac{B}{2}\right)\cot\left(\dfrac{C}{2}\right)$

$$= \sqrt{\frac{s(s-b)}{(s-a)(s-c)}} \times \sqrt{\frac{s(s-c)}{(s-a)(s-b)}}$$

$$= \sqrt{\frac{s(s-b)}{(s-a)(s-c)} \times \frac{s(s-c)}{(s-a)(s-b)}}$$

$$= \sqrt{\frac{s^2}{(s-a)^2}}$$

$$= \frac{s}{s-a}$$

$$= \frac{2s}{2s-2a}$$

$$= \frac{a+b+c}{a+b+c-2a}$$

$$= \frac{4a}{4a-2a}$$

$$= 2$$

Ex-3. In a ΔABC, if $\cot\frac{A}{2}, \cot\frac{B}{2}, \cot\frac{C}{2}$ are in A.P., then prove that a, b, c are in A.P.

Soln. Given $\cot\frac{A}{2}, \cot\frac{B}{2}, \cot\frac{C}{2}$ are in A.P

$$\Rightarrow \quad 2\cot\left(\frac{B}{2}\right) = \cot\left(\frac{A}{2}\right) + \cot\left(\frac{C}{2}\right)$$

$$\Rightarrow \quad 2\sqrt{\frac{s(s-b)}{(s-a)(s-c)}}$$

$$= \sqrt{\frac{s(s-a)}{(s-b)(s-c)}} + \sqrt{\frac{s(s-c)}{(s-a)(s-b)}}$$

$$\Rightarrow \quad 2\sqrt{\frac{(s-b)}{(s-a)(s-c)}}$$

$$= \sqrt{\frac{(s-a)}{(s-b)(s-c)}} + \sqrt{\frac{(s-c)}{(s-a)(s-b)}}$$

$$\Rightarrow \quad 2\sqrt{\frac{(s-b)^2}{(s-b)(s-a)(s-c)}}$$

$$= \sqrt{\frac{(s-a)^2}{(s-a)(s-b)(s-c)}}$$

$$+ \sqrt{\frac{(s-c)^2}{(s-a)(s-b)(s-c)}}$$

$$\Rightarrow \quad 2(s-b) = (s-a)+(s-c)$$

$$\Rightarrow \quad 2(s-b) = (2s-a-c)$$

$$\Rightarrow \quad 2b = a+c$$

$$\Rightarrow \quad a, b, c \in A.P$$

Exercise - 5

(Questions based on Half-Angles Formulae:)

1. In a ΔABC, if $a = 13$, $b = 14$ and $c = 15$, find the value of :

   i) $\sin\frac{A}{2}$ ii) $\cos\frac{B}{2}$

   iii) $\cos A$

2. In a ΔABC, $2ac \sin\left(\frac{A-B+C}{2}\right) =$

   (a) $a^2 + b^2 - c^2$ (b) $c^2 + a^2 - b^2$

   (c) $b^2 - c^2 - a^2$ (d) $c^2 - a^2 - b^2$

3. In a ΔABC, $3a = b + c$, then the value of $\cot\frac{B}{2}\cot\frac{C}{2}$ is

   (a) 1   (b) $\sqrt{3}$

   (c) 2   (d) None

4. In a ΔABC, if $\cos A + \cos C = 4\sin^2\left(\frac{B}{2}\right)$, then a, b, c are in

   (a) A.P   (b) G.P.

   (c) H.P.   (d) A.G.P.

5. In a ΔABC, if $(s-a)(s-b) = s(s-c)$ then the angle C is

   (a) 90°   (b) 45°

   (c) 30°   (d) 60°

6. In a ΔABC, if $\cot\frac{A}{2}, \cot\frac{B}{2}, \cot\frac{C}{2}$ and in A.P., then a, b, c are in

   (a) AP   (b) G.P.

   (c) H.P.   (d) A.G.P.

7. In a ΔABC, $\tan\frac{A}{2}\tan\frac{B}{2}$ is

(a) $(s-b):(s-c)$ (b) $(s-c):s$
(c) $(s-a):(s-b)$ (d) $(s-b):(s-a)$

Answers

2.	(b)	3.	(c)	4.	(a)
5.	(a)	6.	(c)	7.	(b)

6.7 AREA OF TRIANGLE :

Statement :

Prove that the area of ΔABC is given by

i) $\Delta = \frac{1}{2}$ bc sin A

ii) $\Delta = \frac{1}{2}$ ca sin B

iii) $\Delta = \frac{1}{2}$ ab sin C

Proof: (i) We have, $\Delta = \frac{1}{2}\times AB\times CD$

$\Rightarrow \Delta = \frac{1}{2}\times c\times b\sin A$

$\Rightarrow \Delta = \frac{1}{2}bc\sin A$

Similary, we can easily proved that (ii) and (iii).

1) Area of a Triangle (Hero's Formula)

In any ΔABC : $\Delta = \sqrt{s(s-a)(s-b)(s-c)}$

Proof: As, we know that $\Delta = \frac{1}{2}bc\sin A$

$= \frac{1}{2}\times bc\times 2\sin\left(\frac{A}{2}\right)\cos\left(\frac{A}{2}\right)$

$= bc\times\sqrt{\frac{(s-b)(s-c)}{bc}}\times\sqrt{\frac{s(s-a)}{bc}}$

$= \sqrt{s(s-a)(s-b)(s-c)}$

Ex-1. In any ΔABC, if $a=\sqrt{2}$ cm, $b=\sqrt{3}$ cm. and $c=\sqrt{5}$ cm., then find the area of the ΔABC.

Soln. $\Delta ABC = \frac{1}{2}\times\sqrt{2}\times\sqrt{3} = \sqrt{\frac{3}{2}}$ s.u.

Ex-2. In any ΔABC, prove that $\Delta = \frac{a^2-b^2}{2}\times\frac{\sin A\sin B}{\sin(A-B)}$.

Soln. We have, $\frac{a^2-b^2}{2}\times\frac{\sin A\sin B}{\sin(A-B)}$

$= \frac{k^2(\sin^2 A-\sin^2 B)}{2}\times\frac{\sin A.\sin B}{\sin(A-B)}$

$= \frac{k^2\times\sin(A+B)\times\sin(A-B)}{2}\times\frac{\sin A.\sin B}{\sin(A-B)}$

$= \frac{k^2\times\sin(A+B)\times\sin A.\sin B}{2}$

$= \frac{k^2\times\sin(\pi-C)\times\sin A.\sin B}{2}$

$= \frac{k^2\times\sin C\times\frac{a}{k}\times\frac{b}{k}}{2}$

$= \frac{1}{2}\times a\,b\sin C$

$= \Delta$

Ex-3. If the angles of a triangle are 30° and 45° and the included side is $(\sqrt{3}+1)$cm., then prove that the area of the triangle is $\frac{1}{2}(\sqrt{3}+1)$ cm^2.

Soln. As we know that $\frac{a}{\sin A} = \frac{b}{\sin B} = \frac{c}{\sin C}$

$\Rightarrow \frac{(\sqrt{3}+1)}{\sin(105^o)} = \frac{b}{\sin 45^o} = \frac{c}{\sin 30^o}$

$\Rightarrow \frac{(\sqrt{3}+1)}{\cos(15^o)} = \frac{b}{\sin 45^o} = \frac{c}{\sin 30^o}$

$\Rightarrow \frac{(\sqrt{3}+1)}{\frac{\sqrt{3}+1}{2\sqrt{2}}} = \frac{b}{\frac{1}{\sqrt{2}}} = \frac{c}{\frac{1}{2}}$

$\Rightarrow 2\sqrt{2} = b\sqrt{2} = 2c$

$\Rightarrow b = 2$ & $c = \sqrt{2}$

Hence, the area of the triangle is $= \frac{1}{2} bc\sin A$

$$= \frac{1}{2}\times 2\times\sqrt{2}\times\sin(105^\circ)$$

$$= \frac{1}{2}\times 2\times\sqrt{2}\times\frac{\sqrt{3}+1}{2\sqrt{2}}$$

$$= \frac{(\sqrt{3}+1)}{2} \text{ sq. u.}$$

Ex-4. In a ΔABC, prove that :

$$\cot\frac{A}{2}+\cot\frac{B}{2}+\cot\frac{C}{2} = \frac{(a+b+c)}{(b+c-a)}\cot\frac{A}{2}$$

Soln. We have, $\cot\frac{A}{2}+\cot\frac{B}{2}+\cot\frac{C}{2}$

$$= \sqrt{\frac{s(s-a)}{(s-b)(s-c)}}+\sqrt{\frac{s(s-b)}{(s-a)(s-c)}}+\sqrt{\frac{s(s-c)}{(s-a)(s-b)}}$$

$$= \sqrt{\frac{s^2(s-a)^2}{s(s-a)(s-b)(s-c)}}+\sqrt{\frac{s^2(s-b)^2}{s(s-b)(s-a)(s-c)}}+\sqrt{\frac{s^2(s-c)^2}{s(s-c)(s-a)(s-b)}}$$

$$= \sqrt{\frac{s^2(s-a)^2}{\Delta^2}}+\sqrt{\frac{s^2(s-b)^2}{\Delta^2}}+\sqrt{\frac{s^2(s-c)^2}{\Delta^2}}$$

$$= \frac{s(s-a)}{\Delta}+\frac{s(s-b)}{\Delta}+\frac{s(s-c)}{\Delta}$$

$$= \frac{s}{\Delta}(s-a+s-b+s-c)$$

$$= \frac{s}{\Delta}(3s-(a+b+c))$$

$$= \frac{s}{\Delta}(3s-2s)$$

$$= \frac{s^2}{\Delta}$$

$$= \frac{s^2}{\sqrt{s(s-a)(s-b)(s-c)}}$$

$$= \frac{s}{(s-a)}\times\sqrt{\frac{s(s-a)}{(s-b)(s-c)}}$$

$$= \frac{2s}{(2s-2a)}\times\sqrt{\frac{s(s-a)}{(s-b)(s-c)}}$$

$$= \frac{(a+b+c)}{(b+c-a)}\times\cot\left(\frac{A}{2}\right)$$

Ex-5. In a ΔABC, prove that :

$$\frac{\cot\left(\frac{A}{2}\right)+\cot\left(\frac{B}{2}\right)+\cot\left(\frac{C}{2}\right)}{\cot A+\cot B+\cot C} = \frac{(a+b+c)^2}{a^2+b^2+c^2}$$

Soln. We have, $\cot A+\cot B+\cot C$

$$= \frac{\cos A}{\sin A}+\frac{\cos B}{\sin B}+\frac{\cos C}{\sin C}$$

$$= \frac{\frac{b^2+c^2-a^2}{2bc}}{ak}+\frac{\frac{c^2+a^2-b^2}{2ac}}{bk}+\frac{\frac{a^2+b^2-c^2}{2ab}}{ck}$$

$$= \frac{b^2+c^2-a^2}{2abck}+\frac{c^2+a^2-b^2}{2abck}+\frac{a^2+b^2-c^2}{2abck}$$

$$= \frac{a^2+b^2+c^2}{2abck}$$

$$= \frac{(a^2+b^2+c^2)}{4\times\left(\frac{1}{2}ab\right)\times\sin C}$$

$$= \frac{(a^2+b^2+c^2)}{4\times\Delta} \quad \text{.............(i)}$$

Also, $\cot\left(\frac{A}{2}\right)+\cot\left(\frac{B}{2}\right)+\cot\left(\frac{C}{2}\right)$

$$= \frac{(a+b+c)}{(b+c-a)}\times\cot\left(\frac{A}{2}\right)$$

$$= \frac{s^2}{\Delta} \quad \text{...................(ii)}$$

Dividing (ii) by (i), we get,

$$\frac{\cot\left(\frac{A}{2}\right)+\cot\left(\frac{B}{2}\right)+\cot\left(\frac{C}{2}\right)}{\cot(A)+\cot(B)+\cot(C)}=\frac{s^2}{4\left(a^2+b^2+c^2\right)}$$

$$\Rightarrow \frac{\cot\left(\frac{A}{2}\right)+\cot\left(\frac{B}{2}\right)+\cot\left(\frac{C}{2}\right)}{\cot(A)+\cot(B)+\cot(C)}=\frac{(a+b+c)^2}{\left(a^2+b^2+c^2\right)}$$

Ex.-6. If in a ΔABC, prove that $\Delta < \frac{s^2}{4}$

Soln. Let a, b, and c are the sides of a triangle and s be the semi perimeter.

Let the four quantities are s, (s - a), (s - b) and (s - c)

Applying, A.M. $\geq$ G. M, we get,

$$\Rightarrow \frac{s+(s-a)+(s-b)+(s-c)}{4}$$

$$\geq \sqrt[4]{s(s-a)(s-b)(s-c)}$$

$$\Rightarrow \frac{4s-(a+b+c)}{4} \geq \sqrt[4]{\Delta^2}$$

$$\Rightarrow \frac{4s-2s}{4} \geq (\Delta)^{\frac{1}{2}}$$

$$\Rightarrow \frac{s}{2} \geq (\Delta)^{\frac{1}{2}}$$

$$\Rightarrow \Delta < \frac{s^2}{4}$$

Ex-7. If α, β, γ are the lengths of the altitudes of a triangle ABC, then prove that

$$\frac{1}{\alpha^2}+\frac{1}{\beta^2}+\frac{1}{\gamma^2}=\frac{1}{\Delta}(\cot A+\cot B+\cot C)$$

Soln. Let AD = α, BE = β and CF = γ

Then,

$$\Delta=\frac{1}{2}\times a\times AD=\frac{1}{2}\times b\times BE=\frac{1}{2}\times c\times CF$$

$$\Rightarrow AD=\frac{2\Delta}{a},\ BE=\frac{2\Delta}{b},\ CF=\frac{2\Delta}{c}$$

$$\Rightarrow \alpha=\frac{2\Delta}{a},\ \beta=\frac{2\Delta}{b},\ \gamma=\frac{2\Delta}{c}$$

Now, $\frac{1}{\alpha^2}+\frac{1}{\beta^2}+\frac{1}{\gamma^2}$

$$=\frac{a^2}{4\Delta^2}+\frac{b^2}{4\Delta^2}+\frac{c^2}{4\Delta^2}$$

$$=\frac{\left(a^2+b^2+c^2\right)}{4\Delta^2}$$

$$=\frac{1}{\Delta}\times\frac{\left(a^2+b^2+c^2\right)}{4\Delta}$$

$$=\frac{1}{\Delta}\times(\cot A+\cot B+\cot C)$$

$$=\frac{(\cot A+\cot B+\cot C)}{\Delta}$$

Hence, the result.

Ex.-8. If p_1, p_2, p_3 are the altitudes of a triangle from the vertices A, B, C and Δ be the area of the triangle ABC, prove that

$$\frac{1}{p_1}+\frac{1}{p_2}-\frac{1}{p_3}=\frac{2ab}{(a+b+c)\times\Delta}\times\cos^2\left(\frac{C}{2}\right).$$

Soln. Let AD = p_1, BE = p_2 and CF = p_3

Then, $\Delta=\frac{1}{2}\times a\times p_1=\frac{1}{2}\times b\times p_2=\frac{1}{2}\times c\times p_3$

$$\Rightarrow p_1=\frac{2\Delta}{a},\ p_2=\frac{2\Delta}{b},\ p_3=\frac{2\Delta}{c}$$

Now, $\frac{1}{p_1}+\frac{1}{p_2}-\frac{1}{p_3}$

$$=\frac{a}{2\Delta}+\frac{b}{2\Delta}-\frac{c}{2\Delta}$$

$$=\frac{(a+b-c)}{2\Delta}$$

$$=\frac{(a+b+c-2c)}{2\Delta}$$

$$=\frac{(2s-2c)}{2\Delta}$$

$$=\frac{(s-c)}{\Delta}$$

$$=\frac{2ab\times s(s-c)}{\Delta\times s}\times\frac{1}{2ab}$$

$$= \frac{2ab}{\Delta \times s} \times \frac{s(s-c)}{2ab}$$

$$= \frac{2\,ab}{(a+b+c)\Delta} \times \cos^2\left(\frac{C}{2}\right)$$

Ex-9. If a, b, c and d are the sides of a quadrilateral, then find the minimum value of $\frac{a^2+b^2+c^2}{d^2}$

Soln. We have, $(a-b)^2+(b-c)^2+(c-d)^2 \geq 0$

$\Rightarrow \quad 2(a^2+b^2+c^2) \geq 2(ab+bc+ca)$

$\Rightarrow \quad 3(a^2+b^2+c^2) \geq (a^2+b^2+c^2) + 2(ab+bc+ca)$

$\Rightarrow \quad 3(a^2+b^2+c^2) > (a+b+c)^2 > d^2$

$\Rightarrow \quad \frac{3(a^2+b^2+c^2)}{d^2} > 1$

$\Rightarrow \quad \frac{(a^2+b^2+c^2)}{d^2} > \frac{1}{3}$

Thus, the minimum value of $\frac{(a^2+b^2+c^2)}{d^2}$ is $\frac{1}{3}$

Ex-10. In a triangle ABC, if $\cos A + \cos B + \cos C = \frac{3}{2}$, then the triangle is equilateral.

Soln. We have, $\cos A + \cos B + \cos C = \frac{3}{2}$

$\Rightarrow \quad 1+4\sin\left(\frac{A}{2}\right)\sin\left(\frac{B}{2}\right)\sin\left(\frac{C}{2}\right)=\frac{3}{2}$

$\Rightarrow \quad \sin\left(\frac{A}{2}\right)\sin\left(\frac{B}{2}\right)\sin\left(\frac{C}{2}\right)=\frac{1}{8}$

It is possible only when

$\sin\left(\frac{A}{2}\right)=\frac{1}{2},\ \sin\left(\frac{B}{2}\right)=\frac{1}{2},\ \sin\left(\frac{C}{2}\right)=\frac{1}{2}$

$\Rightarrow \quad \frac{A}{2}=\frac{\pi}{6},\ \frac{B}{2}=\frac{\pi}{6},\ \frac{C}{2}=\frac{\pi}{6}$

$\Rightarrow \quad A=\frac{\pi}{3},\ B=\frac{\pi}{3},\ C=\frac{\pi}{3}$

$\Rightarrow \quad \Delta$ is an equilateral .

Ex-11 In a ΔABC, $c\,(a+b)\cos\frac{B}{2} = b\,(a+c)\cos\frac{C}{2}$, then prove that the triangle is isosceles.

Soln. We have, $c\,(a+b)\cos\frac{B}{2} = b\,(a+c)\cos\frac{C}{2}$

$\Rightarrow \quad c(a+b)\times\sqrt{\frac{s(s-b)}{ac}} = b(a+c)\times\sqrt{\frac{s(s-c)}{ab}}$

$\Rightarrow \quad c(a+b)\times\sqrt{\frac{(s-b)}{c}} = b(a+c)\times\sqrt{\frac{(s-c)}{b}}$

$\Rightarrow \quad c^2(a+b)^2\times\frac{(s-b)}{c} = b^2(a+c)^2\times\frac{(s-c)}{b}$

$\Rightarrow \quad c(a+b)^2\times(s-b) = b(a+c)^2\times(s-c)$

$\Rightarrow \quad c(a+b)^2\times(s-b) = b(a+c)^2\times(s-c)$

$\Rightarrow \quad c(a+b)^2\times(a+c-b) = b(a+c)^2\times(a+b-c)$

$\Rightarrow \quad \frac{(a+c-b)}{b(a+c)^2} = \frac{(a+b-c)}{c(a+b)^2}$

$\Rightarrow \quad \frac{1}{b(a+c)} - \frac{1}{(a+c)^2} = \frac{1}{c(a+b)} - \frac{1}{(a+b)^2}$

$\Rightarrow \quad \frac{1}{b(a+c)} - \frac{1}{c(a+b)} = \frac{1}{(a+c)^2} - \frac{1}{(a+b)^2}$

$\Rightarrow \quad \frac{ac+c^2-ab-b^2}{bc(a+b)(a+c)} = \frac{(a+b)^2-(a+c)^2}{(a+c)^2(a+b)^2}$

$\Rightarrow \quad \frac{a(c-b)+(c^2-b^2)}{bc} = \frac{2a(b-c)+(b^2-c^2)}{(a+c)(a+b)}$

$\Rightarrow \quad \frac{(c-b)(a+b+c)}{bc} = \frac{(b-c)(2a+b+c)}{(a+c)(a+b)}$

$\Rightarrow \quad (c-b)\left(\frac{(a+b+c)}{bc} + \frac{(2a+b+c)}{(a+c)(a+b)}\right) = 0$

$\Rightarrow \quad (c-b)=0 \left(\because \left(\frac{(a+b+c)}{bc} + \frac{(2a+b+c)}{(a+c)(a+b)}\right) \neq 0\right)$

$\Rightarrow \quad b=c$

$\Rightarrow \quad \Delta$ is isosceles .

Exercise - 6

(Questions based on Area of Triangle)

1. In a ΔABC, if $\angle A = 60°$, $b = 4$ cm. and $c = \sqrt{3}$ cm. then the area of the ΔABC is

(a) 3 cm² (b) 5 cm²
(c) 8 cm² (d) None

2. In a ΔABC, If $\frac{\cos A}{a} = \frac{\cos B}{b} = \frac{\cos C}{c}$ and the side $a = 2$, then the area of the triangle is

(a) 1 (b) 2
(c) $\frac{\sqrt{3}}{2}$ (d) $\sqrt{3}$

3. If in a triangle ABC, $a = 6$, $b = 3$ and $\cos(A - B) = 4/5$, then its area in sq. units is

(a) 3 (b) 6
(c) 9 (d) 12

4. If $\cos A + \cos B + \cos C = 3/2$, then triangle ABC is

(a) Rt angled (b) Isoceles
(c) equilateral (d) None.

5. The two adjacent sides of a cyclic quadrilateral are 2 and 3 and the angle between them is 60. If the area of the quadrilateral is $4\sqrt{3}$, then the remaining two sides are

(a) 2, 3 (b) 1, 2
(c) 3, 4 (d) 2, 2

6. If a, b, c , d be the sides of a quadrilateral, then the minimum value of $\frac{a^2+b^2+c^2}{d^2}$ is

(a) 1/2 (b) 1/3
(c) 1/4 (d) 1

Answers

1.(a)	2.(d)	3.(c)
4.(c)	5.(a)	6.(b)

Radii of Circle Connected with a Triangle

6.8 INTRODUCTION :

On this topic, we shall discuss various circles connected with a triangle and the formula for their radii in terms of elements of the triangle.

6.9 CIRCUMCIRCLE OF A TRIANGLE AND ITS RADIUS :

The circle which passess through the angular points of a Δ is called the circumcircle. The centre of this circle is the point of intersection of perpendicular bisectors of the sides and is called the circumcentre and its radius is always denoted by R.

The circumcentre may lie within outside or upon one of the sides of the Δ. In a right angled triangle the circumcentre is the vertex whose right angle is formed.

6.10 CIRCUM-RADIUS :

Statement : Prove that the circum-radius R of a ΔABC is given by

i) $$R = \frac{a}{2\sin A} = \frac{b}{2\sin B} = \frac{c}{2\sin C}$$

ii) $$R = \frac{abc}{4\Delta}$$

Proof :

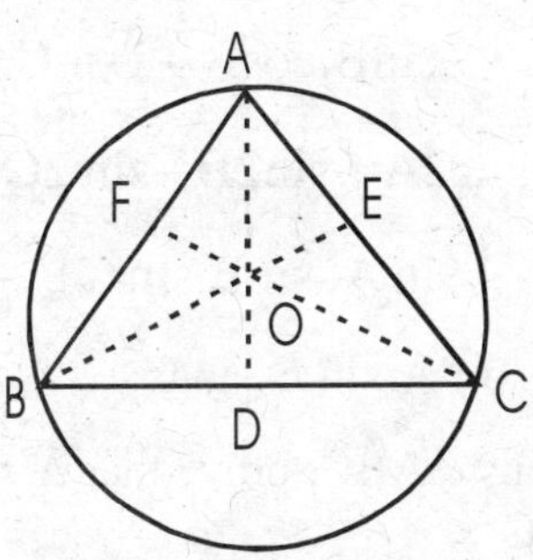

Let ABC be a Δ and let the perpendicular bisectors of its sides BC, CA and AB intersect at O. Then O is the circumcentre such that

$$OA = OB = OC = R$$

i) We have $\angle BOC = 2\angle BAC = 2A$

$\Rightarrow$ $\angle BOD = \angle COD = A$

In ΔBOD, $\sin A = \frac{BD}{OB} = \frac{\frac{a}{2}}{R} = \frac{a}{2R}$

$\Rightarrow$ $R = \frac{a}{2\sin A}$

Similarly, it can be shown that

$$R = \frac{b}{2\sin B} \text{ and } R = \frac{c}{2\sin C}$$

Hence, $R = \frac{a}{2\sin A} = \frac{b}{2\sin B} = \frac{c}{2\sin C}$

ii) As we know that,

$\Delta = \frac{1}{2}$ bc sinA.

$\Rightarrow \sin A = \frac{2\Delta}{bc}$

$\Rightarrow \frac{a}{2R} = \frac{2\Delta}{bc}$

$\Rightarrow 4R\Delta = abc$

$\Rightarrow R = \frac{abc}{4\Delta}$

Hence, the result.

Ex-1. In any ΔABC, prove that a cos A + b cos B + c cos C = 4R sinA sinB sinC

Soln. We have, a cos A + b cos B + c cos C

$= 2R(\sin A.\cos A + \sin B.\cos B + \sin C.\cos C)$

$= \frac{2R}{2}\left(\begin{array}{l}2\sin A.\cos A + \\ 2\sin B.\cos B + 2\sin C.\cos C\end{array}\right)$

$= R(\sin 2A + \sin 2B + \sin 2C)$

$= R(4\sin A.\sin B.\sin C)$

$= 4R.\sin A.\sin B.\sin C$

Ex-2. In any ΔABC, prove that $\Delta = 2R^2 \sin A \sin B \sin C$

Soln. We have, $\Delta = \frac{1}{2} \times a \times b \times \sin C$

$\Rightarrow \Delta = \frac{1}{2} \times 2R\sin A \times 2R\sin B \times \sin C$

$\Rightarrow \Delta = 2R^2 \times \sin A.\sin B.\sin C$

Ex.-3. In any triangle ABC, prove that,

$\frac{\sin A}{a} + \frac{\sin B}{b} + \frac{\sin C}{c} = \frac{3}{2R}$

Soln. We have, $\frac{\sin A}{a} + \frac{\sin B}{b} + \frac{\sin C}{c}$

$= \frac{\sin A}{2R\sin A} + \frac{\sin B}{2R\sin B} + \frac{\sin C}{2R\sin C}$

$= \frac{1}{2R} + \frac{1}{2R} + \frac{1}{2R}$

$= \frac{3}{2R}$

Ex-4. In any triangle ABC, a, b, c are in A.P and p_1, p_2 and p_3 are the altitudes of the given triangle, then prove that, $\frac{1}{p_1} + \frac{1}{p_2} + \frac{1}{p_3} \le \frac{3R}{\Delta}$

Soln. Let AD = p_1, BE = p_2 and CF = p_3.

Then.

$\Delta = \frac{1}{2} \times a \times p_1 = \frac{1}{2} \times b \times p_2 = \frac{1}{2} \times c \times p_3$

$\Rightarrow p_1 = \frac{2\Delta}{a},\ p_2 = \frac{2\Delta}{b},\ p_3 = \frac{2\Delta}{c}$

Now, $\frac{1}{p_1} + \frac{1}{p_2} + \frac{1}{p_3}$

$= \frac{a}{2\Delta} + \frac{b}{2\Delta} + \frac{c}{2\Delta}$

$= \frac{a+b+c}{2\Delta}$

$= \frac{2R(\sin A + \sin B + \sin C)}{2\Delta}$

$\le \frac{3R}{\Delta}$ $(\because \sin A \le 1,\ \sin B \le 1,\ \sin C \le 1)$

Exercise - 7

(Questions based on Circum-radius of a Triangle:)

1. In a ΔABC, if a = 18 cm., b = 24 cm. and c = 30 cm., then R is

(a) 16 cm. (b) 15 cm.
(c) 24 cm. (d) 30 cm.

2. If the length of the sides of a Δ are 3, 4 and 5 units, then the circum-radius R is

(a) 2 (b) 2.5
(c) 3 (d) 3.5

3. In an equilateral triangle of side $2\sqrt{3}$ cm., then the circum-radius R is

(a) 1 cm. (b) $\sqrt{3}$ cm.
(c) 2 cm. (d) $2\sqrt{3}$ cm.

4. If $8R^2 = a^2 + b^2 + c^2$, then the Δ is

(a) Right angled (b) Isosceles
(c) Equilateral (d) None

5. If p_1, p_2 and p_3 are the altitudes of a triangle ABC, from the vertices A, B and C respectively. then

the value of $\frac{\cos A}{p_1}+\frac{\cos B}{p_2}+\frac{\cos C}{p_3}$ is

(a)* $\frac{1}{R}$ (b) $\frac{2}{R}$

(c) $\frac{3}{R}$ (d) $\frac{4}{R}$

Answers

1.(b) 2.(b) 3.(c)
4.(a) 5.(a)

Inscribed Circle and its Radius

On this topic, we shall discuss the formula for the circle which can be inscribed within a Δ that touches each of the sides. Such a circle is called its inscribed circle or incircle.

The centre of this circle is the point of intersection of the bisecter of the angles of the triangle . The radius of this circle is a always denoted by 'r' and is equal to the length of the perpendicular from its centre to any of the sides of the Δ.

6.11 IN-RADIUS :

Statement : Prove that the in-radius r of the inscribed circle of a triangle ABC is given by

i) $r=\frac{\Delta}{s}$

ii) $r=(s-a)\tan\frac{A}{2}$; $r=(s-b)\tan\frac{B}{2}$;

$r=(s-c)\tan\frac{C}{2}$

iii) $r=\frac{a\sin\left(\frac{B}{2}\right)\sin\left(\frac{C}{2}\right)}{\cos\left(\frac{A}{2}\right)}$;

$r=\frac{b\sin\left(\frac{A}{2}\right)\sin\left(\frac{C}{2}\right)}{\cos\left(\frac{B}{2}\right)}$; $r=\frac{c\sin\left(\frac{B}{2}\right)\sin\left(\frac{A}{2}\right)}{\cos\left(\frac{C}{2}\right)}$

iv) $r=4R\sin\left(\frac{A}{2}\right)\sin\left(\frac{B}{2}\right)\sin\left(\frac{C}{2}\right)$

Proof :

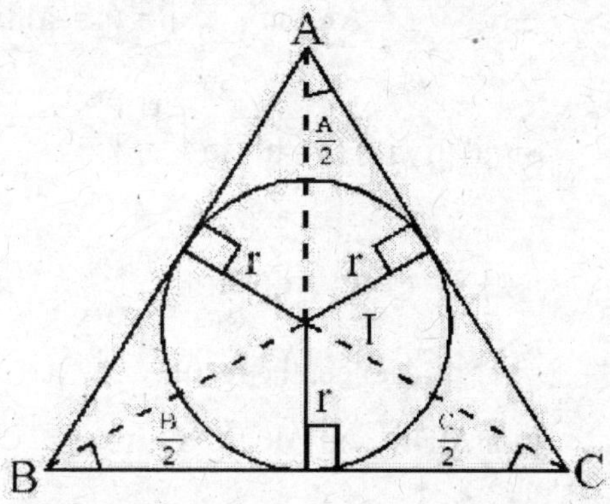

Let ABC be a triangle such that the internal bisectors of the angles of the triangle intersect at I. Suppose the incircle touches the sides BC, CA and AB at D, E and F resp. Then ID, IE and IF are perpendicular to these sides and

ID = IE = IF = r.

i) We have, ar (ΔABC) = ar (ΔIBC) + ar (ΔICA) + ar (ΔIAB)

$\Rightarrow \quad \Delta=\frac{1}{2}ar+\frac{1}{2}br+\frac{1}{2}cr$

$\Rightarrow \quad \Delta=\frac{1}{2}r(a+b+c)$

$=\frac{1}{2}r\times 2s$

$=rs$

$\Rightarrow \quad r=\frac{\Delta}{s}$

ii) We know that the length of the tangents to a circle from a given point are equal.

∴ AE = AF, BD = BF and CD = CE

Now, 2 s = a + b+ c

= BC + CA + AB

= (BD + DC) + (CE + EA) + (AF + FB)

= 2 (BD + AE + CD)

= 2 (BC + AE)

= 2 (a + AE)

$\Rightarrow$ s = a + AE.

$\Rightarrow$ AE = s – a

Now, in ΔIAE, $\tan\frac{A}{2}=\frac{r}{AE}$

$\Rightarrow$ $r = AE \tan\left(\frac{A}{2}\right) = (s-a)\tan\left(\frac{A}{2}\right)$

Similarly $r = (s-b)\tan\left(\frac{B}{2}\right)$,

$r = (s-c)\tan\left(\frac{C}{2}\right)$.

Hence, the result.

iii) In a ΔIBD and ΔICD, we have

$$\tan\left(\frac{B}{2}\right) = \frac{r}{BD} \text{ and } \tan\left(\frac{C}{2}\right) = \frac{r}{CD}$$

$$\Rightarrow \quad BD = \frac{r}{\tan\left(\frac{B}{2}\right)},\ CD = \frac{r}{\tan\left(\frac{C}{2}\right)}$$

Now, $a = BD + CD$

$$= \frac{r}{\tan\left(\frac{B}{2}\right)} + \frac{r}{\tan\left(\frac{C}{2}\right)}$$

$$= r\left[\frac{\cos\left(\frac{B}{2}\right)}{\sin\left(\frac{B}{2}\right)} + \frac{\cos\left(\frac{C}{2}\right)}{\sin\left(\frac{C}{2}\right)}\right]$$

$$= r\left[\frac{\sin\left(\frac{C}{2}\right)\cos\left(\frac{B}{2}\right) + \cos\left(\frac{C}{2}\right)\sin\left(\frac{B}{2}\right)}{\sin\left(\frac{B}{2}\right)\sin\left(\frac{C}{2}\right)}\right]$$

$$= r\frac{\sin\left(\frac{C}{2}+\frac{B}{2}\right)}{\sin\left(\frac{B}{2}\right)\sin\left(\frac{C}{2}\right)}$$

$$= r\frac{\sin\left(\frac{\pi}{2}-\frac{A}{2}\right)}{\sin\left(\frac{B}{2}\right)\sin\left(\frac{C}{2}\right)}$$

$$= r\frac{\cos\left(\frac{A}{2}\right)}{\sin\left(\frac{B}{2}\right)\sin\left(\frac{C}{2}\right)}$$

$$\Rightarrow \quad r = a\frac{\sin\left(\frac{B}{2}\right)\sin\left(\frac{C}{2}\right)}{\cos\left(\frac{A}{2}\right)}$$

Similarly, $r = \dfrac{b\sin\left(\frac{C}{2}\right)\sin\left(\frac{A}{2}\right)}{\cos\left(\frac{B}{2}\right)}$,

$$r = \frac{c\sin\left(\frac{A}{2}\right)\sin\left(\frac{B}{2}\right)}{\cos\left(\frac{C}{2}\right)}$$

iv) We have, $r = \dfrac{a\sin\left(\frac{B}{2}\right)\sin\left(\frac{C}{2}\right)}{\cos\left(\frac{A}{2}\right)}$

$$= 2R\sin A \times \frac{\sin\left(\frac{B}{2}\right)\sin\left(\frac{C}{2}\right)}{\cos\left(\frac{A}{2}\right)}$$

$$= \frac{4R\sin\left(\frac{A}{2}\right)\cos\left(\frac{A}{2}\right)\sin\left(\frac{B}{2}\right)\sin\left(\frac{C}{2}\right)}{\cos\left(\frac{A}{2}\right)}$$

$$= 4R\sin\left(\frac{A}{2}\right)\sin\left(\frac{B}{2}\right)\sin\left(\frac{C}{2}\right)$$

Hence, the result.

Ex-1. In any ΔABC, prove that $a\cos A + b\cos B + c\cos C = 4R\sin A\sin B\sin C$.

Soln. We have, $a\cos A + b\cos B + c\cos C$

$= 2R(\sin A\cos A + \sin B.\cos B + \sin C.\cos C)$

$= R\begin{pmatrix}2\sin A\cos A + 2\sin B.\cos B \\ +2\sin C.\cos C\end{pmatrix}$

$= R(\sin 2A + \sin 2B + \sin 2C)$

$= R(4\sin A.\sin B.\sin C)$

$= 4(R\sin A.\sin B.\sin C)$

Ex-2. In a ΔABC, prove that

$$\frac{1}{ab} + \frac{1}{bc} + \frac{1}{ca} = \frac{1}{2rR}$$

Soln. We have, $\frac{1}{ab}+\frac{1}{bc}+\frac{1}{ca}$

$$= \frac{a+b+c}{abc}$$

$$= \frac{2s}{4\Delta R}$$

$$= \frac{\frac{\Delta}{r}}{2\Delta R}$$

$$= \frac{1}{2rR}$$

Ex-3. In a ΔABC, prove that $\cos A + \cos B + \cos C = \left(1+\frac{r}{R}\right)$

Soln. We have, $\cos A + \cos B + \cos C$

$$= 2\cos\left(\frac{A+B}{2}\right)\cos\left(\frac{A-B}{2}\right)+\cos C$$

$$= 2\cos\left(\frac{\pi}{2}-\frac{C}{2}\right)\cos\left(\frac{A-B}{2}\right)+\cos C$$

$$= 2\sin\left(\frac{C}{2}\right)\cos\left(\frac{A-B}{2}\right)+\cos C$$

$$= 2\sin\left(\frac{C}{2}\right)\cos\left(\frac{A-B}{2}\right)+1-2\sin^2\left(\frac{C}{2}\right)$$

$$= 1+2\sin\left(\frac{C}{2}\right)\left(\cos\left(\frac{A-B}{2}\right)-\sin\left(\frac{C}{2}\right)\right)$$

$$= 1+2\sin\left(\frac{C}{2}\right)\left(\cos\left(\frac{A-B}{2}\right)-\cos\left(\frac{A+B}{2}\right)\right)$$

$$= 1+2\sin\left(\frac{C}{2}\right)\left(2\sin\left(\frac{A}{2}\right)\sin\left(\frac{B}{2}\right)\right)$$

$$= 1+4\sin\left(\frac{A}{2}\right)\sin\left(\frac{B}{2}\right)\sin\left(\frac{C}{2}\right)$$

$$= \left(1+\frac{r}{R}\right)$$

Ex-4. In a ΔABC, prove that $\sin A + \sin B + \sin C = \frac{s}{R} = \frac{\Delta}{Rr}$

Soln. We have, $\sin A + \sin B + \sin C$

$$= \frac{a}{2R}+\frac{b}{2R}+\frac{c}{2R}$$

$$= \frac{a+b+c}{2R}$$

$$= \frac{2s}{2R}$$

$$= \frac{s}{R}$$

$$= \frac{\frac{\Delta}{r}}{R}$$

$$= \frac{\Delta}{rR}$$

Ex-5. In any ΔABC, prove that $a \cot A + b \cot B + c \cot C = 2(r + R)$

Soln. We have, $a \cot A + b \cot B + c \cot C$

$$= \left(2R\sin A \times \frac{\cos A}{\sin A} + 2R\sin B \times \frac{\cos B}{\sin B} + 2R\sin C \times \frac{\cos C}{\sin C}\right)$$

$$= 2R(\cos A+\cos B+\cos C)$$

$$= 2R\left(1+\frac{r}{R}\right)$$

$$= 2(R+r)$$

Ex-6. In any ΔABC, prove that

$$\cos^2\frac{A}{2} + \cos^2\frac{B}{2} + \cos^2\frac{C}{2} = 2 + \frac{r}{2R}.$$

Soln. We have, $\cos^2\frac{A}{2} + \cos^2\frac{B}{2} + \cos^2\frac{C}{2}$

$$= \frac{1}{2}\left(2\cos^2\left(\frac{A}{2}\right)+2\cos^2\left(\frac{B}{2}\right)+2\cos^2\left(\frac{C}{2}\right)\right)$$

$$= \frac{1}{2}(1+\cos(A)+1+\cos B+1+\cos C)$$

$$= \frac{1}{2}(3+\cos(A)+\cos B+\cos C)$$

$$= \frac{1}{2}\left(3+\left(1+\frac{r}{R}\right)\right)$$

$$= \frac{1}{2}\left(4+\frac{r}{R}\right)$$

$$= \left(2+\frac{r}{2R}\right)$$

Ex-7. If the distances of the sides of a triangle ABC from a circum-center be x, y and z respectively, then prove that $\frac{a}{x}+\frac{b}{y}+\frac{c}{z}=\frac{abc}{4xyz}$.

Soln. Let O is the circum centre and OD = x, OE = y, OF = z. respectively.

Also, OA = R = OB = OC

We have, x = OD = R cos A

$$= \frac{a}{2\sin A}.\cos A = \frac{a}{2\tan A}$$

$$\Rightarrow \quad \tan A = \frac{a}{2x}$$

Similarly, $\tan B = \frac{b}{2y}$ & $\tan C = \frac{c}{2z}$

As we know that in a triangle ABC, $\tan A + \tan B + \tan C = \tan A.\tan B.\tan C$

$$\Rightarrow \quad \frac{a}{2x}+\frac{b}{2y}+\frac{c}{2z}=\frac{a}{2x}.\frac{b}{2y}.\frac{c}{2z}$$

$$\Rightarrow \quad \frac{a}{x}+\frac{b}{y}+\frac{c}{z}=\frac{a.b.c}{4.x.y.z}.$$

Ex-8. If in a triangle Δ ABC, O is the circum center and R is the circum - radius and R_1, R_2, R_3 are the circum radii of the traingles Δ OBC, Δ OCA and ΔOAB respectively, then prove that

$$\frac{a}{R_1}+\frac{b}{R_2}+\frac{c}{R_3}=\frac{abc}{R^3}.$$

Soln. We have,

$$R_1 = \frac{OB.OC.BC}{4\Delta OBC} = \frac{R.R.a}{4\Delta_1} = \frac{R^2.a}{4\Delta_1}$$

$$\Rightarrow \quad \frac{a}{R_1} = \frac{4\Delta_1}{R^2}$$

Similarly, $\frac{b}{R_2} = \frac{4\Delta_2}{R^2}$ and $\frac{c}{R_3} = \frac{4\Delta_3}{R^2}$

Thus, $$\frac{a}{R_1}+\frac{b}{R_2}+\frac{c}{R_3} = \frac{4\Delta_1}{R^2}+\frac{4\Delta_2}{R^2}+\frac{4\Delta_3}{R^2}$$

$$= \frac{4(\Delta_1+\Delta_2+\Delta_3)}{R^2}$$

$$= \frac{4\Delta}{R^2}$$

$$= \frac{4\Delta}{R^2}$$

Ex-9. If p_1, p_2, p_3 are respectively the perpendiculars from the vertices of a Δ to the opposite sides, then prove that $p_1.p_2.p_3 = \frac{(abc)^2}{8R^3}$.

Soln. Let AD = p_1, BE = p_2 and CF = p_3

Then, $$\Delta = \frac{1}{2}a\,p_1 = \frac{1}{2}b\,p_2 = \frac{1}{2}c\,p_3$$

$$\Rightarrow \quad p_1 = \frac{2\Delta}{a}, p_2 = \frac{2\Delta}{b}, p_3 = \frac{2\Delta}{c}$$

We have, $$p_1.p_2.p_3 = \frac{2\Delta}{a}.\frac{2\Delta}{b}.\frac{2\Delta}{c}$$

$$= \frac{8\Delta^3}{abc}$$

$$= \frac{8\left(\frac{abc}{4R}\right)^3}{abc}$$

6.12 ADVANCE LEVEL :

1. The bisectors of the angles of a ΔABC :

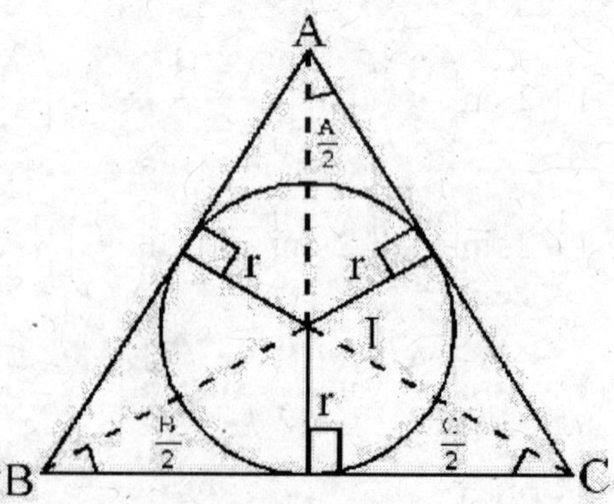

Since IA is the internal angle bisctor of $\angle$A, so we can write

$$\frac{AB}{AC} = \frac{BD}{DC}$$

$$\Rightarrow \quad \frac{BD}{DC} = \frac{AB}{AC} = \frac{c}{b}$$

$$\Rightarrow \quad \frac{DC}{BD}+1 = \frac{b}{c}+1$$

$$\Rightarrow \quad \frac{BC}{BD} = \frac{b+c}{c}$$

$$\Rightarrow \quad BD = \frac{ac}{b+c}$$

In , ΔABD, $\dfrac{BD}{\sin\left(\frac{A}{2}\right)} = \dfrac{AD}{\sin B}$

$$\Rightarrow \quad AD = \frac{BD \sin B}{\sin\left(\frac{A}{2}\right)}$$

$$= \frac{ac \sin B}{(b+c)\sin\left(\frac{A}{2}\right)} = \frac{2\Delta}{(b+c)\sin(A/2)}$$

similarly, $BE = \dfrac{2\Delta}{(c+a)\sin\left(\frac{B}{2}\right)}$

& $CF = \dfrac{2\Delta}{(a+b)\sin\left(\frac{C}{2}\right)}$

Exercise - 8

(Questions based on in-radius and circum-radius)

1. In a ΔABC, if $a = 4$ cm., $b = 6$ cm. and $c = 8$ cm., then r is

(a) $\sqrt{\frac{5}{3}}$ (b) $\sqrt{\frac{3}{5}}$

(c) $\frac{\sqrt{5}}{3}$ (d) None

2. If the sides of a triangle are 3 : 7 : 8, then R : r is

(a) 2 : 7 (b) 7 : 2

(c) 3 : 7 (d) 7 : 3

3. In an equilateral Δ, the in-radius and the circum-radius are connected by

(a) $r = 4R$ (b) $r = \frac{R}{2}$

(c) $r = \frac{R}{3}$ (d) None

4. If a, b, c are sides of a triangle ABC , then the value of $\frac{1}{ab}+\frac{1}{bc}+\frac{1}{ca}$ is

(a) $\frac{1}{r}$ (b) $\frac{1}{R}$

(c) $\frac{1}{2rR}$ (d) $\frac{2}{rR}$

5. In a triangle ABC , the value of

$$\frac{a\sec A + b\sec B + c\sec C}{\tan A.\tan B.\tan C} \text{ is}$$

(a) R (b) 2R

(c) R/2 (d) r

6. If the sides of a triangle be 18, 24 , 30 cms, then the in-radius is

(a) 2 (b) 4

(c) 6 (d) 9.

Answers

1.(a)	2.(b)	3.(a)
4.(c)	5.(a)	6.(c)

6.13 ESCRIBED CIRCLE OF A TRIANGLE AND THEIR RADII

In circle which touches the side BC and two sides AB and AC produced of a triangle ABC is called the Escribed circle opposite to the angle A. Its radius is denoted by r_1. Similarly, r_2 and r_3 denote the radii of escribed circle opposite to the angle B and C respectively.

The centres of the escribed circle are called the Ex-centres. The centre of the escribed circle opposite to the angle A is the point of intersection of the external bisectors of angles B and C. The internal bisector of angle A also passes through the same point. The centre is generally denoted by I_1.

6.14 RADII OF EX-CIRCLES :

Statement : In any ΔABC, prove that the ex-radii are given by

i) $r_1 = \dfrac{\Delta}{s-a}$; $r_2 = \dfrac{\Delta}{s-b}$; $r_3 = \dfrac{\Delta}{s-c}$

ii) $r_1 = s\tan\dfrac{A}{2}$; $r_2 = s\tan\dfrac{B}{2}$; $r_3 = s\tan\dfrac{C}{2}$

iii) $r_1 = \dfrac{a\cos\frac{B}{2}\cos\frac{C}{2}}{\cos\frac{A}{2}}$; $r_2 = \dfrac{b\cos\frac{C}{2}\cos\frac{A}{2}}{\cos\frac{B}{2}}$

$$r_3 = \frac{c\cos\frac{A}{2}\cos\frac{B}{2}}{\cos\frac{C}{2}}$$

iv) $r_1 = 4R \sin\frac{A}{2}\cos\frac{B}{2}\cos\frac{C}{2}$

$r_2 = 4R \sin\frac{B}{2}\cos\frac{A}{2}\cos\frac{C}{2}$

$r_3 = 4R \sin\frac{C}{2}\cos\frac{A}{2}\cos\frac{B}{2}$

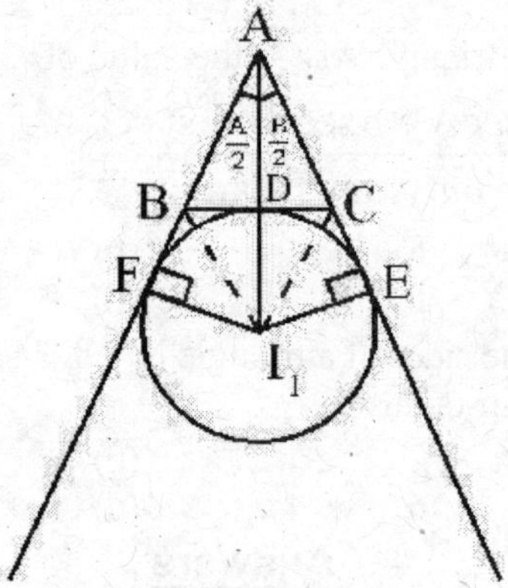

Let I_1 be the point of Intrersection of external bisectors of angles B and C of Δ ABC. Suppose the circle touches the side BC at D and sides AB and AC produced at F and E respectively.

Clearly $I_1D = I_1E = I_1F = r_1$.

i) ar (ΔABC)

= ar (ΔI_1AC) + ar (ΔI_1AB) – ar (ΔI_1BC)

$\Rightarrow \quad \Delta = \frac{1}{2}r_1b + \frac{1}{2}r_1c - \frac{1}{2}r_1a$

$= \frac{1}{2}r_1(b + c - a)$

$\Rightarrow \quad \Delta = \frac{r_1}{2}(a + b + c - 2a)$

$= \frac{r_1}{2}(2s - 2a)$

$= r_1(s - a)$

$\Rightarrow \quad r_1 = \frac{\Delta}{s-a}$

Similarly, it can be shown that

$r_2 = \frac{\Delta}{s-b}$ and $r_3 = \frac{\Delta}{s-c}$

ii) We know that the lengths of the tangents to a circle from an external point are equal.

$\therefore$ AE = AF, BD = BF and CD = CE

Now, AE + AF = (AC + CE) + (AB + BF)

= (AC + CD) + (AB + BD)

= AC + AB + (BD + CD)

= AC + AB + BC

= b + c + a

$\Rightarrow \quad 2AF = 2s$

$\Rightarrow \quad AF = s$

Now, ΔI_1AF, $\tan\frac{A}{2} = \frac{I_1F}{AF} = \frac{r_1}{AF} = \frac{r_1}{s}$

$\Rightarrow \quad r_1 = s\tan\frac{A}{2}$

Similarly, it can be shown that $r_2 = s\tan\frac{B}{2}$ and $r_3 = s\tan\frac{C}{2}$.

Hence, the result.

iii) In ΔI_1BD, we have

$\tan\left(\frac{\pi - B}{2}\right) = \frac{I_1D}{BD} = \frac{r_1}{BD}$

$\Rightarrow \quad \tan\left(\frac{\pi}{2} - \frac{B}{2}\right) = \frac{r_1}{BD}$

$\Rightarrow \quad \cot\frac{B}{2} = \frac{r_1}{BD}$

$\Rightarrow \quad BD = r_1\tan\frac{B}{2}$

Similarly, in ΔI_1CD, we have, $CD = r_1\tan\frac{C}{2}$.

Now, a = BC

= BD + DC

$= r_1\tan\frac{B}{2} + r_1\tan\frac{C}{2}$

$= r_1\left(\tan\frac{B}{2} + \tan\frac{C}{2}\right)$

$= r_1\frac{\sin\left(\frac{B+C}{2}\right)}{\cos\frac{B}{2}\cos\frac{C}{2}}$

$= r_1\frac{\cos\frac{A}{2}}{\cos\frac{B}{2}.\cos\frac{C}{2}}$

$$\Rightarrow \quad r_1 = \frac{a\cos\frac{B}{2}\cos\frac{C}{2}}{\cos\frac{A}{2}}$$

Similarly, $r_2 = \dfrac{b\cos\frac{A}{2}\cos\frac{C}{2}}{\cos\frac{B}{2}}$ and

$$r_3 = \frac{c\cos\frac{A}{2}\cos\frac{B}{2}}{\cos\frac{C}{2}}$$

iv) We have, $r_1 = \dfrac{a\cos\frac{B}{2}\cos\frac{C}{2}}{\cos\frac{A}{2}}$ and

$$R = \frac{a}{2\sin A}$$

$$\Rightarrow \quad r_1 = \frac{2R\sin A\cos\frac{B}{2}\cos\frac{C}{2}}{\cos\frac{A}{2}}$$

$$= R\frac{\sin\frac{A}{2}\cos\frac{A}{2}\cos\frac{B}{2}\cos\frac{C}{2}}{\cos\frac{A}{2}}$$

$$= 4R\sin\frac{A}{2}\cos\frac{B}{2}\cos\frac{C}{2}$$

Similarly, $r_2 = 4R\sin\frac{B}{2}\cos\frac{A}{2}\cos\frac{C}{2}$

and $r_3 = 4R\sin\frac{C}{2}\cos\frac{A}{2}\cos\frac{B}{2}$.

Hence, the result

This completes the proof of the statement.

Ex-1. In a ΔABC, if $a = 18$ cm., $b = 24$ cm., and $c = 30$ cm., then find the value of r_1, r_2 and r_3

Soln. Now $2s = a + b + c = 18 + 24 + 30 = 72$

$\Rightarrow \quad 2s = 72 \Rightarrow \quad s = 36$

We have, $\Delta = \sqrt{s(s-a)(s-b)(s-c)}$

$= \sqrt{36(36-18)(36-24)(36-30)}$

$= \sqrt{36\times18\times12\times6}$

$= \sqrt{36\times9\times12\times12}$

$= 6\times3\times12$

$= 216$

Thus, $r_1 = \dfrac{\Delta}{s-a} = \dfrac{216}{36-18} = \dfrac{216}{18} = 12$

$r_2 = \dfrac{\Delta}{s-b} = \dfrac{216}{36-24} = \dfrac{216}{12} = 18$

and $r_3 = \dfrac{\Delta}{s-c} = \dfrac{216}{36-30} = \dfrac{216}{6} = 36$

Ex-2. In a triangle ABC if $\dfrac{s-c}{s-a} = \dfrac{b-c}{a-b}$, then prove that a, b, c are in A.P

Soln. Given $\dfrac{s-c}{s-a} = \dfrac{b-c}{a-b}$

$\Rightarrow \quad \dfrac{2s-2c}{2s-2a} = \dfrac{b-c}{a-b}$

$\Rightarrow \quad \dfrac{a+b-c}{b+c-a} = \dfrac{b-c}{a-b}$

$\Rightarrow \quad \dfrac{a+b-c}{b-c} = \dfrac{b+c-a}{a-b}$

$\Rightarrow \quad \dfrac{a}{b-c} + 1 = \dfrac{c}{a-b} - 1$

$\Rightarrow \quad \dfrac{c}{a-b} + \dfrac{a}{b-c} = 2$

$\Rightarrow \quad \dfrac{c(b-c)+a(a-b)}{(a-b)(b-c)} = 2$

$\Rightarrow \quad \dfrac{bc - c^2 + a^2 - ab}{(a-b)(b-c)} = 2$

$\Rightarrow \quad bc - c^2 + a^2 - ab = 2(ab - ac - b^2 + bc)$

$\Rightarrow \quad 2b^2 - bc - c^2 + a^2 - 3ab + 2ac = 0$

Ex-3. If r_1, r_2, r_3 are in H.P, then prove that a, b, c are in A,P

Soln. Given r_1, r_2, r_3 are in H.P

$\Rightarrow \quad r_2 = \dfrac{2r_1r_3}{r_1+r_3}$

$$\Rightarrow \quad \frac{\Delta}{s-b} = \frac{2.\frac{\Delta}{s-a}.\frac{\Delta}{s-c}}{\frac{\Delta}{s-a}+\frac{\Delta}{s-c}}$$

$$\Rightarrow \quad \frac{1}{s-b} = \frac{2.\frac{1}{s-a}.\frac{1}{s-c}}{\frac{1}{s-a}+\frac{1}{s-c}}$$

$$\Rightarrow \quad \frac{1}{s-b}\left(\frac{1}{s-a}+\frac{1}{s-c}\right) = 2.\frac{1}{s-a}.\frac{1}{s-c}$$

$$\Rightarrow \quad \frac{1}{s-b}\left(\frac{s-c+s-a}{(s-a)(s-c)}\right) = \frac{2}{(s-a)(s-c)}$$

$$\Rightarrow \quad (2s-a-c) = 2(s-b)$$

$$\Rightarrow \quad a+c = 2b$$

$$\Rightarrow \quad a, b, c \in A.P$$

Ex-4. In a triangle ABC, if a, b, c are in A.P as well as in G.P then prove that the value of

$\left(\frac{r_1}{r_2} - \frac{r_2}{r_3} + 10\right)$ is 10.

Soln. Since a, b, c are in A.P as well as in G.P, so

$a = b = c$

Now, $r_1 = \frac{\Delta}{s-a} = r_2 = r_3$

Thus, $\left(\frac{r_1}{r_2} - \frac{r_2}{r_3} + 10\right) = 1 - 1 + 10 = 10$

Ex-5. If $r_1 < r_2 < r_3$ and the ex-radii of a right angled triangle and $r_1 = 1,\ r_2 = 2$, then prove that $r_3 = \frac{3+\sqrt{17}}{2}$.

Soln. We have, $r_1 = \frac{\Delta}{s-a} = 1,\ r_2 = \frac{\Delta}{s-b} = 2$ and $r_3 = \frac{\Delta}{s-c}$

$$\Rightarrow \quad s-a = \Delta,\ s-b = \frac{\Delta}{2} \text{ and } s-c = \frac{\Delta}{r_3}$$

$$\Rightarrow \quad c = \Delta\left(1+\frac{1}{2}\right),\ a = \Delta\left(\frac{1}{2}+\frac{1}{r_3}\right),\ b = \Delta\left(1+\frac{1}{r_3}\right)$$

Since triangle is right angled, so $a^2 + b^2 = c^2$

$$\Rightarrow \quad \Delta^2\left(\frac{3}{2}\right)^2 = \frac{\Delta^2(r_3+2)^2}{4(r_3)^2} + \Delta^2\left(\frac{r_3+1}{r_3}\right)^2$$

$$\Rightarrow \quad \left(\frac{3}{2}\right)^2 = \frac{(r_3+2)^2}{4(r_3)^2} + \left(\frac{r_3+1}{r_3}\right)^2$$

$$\Rightarrow \quad 9r_3^2 = (r_3+2)^2 + 4(r_3+1)^2$$

$$\Rightarrow \quad 4r_3^2 - 12r_3 - 8 = 0$$

$$\Rightarrow \quad r_3^2 - 3r_3 - 2 = 0$$

$$\Rightarrow \quad r_3 = \frac{3 \pm \sqrt{17}}{2} = \frac{3+\sqrt{17}}{2}$$. as r_3 is positive.

Ex-6. Two sides of a triangle are the roots of $x^2 - 5x + 3 = 0$. If the angle between the sides is $\frac{\pi}{3}$. then prove that the value of $r.R$ is 2/3.

Soln. Let a, b be the sides of a triangle. Then $a + b = 5$ and $ab = 3$

Now, $\cos C = \frac{a^2+b^2-c^2}{2ab}$

$$\Rightarrow \quad \cos\left(\frac{\pi}{3}\right) = \frac{19-c^2}{6}$$

$$\Rightarrow \quad \frac{1}{2} = \frac{19-c^2}{6}$$

$$\Rightarrow \quad 19 - c^2 = 3$$

$$\Rightarrow \quad c = 4$$

Thus, $r.R = \frac{\Delta}{s} \times \frac{abc}{4\Delta}$

$$= \frac{abc}{4s} = \frac{ab.c}{2(a+b+c)} = \frac{3.4}{2(5+4)} = \frac{12}{18} = \frac{2}{3}$$

Ex-7. In an isosceles triangle of which one angle is 120°, circle of radius $\sqrt{3}$ is inscribed, then prove that the area of the triangle is $(12 + 7\sqrt{3})$ sq. u.

Soln. By sine rule, $\frac{a}{\sin(120^\circ)} = \frac{b}{\sin(30^\circ)}$

$$\Rightarrow \quad \frac{a}{\sqrt{3}/2} = \frac{b}{1/2}$$

$$\Rightarrow \quad a = b\sqrt{3}$$

Also, from the above figure, $r = \sqrt{3}$

$$\Rightarrow \quad \frac{\sqrt{3}}{a/2} = \tan(15^\circ)$$

$$\Rightarrow \quad \frac{2\sqrt{3}}{a} = \left(2-\sqrt{3}\right)$$

$$\Rightarrow \quad a = \frac{2\sqrt{3}}{\left(2-\sqrt{3}\right)}$$

Now, $b = \frac{a}{\sqrt{3}} = \frac{2\sqrt{3}}{\left(2-\sqrt{3}\right)} \times \frac{1}{\sqrt{3}} = \frac{2}{\left(2-\sqrt{3}\right)}$

Thus, the required area $= \frac{1}{2} \times ab \times \sin\left(30^{\circ}\right)$

$$= \frac{1}{2} \times \frac{2\sqrt{3}}{\left(2-\sqrt{3}\right)} \times \frac{2}{\left(2-\sqrt{3}\right)} \times \frac{1}{2}$$

$$= \sqrt{3} \times \left(2+\sqrt{3}\right)^2$$

$$= \sqrt{3} \times \left(7+4\sqrt{3}\right)$$

$$= \left(12+7\sqrt{3}\right) \text{ sq. u.}$$

Ex-8. If a triangle of maximum area is inscribed within a circle of radius R, then prove that

$$\frac{1}{r_1} + \frac{1}{r_2} + \frac{1}{r_3} = \frac{\sqrt{2}+1}{R}$$

Soln. Let ABC be a right angled triangle in a circle of radius R

Therefore, BC = 2R = Diameter.

Now, $\Delta = \frac{1}{2} \times AB \times AC \times \sin\left(90^{\circ}\right)$

$$= \frac{1}{2} \times AB \times AC$$

It wil maximum, when AB = AC

Thus, $AB^2 + AC^2 = BC^2 = 4R^2$

$$\Rightarrow \quad 2AB^2 = 4R^2$$

$$\Rightarrow \quad AB^2 = 2R^2$$

Now, $2s = AB + BC + CA = 2AB + BC$

$$= 2R\left(1+\sqrt{2}\right)$$

and $r = \frac{\Delta}{s} = \frac{R^2}{R\left(\sqrt{2}+1\right)} = \frac{R}{\left(\sqrt{2}+1\right)}$

Therefore, $\frac{1}{r_1} + \frac{1}{r_2} + \frac{1}{r_3} = \frac{1}{r} = \frac{\left(\sqrt{2}+1\right)}{R}$

Ex-9. If in a triangle $r = r_1 - r_2 - r_3$, then prove that the triangle is right angled.

Soln. Given $r = r_1 - r_2 - r_3$

$$\Rightarrow \quad r_1 - r = r_2 + r_3$$

$$\Rightarrow \quad \frac{\Delta}{s-a} - \frac{\Delta}{s} = \frac{\Delta}{s-b} + \frac{\Delta}{s-c}$$

$$\Rightarrow \quad \frac{1}{s-a} - \frac{1}{s} = \frac{1}{s-b} + \frac{1}{s-c}$$

$$\Rightarrow \quad \frac{(s-a-s)}{(s-a)} = \frac{(s-c+s-b)}{(s-b)(s-c)}$$

$$\Rightarrow \quad \frac{a}{(s-a)} = \frac{a}{(s-b)(s-c)}$$

$$\Rightarrow \quad \frac{(s-b)(s-c)}{(s-a)} = 1$$

$$\Rightarrow \quad \tan^2\left(\frac{A}{2}\right) = 1$$

$$\Rightarrow \quad \tan\left(\frac{A}{2}\right) = 1$$

$$\Rightarrow \quad \tan\left(\frac{A}{2}\right) = \tan\left(\frac{\pi}{4}\right)$$

$$\Rightarrow \quad A = \frac{\pi}{2}$$

Thus, the triangle is right angled.

Ex-10. In a ΔABC, prove that $r.r_1.r_2.r_3 = \Delta^2$

Soln. We have, $r.r_1.r_2.r_3$

$$= \frac{\Delta}{s} \cdot \frac{\Delta}{(s-a)} \cdot \frac{\Delta}{(s-b)} \cdot \frac{\Delta}{(s-c)}$$

$$= \frac{\Delta^4}{s(s-a)(s-b)(s-c)}$$

$$= \frac{\Delta^4}{\Delta^2}$$

$$= \Delta^2$$

Ex.11. In any ΔABC, prove that $\frac{r_1}{bc} + \frac{r_2}{ca} + \frac{r_3}{ab} = \frac{1}{r} - \frac{1}{2R}$

Soln. We have $\dfrac{r_1}{bc}$

$$= \frac{4R\sin\left(\frac{A}{2}\right)\cos\left(\frac{B}{2}\right)\cos\left(\frac{C}{2}\right)}{(2R\sin B)(2R\sin C)}$$

$$= \frac{\sin\left(\frac{A}{2}\right)}{4R\sin\left(\frac{B}{2}\right)\sin\left(\frac{C}{2}\right)}$$

$$= \frac{\sin^2\left(\frac{A}{2}\right)}{4R\sin\left(\frac{A}{2}\right)\sin\left(\frac{B}{2}\right)\sin\left(\frac{C}{2}\right)}$$

$$= \frac{\sin^2\left(\frac{A}{2}\right)}{r}$$

Similarly, $\dfrac{r_2}{ca} = \dfrac{\sin^2\left(\frac{B}{2}\right)}{r}$ and $\dfrac{r_3}{ab} = \dfrac{\sin^2\left(\frac{C}{2}\right)}{r}$

Now, $\dfrac{r_1}{bc} + \dfrac{r_2}{ca} + \dfrac{r_3}{ab}$

$$= \frac{1}{r}\left(\sin^2\left(\frac{A}{2}\right) + \sin^2\left(\frac{B}{2}\right) + \sin^2\left(\frac{C}{2}\right)\right)$$

$$= \frac{1}{2r}\left(2\sin^2\left(\frac{A}{2}\right) + 2\sin^2\left(\frac{B}{2}\right) + 2\sin^2\left(\frac{C}{2}\right)\right)$$

$$= \frac{1}{2r}(1-\cos(A)+1-\cos(B)+1-\cos(C))$$

$$= \frac{1}{2r}(3-(\cos(A)+\cos(B)+\cos(C)))$$

$$= \frac{1}{2r}\left(3-\left(1+\frac{r}{R}\right)\right)$$

$$= \frac{1}{2r}\left(2-\frac{r}{R}\right)$$

$$= \left(\frac{1}{r}-\frac{1}{2R}\right)$$

Ex-12. In any ΔABC prove that $\dfrac{bc}{r_1} + \dfrac{ca}{r_2} + \dfrac{ab}{r_3}$

$$= 2R\left[\left(\frac{a}{b}+\frac{b}{a}\right)+\left(\frac{b}{c}+\frac{c}{b}\right)+\left(\frac{c}{a}+\frac{a}{c}\right)-3\right]$$

Soln. We have, $\dfrac{bc}{r_1} + \dfrac{ca}{r_2} + \dfrac{ab}{r_3}$

$$= abc\left(\frac{1}{a r_1}+\frac{1}{b r_2}+\frac{1}{c r_3}\right)$$

$$= abc\left(\frac{s-a}{a\Delta}+\frac{s-b}{b\Delta}+\frac{s-c}{c\Delta}\right)$$

$$= \frac{abc}{\Delta}\left(\frac{s-a}{a}+\frac{s-b}{b}+\frac{s-c}{c}\right)$$

$$= \frac{abc}{2\Delta}\left(\frac{2s-2a}{a}+\frac{2s-2b}{b}+\frac{2s-2c}{c}\right)$$

$$= \frac{abc}{2\Delta}\left(\frac{b+c}{a}-1+\frac{c+a}{b}-1+\frac{a+b}{c}-1\right)$$

$$= \frac{abc}{2\Delta}\left(\left(\frac{a}{b}+\frac{b}{a}\right)+\left(\frac{b}{c}+\frac{c}{b}\right)+\left(\frac{a}{c}+\frac{c}{a}\right)-3\right)$$

$$= 2R\left(\left(\frac{a}{b}+\frac{b}{a}\right)+\left(\frac{b}{c}+\frac{c}{b}\right)+\left(\frac{a}{c}+\frac{c}{a}\right)-3\right)$$

Ex-13. In a ΔABC, prove that $(r+r_1)\tan\left(\frac{B-C}{2}\right) + (r+r_2)\tan\left(\frac{C-A}{2}\right) + (r+r_3)\tan\left(\frac{A-B}{2}\right) = 0.$

Soln. We have, $(r+r_1)\tan\left(\frac{B-C}{2}\right)$

$$= \left(\frac{\Delta}{s}+\frac{\Delta}{s-a}\right)\times\left(\frac{b-c}{b+c}\right)\cot\left(\frac{A}{2}\right)$$

$$= \Delta\left(\frac{s-a+s}{s(s-a)}\right)\times\left(\frac{b-c}{b+c}\right)\times\sqrt{\frac{s(s-a)}{(s-b)(s-c)}}$$

$$= \Delta\left(\frac{b+c}{\sqrt{s(s-a)}}\right)\times\left(\frac{b-c}{b+c}\right)\times\sqrt{\frac{1}{(s-b)(s-c)}}$$

$$= \Delta\times(b+c)\times\left(\frac{b-c}{b+c}\right)\times\frac{1}{\sqrt{s(s-a)(s-b)(s-c)}}$$

$$= \Delta\times(b-c)\times\frac{1}{\Delta}$$

$$= b-c$$

Similarly, $(r+r_2)\tan\left(\frac{C-A}{2}\right)=c-a$

and $(r+r_3)\tan\left(\frac{A-B}{2}\right)=a-b$

Now, $(r+r_1)\tan\left(\frac{B-C}{2}\right)+(r+r_2)\tan\left(\frac{C-A}{2}\right)+$

$(r+r_3)\tan\left(\frac{A-B}{2}\right)$

$= b-c+c-a+a-b$

$= 0$

Ex-14. Prove that

$$\frac{(r_1+r_2)}{1+\cos C}=\frac{(r_2+r_3)}{1+\cos A}=\frac{(r_3+r_1)}{1+\cos B}.$$

Soln. We have, $\frac{r_1+r_2}{1+\cos C}$

$$=\frac{\frac{\Delta}{s-a}+\frac{\Delta}{s-b}}{2\cos^2\left(\frac{C}{2}\right)}$$

$$=\frac{\frac{\Delta(s-a+s-b)}{(s-a)(s-b)}}{2\cos^2\left(\frac{C}{2}\right)}$$

$$=\frac{\Delta(2s-a-b)}{(s-a)(s-b)\times 2\times\frac{s(s-c)}{ab}}$$

$$=\frac{\Delta\times abc}{s(s-a)(s-b)(s-c)}$$

$$=\frac{\Delta\times abc}{\Delta^2}=\frac{abc}{\Delta}$$

Similarly, we can easily prove that $\frac{(r_2+r_3)}{1+\cos A}=\frac{abc}{\Delta}$ & $\frac{(r_3+r_1)}{1+\cos B}=\frac{abc}{\Delta}$

Thus, $\frac{(r_2+r_3)}{1+\cos A}=\frac{abc}{\Delta}$ & $\frac{(r_3+r_1)}{1+\cos B}=\frac{abc}{\Delta}$

Ex-15. Prove that $\left(\frac{1}{r}-\frac{1}{r_1}\right)\left(\frac{1}{r}-\frac{1}{r_2}\right)\left(\frac{1}{r}-\frac{1}{r_3}\right)$

$$=\frac{16R}{r^2(a+b+c)^2}$$

Soln. We have, $\left(\frac{1}{r}-\frac{1}{r_1}\right)\left(\frac{1}{r}-\frac{1}{r_2}\right)\left(\frac{1}{r}-\frac{1}{r_3}\right)$

$$=\left(\frac{s}{\Delta}-\frac{(s-a)}{\Delta}\right)\left(\frac{s}{\Delta}-\frac{(s-b)}{\Delta}\right)\left(\frac{s}{\Delta}-\frac{(s-c)}{\Delta}\right)$$

$$=\frac{1}{\Delta^3}(s-s+a)(s-s+b)(s-s+c)$$

$$=\frac{1}{\Delta^3}\times abc$$

$$=16\times\frac{abc}{4\Delta}\times\frac{s^2}{\Delta^2(2s)^2}.$$

$$=\frac{16R}{r^2(a+b+c)^2}$$

Ex-16. Prove that $\left(\frac{1}{r_1}+\frac{1}{r_2}\right)\left(\frac{1}{r_2}+\frac{1}{r_3}\right)\left(\frac{1}{r_3}+\frac{1}{r_1}\right)$

$$=\frac{64R^3}{(abc)^2}$$

Soln. We have $\left(\frac{1}{r_1}+\frac{1}{r_2}\right)\left(\frac{1}{r_2}+\frac{1}{r_3}\right)\left(\frac{1}{r_3}+\frac{1}{r_1}\right)$

$$=\left(\frac{s-a}{\Delta}+\frac{s-b}{\Delta}\right)\left(\frac{s-b}{\Delta}+\frac{s-c}{\Delta}\right)\left(\frac{s-c}{\Delta}+\frac{s-a}{\Delta}\right)$$

$$=\frac{1}{\Delta^3}\times abc$$

$$=\frac{1}{\left(\frac{abc}{4R}\right)^3}\times abc$$

$$=\frac{64R^3}{(abc)^2}$$

Exercise - 9

(Questions based on escribed circles)

1. In a triangle, the value of $\frac{1}{r_1}+\frac{1}{r_2}+\frac{1}{r_3}$ is

(a) $\frac{1}{r}$ (b) $\frac{1}{R}$

(c) $\frac{2}{R}$ (d) r

2. In a triangle, the value of $r_1 + r_2 + r_3 - r$ is
(a) 4R (b) R
(c) 2R (d) 3R

3. The value of $r_1 r_2 + r_2 r_3 + r_3 r_1$ is
(a) $2s^2$ (b) s^2
(c) $4s^2$ (d) $3s^2$

4. The value of $r - r_1 + r_2 + r_3$ is
(a) 4R cos A (b) 4R cos B
(c) 4R cosC (d) 4R

5. In a triangle if $\left(1-\frac{r_1}{r_2}\right)\left(1-\frac{r_2}{r_3}\right) = 2$, then the triangle is a/an
(a) Equilateral (b) Isosceles
(c) rt. angled (d) Scalene

Answers

1.(a) 2.(a) 3.(a)
4.(a) 5.(c)

6.15. REGULAR POLYGON :

If a polygon has all its sides equal in length and also all its angles equal then the polygon is called a regular polygon.

The circle inscribed in the regular polygon and touching all the sides of the regular polygon is ...... **inscribed circle.**

The circle which pass through all the vertices of the regular polygon is called its **circum scribed circle.**

If the polygon has n-sides, then the sum of the interrior angles is $(n-2) \times \pi$ and each angle is $\frac{(n-2)\times\pi}{n}$.

6.16.

Statement : In a regular polygon of $A_1A_2A_3....A_n$ of n-sides of each length a is given by

i) $R = \frac{a}{2}\ \text{cosec}\left(\frac{\pi}{n}\right)$, where R = circum-radius.

ii) $r = \frac{a}{2}\cot\left(\frac{\pi}{n}\right)$, where r = in-radius.

iii) $\Delta = \frac{a^2 n}{4}\cot\left(\frac{\pi}{n}\right)$, where Δ = area of the regular polygon

iv) $\Delta = \frac{nR^2}{2}\sin\left(\frac{2\pi}{n}\right)$

v) $\Delta = nr^2\tan\left(\frac{\pi}{n}\right)$

Proof :

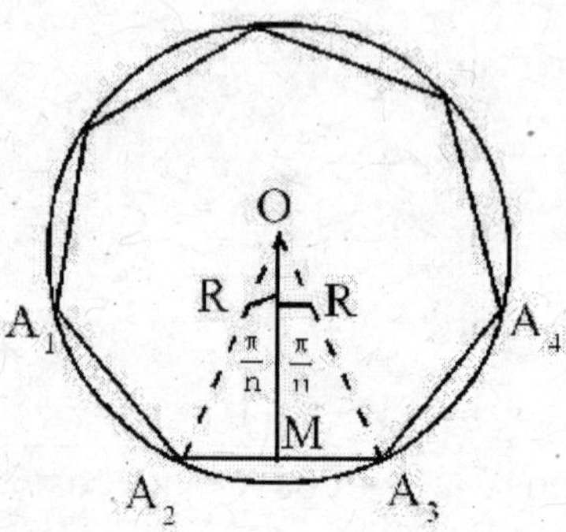

i) Let A_1A_2, A_2A_3 and A_3A_4 be three consecutive sides of one regular polygon.

$A_1A_2 = A_2A_3 = A_3A_4 = a.$

Let the bisectors OA_2 and OA_3 of the angles $\angle A_1A_2A_3$ and $\angle A_2A_3A_4$ respectively meet at O.

Clearly, O is the centre of both circumscribed circle and inscribed circle.

Let R and r be the radii of circumcribed and inscribed circle, then clearly

$OA_2 = OA_3 = R$ and

$OM = r$, where OM is perpendicular from O on A_2A_3.

$\therefore \quad A_2M = \frac{1}{2}a = A_3M.$

We know that, the whole angle subtended at the centre 'O' by the regular polygon is 2π radian

$$\therefore \quad \angle A_2 O A_3 = \frac{2\pi}{n}$$

$$\Rightarrow \quad \angle A_2 O M = \angle A_3 OM = \frac{\pi}{n}.$$

From ΔA_2OM, we have,

$$\sin\left(\frac{\pi}{n}\right) = \frac{A_2M}{OA_2} = \frac{\frac{1}{2}a}{R} = \frac{a}{2R}$$

$$\Rightarrow \quad R = \frac{a}{2\sin\left(\frac{\pi}{n}\right)} = \frac{a}{2}\ \text{cosec}\left(\frac{\pi}{n}\right)$$

again, $\tan\left(\frac{\pi}{n}\right) = \frac{A_2M}{OM} = \frac{a}{2r}$

$\Rightarrow \quad r = \frac{1}{2} a \cot\left(\frac{\pi}{n}\right)$

ii) Let the area of the regular polygon be denoted by Δ.

Join O to the n-vertices $A_1, A_2, \ldots, A_n$ of the regular polygon.

The regular polygon is divided into n-isosceles triangle of equal area.

Therefore, the area Δ of regular polygon is given by Δ = n-times the area of isosceles $\Delta A_2 OA_3$

$$\Rightarrow \quad \Delta = n \times \frac{1}{2} \times A_2A_3 \times OM$$

$$\Rightarrow \quad \Delta = n \times \frac{1}{2} \times a \times OM$$

$$\Rightarrow \quad \Delta = n \times \frac{1}{2} \times a \times \frac{a}{2} \cot\left(\frac{\pi}{n}\right)$$

$$\Rightarrow \quad \Delta = \frac{na^2}{4} \cot\left(\frac{\pi}{n}\right)$$

iii) In ΔOA_2M, we have,

$$\cos\left(\frac{\pi}{n}\right) = \frac{OM}{OA_2}$$

$$\Rightarrow \quad OM = R\cos\left(\frac{\pi}{n}\right)$$

Also, $\sin\left(\frac{\pi}{n}\right) = \frac{A_2M}{OA_2} = \frac{\frac{1}{2}a}{R}$

$$\Rightarrow \quad a = 2R\sin\left(\frac{\pi}{n}\right)$$

$$\therefore \quad \Delta = \frac{na}{2} \times OM$$

$$= \frac{n}{2} \times 2R\sin\left(\frac{\pi}{n}\right) \times R\cos\left(\frac{\pi}{n}\right)$$

$$= \frac{nR^2}{2} \sin\left(\frac{2\pi}{n}\right)$$

iv) In ΔOA_2M, we have OM = r

$$\therefore \quad \tan\left(\frac{\pi}{n}\right) = \frac{A_2M}{OM}$$

$$\Rightarrow \quad \tan\left(\frac{\pi}{n}\right) = \frac{\frac{a}{2}}{r} = \frac{a}{2r}$$

$$\Rightarrow \quad a = 2r\tan\left(\frac{\pi}{n}\right)$$

$$\therefore \quad \Delta = \frac{na}{2} \times OM$$

$$= \frac{n}{2} \times 2r\tan\left(\frac{\pi}{n}\right) \times r$$

$$= nr^2 \tan\left(\frac{\pi}{n}\right)$$

Hence, the result

This completes the proof of the statement.

6.17 SOME SOLVED EXAMPLES :

Ex-1. If $A_0, A_1, \ldots A_5$ be the consecutive vertices of a regular hexagon inscribed in a unit circle. Then find the product of length of A_0A_1, A_0A_2 and A_0A_4.

Soln. Here, $OA_0 = OA_1 = OA_2 = \ldots\ldots\ldots = OA_5 = 1$.

and

$$\angle A_0OA_1 = \frac{2\pi}{6} = \angle A_1OA_2 = \ldots\ldots\ldots = \angle A_4OA_5$$

Now,

$$\cos\left(\frac{\pi}{3}\right) = \frac{OA_0^{\ 2} + OA_1^{\ 2} - A_1A_2}{2OA_0.OA_1} = \frac{1+1-A_1A_2}{2.1.1}$$

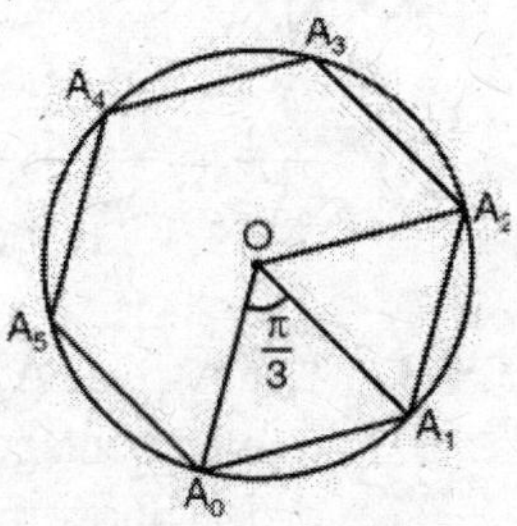

$$\Rightarrow \quad \frac{1}{2} = \frac{2 - A_0A_1^{\ 2}}{2}$$

$$\Rightarrow \quad A_0A_1^{\ 2} = 1$$

$$\Rightarrow \quad A_0A_1 = 1$$

Also, $\cos\left(\frac{2\pi}{3}\right) = \frac{OA_0^{\ 2} + OA_2^{\ 2} - A_0A_2^{\ 2}}{2.OA_0.OA_2}$

$$\Rightarrow \quad -\frac{1}{2} = \frac{1+1-A_0A_2^{\ 2}}{2.1.1}$$

$$\Rightarrow \quad -\frac{1}{2} = \frac{2-A_0A_2^{\ 2}}{2}$$

$$\Rightarrow \quad A_0A_2 = \sqrt{3}$$

Again,

$$\cos\left(\frac{4\pi}{3}\right)=\frac{OA_0^{\ 2}+OA_4^{\ 2}-A_0A_4^{\ 2}}{2.OA_0.OA_4}$$

$$=\frac{1+1-A_0A_4^{\ 2}}{2.1.1}$$

$$=\frac{2-A_0A_4^{\ 2}}{2}$$

$$\Rightarrow\quad -\frac{1}{2}=\frac{2-A_0A_4^{\ 2}}{2}$$

$$\Rightarrow\quad A_0A_4=\sqrt{3}$$

Hence, the value of

$A_0A_1.A_0A_2.A_0A_4=1.\sqrt{3}.\sqrt{3}=3$.

Ex-2. If the Area of circle is A_1 and area of regular pentagon inscribed in the circle is A_2.

Find the ratio of area of two.

Soln. Let R be the radius of the circle.

Then, $A_1=\pi R^2$

$$\Delta=\frac{nR^2}{2}\sin\left(\frac{2\pi}{n}\right)$$

and $A_2=\dfrac{5.R^2}{2}\sin\left(\dfrac{360^o}{5}\right)$

$$=\frac{5}{2}R^2\sin(72^o)=\frac{5R^2}{2}\times\cos(18^o)$$

Now,

$$\frac{A_1}{A_2}=\frac{\pi R^2}{\frac{5}{2}R^2\cos(18^o)}=\frac{2\pi}{5}\times\sec(18^o)$$

Ex-3. Let A_1, A_2, A_3, A_4 and A_5 be the vertices of a regular pentagon inscribed in a unit circle taken in order. Show that :

$A_1A_2\times A_1A_3=\sqrt{5}$.

Soln. Here,

$OA_1=OA_2=OA_3=OA_4=OA_5=1$ and

$$\angle A_1OA_2=\frac{2\pi}{5}=\angle A_2OA_3=.........=\angle A_4OA_5$$

Now, $\cos\left(\dfrac{2\pi}{5}\right)=\dfrac{OA_1^{\ 2}+OA_2^{\ 2}-A_1A_2^{\ 2}}{2.OA_1.OA_2}$

$$\Rightarrow\quad \sin(18^o)=\frac{1+1-A_1A_2^{\ 2}}{2.1.1}$$

$$\Rightarrow\quad \frac{\sqrt{5}-1}{4}=\frac{2-A_1A_2^{\ 2}}{2}$$

$$\Rightarrow\quad A_1A_2^{\ 2}=2-\frac{\sqrt{5}-1}{2}=\frac{5-\sqrt{5}}{2}$$

$$\Rightarrow\quad A_1A_2=\sqrt{\frac{5-\sqrt{5}}{2}}$$

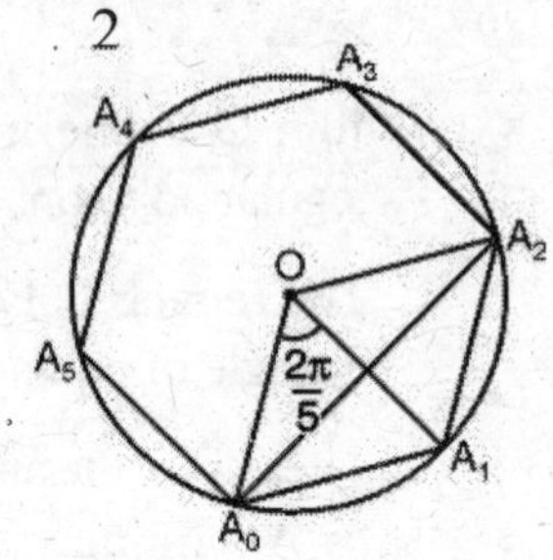

Similarly, $A_1A_3=\sqrt{\dfrac{5+\sqrt{5}}{2}}$

Thus, $A_1A_2\times A_1A_3$

$$=\sqrt{\left(\frac{5-\sqrt{5}}{2}\right)\times\left(\frac{5+\sqrt{5}}{2}\right)}=\sqrt{\frac{25-5}{4}}=\sqrt{\frac{20}{4}}=\sqrt{5}$$

Ex-4. The sides of a regular do-decagon is 2 ft. Find the radius of the circumscribed circle.

Soln. $R=\dfrac{a}{2}\operatorname{cosec}\left(\dfrac{\pi}{n}\right)$,

As we know that, the circum-radius of n sided regular polygon

$=\dfrac{a}{2}\operatorname{cosec}\left(\dfrac{\pi}{n}\right)$, where a = side and n = number of sides

$$=\quad 6\times\operatorname{cosec}\left(\frac{\pi}{12}\right)$$

$$=\quad 6\times\operatorname{cosec}(15^o)$$

$$=\quad \frac{6\times2\sqrt{2}}{\sqrt{3}-1}$$

$$=\quad 6\sqrt{2}\left(\sqrt{3}+1\right)$$

Ex-5. A regular pentagon and a regular decagon have the same perimeter. Find the ratio of its area.

Soln. Let the perimeter of the pentagon and the decagon be 10 x.

Then each side of the pentagon is 2x and the decagon is x.

Let A_1 = the area of the pentagon

$$=5x^2\cot\left(\frac{\pi}{5}\right)$$

and A_2 = the area of the decagon

$$= \frac{5}{2}x^2\cot\left(\frac{\pi}{10}\right)$$

Now, $$\frac{A_1}{A_2} = \frac{5x^2\cot\left(\frac{\pi}{5}\right)}{\frac{5}{2}x^2\cot\left(\frac{\pi}{10}\right)} = \frac{2\cot\left(\frac{\pi}{5}\right)}{\cot\left(\frac{\pi}{10}\right)}$$

$$= \frac{2\cot(36^o)}{\cot(18^o)} = \frac{2\cos(36^o)\sin(18^o)}{\sin(36^o)\cos(18^o)}$$

$$= \frac{2\cos(36^o)\sin(18^o)}{2\sin(18^o)\cos^2(18^o)}$$

$$= \frac{2\cos(36^o)}{(1+\cos(36^o))}$$

$$= \frac{2\left(\frac{\sqrt{5}+1}{4}\right)}{\left(1+\frac{\sqrt{5}+1}{4}\right)}$$

$$= \frac{2(\sqrt{5}+1)}{(\sqrt{5}+5)}$$

$$= \frac{2(\sqrt{5}+1)}{\sqrt{5}(\sqrt{5}+1)}$$

$$= \frac{2}{\sqrt{5}}$$

Ex-6. If 2a be the sides of a regular polygon of n - sides . R and r be the circum radius and inradius , then prove that $r + R = a\cot\left(\frac{\pi}{2n}\right)$

Soln. We have,

$$R = \frac{2a}{2}\text{cosec}\left(\frac{\pi}{n}\right) = a\,\text{cosec}\left(\frac{\pi}{n}\right)$$

and $$r = \frac{2a}{2}\cot\left(\frac{\pi}{n}\right) = a\cot\left(\frac{\pi}{n}\right)$$

Now, $r + R$

$$= a\cot\left(\frac{\pi}{n}\right) + a\,\text{cosec}\left(\frac{\pi}{n}\right)$$

$$= a\left(\cot\left(\frac{\pi}{n}\right) + \text{cosec}\left(\frac{\pi}{n}\right)\right)$$

$$= a\left(\frac{1+\cos\left(\frac{\pi}{n}\right)}{\sin\left(\frac{\pi}{n}\right)}\right)$$

$$= a\left(\frac{2\cos^2\left(\frac{\pi}{2n}\right)}{2\sin\left(\frac{\pi}{2n}\right)\cos\left(\frac{\pi}{2n}\right)}\right)$$

$$= a\cot\left(\frac{\pi}{2n}\right)$$

Ex-7. A regular pentagon and a regular decagon have the same area , then find the ratio of their perimeter.

Soln. Let A_1 be the area of the regular pentagon A_2 be the area of the regular decagon .

Therefore, $A_1 = A_2$

$$\Rightarrow \frac{5a^2}{4}\cot\left(\frac{\pi}{5}\right) = \frac{6b^2}{4}\cot\left(\frac{\pi}{6}\right)$$

$$\Rightarrow 5a^2\cot\left(\frac{\pi}{5}\right) = 6b^2\cot\left(\frac{\pi}{6}\right)$$

$$\Rightarrow 5a^2\cot(36^o) = 6b^2\cot(30^o)$$

$$\Rightarrow 5a^2\cot(36^o) = 6\sqrt{3}\,b^2$$

$$\Rightarrow \frac{a^2}{6\sqrt{3}} = \frac{b^2}{5\cot(36^o)} = \lambda$$

Hence, the ratio of their perimeters

$$= \frac{5a}{6b}$$

$$= \frac{5}{6}\times\frac{\sqrt{\lambda 6\sqrt{3}}}{\sqrt{5\lambda\lambda c d(36^o)}} = \sqrt{\frac{5\sqrt{3}}{6}\tan(36^o)}$$

Ex-8. If the number of sides of two regular polygon having the same perimeter be n and 2n respectively , prove that their areas are in the ratio $2\cos(\pi/n):(1+\cos(\pi/n))$.

Soln. Let the perimeter of the two polygons are n x and 2 n x respectively.

Then each side of the polygons are $2x$ and x.

Let A_1 = the area of the polygon of n sides

$$= n x^2 \cot\left(\frac{\pi}{n}\right)$$

and A_2 = the area of the decagon

$$= \frac{5}{2} n^2 \cot\left(\frac{\pi}{2n}\right),$$

$$\frac{A_1}{A_2} = \frac{5n^2\cot\left(\frac{\pi}{n}\right)}{\frac{5}{2}n^2\cot\left(\frac{\pi}{2n}\right)} = \frac{2\cot\left(\frac{\pi}{n}\right)}{\cot\left(\frac{\pi}{2n}\right)}$$

$$= \frac{2\cos\left(\frac{\pi}{n}\right)\sin\left(\frac{\pi}{2n}\right)}{2\sin\left(\frac{\pi}{2n}\right)\cos^2\left(\frac{\pi}{2n}\right)}$$

$$= \frac{2\cos\left(\frac{\pi}{n}\right)}{1+\cos\left(\frac{\pi}{n}\right)}$$

Ex-9. Let $A_1, A_2, A_3, \ldots\ldots\ldots\ldots, A_n$ be the vertices of an n-sided regular polygon such that

$$\frac{1}{A_1A_2} = \frac{1}{A_1A_3} + \frac{1}{A_1A_4},$$ then find the value of n.

Soln. Let O be the centre and $A_1A_2\ldots\ldots.A_n$ be the regular polygon of n-sides.

Let $OA_1 = OA_2 = \ldots\ldots\ldots\ldots\ldots\ldots.= OA_n = r$

and $\angle A_1OA_2 = \angle A_2OA_3$

$$= \ldots\ldots\ldots = \angle A_nOA_1 = \frac{2\pi}{n}$$

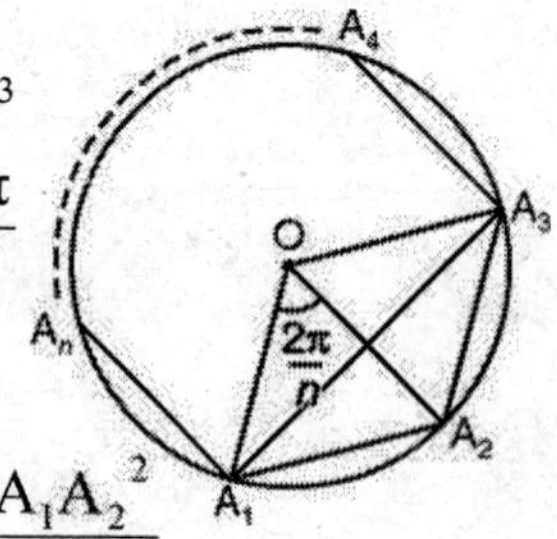

From the triangle OA_1A_2,

$$\cos\left(\frac{2\pi}{n}\right) = \frac{OA_1^2 + OA_2^2 - A_1A_2^2}{2.OA_1.OA_2}$$

$$= \frac{r^2 + r^2 - A_1A_2^2}{2.r.r}$$

$$\Rightarrow \quad A_1A_2^2 = 2r^2 - 2r^2\cos\left(\frac{2\pi}{n}\right)$$

$$\Rightarrow \quad A_1A_2^2 = 2r^2\left(1-\cos\left(\frac{2\pi}{n}\right)\right)$$

$$\Rightarrow \quad A_1A_2^2 = 2r^2.2\sin^2\left(\frac{2\pi}{n}\right) = 4r^2.\sin^2\left(\frac{2\pi}{n}\right)$$

$$\Rightarrow \quad A_1A_2 = 2r.\sin\left(\frac{2\pi}{n}\right).$$

Similarly, $A_1A_3 = 2r.\sin\left(\frac{4\pi}{n}\right)$

& $A_1A_4 = 2r.\sin\left(\frac{6\pi}{n}\right)$

Given, $$\frac{1}{A_1A_2} = \frac{1}{A_1A_3} + \frac{1}{A_1A_4}$$

$$\Rightarrow \quad \frac{1}{2r.\sin\left(\frac{2\pi}{n}\right)} = \frac{1}{2r.\sin\left(\frac{4\pi}{n}\right)} + \frac{1}{2r.\sin\left(\frac{6\pi}{n}\right)}$$

$$\Rightarrow \quad \frac{1}{\sin\left(\frac{\pi}{n}\right)} = \frac{1}{\sin\left(\frac{2\pi}{n}\right)} + \frac{1}{\sin\left(\frac{3\pi}{n}\right)}$$

$$\Rightarrow \quad \frac{1}{\sin\left(\frac{\pi}{n}\right)} - \frac{1}{\sin\left(\frac{3\pi}{n}\right)} = \frac{1}{\sin\left(\frac{2\pi}{n}\right)}$$

$$\Rightarrow \quad \frac{\sin\left(\frac{3\pi}{n}\right) - \sin\left(\frac{\pi}{n}\right)}{\sin\left(\frac{\pi}{n}\right)\sin\left(\frac{3\pi}{n}\right)} = \frac{1}{\sin\left(\frac{2\pi}{n}\right)}.$$

$$\Rightarrow \quad \frac{2\cos\left(\frac{2\pi}{n}\right)\sin\left(\frac{\pi}{n}\right)}{\sin\left(\frac{\pi}{n}\right)\sin\left(\frac{3\pi}{n}\right)} = \frac{1}{\sin\left(\frac{2\pi}{n}\right)}$$

$$\Rightarrow \quad 2\cos\left(\frac{2\pi}{n}\right)\sin\left(\frac{2\pi}{n}\right) = \sin\left(\frac{3\pi}{n}\right)$$

$$\Rightarrow \quad \sin\left(\frac{4\pi}{n}\right) = \sin\left(\frac{3\pi}{n}\right)$$

$$\Rightarrow \quad \sin\left(\frac{4\pi}{n}\right) = \sin\left(\pi - \frac{3\pi}{n}\right)$$

$$\Rightarrow \quad \left(\frac{4\pi}{n}\right) = \left(\pi - \frac{3\pi}{n}\right)$$

$\Rightarrow \quad \left(\frac{7\pi}{n}\right) = \pi$

$\Rightarrow \quad n = 7$.

Ex-10. If If A, A_1, A_2, A_3 are the areas of incircle and the ex-circles of a triangle , then prove that $\frac{1}{\sqrt{A_1}} + \frac{1}{\sqrt{A_2}} + \frac{1}{\sqrt{A_3}} = \frac{1}{\sqrt{A}}$.

Soln. Let r be the radius of the in-circle and r_1, r_2 and r_3 are the ex-radii of the given triangle

Then $\frac{1}{\sqrt{A_1}} + \frac{1}{\sqrt{A_2}} + \frac{1}{\sqrt{A_3}}$

$= \frac{1}{\sqrt{\pi r_1^2}} + \frac{1}{\sqrt{\pi r_1^2}} + \frac{1}{\sqrt{\pi r_1^2}}$

$= \frac{1}{\sqrt{\pi}}\left(\frac{1}{r_1} + \frac{1}{r_2} + \frac{1}{r_3}\right)$

$= \frac{1}{\sqrt{\pi}} \times \frac{1}{r}$

$= \frac{1}{\sqrt{\pi r^2}}$

$= \frac{1}{\sqrt{A}}$

Hence, the result.

Exercise - 10

(Questions based on Regular Polygons)

1. If R is the radius of the circumscribing circle of a regular polygon of n-sides with side length 'a' , then R is

(a) $\frac{a}{2}\sin\left(\frac{\pi}{n}\right)$ (b) $\frac{a}{2}\sin\left(\frac{\pi}{2n}\right)$

(c) $\frac{a}{2}\text{cosec}\left(\frac{\pi}{2n}\right)$ (d)* $\frac{a}{2}\text{cosec}\left(\frac{\pi}{n}\right)$

2 If A is the area and s be the semi perimeter of a triangle, then

(a) $A \le \frac{s^2}{3\sqrt{3}}$ (b) $A \le \frac{s^2}{2}$

(c) $A > \frac{s^2}{3}$ (d) $A > \frac{s^2}{4}$

3 If A_1, A_2, A_3 are the areas of an inscribed poly gon of 2n sides and n -sides and circumscribed polygon of n -sides , then A_2, A_1 , A_3 are in

(a) A.P (b) G.P

(c) H.P (d) A.G.P.

4 If the perimeter of the circle and the perimeter of the polygon of n-sides are same , then the ratio of the area of the circle and the area of the poly gon of n - sides is

(a) $\tan\left(\frac{\pi}{n}\right) : \frac{\pi}{n}$ (b) $\cos\left(\frac{\pi}{n}\right) : \frac{\pi}{n}$

(c) $\sin\left(\frac{\pi}{n}\right) : \frac{\pi}{n}$ (d) $\cot\left(\frac{\pi}{n}\right) : \frac{\pi}{n}$

5. The sum of the radii of the circle , which are respectively inscribed in and circum-scribed about a regular polygon of n-sides , is

(a) $\frac{a}{2}\cot\left(\frac{\pi}{n}\right)$ (b) $\frac{a}{2}\cot\left(\frac{\pi}{2n}\right)$

(c) $\frac{a}{2}\tan\left(\frac{\pi}{2n}\right)$ (d) $\frac{a}{2}\tan\left(\frac{\pi}{n}\right)$.

Answers

1.(d) 2.(a) 3.(b) 4.(a) 5.(b)

(Advance Level)

1. **Orthocentre and Pedal triangle of any triangle.**

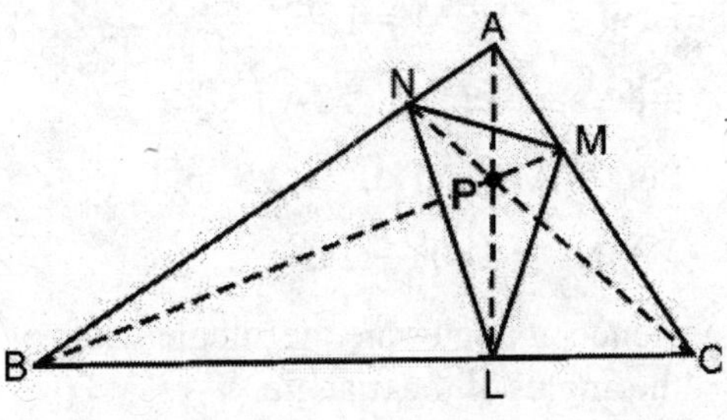

Let ABC be any triangle and let AL, BM and CN be the perpendiculars from A, B and C upon the opposite sides of the triangle. They meet at a point P. This point P is called the orthocentre of the triangle

The triangle LMN, which is formed by joining the feet of the perpendiculars, is called the pedal triangle.

(i) Distances of the Orthocentre from the angular points of the pedal triangle.

We have , PL = LB tan (PBL)

$= \text{LB}\ \tan(90^{\circ} - C)$

= AB cos B cot C

$= \dfrac{c}{\sin C} \times \cos B \cos C$

= 2RcosB cosC.

Similarly , PM = 2R cos A cos C,

PN = 2R cos A cos B

Again, AP = AM sec(LAC) = c cosA cosec

$C = \dfrac{c}{\sin C} \times \cos A = 2R \cos A.$

Similarly, BP = 2R cos B and CP = 2Rcos C.

(ii). The sides and the angles of the pedal triangle

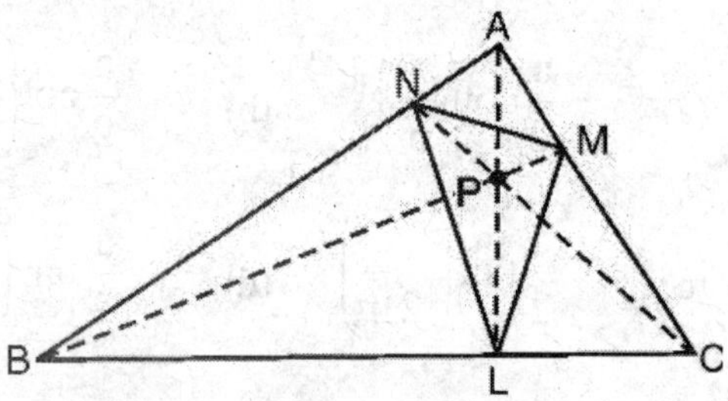

Since the angles PLC , PMC are right angles, so the points P, L, C and M lie on a circle.

Thus, $\angle PLM = \angle PCM = 90^{\circ} - A$

Similarly, P, L, B and M lie on a circle and therefore $\angle PLM = \angle PBN = 90^{\circ} - A$

Hence, $\angle NLM = 180^{\circ} - 2A$

= the supplement of 2A.

So, $\angle LMN = 180^{\circ} - 2B$ and

$\angle MNL = 180^{\circ} - 2C$

Hence, it angles are the supplements of twice the angles of the triangle.

Again, from the triangle AMN, we have,

$$\frac{MN}{\sin A} = \frac{AM}{\sin(ANM)} = \frac{AB\cos A}{\cos(PNM)}$$

$$= \frac{c\cos A}{\cos(PAM)} = \frac{c\cos A}{\sin C}$$

$$\Rightarrow\ MN = \frac{c}{\sin C}\ \sin A\ \cos A = a\cos A$$

So, NL = b cos B and LM = c cos C.

Hence, the sides of the pedal triangles are a cosA, b cosB and c cosC respectively.

(iii) Area of a pedal triangle LMN of a triangle ABC is $2\Delta \cos A \cos B \cos C$

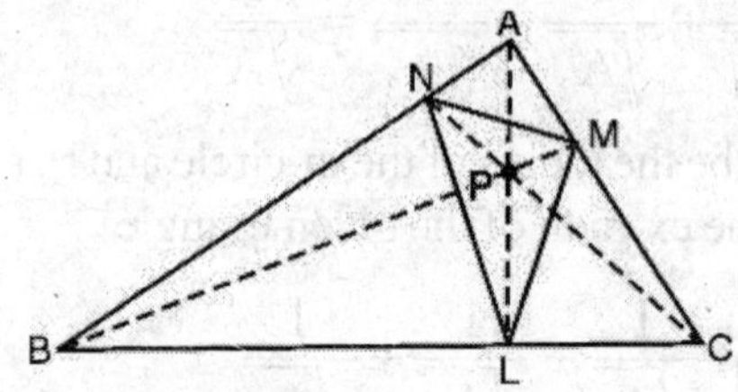

Here, PL = 2R cos B cos C, OM = 2R cos C. cosA, ON = 2R cos A. cos B

$$\text{ar}(\Delta LMN)$$

$$= \frac{1}{2} \times (R\sin 2A) \times (R\sin 2B) \times (\sin 2C)$$

$$= \frac{1}{2} \times R^2 \times (\sin 2A.\sin 2B.\sin 2C)$$

$$= \frac{1}{2} \times R^2 \times (8\sin A.\sin B.\sin C)$$

$$\times (\cos A.\cos B.\cos C)$$

$$= \frac{1}{2} \times R^2 \times \left(8.\frac{a}{2R}.\frac{b}{2R}.\frac{c}{2R}\right)$$

$$\times (\cos A.\cos B.\cos C)$$

$$= \frac{1}{2} \times \frac{abc}{R} \times (\cos A.\cos B.\cos C)$$

$$= 2\Delta \times (\cos A.\cos B.\cos C)$$

(iv) The circum-radius of a pedal triangle LMN of a triangle ABC is $\dfrac{R}{2}$

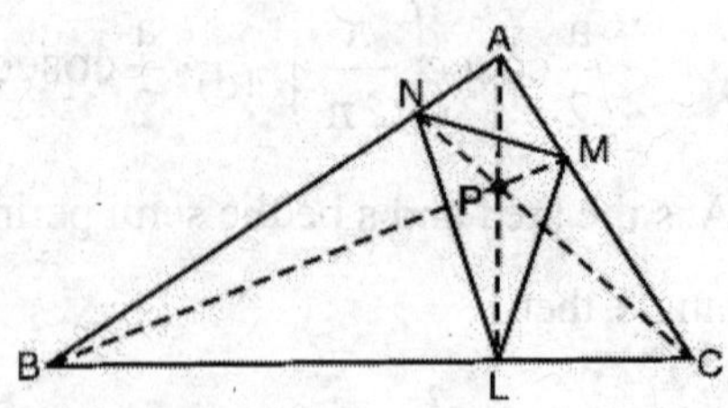

Circum-radius

$$\frac{MN}{2\sin(MLN)} = \frac{R\sin 2A}{2\sin(180^{\circ} - 2A)} = \frac{R\sin 2A}{2\sin 2A} = \frac{R}{2}$$

(v) The in-radius of a pedal triangle LMN of a triangle ABC is 2RcosAcosBcosC.

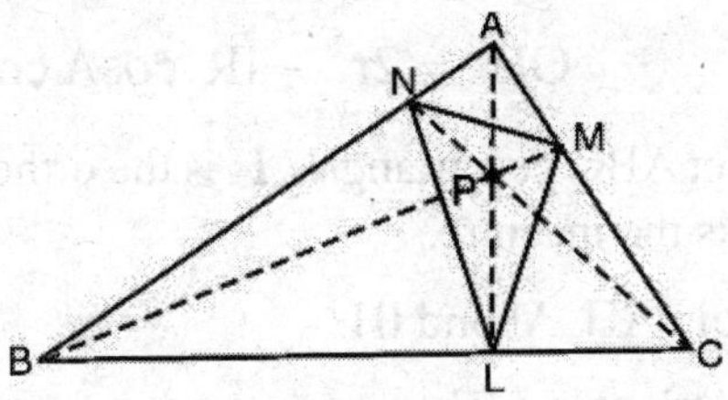

In-radius

$$= \frac{ar(\Delta LMN)}{\text{semi}-\text{perimeter}(\Delta LMN)}$$

$$= \frac{\frac{1}{2}R^2.\sin\ 2A.\sin\ 2B.\sin\ 2C}{2R.\sin\ A.\sin\ B.\sin\ C}$$

$$= 2R.\cos A.\cos B.\cos C$$

2. Distance between the Circumcentre and Orthocentre :

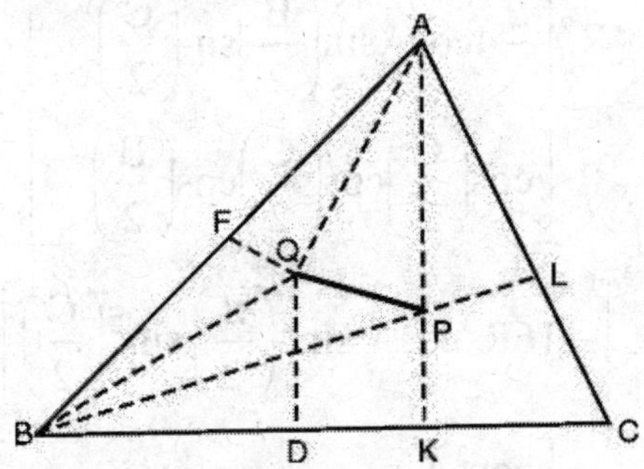

If OF perpendicular to AB, we have,

$\angle OAF = 90^o - \angle AOF = 90^o - C$

Also, $\angle PAL = 90^o - C$

Thus, $\angle OAP$

$$= A - \angle OAF - \angle PAL$$

$$= A - 2(90^o - C)$$

$$= A + 2C - 180^o$$

$$= A + 2C - (A+B+C) = C - B$$

Also, OA = R and PA = 2R cosA

Thus, $OP^2 = OA^2 + PA^2 - 2.OA.PA.\cos(OAP)$

$$= R^2 + 4R^2\cos^2A - 4R^2\cos A\cos(C-B)$$

$$= R^2 + 4R^2\cos A[\cos A - \cos(C-B)]$$

$$= R^2 - 4R^2\cos A[\cos(B+C) - \cos(B-C)]$$

$$= R^2 - 8R^2\cos A\cos B\cos C$$

$$= R^2(1 - 8\cos A\cos B\cos C)$$

$$\Rightarrow OP = R\sqrt{(1-8\cos A\cos B\cos C)}$$

3. Distance between the circumcentre and the Incentre.

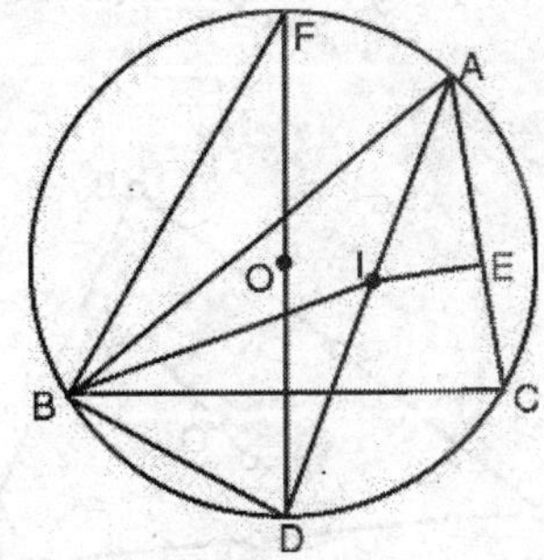

Let O be the circumcentre and OF be perpendicular to AB.

Let I be the incentre and IE perpendicular to AC.

Then $\angle OAF = 90^o - C$

$\angle OAI = \angle IAF - \angle OAF$

$$= \frac{A}{2} - (90^o - C)$$

$$= \frac{A}{2} + C - \frac{A+B+C}{2}$$

$$= \frac{C-B}{2}$$

Also, $AI = \dfrac{IE}{\sin\left(\frac{A}{2}\right)} = \dfrac{r}{\sin\left(\frac{A}{2}\right)}$

$$= 4R\sin\left(\frac{B}{2}\right)\sin\left(\frac{C}{2}\right)$$

$$= 1 - 8\sin\left(\frac{B}{2}\right)\sin\left(\frac{C}{2}\right)\left(\cos\left(\frac{B}{2} + \frac{C}{2}\right)\right)$$

$$= 1 - 8\sin\left(\frac{B}{2}\right)\sin\left(\frac{C}{2}\right)\sin\left(\frac{A}{2}\right)$$

$$\Rightarrow OI = R\sqrt{1 - 8\sin\left(\frac{B}{2}\right)\sin\left(\frac{C}{2}\right)\sin\left(\frac{A}{2}\right)}$$

Also, $OI^2 = R^2 - 2R \times 4R\sin\left(\frac{A}{2}\right)\sin\left(\frac{B}{2}\right)\sin\left(\frac{C}{2}\right)$

$$= R^2 - 2Rr$$

$$\Rightarrow OI = \sqrt{R^2 - 2Rr}$$

Hence, the result

4. Distance between the Circumcentre and Centroid :

i.e. $OG^2 = R^2 - \frac{1}{9}(a^2+b^2+c^2)$.

Soln.

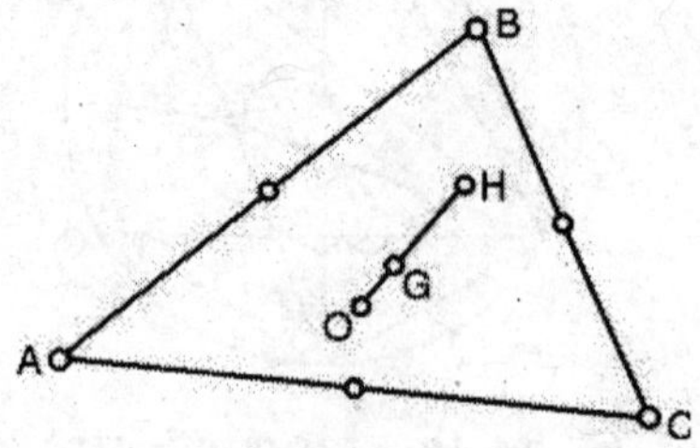

As we know that, Centroid divides the orthocentre and circumcentre in the ratio 2:1

Thus, $OG = \frac{1}{3}.OH$

$\Rightarrow \quad OG^2 = \frac{1}{9}.OH^2$

$= \frac{1}{9}(R^2 - 8R^2\cos A.\cos B.\cos C)$

$\Rightarrow \quad OG^2 = \frac{R^2}{9} \times$

$(1-4\{\cos(A+B)+\cos(A-B)\}.\cos C)$

$\Rightarrow \quad OG^2 = \frac{R^2}{9} \times$

$(1+4\cos^2 C+4\cos(A-B).\cos(A+B))$

$\Rightarrow \quad OG^2 = \frac{R^2}{9}(1+4\cos^2 C+4\cos^2 A-4\sin^2 B)$

$\Rightarrow \quad OG^2 = \frac{R^2}{9}(1+4-4\sin^2 C+4-4\sin^2 A-4\sin^2 B)$

$\Rightarrow \quad OG^2 = \frac{R^2}{9}(9-4(\sin^2 A+\sin^2 B+\sin^2 C))$

$\Rightarrow \quad OG^2 = R^2 - \frac{1}{9} \times$

$\{(2R\sin A)^2+(2R\sin A)^2+(2R\sin A)^2\}$

$\Rightarrow \quad OG^2 = R^2 - \frac{1}{9}(a^2+b^2+c^2)$

5. Distance between Incentre and Orthocentre :

i.e. $OH^2 = 2r^2 - 4R^2\cos A\cos B\cos C$

Let ABC be a triangle , H is the orthocentre and I is the incentre.

Join AH, AI and IH.

In triangle AIH,

$IH^2 = AH^2 + AI^2 - 2.AH.AI.\cos(\angle IAH)$

$\angle IAH = \frac{A}{2} - \angle HAC$

$= \frac{A}{2} - (90^\circ - C) = \frac{1}{2}(C-B)$

Thus,

$$= 4R^2\begin{bmatrix}\cos^2 A + 4\sin^2\left(\frac{B}{2}\right)\sin\left(\frac{C}{2}\right) \\ -4\cos A\sin\left(\frac{B}{2}\right)\sin\left(\frac{C}{2}\right) \\ \cos\left(\frac{C}{2}\right)\cos\left(\frac{C}{2}\right)\cos\left(\frac{B}{2}\right)\end{bmatrix}$$

$$-\left[-16R^2\cos A.\sin^2\left(\frac{B}{2}\right).\sin^2\left(\frac{C}{2}\right)\right]$$

$$= 4R^2\begin{bmatrix}\cos^2 A + \\ 4\sin^2\left(\frac{B}{2}\right)\sin^2\left(\frac{C}{2}\right)(1-\cos A) \\ -\cos A.\cos B.\sin C\end{bmatrix}$$

$$= 4R^2\begin{bmatrix}\cos^2 A + \\ 8\sin^2\left(\frac{A}{2}\right)\sin^2\left(\frac{B}{2}\right)\sin^2\left(\frac{C}{2}\right) \\ -\cos A.\sin B.\sin C\end{bmatrix}$$

$$= 32R^2\sin^2\left(\frac{A}{2}\right)\sin^2\left(\frac{B}{2}\right)\sin^2\left(\frac{C}{2}\right)$$

$$+4R^2\cos A.(\cos A - \sin B.\sin C)$$

$$= 2\left[16R^2\sin^2\left(\frac{A}{2}\right)\sin^2\left(\frac{B}{2}\right)\sin^2\left(\frac{C}{2}\right)\right]$$

$$-4R^2\cos A.[\cos(B+C)+\sin B.\sin C]$$

$= 2r^2 - 4R^2 \cos A.\cos B.\cos C.$

Hence, the result.

6. Excentral Triangle.

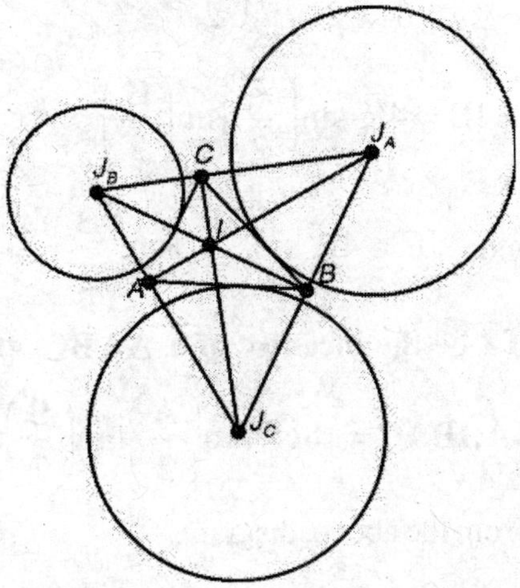

The triangle formed by joining the three excentres I_A, I_B, I_C of a ΔABC is called an excentral of excentric triangle

(i) ΔABC is the pedal triangle of $\Delta\ I_A I_B I_C$

(ii) Its angles are $\left(\frac{\pi}{2}-\frac{A}{2}\right), \left(\frac{\pi}{2}-\frac{B}{2}\right), \left(\frac{\pi}{2}-\frac{C}{2}\right)$.

(iii) Its sides are

$$I_B I_C = 4R\cos\left(\frac{A}{2}\right),$$

$$I_A I_C = 4R\cos\left(\frac{B}{2}\right),$$

and $\quad I_A I_B = 4R\cos\left(\frac{C}{2}\right)$

(iv) $II_A = 4R\sin\left(\frac{A}{2}\right);$

$$II_B = 4R\sin\left(\frac{B}{2}\right);$$

$$II_C = 4R\sin\left(\frac{C}{2}\right)$$

(v) Incentre I of ΔABC is the orthocentre of the excentric triangle $\Delta\ I_A I_B I_C$.

(vi) $ar(I_A I_B I_C) = \frac{1}{2} \times \left(4R\cos\left(\frac{B}{2}\right)\right)$

$$\times \left(4R\cos\left(\frac{B}{2}\right)\right) \times \sin\left(90^o - \frac{A}{2}\right)$$

$$= 8.R^2.\cos\left(\frac{A}{2}\right).\cos\left(\frac{B}{2}\right).\cos\left(\frac{C}{2}\right)$$

(vii) Circum-radius $= \dfrac{I_2 I_3}{2\sin(\angle I_2 I_1 I_3)}$

$$= \frac{I_2 I_3}{2\sin\left(90^o - \frac{A}{2}\right)}$$

$$= \frac{4R\cos\left(\frac{A}{2}\right)}{2\cos\left(\frac{A}{2}\right)} = 2R$$

Soln. We know that,

$$II_A = a\sec\left(\frac{A}{2}\right),$$

$$II_B = a\sec\left(\frac{B}{2}\right),$$

$$II_C = a\sec\left(\frac{C}{2}\right)$$

Also,

$$I_A I_B = c\,\mathrm{cosec}\left(\frac{A}{2}\right),\ I_B I_C = c\,\mathrm{cosec}\left(\frac{B}{2}\right),$$

$$I_C I_A = c\,\mathrm{cosec}\left(\frac{C}{2}\right)$$

Thus, $II_A.II_B.II_C$

$$= abc \times \sec\left(\frac{A}{2}\right)\sec\left(\frac{B}{2}\right)\sec\left(\frac{C}{2}\right) \quad ....(i)$$

Also, $a = 2R\sin A$, $b = 2R\sin B$ and $c = 2R\sin C$

From (i), we get, $II_A.II_B.II_C$

$$= (2R\sin A)(2R\sin B)(2R\sin C)$$

$$\times \sec\left(\frac{A}{2}\right)\sec\left(\frac{B}{2}\right)\sec\left(\frac{C}{2}\right)$$

$$= 8R^3 \times \frac{\left(2\sin\frac{A}{2}\cos\frac{A}{2}\right)\left(2\sin\frac{B}{2}\cos\frac{B}{2}\right)\left(2\sin\frac{C}{2}\cos\frac{C}{2}\right)}{\cos\frac{A}{2}\cos\frac{B}{2}\cos\frac{C}{2}}$$

$$= 64R^3\sin\frac{A}{2}\sin\frac{B}{2}\sin\frac{C}{2}$$

$$= 16R^2 \times 4R\sin\frac{A}{2}\sin\frac{B}{2}\sin\frac{C}{2}$$

$$= 16R^2 r$$

Hence, the result.

(vi) If I is the incentre and I_A, I_B, I_C are the excentres of the triangle **ΔABC**, then prove that $II_A.II_B.II_C = 16rR^2$

7. The distance between the incentre and the angular point of a triangle ABC.

i.e. $IA = 4R\sin\left(\frac{B}{2}\right)\sin\left(\frac{C}{2}\right)$,

$IB = 4R\sin\left(\frac{A}{2}\right)\sin\left(\frac{C}{2}\right)$,

$IC = 4R\sin\left(\frac{A}{2}\right)\sin\left(\frac{B}{2}\right)$

Soln.

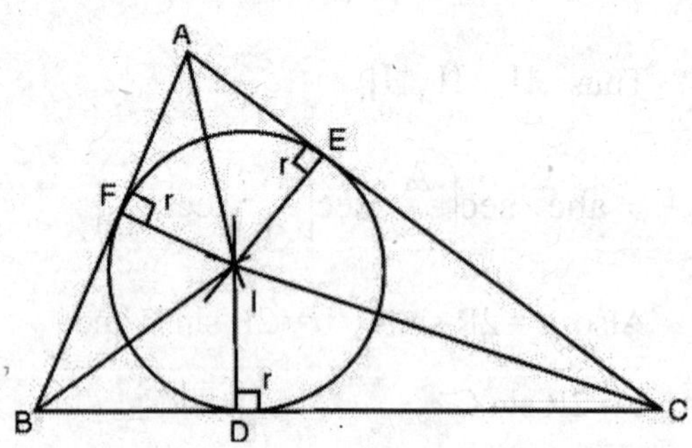

We have, $\sin\left(\frac{A}{2}\right) = \frac{IF}{IA}$

$\Rightarrow \quad r = IA.\sin\left(\frac{A}{2}\right)$

$\Rightarrow \quad 4R\sin\left(\frac{A}{2}\right)\sin\left(\frac{B}{2}\right)\sin\left(\frac{C}{2}\right) = IA.\sin\left(\frac{A}{2}\right)$

$\Rightarrow \quad IA = 4R\sin\left(\frac{B}{2}\right)\sin\left(\frac{C}{2}\right)$,

Similarly, $IB = 4R\sin\left(\frac{A}{2}\right)\sin\left(\frac{C}{2}\right)$

and $IC = 4R\sin\left(\frac{A}{2}\right)\sin\left(\frac{B}{2}\right)$

8. If I be the incentre of a ΔABC, then

$IA.IB.IC = abc \times \tan\left(\frac{A}{2}\right)\tan\left(\frac{B}{2}\right)\tan\left(\frac{C}{2}\right)$

Soln. From the above, diagram,

$$IA = \frac{r}{\sin\left(\frac{A}{2}\right)},\ IB = \frac{r}{\sin\left(\frac{B}{2}\right)},$$

$$IC = \frac{r}{\sin\left(\frac{C}{2}\right)}$$

Thus, $IA.IB.IC = \frac{r^3}{\sin\left(\frac{A}{2}\right)\sin\left(\frac{B}{2}\right)\sin\left(\frac{C}{2}\right)}$.

$$= \frac{r^3 \times 4R}{4R\sin\left(\frac{A}{2}\right)\sin\left(\frac{B}{2}\right)\sin\left(\frac{C}{2}\right)}$$

$$= \frac{r^3.4R}{r} = 4r^2.R = 4.\frac{\Delta^2}{s^2}.\frac{abc}{4\Delta} = abc.\frac{\Delta}{s^2}$$

$$= abc \times \frac{\sqrt{s(s-a)(s-b)(s-c)}}{s^2}$$

$$= abc \times \sqrt{\frac{s(s-a)(s-b)(s-c)}{s^4}}$$

$$= abc \times \sqrt{\frac{(s-a)}{s}.\frac{(s-b)}{s}.\frac{(s-c)}{s}}$$

$$= abc \times \sqrt{\frac{(s-a)(s-b)}{s(s-c)}}.\sqrt{\frac{(s-b)(s-c)}{s(s-a)}}.$$

$$\sqrt{\frac{(s-c)(s-a)}{s(s-b)}}.$$

$= \quad abc \times \tan\left(\frac{C}{2}\right)\tan\left(\frac{A}{2}\right)\tan\left(\frac{B}{2}\right)$

$= \quad abc \times \tan\left(\frac{A}{2}\right)\tan\left(\frac{B}{2}\right)\tan\left(\frac{C}{2}\right)$

Hence, the result.

9. O is the circumcentre of a triangle ΔABC and R_1, R_2 & R_3 are respectively the radii of the circumference of the Δ's OBC, OCA and OAB respectively, then $\frac{a}{R_1}+\frac{b}{R_2}+\frac{c}{R_3}=\frac{abc}{R^3}$.

Soln.

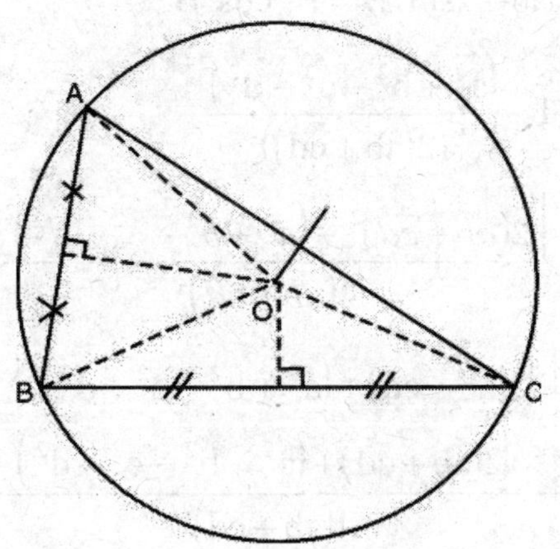

As we know that $R=\frac{abc}{4\Delta}$

Let $\Delta OBC=\Delta_1$, $\Delta OCA=\Delta_2$, $\Delta OAB=\Delta_3$

In ΔOBC,

$R_1=\frac{0B.0C.BC}{4\Delta_1}=\frac{R.R.a}{4\Delta_1}=\frac{R^2.a}{4\Delta_1}$

$\Rightarrow \quad \frac{a}{R_1}=\frac{4\Delta_1}{R^2}$

similarly, $\frac{b}{R_2}=\frac{4\Delta_2}{R^2}$ & $\frac{c}{R_3}=\frac{4\Delta_3}{R^2}$

Thus, $\frac{a}{R_1}+\frac{b}{R_2}+\frac{c}{R_3}$

$= \quad \frac{4}{R^2}(\Delta_1+\Delta_2+\Delta_3)$

$= \quad \frac{4}{R^2}\times\Delta=\frac{4\Delta}{R^2}=\frac{abc}{R^3}$

Hence, the result.

10. The distance between the centre of the nine - point circle from the angle A is $\frac{R}{2}\sqrt{(1+8\cos A.\cos B.\cos C)}$.

Soln.

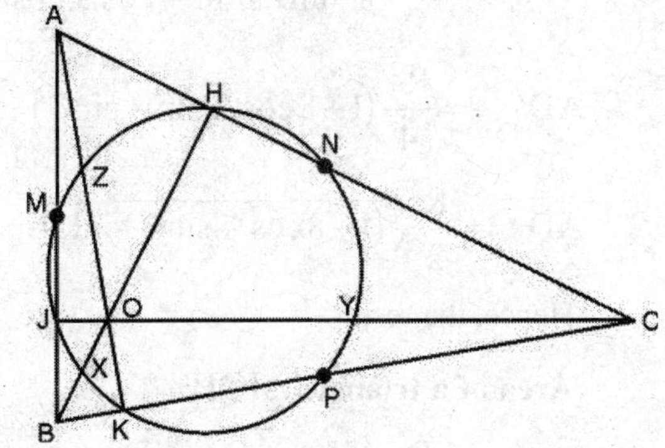

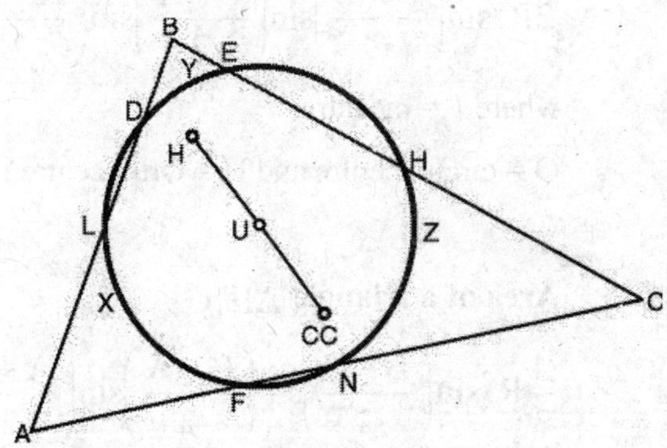

Let ABC be a triangle, H = orthocentre, O = circumcentre, D = nine-point centre

As we know that nine point centre is the mid - point of the orthocentre and circumcentre of a triangle.

Thus AOH be a triangle , where, AD is the median.

In ΔAOH, $2(AD^2+DO^2)=AH^2+AO^2$

$\Rightarrow \quad 2AD^2=AH^2+AO^2-\frac{1}{2}OH^2$ ....(i)

Now from the triangle ABC, we can write,

$AH=2R\cos A$, $OA=R$,

$OH=R\sqrt{(1-8\cos A.\cos B.\cos C)}$.

From (i), we get, $2AD^2$

$= \quad R^2(4\cos^2 A+1)$

$-\frac{R^2}{2}(1-8\cos A.\cos B.\cos C)$

$= \quad \frac{R^2}{2}\left[1+8\cos A.\left\{\begin{matrix}\cos(180^\circ-B+C)\\ +\cos B.\cos C\end{matrix}\right\}\right]$

$$\Rightarrow \quad AD^2 = \frac{R^2}{4}\left[1+8\cos A.\begin{Bmatrix}-\cos B.\cos C\\ +\sin B.\sin C+\cos B.\cos C\end{Bmatrix}\right]$$

$$\Rightarrow \quad AD^2 = \frac{R^2}{4}(1+8\cos A.\sin B.\sin C)$$

$$\Rightarrow \quad AD = \frac{R}{2}\sqrt{(1+8\cos A.\sin B.\sin C)}$$

Hence, the result.

11. **Area of a triangle ΔIOH**

$= 2R^2\sin\left(\frac{B-C}{2}\right)\sin\left(\frac{C-A}{2}\right)\sin\left(\frac{A-B}{2}\right)$,

where I = Incentre,

O = circumcentre and H = Orthocentre.:

12. **Area of a triangle ΔIHG**

$= \frac{4}{3}R^2\sin\left(\frac{B-C}{2}\right)\sin\left(\frac{C-A}{2}\right)\sin\left(\frac{A-B}{2}\right)$

where I = Incentre,

H = Orthocentre., G = Centroid.

Quadrilateral

1. **Area of a quadrilateral, which is inscribed in a circle.**

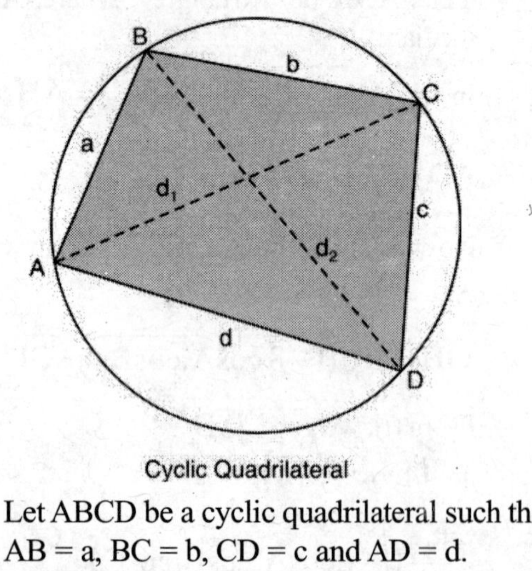

Cyclic Quadrilateral

Let ABCD be a cyclic quadrilateral such that AB = a, BC = b, CD = c and AD = d.

$$ar(ABCD) = ar(\Delta ABC) + ar(\Delta ADC)$$

$$= \frac{1}{2}ab\sin B + \frac{1}{2}cd\sin D$$

$$= \frac{1}{2}ab\sin B + \frac{1}{2}cd\sin(\pi - B)$$

$$= \frac{1}{2}ab\sin B + \frac{1}{2}cd\sin B$$

$$= \frac{1}{2}(ab+cd)\sin B$$

From $\Delta's$, BAC and BCD, we have,

$$a^2+b^2-2ab\cos B = c^2+d^2-2cd\cos D$$

$$\Rightarrow \quad a^2+b^2-2ab\cos B = c^2+d^2+2cd\cos B$$

$$\Rightarrow \quad \cos B = \left(\frac{a^2+b^2-c^2-d^2}{2(ab+cd)}\right)$$

Now, $\sin^2 B = 1-\cos^2 B$

$$= 1-\frac{(a^2+b^2-c^2-d^2)^2}{(2(ab+cd))^2}$$

$$= \frac{\left[\{2(ab+cd)^2-(a^2+b^2-c^2-d^2)^2\}\right]}{4(ab+cd)^2}$$

$$= \frac{\left[\begin{matrix}\{2(ab+cd)+(a^2+b^2-c^2-d^2)^2\}\\ \times\{2(ab+cd)+(a^2+b^2-c^2-d^2)^2\}\end{matrix}\right]}{4(ab+cd)^2}$$

$$= \frac{\begin{matrix}\{(a^2+b^2+2ab)-(c^2-2cd+d^2)\}\\ \times\{(c^2+2cd+d^2)-(a^2+b^2-2ab)\}\end{matrix}}{4(ab+cd)^2}$$

$$= \frac{\begin{matrix}\{(a+b)^2-(c-d)^2\}\\ \times\{(c+d)^2-(a-b)^2\}\end{matrix}}{4(ab+cd)^2}$$

$$= \frac{\begin{matrix}\{(a+b+c-d)(a+b-c+d)\}\\ \times\{(c+d+a-b)(c+d-a+b)\}\end{matrix}}{4(ab+cd)^2}$$

Let $a+b+c+d = 2s$

Thus, $(a+b+c-d)$

$= (a+b+c+d-2d) = 2(s-d)$

similarly, $(a+b+d-c) = 2(s-c)$,

$(a+c+d-b) = 2(s-b)$,

$(b+c+d-a) = 2(s-a)$

$\Rightarrow \quad \sin^2 B$

$$= \frac{2(s-d)\times 2(s-c)\times 2(s-b)\times 2(s-a)}{4(ab+cd)^2}$$

$\Rightarrow \quad \sin^2 B$

$$= \frac{16\times(s-a)\times(s-b)\times(s-c)\times(s-d)}{4(ab+cd)^2}$$

$\Rightarrow \quad (ab+cd)\sin B$

$$= 2\times\sqrt{(s-a)(s-b)(s-c)(s-d)}$$

Hence, the area of the quadrilateral

$$= \frac{1}{2}(ab+cd)\sin B$$

$$= \sqrt{(s-a)(s-b)(s-c)(s-d)}$$

2. The radius of the circle circumscribing the quadrilateral ABCD.

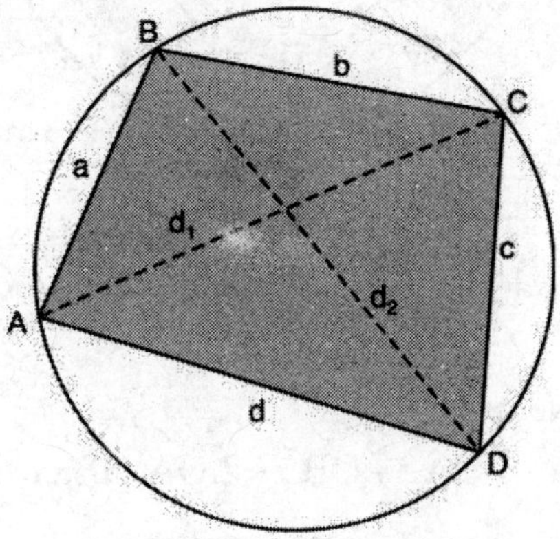

Cyclic Quadrilateral

We have, $\cos B = \dfrac{a^2+b^2-c^2-d^2}{2(ab+cd)}$

and $AC^2 = a^2+b^2-2\,ab\cos B$

$$= a^2+b^2-2\,ab\times\left(\frac{a^2+b^2-c^2-d^2}{ab+cd}\right)$$

$$= \frac{(a^2+b^2)cd+(c^2+d^2)ab}{(ab+cd)}$$

$$= \frac{(ac+bd)(ad+bc)}{(ab+cd)}.$$

In ΔABC, R

$$= \frac{1}{2}\times\frac{AC}{\sin B}$$

$$= \sqrt{\left(\frac{(ac+bd)(ad+bc)}{(ab+cd)}\right)}$$

$$\div 4\sqrt{\frac{(s-a)(s-b)(s-c)(s-d)}{(ab+cd)^2}}$$

$$= \frac{1}{2}\times\frac{AC}{\sin B} = \frac{1}{4}\sqrt{\left(\frac{(ab+cd)(ac+bd)(ad+bc)}{(s-a)(s-b)(s-c)(s-d)}\right)}$$

3. Area of a quadrilateral ABCD, when it is not inscribed.

$$ar(ABCD) = ar(\Delta ABC)+ar(\Delta ACD)$$

$$= \frac{1}{2}ab\sin B+\frac{1}{2}cd\sin D$$

$\Rightarrow \quad 4\Delta = 2\,ab\ \sin B+2\,cd\sin D$ ....(i)

Also, $a^2+b^2-2ab\cos B = c^2+d^2-2cd\cos D$

$\Rightarrow \quad 2ab\cos B - 2cd\cos D = a^2+b^2-c^2-d^2$ ......(ii)

Squaring (i) and (ii) and adding, we get,

$16\Delta^2+(a^2+b^2-c^2-d^2)^2$

$$= 4a^2b^2+4c^2d^2 - 8\,abcd(\cos B\cos D-\sin B\sin D)$$

$$= 4(a^2b^2+c^2d^2)-8\,abcd\cos(B+D)$$

$$= 4(a^2b^2+c^2d^2)-8\,abcd\cos 2\alpha$$

$$= 4(a^2b^2+c^2d^2)-8\,abcd(2\cos^2\alpha-1)$$

$$= 4(ab+cd)^2-16\,abcd\cos^2\alpha$$

$16\Delta^2$

$$= 4(ab+cd)^2 - (a^2+b^2-c^2-d^2)^2-16\,abcd\cos^2\alpha$$ ....(iii)

Thus, $16\Delta^2$

$$= 2(s-a).2(s-b).2(s-c).2(s-d) - 16abcd\cos^2\alpha$$

$\Rightarrow \quad \Delta^2$

$$= (s-a).(s-b).(s-c).(s-d)-abcd\cos^2\alpha$$

$\Rightarrow \quad \Delta$

$$= \sqrt{(s-a).(s-b).(s-c).(s-d)-abcd\cos^2\alpha}$$

4. Area of a quadilateral which can have a circle inscribed in it.

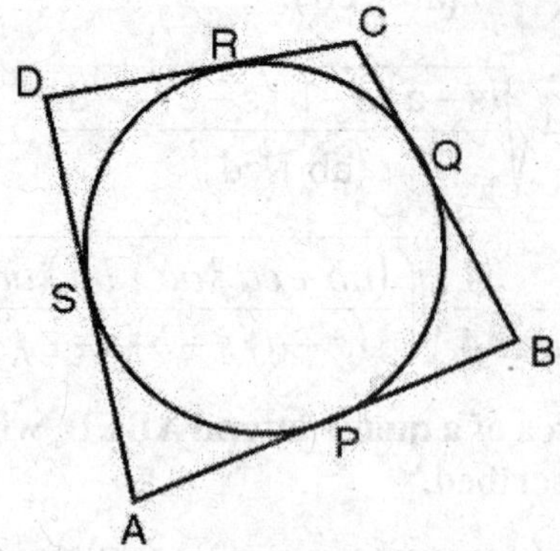

We have, AP = AS, BP = BQ , CQ = CR and DR = RS

$$AP + BP + CR + DR = AS + BQ + CQ + DS$$

$$\Rightarrow \quad AB + CD = BC + DA$$

$$\Rightarrow \quad a + c = b + d$$

Hence, $s = \dfrac{a+b+c+d}{2} = a + c = b + d$

Thus, $s - a = c$, $s - b = d$, $s - c = a$, $s - d = b$

As we know that,

$\Delta^2 = abcd - abcd\cos^2\alpha = abcd\sin^2\alpha$

$\Rightarrow \quad \Delta = \sqrt{abcd}\ \sin\alpha$

5. The area of a quadrilateral, which can be both inscribed in a circle and circumscribed on another circle and the radius of the later circle is $\dfrac{2\sqrt{abcd}}{a+b+c+d}$

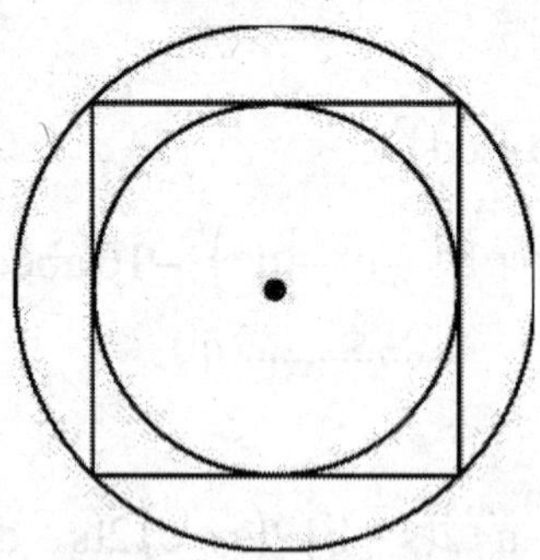

In a quadrilateral ABCD, $\angle B + \angle D = 180^\circ$

$$\Rightarrow \quad 2\alpha = 180^\circ$$

$$\Rightarrow \alpha = 90^\circ$$

Hence, the area of the quadrilateral, which is inscribed in a circle and circumscribed on another circle is

$= \sqrt{a\,b\,c\,d}$.

We have, $r = \dfrac{\Delta}{s} = \dfrac{2\Delta}{2s} = \dfrac{2\sqrt{abcd}}{a+b+c+d}$

6. a, b, c and d are the sides of a quadrilateral taken in order, and θ is the angle between the diagonals opposite to b or d, then the area of the quadrilateral is $\dfrac{1}{4}\left(a^2 + c^2 - b^2 - d^2\right)\tan\theta$

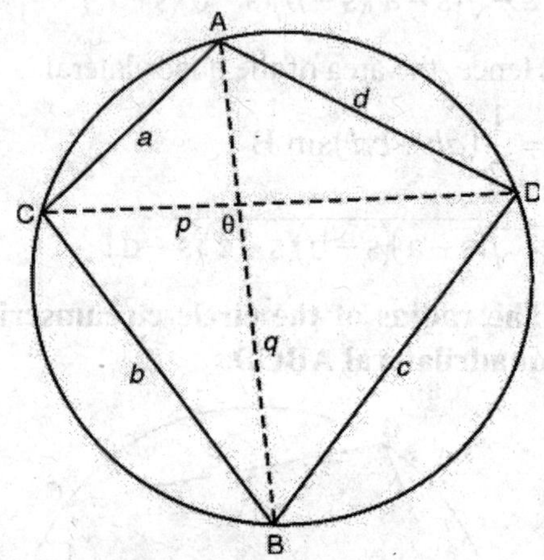

Area of a quadrilateral ABCD

$= \dfrac{1}{2} \times AC \times BD \times \sin\theta$ ....(i)

Now,

$a^2 = OA^2 + OB^2 - 2.OA.OB.\cos\left(180^\circ - \theta\right)$

$b^2 = OC^2 + OB^2 - 2.OC.OB.\cos\theta$,

$c^2 = OC^2 + OD^2 - 2.OC.OD.\cos\left(180^\circ - \theta\right)$

$d^2 = OA^2 + OD^2 - 2.OA.OD.\cos\theta$.

We have, $a^2 - b^2 + c^2 - d^2$

$= \quad 2\cos\theta \begin{pmatrix} OA.OB + OB.OC \\ + OC.OD + OA.OD \end{pmatrix}$

$= \quad 2.\cos\theta.AC.BD$

From (i), we get, area of the quadrilateral

$ABCD = \dfrac{1}{2}\left(a^2 - b^2 + c^2 - d^2\right)\tan\theta$.

7. a, b, c and d are the sides of a quadrilateral and p and q be its diagonals , then its area is

$$\frac{1}{4} \times \sqrt{\left(4p^2q^2 - \left(a^2 + c^2 - b^2 - d^2\right)^2\right)}$$

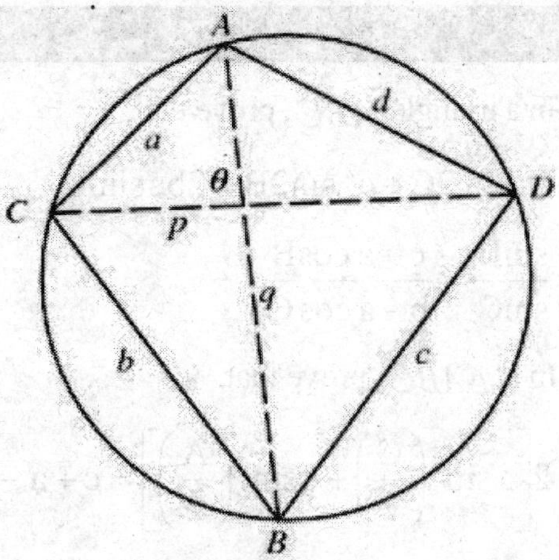

As we know that,

$$a^2 - b^2 + c^2 - d^2 = 2pq\cos\theta$$

Area of a quadrilateral ABCD

$$= \frac{1}{2} pq \sin\theta$$

$$= \sqrt{\frac{1}{4}p^2q^2\sin^2\theta}$$

$$= \sqrt{\frac{1}{4}p^2q^2(1-\cos^2\theta)}$$

$$= \sqrt{\frac{1}{4}\left(4p^2q^2-(2pq\cos\theta)^2\right)}$$

$$= \frac{1}{2}\sqrt{\left(4p^2q^2-\left(a^2-b^2+c^2-d^2\right)^2\right)}$$

Hence, the result.

8. If a quadrilateral can be inscribed in a circle, then the angle between its diagonals is

$$\sin^{-1}\left(\frac{2\sqrt{(s-a)(s-b)(s-c)(s-d)}}{(ac+bd)}\right)$$

Soln. Let ABCD be a quadrilateral, whose sides are a, b, c and d resp. and its diagonals are p and q.

Let the angle between the diagonals be θ.

Then, $pq = ac + bd$.

Area of a quadrilateral ABCD $= \frac{1}{2} \times pq\sin\theta$

$$\Rightarrow \quad \sin\theta = \frac{2\sqrt{(s-a)(s-b)(s-c)(s-d)}}{(ac+bd)}$$

$$\Rightarrow \quad \theta = \sin^{-1}\left(\frac{2\sqrt{(s-a)(s-b)(s-c)(s-d)}}{(ac+bd)}\right)$$

Hence, the result.

9. If a quadrilateral can be inscribed in a circle as well as circumscribed about another circle, then the angle between its diagonal is $\cos^{-1}\left(\frac{ac-bd}{ac+bd}\right)$.

Since the quadrilateral be circumscribed, then we can write,

$$\frac{1}{2}pq\sin\theta = \sqrt{abcd}$$

$$\Rightarrow \quad \sin\theta = \left(\frac{2\sqrt{abcd}}{pq}\right)$$

Therefore, $\cos\theta = \sqrt{1-\sin^2\theta} =$

$$\cos\theta = \sqrt{1-\left(\frac{4abcd}{(ac+bd)^2}\right)} = \left(\frac{ac-bd}{ac+bd}\right)$$

$$\Rightarrow \quad \theta = \cos^{-1}\left(\frac{ac-bd}{ac+bd}\right)$$

Hence, the result.

10. If a, b, c and d be the sides of a quadrilateral, taken in order such that α, β & γ be the angles between the sides a and b , b and c , c and a resp, then

$$d^2 = a^2 + b^2 + c^2 - 2ab\cos\alpha$$
$$- 2bc\cos\beta - 2ca\cos\gamma$$

Level– I QUESTIONS BASED ON FUNDAMENTALS

1. In any triangle ABC, prove that
$(b^2-c^2)\cot A+(c^2-a^2)\cot B+(a^2-b^2)\cot C=0$

2. In a tringle Δ ABC, prove that

(i) $a\sin(B-C)+b\sin(C-A)+c\sin(A-B)=0$

(ii) $\dfrac{a^2\sin(B-C)}{\sin A}+\dfrac{b^2\sin(C-A)}{\sin B}+\dfrac{c^2\sin(C-A)}{\sin C}=0$

3. In a triangle ABC, prove that for any angle θ,
$b\cos(A-\theta)+c\cos(B+\theta)=c\cos\theta$

4. If $\cos A+\cos B=4\sin^2\left(\dfrac{C}{2}\right)$, prove that the sides , a, c , b of the triangle ABC are in A.P

5. If in a ΔABC $\dfrac{\sin A}{\sin C}=\dfrac{\sin(A-B)}{\sin(B-C)}$, then prove that, a^2, b^2, c^2 are in A.P

6. In a triangle ABC, prove that,

(i) $2(bc\cos A+ca\cos B+ab\cos C)=a^2+b^2+c^2$

(ii) $(a-b)^2\cos^2\left(\dfrac{C}{2}\right)+(a+b)^2\sin^2\left(\dfrac{C}{2}\right)=c^2$

7. With usual notation, if in a triangle ABC, $\dfrac{b+c}{11}=\dfrac{c+a}{12}=\dfrac{a+b}{13}$, then prove that, $\dfrac{\cos A}{7}=\dfrac{\cos B}{19}=\dfrac{\cos C}{25}$

8. Let a, b and c be the sides of a ΔABC. If $a^2, b^2 \& c^2$ are the roots of the equation $x^3-Px^2+Qx-R=0$,
where P, Q and R are costants, then find the value of $\dfrac{\cos A}{a}+\dfrac{\cos B}{b}+\dfrac{\cos C}{c}$, in terms of P, Q and R.

9. In a triangle ABC, prove that,

(i) $b^2\sin 2C+c^2\sin 2B=2bc\sin A$

(ii) $\dfrac{\sin B}{\sin C}=\dfrac{c-a\cos B}{b-a\cos C}$

10. In a ΔABC, prove that,

(i) $2\left[a\sin^2\left(\dfrac{C}{2}\right)+c\sin^2\left(\dfrac{A}{2}\right)\right]=c+a-b$

(ii) $\dfrac{\cos^2\left(\frac{A}{2}\right)}{a}+\dfrac{\cos^2\left(\frac{B}{2}\right)}{b}+\dfrac{\cos^2\left(\frac{C}{2}\right)}{c}=\dfrac{s^2}{abc}$

11. In a ΔABC, prove that
$\left(bc\cos^2\left(\dfrac{A}{2}\right)+ca\cos^2\left(\dfrac{B}{2}\right)+ab\cos^2\left(\dfrac{C}{2}\right)\right)=\dfrac{1}{4}(a+b+c)^2$

12. In a ΔABC, prove that,
$(b-c)\cot\left(\dfrac{A}{2}\right)+(c-a)\cot\left(\dfrac{B}{2}\right)+(a-b)\cot\left(\dfrac{C}{2}\right)=0$

13. If the sides a, b, c of a triangle are in A.P, then find the value of $\tan\left(\dfrac{A}{2}\right)+\tan\left(\dfrac{C}{2}\right)$ in terms of $\cot\left(\dfrac{B}{2}\right)$

14. If D is mid-point of CA in triangle ABC and Δ is the area of triangle, then prove that $\tan(\angle ADB)=\dfrac{4\Delta}{a^2-c^2}$

15. In any triangle ABC, prove that, $4\Delta(\cot A+\cot B+\cot C)=a^2+b^2+c^2$

16. In any triangle ABC, prove that, $\left(\dfrac{2abc}{a+b+c}\right).\cos\left(\dfrac{A}{2}\right)\cos\left(\dfrac{B}{2}\right)\cos\left(\dfrac{C}{2}\right)=\Delta$

17. If in a triangle ABC, a = 6, b = 3 and $\cos(A-B)=\dfrac{4}{5}$, then find its area.

18. If in a triangle ABC, $\angle A = 30^{o}$ and the area of the triangle is $\frac{a^2\sqrt{3}}{4}$, then prove that either

$B = 4\,C$ or $C = 4\,B$

19. In any triangle ΔABC, prove that, $Rr(\sin A + \sin B + \sin C) = \Delta$.

20. In any triangle ΔABC, prove that,

$$a\cos B\cos C + b\cos C\cos A + c\cos A\cos B = \frac{\Delta}{R}$$

21. In any triangle ΔABC, prove that,

$$\frac{1}{bc}+\frac{1}{ca}+\frac{1}{ab}=\frac{1}{2Rr}$$

22. In any triangle ΔABC, prove that,

$$\cos^2\left(\frac{A}{2}\right)+\cos^2\left(\frac{B}{2}\right)+\cos^2\left(\frac{C}{2}\right)=2+\frac{r}{2R}$$

23. In any triangle ΔABC, prove that, $a\cot A + b\cot B + c\cot C = 2(R + r)$.

24. In any triangle ΔABC, prove that,

$$r_1 + r_2 - r_3 + r = 4R\cos C$$

25. In any triangle ΔABC, prove that,

$$\frac{1}{r_1^2}+\frac{1}{r_2^2}+\frac{1}{r_3^2}+\frac{1}{r^2}=\frac{a^2+b^2+c^2}{\Delta^2}$$

26. In any triangle ΔABC, prove that, prove that,

$$\left(\frac{1}{r}+\frac{1}{r_1}+\frac{1}{r_2}+\frac{1}{r_3}\right)^2=\frac{4}{r}\left(\frac{1}{r_1}+\frac{1}{r_2}+\frac{1}{r_3}\right)$$

27. In any triangle ΔABC, prove that, prove that,

$$\frac{bc - r_2r_3}{r_1}=\frac{ca - r_1r_3}{r_2}=\frac{ab - r_1r_2}{r_3}$$

28. Prove that the radii of the three escribed circles of a triangle are the roots of

$$x^3 - (r + 4R)x^2 + s^2x - rs^2 = 0$$

29. If $\alpha, \beta \,\&\, \gamma$ are the respective altitudes of a triangle ABC, prove that

(i) $$\frac{1}{\alpha^2}+\frac{1}{\beta^2}+\frac{1}{\gamma^2}=\frac{\cot A + \cot B + \cot C}{\Delta}$$

(ii) $$\frac{1}{\alpha}+\frac{1}{\beta}-\frac{1}{\gamma}=\left(\frac{2ab}{(a+b+c)\Delta}\times\cos^2\left(\frac{C}{2}\right)\right)$$

30. If in a triangle ΔABC, $a^2 + b^2 + c^2 = 8R^2$, then prove that the triangle ABC is right angled triangle.

31. The sides of a triangle are $x^2 + x + 1$, $2x + 1$ and x 2 - 1, prove that the greatest angle is 120°.

32. The sides of a triangle are three consecutive natural numbers and its largest angle is twice the smallest one. Determine the side of the triangle. Ans. 4, 5 and 6. **Ans. 4 , 5 and 6.**

33. ABC is a triangle . Its area is 12 sq. cm and base is 6 cm . The difference of the angles is 60°. Prove that the angle A opposite to the base is given by 8 sin A - 6 cosA = 3.

34. In any triangle ABC, if $\cos\theta = \frac{a}{b+c}$, $\cos\phi = \frac{b}{a+c}$, $\cos\psi = \frac{c}{a+b}$, where $\theta, \varphi \,\&\, \psi$ lie between 0 and π, prove that,

$$\tan^2\left(\frac{\theta}{2}\right)+\tan^2\left(\frac{\varphi}{2}\right)+\tan^2\left(\frac{\psi}{2}\right)=1$$

35. The sides of a triangle are ABC are in A.P If the angles A and C are the greatest and the smallest angles respectively, then prove that,

$$4(1-\cos A)(1-\cos C)=\cos A+\cos C$$

36. In triangle ABC, if a, b, c are in H.P, prove that $\sin^2\left(\frac{A}{2}\right)$, $\sin^2\left(\frac{B}{2}\right)$, $\sin^2\left(\frac{C}{2}\right)$ are also in H.P.

37. In a triangle ABC, if the sides a, b, c be in A.P, prove that $\cos A.\cot\left(\frac{A}{2}\right)$, $\cos B.\cot\left(\frac{B}{2}\right)$, $\cos C.\cot\left(\frac{C}{2}\right)$ are also in A.P.

38. The sides of a triangle are in A.P and its area is 3/5 th of an equilateral triangle of the same perimeter. Prove that the sides are in the ratio 3 : 5 : 7.

39. The ex-radii r_1, r_2, r_3 of a triangle ABC are in H.P, prove that the sides a, b, c are in A.P

40. In usual notation, if $r_1 = r_2 + r_3 + r$, then prove that the triangle is right angled.

41. If A, B, C are the angles of a triangle, then prove that $\cos A + \cos B + \cos C = 1 + \frac{r}{R}$, where r = in-radius and R = circum - radius.

42. In triangle ABC, if $8R^2 = a^2 + b^2 + c^2$, prove that the triangle is right angled.

43. Let $A_1, A_2, A_3, \ldots, A_n$ be the vertices of an n-sided regular polygon such that $\frac{1}{A_1A_2} = \frac{1}{A_1A_3} + \frac{1}{A_1A_4}$, then prove that the value of n is 7.

44. If A, A_1, A_2 & A_3 are the area of the inscribed and escribed circles of a triangle, then prove that $\frac{1}{\sqrt{A}} = \frac{1}{\sqrt{A_1}} + \frac{1}{\sqrt{A_2}} + \frac{1}{\sqrt{A_3}}$.

45. The sides of a quadrilateral are 3, 4, 5 and 6 cms. The sum of a pair of opposite angles is 120°. Prove that the area of the quadrilateral is $3\sqrt{30}$ sq. cm.

46 In triangle ABC, prove that,
$r^2 + r_1^2 + r_2^2 + r_3^2 = 16R^2 - a^2 - b^2 - c^2$

47. In triangle ABC, prove that,
$IA.IB.IC = abc \tan\left(\frac{A}{2}\right).\tan\left(\frac{B}{2}\right).\tan\left(\frac{C}{2}\right)$

48. In a triangle ABC, if $\left(1 - \frac{r_1}{r_2}\right)\left(1 - \frac{r_1}{r_3}\right) = 2$, prove that the triangle is right angled.

49. In triangle ABC, prove that,
$\cot^2 A + \cot^2 B + \cot^2 C \geq 1$

50. If cosA = tan B, cosB = tan C and cosC = tan A, then prove that, sin A = sin B = sin C = 2 sin 18.

Level– II FOR JEE-MAIN AND ADVANCED EXAMS ONLY

1. In ΔABC, $a \geq b \geq c$, if $\frac{a^3 + b^3 + c^3}{\sin^3 A + \sin^3 B + \sin^3 C} = 8$, then the maximum value of a is
   (a) $\frac{1}{2}$ (b) 2
   (c) 8 (d) 64

2. Sides of a triangle ABC are in A.P. If $a < \min\{b, c\}$, then cos A may be equal to
   (a) $\frac{3c - 4b}{2a}$ (b) $\frac{3c - 4b}{2c}$
   (c) $\frac{4c - 3b}{2b}$ (d) $\frac{4c - 3b}{2c}$

3. If a Δ ABC, $a^4 + b^4 + c^4 = 2a^2b^2 + b^2c^2 + 2c^2a^2$, then sin A is
   (a) $\frac{1}{\sqrt{2}}$ (b) $\frac{1}{2}$
   (c) $\frac{\sqrt{3}}{2}$ (d) $\frac{\sqrt{3}+1}{2\sqrt{2}}$

4. In a triangle ABC, $2a^2 + 4b^2 + c^2 = 4ab + 2ac$, then the numerical value of cos B is
   (a) 0 (b) $\frac{3}{8}$
   (c) $\frac{5}{8}$ (d) $\frac{7}{8}$

5. If a, b, c be the sides of ΔABC and if roots of the equation $a(b - c)x^2 + b(c - a)x + c(a - b) = 0$ are equal then $\sin^2\left(\frac{A}{2}\right).\sin^2\left(\frac{B}{2}\right).\sin^2\left(\frac{C}{2}\right)$ are in
   (a) A.P. (b) G.P.
   (c) H.P. (d) A.G.P.

6. In a triangle ABC, $(a + b + c)(b + c - a) = k\,bc$ if
   (a) $k < 0$ (b) $k > 6$
   (c) $0 < k < 4$ (d) $k > 4$

7. $a^3 \cos(B - C) + b^3 \cos(C - A) + c^3 \cos(A - B)$ is equal
   (a) 3 abc (b) (a + b + c)
   (c) abc (a + b + c) (d) 0

8. If $\frac{\cos A}{a} = \frac{\cos B}{b} = \frac{\cos C}{c}$ and a = 2, then the area of the triangle is
   (a) 1 (b) 2
   (c) $\frac{\sqrt{3}}{2}$ (d) $\sqrt{3}$

9. If in a ΔABC, cos A + 2 cos B + cos C = 2, then a, b, c are in

(a) A.P. (b) G.P.
(c) H.P. (d) None

10. If in a ΔABC, $\sin^3 A + \sin^3 B + \sin^3 C = 3 \sin A \sin B \sin C$, then the value of $\begin{vmatrix} a & b & c \\ b & c & a \\ c & a & b \end{vmatrix}$

(a) 0
(b) $(a + b + c)^2$
(c) $(a + b + c)(ab + bc + ca)$
(d) None

11. If $b + c = 3a$, then the value of $\cot\left(\frac{B}{2}\right) \cot\left(\frac{C}{2}\right)$ is

(a) 1 (b) 2
(c) $\sqrt{3}$ (d) $\sqrt{2}$

12. Let $A_0A_1A_2A_3A_4A_5$ be a regular hexagon inscribed in a circle of unit radius. The product of length of the line segment A_0A_1, A_0A_2, A_0A_4 is

(a) $\frac{3}{4}$ (b) $3\sqrt{3}$
(c) 3 (d) $\frac{3\sqrt{3}}{2}$

13. In a ΔABC, the value of $\frac{a\cos A + b\cos B + c\cos C}{a + b + c}$ is

(a) $\frac{R}{r}$ (b) $\frac{R}{2r}$
(c) $\frac{r}{R}$ (d) $\frac{2r}{R}$

14. In a ΔABC, the sides a, b, c are the roots of the equation $x^3 - 11x^2 + 38x - 40 = 0$. Then $\frac{\cos A}{a} + \frac{\cos B}{b} + \frac{\cos C}{c}$ is

(a) 1 (b) $\frac{3}{4}$
(c) $\frac{9}{16}$ (d) None

15. The ex-radii of a Δ r_1, r_2, r_3 are in A.P., then the sides a, b, c are in

(a) A.P. (b) G.P.
(c) H.P. (d) A.G.P.

16. In any ΔABC, $\sum \frac{\sin^2 A + \sin A + 1}{\sin A}$ is always greater than

(a) 9 (b) 3
(c) 27 (d) 36

17. In a triangle $\left(1 - \frac{r_1}{r_2}\right)\left(1 - \frac{r_1}{r_3}\right) = 2$, then the triangle is

(a) right angled (b) isosceles
(c) equilateral (d) None

18. In a ΔABC, $a = 2b$ and $|A - B| = \frac{\pi}{3}$, then $\angle C$ is

(a) $\frac{\pi}{4}$ (b) $\frac{\pi}{3}$
(c) $\frac{\pi}{6}$ (d) None

19. If the median of ΔABC, through A is perpendicular to AB, then

(a) $\tan A + \tan B = 0$
(b) $2\tan A + \tan B = 0$
(c) $\tan A + 2\tan B = 0$
(d) None

20. In a ΔABC, $\cos A + \cos B + \cos C = \frac{3}{2}$, then the Δ

(a) Isosceles (b) right angled
(c) equilateral (d) None

21. If $A_1 \ldots\ldots A_n$ be a regular polygon of n-sides and $\frac{1}{A_1A_2} = \frac{1}{A_1A_3} + \frac{1}{A_1A_4}$, then

(a) $n = 5$ (b) $n = 6$
(c) $n = 7$ (c) None

22. In a ΔABC, $\tan\left(\frac{A}{2}\right) = \frac{5}{6}$ and $\tan\left(\frac{C}{2}\right) = \frac{2}{5}$, then

(a) a, c, b $\in$ A. P (b) a, b, c $\in$ A.P.

(c) b, a, c $\in$ A.P. (d) a, b, c $\in$ G.P.

23. If the angles of a triangle are in the ratio 1 : 2 : 3, then the corresponding sides are in the ratio

(a) $2:3:1$ (b) $\sqrt{3}:2:1$

(c) $2:\sqrt{3}:1$ (d) $1:\sqrt{3}:2$

24. In a ΔABC, $a \cot A + b \cot B + c \cot C$ is

(a) $r+R$ (b) $r-R$

(c) $2(r+R)$ (d) $2(r-R)$

25. If A, A_1, A_2, A_3 are the areas of in circle and the ex circles of a triangle, then

$$\frac{1}{\sqrt{A_1}}+\frac{1}{\sqrt{A_2}}+\frac{1}{\sqrt{A_3}} \text{ is}$$

(a) $\frac{2}{\sqrt{A}}$ (b) $\frac{1}{\sqrt{A}}$

(c) $\frac{1}{2\sqrt{A}}$ (d) $\frac{3}{\sqrt{A}}$

26. In any ΔABC, $\Pi\left(\frac{\sin^2 A+\sin A+1}{\sin A}\right)$ is always greater than

(a) 9 (b) 3

(c) 27 (d) 36

27. In an equilateral triangle, $R:r:r_2$ is

(a) $1:1:1$ (b) $1:2:3$

(c) $2:1:3$ (d) $3:2:4$

28. In a ΔABC, $\tan A \tan B \tan C = 9$. For such triangles, if $\tan^2A + \tan^2B + \tan^2C = \lambda$, then

(a) $9.\sqrt[3]{3} < \lambda < 27$ (b) $\lambda \not\leq 27$

(c) $\lambda < 9.\sqrt[3]{3}$ (d) $\lambda > 27$

29. In a ΔABC, $a^2 \cos^2 A = b^2 + c^2$, then

(a) $A < \frac{\pi}{4}$ (b) $\frac{\pi}{4} < A < \frac{\pi}{2}$

(c) $A > \frac{\pi}{2}$ (d) $A = \frac{\pi}{2}$

30. In a ΔABC, $A:B:C = 3:5:4$, then $a+b+c\sqrt{2}$ is

(a) 2b (b) 2c

(c) 3b (d) 3a

31. If A, B, C are angles of a triangle such that the angle A is obtuse, then $\tan B \tan C <$

(a) 0 (b) 1

(c) 2 (d) 3

32. In a triangle, if $r_1 > r_2 > r_3$, then

(a) $a > b > c$ (b) $a < b < c$

(c) $a > b$ and $b < c$ (d) $a < b$ and $b > c$

33. $4rR \cos\frac{A}{2} \cos\frac{B}{2} \cos\frac{C}{2}$ is

(a) s (b) s^2

(c) Δ^2 (d) Δ

34. If $(a-b)(s-c) = (b-c)(s-a)$, then r_1, r_2, r_3 are in

(a) H.P. (b) G.P.

(c) A.P. (d) A.G.P.

35. If $c^2 = a^2 + b^2$, $2s = a + b + c$, then $4s(s-a)(s-b)(s-c)$ is

(a) s^4 (b) b^2c^2

(c) c^2a^2 (d) a^2b^2

36. $\frac{1}{r_1^2}+\frac{1}{r_2^2}+\frac{1}{r_3^2}+\frac{1}{r^2}$ is

(a) $\frac{a^2+b^2+c^2}{s^2}$ (b) $\frac{\Sigma a^2}{\Delta^2}$

(c) 4R (d) 4r

37. $(r_1-r)(r_2-r)(r_3-r)$ is

(a) $\frac{R}{r}$ (b) $4R^2r$

(c) $4Rr^2$ (d) 4R

38. If the sides be 13, 14, 15, then

$$\frac{b-c}{r_1}+\frac{c-a}{r_2}+\frac{a-b}{r_3} \text{ is}$$

(a) 5 (b) 4

(c) 0 (d) 1

39. $\frac{r_1}{bc}+\frac{r_2}{ca}+\frac{r_3}{ab}$ is

(a) $\frac{1}{2R}-\frac{1}{r}$ (b) $2R-r$

(c) $r-2R$ (d) $\frac{1}{r}-\frac{1}{2R}$

40. $r_1+r_2=$

(a) $c\tan\left(\frac{C}{2}\right)$ (b) $c\cot\left(\frac{C}{2}\right)$

(c) $c\sin\left(\frac{C}{2}\right)$ (d) $c\cos\left(\frac{C}{2}\right)$

41. $16R^2 r r_1 r_2 r_3$ is

(a) abc (b) $a^3b^3c^3$

(c) $a^2b^2c^2$ (d) $a^2b^3c^4$

42. If $\frac{r}{r_1}=\frac{r_2}{r_3}$, then

(a) $A=90°$ (b) $B=90°$

(c) $=90°$ (d) None

43. In a ΔABC, the value of $r r_1 r_2 r_3$ is

(a) Δ (b) Δ^2

(c) Δ^3 (d) Δ^4

44. If $r_1=r_2+r_3+r$, then the Δ is

(a) Equilateral (b) Isosceles

(c) Right angled (d) None

45. $(r_1+r_2)(r_2+r_3)(r_3+r_1)$ is

(a) Rs^2 (b) $2Rs^2$

(c) $3Rs^2$ (d) $4Rs^2$

46. The diameter of the circum circle of a triangle with sides 5 cm, 6 cm., 7 cm is

(a) $\frac{3\sqrt{6}}{2}$ cm. (b) $2\sqrt{6}$ cm

(c) $\frac{35}{48}$ cm (d) $\frac{35}{2\sqrt{6}}$

47. In a Δ ABC, the sides are in the ratio 4 : 5 : 6. The ratio of the circum-radius and the in-radius is

(a) 8 : 7 (b) 3 : 2

(c) 7: 3 (d) 16 : 7

48. If in a triangle, R and r are the circum radius and inradius respectively, then the H.M. of the ex-radii of the triangle is

(a) 3r (b) 2R

(c) R + r (d) None

49. If a , b and c are the sides of a triangle ABC and 3a = b + c , the value of Cot (B/2) cot (C/2) is

(a) 3 (b) 2

(c) 4 (d) 1.

50. In a triangle of ABC , if cos A + cos B = 4 sin 2 (C/2) , then a, b and c are in

(a) A.P (b) G.P

(c) H.P (d) None

ANSWER

1. (b)	**2. (d)**	**3. (b)**	**4. (d)**	**5. (c)**	**6. (c)**	**7. (a)**	**8. (d)**	**9 (a)**	**10 (a)**
11 (b)	**12 (c)**	**13 (c)**	**14 (c)**	**15 (a)**	**16 (a)**	**17 (a)**	**18 (b)**	**19 (c)**	**20 (c)**
21 (c)	**22 (b)**	**23 (d)**	**24 (c)**	**25 (b)**	**26 (c)**	**27 (c)**	**28 (b)**	**29 (c)**	**30 (c)**
31 (b)	**32 (a)**	**33 (d)**	**34 (c)**	**35 (d)**	**36 (b)**	**37 (c)**	**38 (c)**	**39 (d)**	**40. (b)**
41. (c)	**42. (c)**	**43. (b)**	**44. (c)**	**45. (d)**	**46 (d)**	**47. (d)**	**48. (a)**	**49. (b)**	**50. (a)**

Level– III FOR JEE ADVANCED EXAMS ONLY

1. If in a triangle ABC,
$$\frac{\cos A + 2\cos C}{\cos A + 2c\cos B} = \frac{\sin B}{\sin C},$$
prove that the triangle ABC is either isisceles or right angled.

2. In a triangle ABC, if
$$a\tan A + b\tan B = (a+b)\tan\left(\frac{A+B}{2}\right),$$
prove that the triangle is isosceles.

3. In any triangle ABC, prove that,
$$(r_2 + r_1)(r_3 + r_2)\sin C$$
$$= 2r_3\sqrt{r_2 r_3 + r_1 r_3 + r_1 r_2}$$

4. In any triangle ABC, prove that,
$$r^2 + r_1^2 + r_2^2 + r_3^2 = 16R^2 - (a^2 + b^2 + c^2)$$

5. In any triangle ABC, prove that,
$$\left((r + r_1)\tan\left(\frac{B-C}{2}\right)\right)$$
$$+\left((r + r_2)\tan\left(\frac{C-A}{2}\right)\right)$$
$$+\left((r + r_3)\tan\left(\frac{C-A}{2}\right)\right) = 0$$

6. In any triangle ABC, prove that,
$$\frac{\tan\left(\frac{A}{2}\right)}{(a-b)(a-c)} + \frac{\tan\left(\frac{B}{2}\right)}{(b-a)(b-c)}$$
$$+\frac{\tan\left(\frac{C}{2}\right)}{(c-a)(c-b)} = \frac{1}{\Delta}$$

7. If $\left(1 - \frac{r_1}{r_2}\right)\left(1 - \frac{r_1}{r_3}\right) = 2$, prove that the triangle is right angled triangle.

8. In triangle ABC, prove that,
$$\frac{\text{area of the incircle}}{\text{area of triangle ABC}}$$
$$= \frac{\pi}{\cot\left(\frac{A}{2}\right).\cot\left(\frac{B}{2}\right).\cot\left(\frac{C}{2}\right)}$$

9. If a, b, c are in A.P, prove that,
$$\cos A.\cot\left(\frac{A}{2}\right),\ \cos B.\cot\left(\frac{B}{2}\right),$$
$$\cos C.\cot\left(\frac{C}{2}\right)$$ are in A.P.

10. If the circumference of the ΔABC lies on its incircle, then prove that,
$$\cos A + \cos B + \cos C = \sqrt{2}$$

11. ABCD is a trapezium such that AB, DC are parallel and BC is perpendicular to them. If angle $\angle ADB = \theta$, BC = p and CD = q, prove that,
$$AB = \frac{(p^2 + q^2)\sin\theta}{p\cos\theta + q\sin\theta}$$

12. Let O be the circumcenter and H be the orthocenter of ΔABC. If Q is the mid-point of OH, then show that
$$AQ = \frac{R}{2}\sqrt{1 + 8\cos A\cos B\cos C}$$

13. If I_1, I_2 & I_3 are the centres of escribed circles of ΔABC, prove that the area of
$$\Delta I_1 I_2 I_3 = \frac{abc}{2r}$$

14. In ΔABC, prove that,
$$a^2(s-a) + b^2(s-b) + c^2(s-c)$$
$$= 4R\Delta\left(1 - 4\sin\left(\frac{A}{2}\right)\sin\left(\frac{B}{2}\right)\sin\left(\frac{C}{2}\right)\right)$$

15. Let O be a point inside a triangle ABC such that $\angle OAB = \angle OBC = \angle OCA = \omega$, then prove that

(i) $\cot A + \cot B + \cot C = \cot\omega$

(ii) $\operatorname{cosec}^2 A + \operatorname{cosec}^2 B + \operatorname{cosec}^2 C = \operatorname{cosec}^2\omega$

16. Find the distance between the circum-center and the mid-points of the sides of a triangle.

17. Find the distance between the incenter and the angular points of a triangle.

18. Prove that the distance between the Circumcentre (O) and the In-center (I) is
$$OI = R \times \sqrt{\left(1 - 8\sin\left(\frac{A}{2}\right)\sin\left(\frac{B}{2}\right)\sin\left(\frac{C}{2}\right)\right)}$$

19. Prove that the ratio of circum-radius and in-radius of an equilateral triangle is 1/2.

20. Prove that the ratio of the area of the in-circle to the area of a triangle is

$\pi : \tan\left(\frac{A}{2}\right).\tan\left(\frac{B}{2}\right).\tan\left(\frac{C}{2}\right)$.

21. Prove that the distance of the orthocenter from the sides and angular points of a triangle is 2 R cos A, 2 R cos B and 2 R cos C.

22. Prove that the distance between the circum-center and the orthocenter of a triangle is OH = $R\sqrt{1-8\cos A.\cos B.\cos C}$.

23. Prove that the area of an Ex-central Triangle is

$8R^2 \cos\left(\frac{A}{2}\right)\times\cos\left(\frac{B}{2}\right)\times\cos\left(\frac{C}{2}\right)$.

24. Prove that the The circum-radius of an ex-central triangle is $\frac{I_2I_3}{2\sin(I_1I_2I_3)}=2R$.

25. Prove that the the distance between the in-center and the ex-centers are $II_1=4R\sin\left(\frac{A}{2}\right)$

$II_2=4R\sin\left(\frac{B}{2}\right)$, $II_3=4R\sin\left(\frac{C}{2}\right)$.

26. If a^2, b^2, c^2 are in A.P, then prove that cot A. cot B, cot C are in A.P

27. In any triangle ABC, prove that,

$a^3\cos(B-C)+b^3\cos(C-A)$
$+c^3\cos(A-B)=3abc$

28. The sides of a triangle are in A.P and the greatest and least angles are θ and ϕ respectively, then prove that

$4(1-\cos\theta)(1-\cos\varphi)=\cos\theta+\cos\varphi$.

28. If in a triangle, the bisector of the side c be perpendicular to the side d, prove that 2 tanA + tan C = 0.

29. In any triangle, if θ be any angle, then prove that, $b\cos\theta=c\cos(A-\theta)+a\cos(C+\theta)$.

30. If AD, BE and CF are the internal bisectors of the angles of a triangle ABC, prove that

$\frac{1}{AD}\cos\left(\frac{A}{2}\right)+\frac{1}{BE}\cos\left(\frac{B}{2}\right)$
$+\frac{1}{CF}\cos\left(\frac{C}{2}\right)=\frac{1}{a}+\frac{1}{b}+\frac{1}{c}$

31. Prove that the triangle having sides 3 x + 4 y, 4 x + 3 y and 5 x + 5 y units respectively, where x, y > 0, is obtuse angled.

32. If Δ be the area and 's' be the semi perimeter of a triangle, then prove that, $\Delta\le\frac{s^2}{3\sqrt{3}}$

33. Let ABC be a triangle having altitudes h_1, h_2 & h_3 from the vertices A, B, C respectively and r be the in-radius, prove that, $\frac{h_1+r}{h_1-r}+\frac{h_2+r}{h_2-r}+\frac{h_3+r}{h_3-r}\ge 6$.

34. Two circles of radii a and b cut each other at an angle θ. Prove that the length of the common chord is $\frac{2ab\sin\theta}{\sqrt{a^2+b^2+2ab\cos\theta}}$

35. If α, β, γ are the distances of the vertices of a triangle from the corresponding points of contact with the in-circle, prove that $r^2=\frac{\alpha\beta\gamma}{\alpha+\beta+\gamma}$

36. Tangents are drawn to the in-circle of a triangle ABC which are parallel to its sides. If x, y, z be the lengths of the tangents and a, b, c be the sides of a triangle, then prove that,

$\frac{x}{a}+\frac{y}{b}+\frac{z}{c}=1$.

37. If t_1, t_2 & t_3 be the lengths of the tangents from the ecntres of escribed circles to the circum circles, prove that $\frac{1}{t_1^2}+\frac{1}{t_1^2}+\frac{1}{t_1^2}=\frac{2s}{abc}$.

38. if x, y, z be the lengths of the perpendiculars from the circumcentre on the sides BC, CA, AB of a triangle ABC, prove that, $\frac{a}{x}+\frac{b}{y}+\frac{c}{z}=\frac{abc}{4xyz}$

39. If p_1, p_2, p_3 are the altitudes of the triangle ABC from the vertices A, B and C respectively, prove that $\frac{\cos A}{p_1}+\frac{\cos B}{p_2}+\frac{\cos C}{p_3}=\frac{1}{R}$

40. The product of the sines of the angles of a triangle is p and the product of their cosines is q. Prove that the tangents of the angles are the roots of $qx^3 - px^2 + (1+q)x - p = 0$.

41. In a triangle ABC, if cos A. cos B + sin A sin B sin C = 1, prove that the sides are in the ratio $1:1:\sqrt{2}$.

42. The base of a triangle is divided into three equal parts. If t_1, t_2 & t_3 be the tangents of the angles subtended by these parts at the opposite vertices, prove that, $\left(\frac{1}{t_1}+\frac{1}{t_2}\right)\left(\frac{1}{t_2}+\frac{1}{t_3}\right) = 4\left(1+\frac{1}{t_2^{\,2}}\right)$.

43. The three medians of a triangle ABC make angle α, β, γ with each other, prove that, $\cot\alpha + \cot\beta + \cot\gamma + \cot A + \cot B + \cot C = 0$

44. Perpendiculars are drawn from the angles A, B, C of an acute angled triangle on the opposite sides and produced to meet the circum scribing circle. If these parts be α, β, γ respectively, then prove that $\frac{a}{\alpha}+\frac{b}{\beta}+\frac{c}{\gamma} = 2(\tan A + \tan B + \tan C)$

45. In a triangle ABC, the vertices A, B, C are at distances of p, q, r from the orthocentre respectively. Prove that $\left(\frac{a}{p}+\frac{b}{q}+\frac{c}{r}\right) = \frac{abc}{pqr}$

46. The internal bisectors of the angles of a triangle ABC meet the sides in D, E and F. Prove that the area of the triangle DEF is $\frac{2\Delta abc}{(a+b)(b+c)(c+a)}$.

47. In a triangle ABC, the measures of the angles A, B, C are $3\alpha\alpha3\beta$ and 3γ respectively. P, Q, and R are the points within the triangle such that $\angle BAR = \angle RAQ = \angle QAC = \alpha$, $\angle CBP = \angle PBR = \angle RBA = \beta$ and $\angle ACQ = \angle QCP = \angle PCB = \gamma$, then prove that $AR = 8R\sin\beta\sin\gamma\cos(30^\circ - \gamma)$

48. If in a triangle ABC, the median AD and the perpendicular AE from the vertex A to the side BC divides the angle A into three equal parts, show that $\cos\left(\frac{A}{3}\right).\sin^2\left(\frac{A}{3}\right) = \frac{3a^2}{32bc}$.

49. If the sides of a triangle are in A.P and if its greatest angle exceeds the least angle by α, show that the sides are in the ratio $(1-x):1:(1+x)$, where $x = \sqrt{\frac{1-\cos\alpha}{7-\cos\alpha}}$

50. Let A, B, C be three angles such that $A = \frac{\pi}{4}$ and tan B tan C = p. Find all possible values of p such that A, B, C are the angles of a triangle.
Ans. $p < 0$ & $p \geq 3+2\sqrt{2}$.

COMPREHENSIVE LINK PASSAGE

In these questions, a passage (paragraph) has been given followed by questions based on each of the passage. You have to answer the questions based on the passage given.

PASSAGE – I

If p_1, p_2, p_3 are the altitudes of a triangle ABC, from the vertices A, B, C respectively and Δ is the area of the triangle and 's' is the semi perimeter of the triangle.

On the basis of the above information, answer the following questions.

1. If $\frac{1}{p_1}+\frac{1}{p_2}+\frac{1}{p_3} = \frac{1}{2}$, then the least value of $p_1 p_2 p_3$ is

(a) 8 (b) 27
(c) 125 (d) 216

2. The value of $\frac{\cos A}{p_1}+\frac{\cos B}{p_2}+\frac{\cos C}{p_3}$ is

(a) $\frac{1}{r}$ (b) $\frac{1}{R}$
(c) $\frac{a^2+b^2+c^2}{2R}$ (d) $\frac{1}{\Delta}$

3. The minimum value of $\frac{b^2 p_1}{c}+\frac{c^2 p_1}{a}+\frac{a^2 p_1}{b}$ is

(a) Δ (b) 2Δ
(c) 3Δ (d) 4Δ

4. The value of $\frac{1}{p_1^3}+\frac{1}{p_2^3}+\frac{1}{p_3^3}$ is

(a) $\frac{(\Sigma a)^2}{4\Delta^2}$ (b) $\frac{(\Pi a)^3}{8\Delta^3}$

(c) $\frac{\Sigma a^2}{4\Delta^2}$ (d) $\frac{\Pi a^2}{8\Delta^2}$

5. In the triangle ABC, the altitudes are in A.P, then

(a) a, b, c, are in A P

(b) a, b, c are in H.P

(c) a, b, c are in G. P

(d) angles A, B, C are in A.P

PASSAGE-II

ABCD be a cyclic quadrilateral inscribed in a circle of radius R.

Then $\cos B=\frac{(a^2+b^2-c^2-d^2)}{2(ab+cd)}$ and the area of the quadrilateral $=\frac{1}{2}(ab+cd)\sin B$

$=\sqrt{(s-a)(s-b)(s-c)(s-d)}$,

where, $s=\frac{a+b+c+d}{2}$.

Also, $AC^2 . BD^2 = (ac + bd)^2$

i.e. AC . BD = AB . CD + BC . AD and

$R=\frac{AC}{2\sin B}$

On the basis of the above information, answer the following questions.

1. The side of a quadrilateral which can be inscribed in a circle are 6, 6, 8 and 8 cm. Then the circum radius is

(a) 5/2 cm (b) 24/7 cm

(c) 11/7 cm (d) None

2. The sides of a quadrilateral, which can be inscribed in a circle are 5, 5, 12 and 12 cm. Then the in-radius is

(a) 15/17 cm (b) 30/17 cm

(c) 60/17 cm (d) None

3. If a quadrilateral with sides a, b, c, d can be inscribed in one circle and circumscribed about another circle, then its area is

(a) $\sqrt{abcd}$ (b) $\sqrt{2(abcd)}$

(c) $2\times\sqrt{(abcd)}$ (d) None

PASSAGE-III

G is the centroid of the triangle ABC. Perpendiculars from vertices A, B, C meet the sides BC, CA, AB at D, E, F respectively. P, Q, R are the feet of perpendiculars from G on sides BC, CA, AB respectively, L, M, N are the mid points of the sides BC, CA, AB respectively.

On the basis of the above information, answer the following questions.

1. Length of the side PG is

(a) $\frac{1}{2}\text{b}\sin\text{C}$ (b) $\frac{1}{2}\text{c}\sin\text{C}$

(c) $\frac{2}{3}\text{b}\sin\text{C}$ (d) $\frac{1}{3}\text{c}\sin\text{B}$

2. $\text{ar}(\Delta\,\text{GPL}):\text{ar}(\Delta\,\text{ALD})$ is

(a) 1/3 (b) 1/9

(c) 2/3 (d) 4/9

3. Area of ΔPQR is

(a) $\frac{1}{9}(a^2+b^2+c^2)\sin A.\sin B.\sin C$

(b) $\frac{1}{18}(a^2+b^2+c^2)\sin A.\sin B.\sin C$

(c) $\frac{2}{9}(a^2+b^2+c^2)\sin A.\sin B.\sin C$

(d) $\frac{1}{3}(a^2+b^2+c^2)\sin A.\sin B.\sin C$

PASSAGE-IV

In a triangle ABC, R be the circum radius such that $R=\frac{abc}{4\Delta}$ and $r_1, r_2 \,\&\, r_3$ are the exradii, where $r_1=\frac{\Delta}{s-a}$, $r_2=\frac{\Delta}{s-a}$ & $r_3=\frac{\Delta}{s-a}$ and r be in-radius such that $r=\frac{\Delta}{s}$.

On the basis of above information, answers the following questions.

1. If $r_1=r+r_2+r_3$, then the triangle is

(a) equilateral (b) isosceles

(c) right angled (d) None

2. The value of cos A + cos B + cos C is

(a) $1+\frac{r}{R}$ (b) $2\left(1+\frac{r}{R}\right)$

(c) $\frac{1}{2}\left(1+\frac{r}{R}\right)$ (d) $\frac{1}{4}\left(1+\frac{r}{R}\right)$

3. If $\left(1-\frac{r_1}{r_2}\right)\left(1-\frac{r_1}{r_3}\right)=2$, then the triangle is

(a) right angled (b) equilateral

(c) isosceles (d) None.

4. The value of $r_1+r_2+r_3-4R$ is

(a) 2r (b) 3r
(c) r (d) 4r

5. The value of $\frac{1}{r_1}+\frac{1}{r_2}+\frac{1}{r_3}$ is

(a) $\frac{1}{r}$ (b) $\frac{2}{r}$

(c) $\frac{1}{2r}$ (d) $\frac{3}{2r}$

6. The value of $\frac{r_1}{bc}+\frac{r_2}{ca}+\frac{r_3}{ab}$ is

(a) $\frac{1}{r}-\frac{1}{R}$ (b) $\frac{1}{2r}-\frac{1}{R}$

(c) $\frac{1}{r}-\frac{1}{2R}$ (d) $\frac{1}{r}-\frac{1}{3R}$

PASSAGE-V

In any triangle ABC, $\cos A=\frac{b^2+c^2-a^2}{2bc}$, $\cos B=\frac{a^2+c^2-b^2}{2ac}$ and $\cos C=\frac{a^2+b^2-c^2}{2ab}$, where a = BC, b = CA and c = AB respectively.

On the basis of the above information, answer the following questions.

1. If the angles A, B, C are in A.P, then $\frac{a+c}{\sqrt{a^2+c^2-ac}}$ is

(a) $2\cos\left(\frac{A-C}{2}\right)$

(b) $2\sin\left(\frac{A-C}{2}\right)$

(c) $\sin\left(\frac{A-C}{2}\right)$

(d) $\cos\left(\frac{A-C}{2}\right)$.

2. If $\frac{1}{a+c}+\frac{1}{b+c}=\frac{1}{a+b+c}$, then

(a) $\angle C=75^o$ (b) $\angle A=75^o$

(c) $\angle A=60^o$ (d) $\angle C=60^o$

3. the value of $\frac{\cos A}{a}+\frac{\cos B}{b}+\frac{\cos C}{c}$ is

(a) $\frac{a^2+b^2+c^2}{2abc}$

(b) $\frac{a^2+b^2+c^2}{abc}$

(c) $\frac{a^2-b^2+c^2}{abc}$

(d) $\frac{a^2+b^2-c^2}{abc}$

4. If $c^4-2(a^2+b^2)c^2+a^4+a^2b^2+b^4=0$, then the angle C is

(a) 60° (b) 30°

(c) 75° (d) 45°

5. The value of $\frac{\cos 2A}{a^2}-\frac{\cos 2B}{b^2}$ is

(a) $\frac{1}{a^2}+\frac{1}{b^2}$ (b) $\frac{1}{a^2}-\frac{1}{b^2}$

(c) $\frac{2}{a^2}-\frac{2}{b^2}$ (d) $\frac{2}{a^2}+\frac{2}{b^2}$

ANSWERS

Passage-I

1.(d) 2.(b) 3.(d) 4.(c) 5.(b)

Passage-II

1. (d) 2. (a) 3.(b)

Passage-III

1. (*) 2. (*) 3. (*)

Passage-IV

1.(c) 2.(a) 3.(a) 4.(c) 5.(a)

6.(c)

Passage-V

1.(a) 2.(d) 3.(a) 4.(a) 5.(b)

MATCH MATRIX

1. **Match the following columns :**

In any triangle ABC,

	Column-I		Column-II
(A)	b cos C + c cos B	(P)	c
(B)	c cos A + a cos C	(Q)	b
(C)	a cos B + b cos A	(R)	a
(D)	(b+c) cos A + (c+a) cos B + (a+b) cos C	(S)	a + b + c

2. **Match the following columns :**

In any triangle ABC, the value of

	Column - I		Column - II
(A)	a sin (B-C) + b sin (C-A) + c sin (A-B) is	(P)	3abc
(B)	$a^2(\cos^2 B-\cos^2 C)+b^2(\cos^2 C-\cos^2 A)+c^2(\cos^2 A-\cos^2 B)$ is	(Q)	abc
(C)	$\left(\frac{b^2-c^2}{a^2}\right)\sin 2A+\left(\frac{c^2-a^2}{b^2}\right)\sin 2B+\left(\frac{a^2-b^2}{c^2}\right)\sin 2C$ is	(R)	0
(D)	$a^3\cos(B-C)+b^3\cos(C-A)+c^3\cos(A-B)$ is	(S)	2 abc.

3. **Match the following columns :**

In any triangle ABC , the value of

	Column- I		Column - II
(A)	tan B cot C is	(P)	$(a^2+b^2+c^2)$
(B)	2 (bc cos A+ ca cos B + ab cos C) is	(Q)	$\frac{(a^2+b^2-c^2)}{(a^2+b^2+c^2)}$
(C)	$\frac{\cot A+\cot B+\cot C}{\cot\left(\frac{A}{2}\right)+\cot\left(\frac{B}{2}\right)+\cot\left(\frac{C}{2}\right)}$ is	(R)	$\frac{a^2+b^2+c^2}{(a+b+c)^2}$
(D)	$\frac{\cot\left(\frac{B}{2}\right)+\cot\left(\frac{C}{2}\right)}{\cot\left(\frac{A}{2}\right)}$ is	(S)	$\left(\frac{2a}{b+c-a}\right)$

4. Match the following columns :

In any triangle ABC, if

Column- I		Column - II	
(A)	cot A, cot B , cot C are in A.P, then	(P)	a, b, c are in A.P
(B)	$\cos A.\cot\left(\frac{A}{2}\right)$, $\cos B.\cot\left(\frac{B}{2}\right)$, $\cos C.\cot\left(\frac{C}{2}\right)$ are in A.P, then	(Q)	a^2, b^2, c^2 are in A.P
(C)	$\sin^2\left(\frac{A}{2}\right)$, $\sin^2\left(\frac{B}{2}\right)$, $\sin^2\left(\frac{C}{2}\right)$ are in A.P, then	(R)	$\frac{1}{a}, \frac{1}{b}, \frac{1}{c}$ are in A.P
(D)	$\tan\left(\frac{A}{2}\right)$, $\tan\left(\frac{B}{2}\right)$, $\tan\left(\frac{C}{2}\right)$ are in A.P, then	(S)	a, b, c are in G.P.

5. Match the following columns :

In any triangle ABC , if

Column- I		Column - II	
(A)	$\cot A + \cot B + \cot C = \sqrt{3}$, then Δ is	(P)	Isoceles.
(B)	$(a^2+b^2)\sin(A-B) = (a^2-b^2)\sin(A+B)$, then Δ is	(Q)	Right angled.
(C)	$2\cos A = \frac{\sin B}{\sin C}$, then Δ is	(R)	Equilateral.
(D)	$a\tan A + b\tan B = (a+b)\tan\left(\frac{A+B}{2}\right)$, then Δ is	(S)	Acute angled.

6. Match the following columns :

In any triangle ABC , if

Column - I		Column - II	
(A)	$\sin A + \sin B + \sin C = \frac{3\sqrt{3}}{2}$, then Δ is	(P)	Isoceles.
(B)	$\tan A + \tan B + \tan C = 3\sqrt{3}$, then Δ is	(Q)	Right angled.
(C)	$8\sin\left(\frac{A}{2}\right)\sin\left(\frac{B}{2}\right)\sin\left(\frac{C}{2}\right) = 1$, then Δ is	(R)	Equilateral.
(D)	$a^2+b^2+c^2 = 8R^2$, then Δ is	(S)	Obtuse angled.

7. Match the following columns :

In any triangle ABC , if r be the inradius and r_1 , r_2 and r_3 be the ex-radii of the given ΔABC, then the value of

Column - I		Column - II	
(A)	$\frac{1}{r_1}+\frac{1}{r_2}+\frac{1}{r_3}$ is	(P)	r
(B)	$\frac{1}{r_1^2}+\frac{1}{r_2^2}+\frac{1}{r_3^2}+\frac{1}{r^2}$ is	(Q)	$\frac{\Delta^2}{a^2+b^2+c^2}$

(C) $r_1 + r_2 + r_3 - 4R$ is (R) $\frac{1}{r}$

(D) $\frac{r_1}{bc} + \frac{r_2}{ac} + \frac{r_3}{ba} + \frac{1}{2R}$ is (S) $\frac{a^2+b^2+c^2}{\Delta^2}$

8. Match the following columns :

In any triangle ABC, if r be the inradius and R be the circum-radius of the given ΔABC, then the value of

	Column - I		Column - II
(A)	cos A + cos B + cos C is	(P)	$\frac{\Delta}{r.R}$
(B)	a cot A + b cot B + c cot C is	(Q)	$\left(1+\frac{r}{R}\right)$
(C)	sin A + sin B + sin C is	(R)	$2(r+R)$
(D)	$\cos^2\left(\frac{A}{2}\right)+\cos^2\left(\frac{B}{2}\right)+\cos^2\left(\frac{C}{2}\right)$	(S)	$\left(2+\frac{r}{2R}\right)$

9. Match the following columns :

In any triangle ABC, if r be the inradius and r_1, r_2 and r_3 be the ex-radii of the given ΔABC, then the value of

	Column - I		Column - II
(A)	$r - r_1 + r_2 + r_3$ is	(P)	4 R cos C
(B)	$r - r_2 + r_1 + r_3$ is	(Q)	4 R cos B
(C)	$r - r_3 + r_1 + r_2$ is	(R)	4 R cos A
(D)	$r_1 + r_2 + r_3 - r$ is	(S)	4 R cos 2 (A + B + C)

10. Match the following columns :

In any triangle ABC, the minimum value of

	Column - I		Column - II
(A)	$\cot^2 A + \cot^2 B + \cot^2 C$ is	(P)	9
(B)	$\tan^2 A + \tan^2 B + \tan^2 C$ is	(Q)	1
(C)	$\text{cosec}\left(\frac{A}{2}\right)+\text{cosec}\left(\frac{B}{2}\right)+\text{cosec}\left(\frac{C}{2}\right)$ is	(R)	6
(D)	cos A + cos B + cos C is	(S)	3.

11. Match the following columns :

In any triangle ABC, then the maximum value of

	Column - I		Column - II
(A)	cos A + cos B + cos C is	(P)	1
(B)	cos A.. cos B . cos C is	(Q)	3 / 2
(C)	$\tan^2\left(\frac{A}{2}\right)+\tan^2\left(\frac{B}{2}\right)+\tan^2\left(\frac{C}{2}\right)$ is	(R)	1 / 4

(D) $\sin\left(\frac{A}{2}\right).\sin\left(\frac{B}{2}\right).\sin\left(\frac{C}{2}\right)$ is (S) 1 / 8.

ANSWERS

1. (A) → (R) ; (B) → (Q) ; (C) → (P) ; (D) → (P)
2. (A) → (R) ; (B) → (R) ; (C) → (R) ; (D) → (P)
3. (A) → (Q) ; (B) → (P) ; (C) → (R) ; (D) → (S)
4. (A) → (Q) ; (B) → (P) ; (C) → (R) ; (D) → (P)
5. (A) → (R) ; (B) → (P, Q) ; (C) → (P) ; (D) → (P, Q)
6. (A) → (R) ; (B) → (R) ; (C) → (R) ; (D) → (Q)
7. (A) → (R) ; (B) → (S) ; (C) → (P) ; (D) → (R)
8. (A) → (Q ; (B) → (R) ; (C) → (P) ; (D) → (S)
9. (A) → (R) ; (B) → (Q) ; (C) → (P) ; (D) → (S)
10. (A) → (Q) ; (B) → (P) ; (C) → (R) ; (D) → (Q)
11. (A) → (Q) ; (B) → (S) ; (C) → (P) ; (D) → (S)

ASSERTION & REASON

Codes:

(A) **Both A and R are individually true and R is the correct explanantion of A.**

(B) **Both A and R are individually true and R is not the correct explanantion of A.**

(C) **A is true and R is false.**

(D) **A is false and R is true.**

1. Assertion (A): If Δ be the area of a triangle and s be the semi - perimeter, then $\Delta^2 \le \frac{s}{4}$

Reson (R): A.M ≥ G.M

(a) A (b) B
(c) C (d) D

2. Assertion (A): In a triangle ABC, if cos A + 2 cos B + cos C = 2, then a, b, c are in A.P

Reason (R): In a triangle ABC,

$\cos A + \cos B + \cos C$

$= 1 + 4\sin\left(\frac{A}{2}\right).\sin\left(\frac{B}{2}\right).\sin\left(\frac{C}{2}\right)$

(a) A (b) B
(c) C (d) D

3. Assertion (A): In a right angled triangle, $a^2 + b^2 + c^2 = 8R^2$, where R is the circum-radius

Reason (R): $a^2 = b^2 + c^2$

(a) A (b) B
(c) C (d) D

4. Assertion (A): If A, B, C and D are the angles of a cyclic quadrilateral, then $\sin A + \sin B + \sin C + \sin D = 0$

Reason (R): If A, B, C and D are the angles of a cyclic quadrilateral, then $\cos A + \cos B + \cos C + \cos D = 0$

(a) A (b) B
(c) C (d) D

5. Assertion (A): In any triangle, $a \cos A + b \cos B + c \cos C \le s$

Reason (R): In any triangle, $\sin\left(\frac{A}{2}\right)\sin\left(\frac{B}{2}\right)\sin\left(\frac{C}{2}\right) \le \frac{1}{8}$

(a) A (b) B
(c) C (d) D

6. Assertion (A): In any triangle ABC, the min value of $\frac{r_1 + r_2 + r_3}{r}$ is 9

Reason (R): In a triangle ABC, if $\frac{\cos A}{a} = \frac{\cos B}{b} = \frac{\cos C}{c}$, then $\frac{r_1 + r_2 + r_3}{r} = 9$

(a) A (b) B
(c) C (d) D

7. Assertion (A) : In a triangle ABC, the harmonic mean of the ex-radii is three times the in-radius.

Reason (R) : In any triangle, ABC,

$r_1 + r_2 + r_3 = 4R$

(a) A (b) B
(c) C (d) D

8. Assertion (A) : If A, A_1, A_2, A_3 are the areas of in-circle and ex-circles of a triangle, then

$$\frac{1}{\sqrt{A_1}} + \frac{1}{\sqrt{A_2}} + \frac{1}{\sqrt{A_3}} = \frac{1}{\sqrt{A}}$$

Reason (R) : In a triangle,

$$\frac{1}{r_1} + \frac{1}{r_2} + \frac{1}{r_3} = \frac{1}{r}$$

(a) A (b) B
(c) C (d) D

9. Assertion (A) : If x, y and z are resptively the distances of the vertices of a triangle ABC from its orthocentre, $\frac{a}{x} + \frac{b}{y} + \frac{c}{z} = \frac{abc}{xyz}$

Reason (R) : In a triangle ABC, $\tan A + \tan B + \tan C = \tan A \cdot \tan B \cdot \tan C$

(a) A (b) B
(c) C (d) D

10. Assertion (A) : If x , y and z are resptively the distances of the vertices of a triangle ABC from its orthocentre , $x + y + z = 2(R + r)$

Reason (R) : In a triangle ABC,

$$r = 4R \sin\left(\frac{A}{2}\right).\sin\left(\frac{B}{2}\right).\sin\left(\frac{C}{2}\right)$$

(a) A (b) B
(c) C (d) D

11. Assertion (A) : In a triangle ABC, $r_1 + r_2 + r_3 - r = 4R$

Reason (R) : In a triangle ABC, $R = \frac{abc}{4\Delta}$

(a) A (b) B
(c) C (d) D

12. Assertion (A) : In any triangle ABC,

$$2\left(a \sin^2\left(\frac{C}{2}\right) + c\sin^2\left(\frac{A}{2}\right)\right) = a + c - b$$

Reason (R) : In any triangle ABC, $b = c \cos A + a \cos C$.

(a) A (b) B
(c) C (d) D

ANSWERS

1.(a) **2.**(b) **3.**(a) **4.**(d) **5.**(a) **6.**(b) **7.**(c) **8.**(a) **9.**(a) **10.**(a) **11.**(a) **12.**(a)

QUESTIONS ASKED IN ROORKEE EXAM

1. In a right angled triangle ABC, The hypotenuse BC of length 'a' is divided into n equal parts (n an odd positive integers) Let α be the acute angle subtending from A by that segment which contains the mid-point of the hypotenuse of the triangle, prove that $\tan\alpha = \frac{4nh}{(n^2-1)a}$

[Roorkee – 1983]

2. No questions asked in 1984, 1985.

3. If a, b, c are the sides of a triangle ABC and 3 a $= b + c$, then prove that $\cot\left(\frac{B}{2}\right).\cot\left(\frac{C}{2}\right) = 2$

[Roorkee – 1986]

4. If in a triangle, $\frac{a^2 - b^2}{a^2 + b^2} = \frac{\sin(A-B)}{\sin(A+B)}$

Prove that it is either a right angled triangle or an isosceles triangle.

[Roorkee – 1987]

5. In any triangle ABC, show that

$$\left(\cot\left(\frac{A}{2}\right)+\cot\left(\frac{B}{2}\right)\right)\left(a\sin^2\left(\frac{B}{2}\right)+b\sin^2\left(\frac{A}{2}\right)\right)=c\cot\left(\frac{C}{2}\right)$$

[Roorkee – 1988]

6. No questions asked in 1989.

7. If x, y, z are the perpendicular distances of the vertices of a triangle ABC from the opposite sides and Δ be the area of the triangle, then prove that

$$\frac{1}{x^2}+\frac{1}{y^2}+\frac{1}{z^2}=\frac{1}{\Delta}(\cot A+\cot B+\cot C).$$

[Roorkee Main – 1990]

8. The two adjacent sides of a cyclic quadrilateral are 2 and 5 and the angle between them is 60°. If the area of the quadrilateral is $4\sqrt{3}$, find the remaining two sides.

[Roorkee Main – 1991]

9. No questions asked in 1992.

10. If the sides a, b, c of a triangle are in A.P, then find the value of $\tan\left(\frac{A}{2}\right)+\tan\left(\frac{C}{2}\right)$ in terms of $\cot\left(\frac{B}{2}\right)$

[Roorkee Main – 1993]

11. No questions asked in 1994.

12. A cyclic quadrilateral ABCD of area $\frac{3\sqrt{3}}{4}$ is inscribed in a unit circle. If one of its sides AB = 1 and the diagonal BD = $\sqrt{3}$. Find the length of the other sides.

[Roorkee Main – 1995]

13. In a triangle ABC, $\angle C=60^o$ and $\angle A=75^o$. If D is a point on AC such that the area of the triangle BAD is $\sqrt{3}$ times the area of the triangle BCD, find the angle $\angle ABD$.

[Roorkee Main – 1996]

14. If in a triangle ABC, a = 6, b = 3 and $\cos(A-B)=\frac{4}{5}$, then find its area.

[Roorkee Main 1997]

15. Two sides of a triangle are of lengths $\sqrt{6}$ and 4 and the angle opposite to smaller side is 30°. How many such triangles are possible? Find the length of their third side and area.

[Roorkee Main – 1998]

16. The radii r_1, r_2, r_3 of escribed circles of a triangle ABC are in the harmonic progression. If its area is 24 sq. cm. and its perimeter is 24 cm, then find the lengths of its sides.

[Roorkee Main – 1999]

17. No questions asked in 2000, 2001.

ANSWERS

8. $x=3, y=2$.

10. $\frac{2}{3}\times\cot\left(\frac{B}{2}\right)$

12. AB = 1, BC = 1, CD = 1 and AD=2.

13. $\angle ABD = 30^o$.

14. 9 sq. units

15. $(2\sqrt{3}+\sqrt{2}), (2\sqrt{3}-\sqrt{2})$ sq. units

16. 8, 6, 6 or $8+\sqrt{29}, 8, 8-\sqrt{29}$

QUESTIONS ASKED IN IIT-JEE

1. In a triangle ΔABC, D is the midlle point of BC. If AD is perpendicular to AC, then prove that

$$\cos A\cos C=\frac{2(c^2-a^2)}{3ac}$$

[IIT-JEE – 1980]

2. Let the angles A, B, C of a triangle ABC be in A.P and let $b:c=\sqrt{3}:\sqrt{2}$. Find the angle A.

[IIT-JEE – 1981]

3. No questions asked in 1982.

4. The ex-radii r_1, r_2, r_3 of ΔABC are in H.P, then prove that the sides a, b, c are in A.P

[IIT-JEE – 1983]

5. For a triangle ABC, it is given that $\cos A+\cos B+\cos C=\frac{3}{2}$, then prove that ΔABC is equilateral.

[IIT-JEE – 1984]

6. With usual notation, if in a triangle ABC, $\frac{b+c}{11}=\frac{c+a}{12}=\frac{a+b}{13}$, then prove that $\frac{\cos A}{7}=\frac{\cos B}{19}=\frac{\cos C}{25}$.

[IIT-JEE – 1984]

7. A ladder rests against a wall at an angle α to the horizontal . Its foot is pulled away from the wall through a distance a , so that it slides a distance 'b' down the wall making an angle β with the horizontal . Show that

$$a=b\tan\left(\frac{\alpha+\beta}{2}\right)$$

[IIT-JEE – 1985]

8. In ΔABC, the median to the side BC is of length $\frac{1}{\sqrt{11-6\sqrt{3}}}$ and divides angle A into angles 30° and 45°, then find the length of the side BC

[IIT-JEE – 1985]

9. Find the set of all values of a such that a^2+2a, $2a+3$ and a^2+3a+8 are the sides of a triangle

[IIT-JEE – 1985]

10. In a triangle ABC, if cot A , cot B, cot C are in A.P, then a^2, b^2, c^2 are in ....Progression.

[IIT-JEE – 1985]

11. There exist a triangle ABC satisfying the conditions,

(a) b sin A = a , A < $\pi/2$

(b) b sin A > a , A > $\pi/2$

(c) b sin A > a , A < $\pi/2$

(d) b sin A < a , A > $\pi/2$, b > a.

[IIT-JEE – 1986]

12. If in a triangle ABC, cos A cos B + sin A sin B sin C = 1, then show that $a:b:c=1:1:\sqrt{2}$. **[IIT-JEE – 1986]**

13. In a triangle, the lengths of the two larger sides are 10 and 9 respectively . If the angles are in A. P, then the length of the third side can be

(a) $5-\sqrt{6}$ (b) $3\sqrt{3}$

(c). 5 (d) $5+\sqrt{6}$

[IIT-JEE – 1987]

14. No questions asked in 1988, 1989.

15. ABC is a triangle such that, $\sin(2A+B)=\sin(C-A)=-\sin(B+2C)=\frac{1}{2}$.

If A, B and C are in A.P, determine the value of A, B and C.

[IIT-JEE – 1990]

16. The sides of a triangle are three consecutive natural numbers and its largest angle is twice the smallest one. Determine the sides of the triangle.

[IIT-JEE – 1991]

17. In a triangle of base a , the ratio of the other two sides is r (< 1) . Show that the altitude of the triangle is less than or equal to $\frac{ar}{1-r^2}$.

[IIT-JEE – 1991]

18. No questions asked in 1992.

19. In a triangle ABC, D and E are points on BC and AC respectively, such that BD = 2 DC and AE = 3 EC.

Let P be the point of intersection of AD an BE. Find BP/PE using vector methods.

[IIT-JEE – 1993]

20. If in a triangle ABC, $\frac{2\cos A}{a}+\frac{\cos B}{b}+\frac{2\cos C}{c}=\frac{1}{bc}+\frac{1}{ca}$, then the value of the angle A is...............degrees.

[IIT-JEE – 1993]

21. In ΔABC, AD is an altitude from A. Given b > c, $\angle C=23^o$ and $AD=\frac{abc}{b^2-c^2}$, then $\angle B$ is.......

[IIT-JEE – 1994]

22. Let $A_1A_2A_3..........A_n$ be the vertices of n sided regular polygon such that $\frac{1}{A_1A_4}+\frac{1}{A_1A_3}=\frac{1}{A_1A_2}$, then find the value of n.

[IIT-JEE – 1994]

23. Consider the following statement. Consider a triangle ABC

(i) The sides a, b, c and area of triangle is rational.

(ii) $a, \tan\left(\frac{B}{2}\right), \tan\left(\frac{C}{2}\right)$ are rational

(iii) a, sin A, sin B, sin C are rational.

Then prove that $(i) \Rightarrow (ii) \Rightarrow (iii) \Rightarrow (i)$.

[IIT-JEE – 1994]

24. In a triangle ABC, $\angle B = \frac{\pi}{3}$ and $\angle C = \frac{\pi}{4}$. Let D divide BC internally in the ratio 1 : 3 then $\frac{\sin \angle BAD}{\sin \angle CAD}$ is equal to

[IIT-JEE – 1995]

(a) $1/\sqrt{6}$ (b) 1/3

(c) $1/\sqrt{3}$ (d) $\sqrt{2/3}$

25. In a triangle ABC, a : b : c = 4 : 5 : 6. The ratio of the radius of the circum-circle to that of the in-circle is..................

[IIT-JEE --1996]

26. Let A, B, C be three angle such that $\angle A = \frac{\pi}{4}$ and $\tan B \tan C = p$. Find all possible values of p such that A, B, C are the angles of triangle.

[IIT-JEE – 1997]

27. If in a triangle PQR, sin P, sin Q, sin R are in A.P, then

(a) the altitudes are in A.P

(b) the altitudes are in H.P.

(c) the altitudes are in G.P

(d) the medians are in A.P

[IIT-JEE – 1998]

28. Prove that , a triangle ABC is equilateral if and only if $\tan A + \tan B + \tan C = 3\sqrt{3}$

[IIT-JEE – 1998]

29. Let ABC be a triangle having O and I as its circumcentre and incenter respectively. If R and *r* are the circum-radius and the in-radius respectively, then prove that $(IO)^2 = R^2 - 2Rr$, further show that the triangle BIO is a right angles triangle if and only if b is the arithmetic mean of a and c.

[IIT-JEE – 1999]

30. In any triangle ABC, prove that

$$\cot\left(\frac{A}{2}\right) + \cot\left(\frac{B}{2}\right) + \cot\left(\frac{C}{2}\right) = \cot\left(\frac{A}{2}\right)\cot\left(\frac{B}{2}\right)\cot\left(\frac{C}{2}\right)$$

[IIT-JEE – 2000]

31. In a triangle ABC, $\sin\left(\frac{A-B+C}{2}\right) =$

[IIT-JEE – 2000]

(a) $a^2 + b^2 - c^2$ (b) $c^2 + a^2 - b^2$

(c) $b^2 - c^2 - a^2$ (d) $c^2 - a^2 - b^2$

32. In a triangle ABC, let $\angle C = \pi/2$. If r is the in radius and R is the circumradius of the triangle, then $2(r+R)$ is

[IIT-JEE – 2000]

(a) a + b (b) b + c

(c) c + a (d) a + b + c.

33. The in center of the triangle with vertices $(1, \sqrt{3})$, (0, 0) and (2, 0) is

[IIT-JEE – 2000]

(a) $\left(1, \sqrt{\frac{3}{2}}\right)$ (b) $\left(\frac{2}{3}, \frac{1}{\sqrt{3}}\right)$

(c) $\left(\frac{2}{3}, \frac{\sqrt{3}}{2}\right)$ (d) $\left(1, \frac{1}{\sqrt{3}}\right)$

34. Let ABC be a triangle with incenter I and inradius *r*. Let D, E, F be the feet of the perpendicular from I to the sides BC, CA and AB respectively. If r_1, r_2 and r_3 are the radii of circles inscribed in the quadrilaterals AFIE, BDIF and CEIF respectively, prove that

$$\frac{r_1}{r-r_1} + \frac{r_2}{r-r_2} + \frac{r_3}{r-r_3} = \frac{r_1 r_2 r_3}{(r-r_1)(r-r_2)(r-r_3)}$$

[IIT-JEE – 2000]

35. No questions asked in 2001.

36. Which of the following pieces of data does not uniquely determine an acute angled triangle ABC (R being the radius of the circum-circle)?

(a) a, sin A , sin B (b) a, b, c

(c) a, sin B, R (d) a, sin A , R.

[IIT-JEE – 2002]

37. If the angles of a triangle are in the ratio 4 : 1 : 1, then the ratio of the longest side to the perimeter is

(a) $\sqrt{3}:(2+\sqrt{3})$ (b) $1:6$

(c) $1:(2+\sqrt{3})$ (d) $2:3.$

[IIT-JEE – 2003]

38. The sides of a triangle are in the ratio $1:\sqrt{3}:2$, then the angles of the triangle are in the ratio

(a) $1:3:5$ (b) $2:3:4$

(c) $3:2:1$ (d) $1:2:3$

IIT-JEE – 2004]

39. Which of the following is true in a triangle ABC?

(a) $(b+c)\cos\left(\frac{A}{2}\right)=2a\sin\left(\frac{B-C}{2}\right)$

(b) $(b-c)\cos\left(\frac{B-C}{2}\right)=2a\cos\left(\frac{A}{2}\right)$

(c) $(b+c)\sin\left(\frac{B-C}{2}\right)=2a\cos\left(\frac{A}{2}\right)$

(d) $(b-c)\cos\left(\frac{A}{2}\right)=a\sin\left(\frac{B-C}{2}\right)$

[IIT-JEE - 2005]

40. One angle of an isosceles Δ is 120° and radius of its incircle = $\sqrt{3}$. Then the area of the triangle in sq. units is

(a) $7+12\sqrt{3}$ (b) $12-7\sqrt{3}$

(c) $12+7\sqrt{3}$ (d) 4π

[IIT-JEE – 2006]

41. Let O (0, 0), P(3, 4), Q(6, 0) be the vertices of the triangle OPQ. The point R inside the triangle OPQ is such that the triangles OPR, PQR, OQR are of equal area. The co-ordinates of R are

(a) $\left(\frac{4}{3},3\right)$ (b) $\left(3,\frac{2}{3}\right)$

(c) $\left(3,\frac{4}{3}\right)$ (d) $\left(\frac{4}{3},\frac{2}{3}\right)$

[IIT-JEE -2006]

42 In Δ ABC, internal angle bisector of $\angle$ A meets side BC in D. $DE \perp AD$ meets AC in E and AB in F, then

(a) AE is H.M. of b and c

(b) $AD=\frac{bc}{b+c}\cos\left(\frac{A}{2}\right)$

(c) $EF=\frac{4bc}{b+c}\sin\left(\frac{A}{2}\right)$

(d) ΔAEF is isosceles.

[IIT-JEE – 2006]

43. In a triangle ABC with fixed base BC, the vertex A moves such that

$$\cos B+\cos C=4\sin^2\left(\frac{A}{2}\right)$$

If a, b, and c denote the lengths of the sides of the triangle opposite to the angles A, B and C respectively, then

(a) b + c = 4a

(b) b +c = 2 a

(c) locus of a point A is an ellipse

(d) locus of point A is a pair of striaght line

[IIT-JEE-2009]

44. No questions asked in 2007 and 2008.

45. Let ABC and ABC' be two non congruent triangles with sides AB = 4, AC = AC' = $2\sqrt{2}$ and angle $B=30^o$. The absolute vale of the difference between the areas of the triangles is..................

[IIT-JEE-2009]

46. The centres of two circles C_1 and C_2 each of unit radius are at a distance of 6 units from each other. Let P be the mid-point of the line segment joining the centres of C_1 and C_2 and C be a circle touching circles C_1 and C_2 externally. If a common tangent to C_1 and C_2 passing through P is also a common tangent to C_2 and C, then the radisu of the circle C is................

[IIT-JEE-2009]

47. Let ABC be a triangle such that $\angle ACB = \frac{\pi}{6}$ denote the lengths of the sides opposite to A, B, C respectively. The values of x for which $a = x^2 + x + 1,\ b = x^2 - 1\ \&\ c = 2x + 1$ is (are)

(a) $-(2+\sqrt{3})$ (b) $(1+\sqrt{3})$

(c) $(2+\sqrt{3})$ (d) $4\sqrt{3}$

[IIT-JEE-2010]

48. Consider a triangle ABC and let a, b and c denote the lengths of the sides opposite to vertices A, B and C respectively . Suppose a = 6, b = 10 and the area of the triangle is $15\sqrt{3}$. If $\angle ABC$ is obtuse and if r denotes the radius of the incircle of the triangle , then r^2 is equal to ....

[IIT-JEE-2010]

49. The positive integral value of n > 3 satisfying the equation $\frac{1}{\sin\left(\frac{\pi}{n}\right)} = \frac{1}{\sin\left(\frac{2\pi}{n}\right)} + \frac{1}{\sin\left(\frac{3\pi}{n}\right)}$ is........

[IIT-JEE-2011]

50. Let PQR be a triangle with a = 2, b = 7/2 and c = 5/2, where a, b and c are the lengths of the sides of the triangle opposite to the angles P, Q and R respectively, then $\frac{2\sin P - \sin 2P}{2\sin P + \sin 2P}$ is

(a) $\frac{3}{4\Delta}$ (b) $\frac{45}{4\Delta}$

(c) $\left(\frac{3}{4\Delta}\right)^2$ (d) $\left(\frac{45}{4\Delta}\right)^2$

[IIT-JEE-2012]

ANSWERS

2. 75°	22. n = 7
8. BC = 2	25. 16 : 7.
9. a = $(5,\ \infty)$	26. $p \in (-\infty,\ 3-2r2] \cup [3+2\sqrt{2},\ \infty]$
10. A.P	42. (a, b, c , d)
15. $A = 45^o,\ B = 60^o,\ C = 75^o$.	45. 4
19. 8 : 3	46. 8
20. 90°	48. 3
21. 113°	49. n = 7

SELF ASSESMENT-I
(Solve by yourself only)

Time : 3 Hrs. **Ch: Properties of Triangles** **M.M : 100**

Give answer of the following questions.

1. If in a triangle ABC , tan A = 2 , tan B = 3/2 and c = $2\sqrt{65}$, then find the circumradius of the triangle.

2. If in a triangle ABC , cos 3A + cos 3B + cos 3C = 1, then prove that the measure of one of the angles must be $\frac{2\pi}{3}$.

3. If in a triangle ABC, the sides a, b , c are in A.P , then prove that $\cot\left(\frac{A}{2}\right) + \cot\left(\frac{C}{2}\right) = 2\cot\left(\frac{B}{2}\right)$.

4. If in a triangle of ABC, A = 3 B, then prove that $\sin B = \frac{1}{2}\sqrt{\frac{3b-a}{b}}$.

5. Let Δ_1, Δ_2 denote the areas of a triangle and that of the incircle . Prove that

$$\frac{\Delta_1}{\Delta_2} = \left(\cot\frac{A}{2}.\cot\frac{B}{2}.\cot\frac{C}{2}\right) : \pi$$

6. If the exradii r_1, r_2, r_3 of a triangle ABC are in H.P, then prove that the sides a, b, c are in A.P.

7. The lenghts of the sides of a triangle are three consecutive natural numbers and its largest angle is twice the smallest one. Determine the sides of the triangle. (Ans. 4, 5, 6.)

8. Let O be a point in the triangle ABC such that $\angle OAC = \angle OCB = \angle OBA = \alpha$, then prove that cot $\alpha = \cot A + \cot B + \cot C$.

9. If in a triangle ABC, two sides are a = 6, b = 3 and cos (A-B) = $\frac{4}{5}$, then prove that the area of the triangle is 9.

10. In a triangle ABC, the angles A, B, C are in A.P, then prove that

$$2\cos\left(\frac{A-C}{2}\right) = \frac{a+c}{\sqrt{a^2 - ac + c^2}}.$$